The Lotus Chronicles:

One man's sordid tale of passion and madness
while resurrecting a 40-year-dead
Lotus Europa Twin Cam Special

ROB SIEGEL

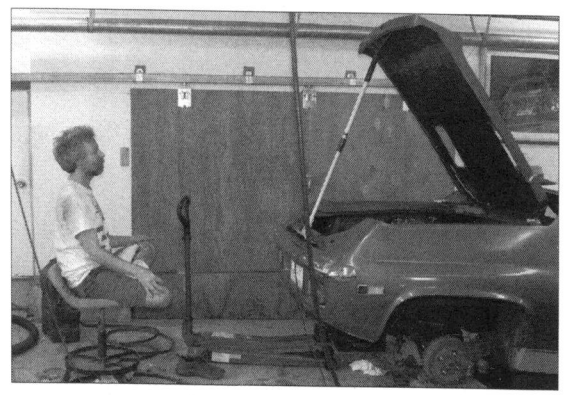

HACK MECHANIC PRESS

The Lotus Chronicles: One man's sordid tale of passion and madness while resurrecting a 40-year-dead Lotus Europa Twin Cam Special

By Rob Siegel

Copyright ©2020 by Rob Siegel

All rights reserved. No part of this book may be used or reproduced in any manner whatsoever without written permission from the publisher, except for brief quotations in critical articles and reviews.

ISBN: 978-0-9989507-3-0

Library of Congress Control Number: 2020916171

The author and publisher recognize that some words, model names, and designations mentioned in this book are the property of the trademark holders. They are used only for identification purposes.

This is not a repair manual. The author is not a professional mechanic. Neither the author nor the publisher is responsible if you injure yourself while working on or driving your car. If you have any doubt as to your ability to do some of the things described in the book, or as to whether your car is safe to drive after doing them, don't do them, and seek the services of a professional mechanic instead.

All photography by Rob Siegel except where noted
Design by King+Sons Design Garage

Printed in the United States

The Hack Mechanic™ is a registered trademark of Rob Siegel

Hack Mechanic Press
19 Mague Place
West Newton, MA 02465

hackmechanicpress.com

Acknowledgements

Thanks to Ray Psulkowski at R.D. Enterprises, Ken Gray at Dave Bean Engineering, and the forums lotuseuropa@groups.io, lotuseuropa.org, the Facebook Lotus Europa group, and lotuselan.net for their expertise.

Thanks to Joel Artenstein who owned Sound Ideas in The Alley in Amherst where I worked briefly when I was 13, and came in one day in 1972 driving that red Lotus Europa that kick-started the whole thing.

Thanks to Joel's employee whose name is lost to history, the forty-something guy who looked at the Lotus and said to me those ten words that haunted me for decades: "A car like that, you can get *sex* out of."

Thanks to all my Facebook friends and followers for being such a snappy quick-witted bunch.

Thanks to Russell Baden Musta who, by way of a chance vehicular encounter on I-81 south, provided the spark that re-lit the long-dormant project.

Big thanks to Ed MacVaugh for proofreading the book on such short notice.

Huge thanks to book designer Eric King for taking on the challenging task of making the book "look like Facebook" and hitting it out of the park.

Thanks to machinist Eric Lindskog at Century Automotive Machine in Waltham MA for helping me get the engine together in something resembling a cost-effective fashion.

Huge thanks to Bill McCurdy (a.k.a. "The Lotus Engine God") at Williams Racing in Harvard MA for the depth of his knowledge, his humor, his endless patience with me, and not throwing me out on my ear when I crawled back to him with my tail between my legs.

But deepest thanks go to my honey my baby, the one and only Maire Anne Diamond, who never once said "WHAT THE FUCK ARE YOU DOING??" Or rolled her eyes at the endless stream of Lolita-related adolescent sexual innuendos. At least not while I was watching.

Contents

Preface 7

Introduction 11

Chapter I: Shit Gets Real 15

Chapter II: The Project Runs Into Molasses 51

Chapter III: The Doldrums 83

Chapter IV: The Block Gets Reassembled 103

Chapter V: Full Engine Reassembly 139

Chapter VI: Under Pressure 195

Chapter VII: Putting Lolita Back Together 255

Chapter VIII: Sploosh 351

Chapter IX: Barely Legal 407

Chapter X: A Mini Road Trip? 427

Chapter XI: Life After LOG 451

Chapter XII: The Winter Projects 465

Chapter XIII: Emerging Into Covid-19 471

Chapter XIV: Adding Up The Costs 495

Epilogue 503

End Notes 509

Preface

I've had a very active presence on Facebook since joining it in 2008. It instantly seemed to be a natural fit for my bred-in-the-bone wise-assery. Although much of my posting revolves around stupid cars, hack repairs, and highly questionable road trips, there's also quite a bit about music, and occasional political rants. Some people who follow me specifically for the car stuff get hot under the collar when I post something left-leaning. I've had folks stay "stick to cars" (ha!) or ask me to set up a separate car-related Facebook page. I've always felt that most of us are more than one thing, and that it's not only okay to present ourselves multi-dimensionally, but it's unhealthy *not* doing so. (Note: Y'all can chill; this book is almost completely apolitical.)

In addition, when I was still an engineer working in my erstwhile profession of unexploded ordnance detection (surveying terrestrial and marine sites to find bombs left over from military training activities), there were posts from the field when I was off on a geophysical survey. In 2012, I was on a three-month operation in the Kisatchie National Forest in a remote part of Louisiana. The nearest town we could drive to for sandwiches and supplies was called Dry Prong, which, you have to admit, is funny, particularly to anyone of the male gender (all men, no matter what our chronological age, remain 13-year-old boys who see dick jokes everywhere). I began doing a series of Facebook posts about the survey, the geophysics, the jury-rigged repairs, the sanitary preparations when working in the backwoods (I bought a "Luggable Loo"—a bucket with a toilet seat—and it became an Internet celebrity), the colorful locals, and the food (the fact that I actually ate alligator sushi was of great interest). To frame the posts in a way that made them instantly identifiable and set them off from anything else I might be writing, I began titling each of them "The Dry Prong Chronicles, Day XXX." The series of posts developed quite a following. Many folks said they were sad when it ended shortly before Thanksgiving 2012.

So, when I bought the '74 Lotus Europa Twin Cam Special (a.k.a. "Lolita") in June of 2013 that had been sitting since 1979 and had a seized engine, and began posting about it on Facebook, it seemed natural to do what I did with the Dry Prong posts—title the series "The Lotus Chronicles" and give each post a day number.

I never expected it to go on for seven years.

That it did was a combination of the fact that the project itself went on and on, and that I couldn't really write about it in my monthly column *The Hack Mechanic* for BMW CCA *Roundel* Magazine, at least not repeatedly and directly. Once I began writing weekly pieces for Hagerty's website, though, the Lotus became fair game, and I wrote quite a bit about it there, but did so in parallel with continuing the shorter punchier "Lotus Chronicles" posts on Facebook.

I never really had any doubt that that, once the car's fate was determined and the story had some sort of defined end, I'd write a book about the Lotus. The question was whether I'd use the Facebook posts as the starting point for content or put together a book of the actual posts. As I re-read them all, it was clear: While my posts themselves had some punch and snap to them, it was the comments from readers that were often priceless. I've long said that any shared interest provides a great conduit for human connection (my wife is an avid quilter, and experiences much of the connection there that I have in the car world), it just happens that the vintage BMW community is an exceptional one, filled with people who, in addition to being interesting, intelligent, experienced, and resourceful, can be laugh-out-loud funny. And, for some reason, taking that community and mixing it online with my singer-songwriter friends, a few relatives, folks I knew from Amherst, Lexington, and UMass, and random Facebook friends picked up along the way, created something magical. The responses were supportive, sarcastic, hilarious, helpful, acerbic, thoughtful, and more. Plus, from a practical standpoint, while I have a lot of experience with 1970s-era BMWs, this was my first vintage Lotus (and my first British car in 35 years), so my familiarity with this corner of the automotive landscape was limited. The vast majority of the comments were intelligent and constructive, and there were times when solutions to my problems were essentially crowd-sourced. The back-and-forth was so good that many non-car-friends said they looked forward to every post.

Trying to create a book out of a seven-year-long series of Facebook posts and their comments, though, proved quite challenging. First there was the mechanics of accessing the content itself. Facebook does have an option to download your information, and my assumption was that I could simply do that, search for "The Lotus Chronicles" and then pull each post and its comments into a Word manuscript, maybe even automating the process. Perhaps, if my software skills were still sharp, that might have been possible, but for a number of reasons, I wound up having to scroll through a year's worth of posts at a time and individually click, drag, copy, and paste every comment into Word. That alone took several weeks.

Next there was the issue of volume. Although my original intent was to include every post and every response, once I had everything in a Word document, it was clear that I couldn't include them all, as well as other needed backstory material, without generating a *War and Peace*-sized volume. So as much as I hated to do it, I needed to trim. Obviously, posts that simply said "Congrats" weren't James Joyce-level genius that needed to be preserved for the ages. I'll admit that I took particular glee in culling nearly all the calls to "LS1 it" (install a Corvette engine) or "M42 it" (install a BMW 4-cylinder twin-cam engine). And the joke about the British, warm beer, and Lucas refrigerators only needed to be told once. Some folks' stories of, say, rebuilding the engine in their MGA had to go. Editing out a digression between me, a junior high school friend, and someone else from UMass Amherst about The Drake, a notorious and long-gone bar, though, was difficult. But some responses, even though lengthy and meandering, were just too precious. I tried to maintain as light an editing hand as possible to preserve the synergy.

Then there was the cleanup. The formatting and grammar in short Facebook responses are a cut above cell phone texts, but still pretty loose. I corrected spelling and did some light editing without mandating that each one be as flawlessly constructed as a Salinger short story. Oddly, most challenging was making the usage of "Twin Cam" consistent. My rule, supported (I think) by the literature, was that the car is a "Lotus Europa Twin Cam Special," the engine is a "Lotus Twin-Cam," and any head with two overhead camshafts, including this one, is a "twin-cam head." Fight me.

Book designer Eric King and I then faced the question of how to evoke Facebook's look and feel. Initially I wanted it to literally look like Facebook, but this turned out not to be feasible. With Facebook's many changes, there are 20 separate text and graphic elements in a FB post and its responses, and all of them, such as the fingernail-sized photo of the responder and the time and date of each response, clearly aren't necessary in a book. I wanted to include the number of likes each post got, as seeing which ones went DING DING DING was part of the fun, but in 2016 Facebook changed from a single thumbs-up like icon to a choice of six "like" icons ranging from heart to angry face. So, if a response got 15 likes, was it 15 likes or was it eight hearts, five laughs, and two thumbs? Obtaining that information required manually clicking on each icon in each reply and typing it into the Word document. And even if I did that (which I didn't), reproducing the happy colorful feel of the FB icons in a black-and-white book wasn't possible. Eric came up with a readable format that evoked the Facebook look and feel without actively copying it.

Even the day counts themselves posed a problem. When "The Lotus Chronicles" posts first began, I was counting from June 20th 2013, the day the car arrived at my house in West Newton. But after work on the car stopped and there was a pause of several months between posts, when I restarted them, I began numbering from the date the bill of sale and title were signed, which was June 10th. However, I saw that the date that Ben Thongsai looked at the car for me and gave the seller a deposit on my behalf was June 8th, so for a while I used *that* date. In the end I left all the day counts alone. It's not like future generations will be lead astray by them or anything.

I added some supporting sections such as email conversations and interjections that provide context, but really, I just wanted to tell the story, and the Facebook posts and their responses and the in-process snapshot they provided were the best way to tell it.

Ladies and gentlemen: Start your Lotus engines.

Introduction

In 2010 I received an unsolicited email from a New York literary agent who'd read one of my columns. "You're very funny and you write very well," she said. "Have you ever thought about writing a book?" I replied "Well, I am *now*." Having skipped the whole process of pitching manuscripts to try to get an agent's interest, I felt like I'd won the lottery. I labored for about a year on a manuscript with the draft title "Gearhead," something we agreed would be sort of the car-guy version of *Men Are From Mars, Women Are From Venus*. Central to that was a chapter I hoped to be the showpiece of the book. I titled the chapter "So why DO men love cars, REALLY?" In it, I explained that most men are total sensation junkies, and that we spend much of our lives seeking sensation, whether it's through sex, fast cars, drug use, skydiving, gambling, etc. I wrote about how a car, particularly a beautiful fast car with arresting lines, a powerful engine, and taut handling, becomes an object of desire to obtain and an object of passion to experience. I explained how owning and driving such a passion-car, appreciating its looks, nailing the accelerator, tossing it hard into curves, making it squeal and scream, is an experience laden with so much sensation that it borders on the sexual.

And explained how I totally didn't understand this when I was 13, and told the story of a guy who I worked for who was a serial consumer of sports cars, and who one day came to work driving a red Lotus Europa. I'd never before seen a car in the flesh that was so low and sexy, and was completely captivated by it. Another guy who also worked in the store, maybe in his early 40s, looked at the little Lotus and said something that I've remembered ever since: *"A car like that, you can get sex out of."* As a 13-year old kid, I thought, "Yeah, I get it, girls like guys with cool cars, it must almost be unfair to own something like this." But I didn't get it. Not for decades. It wasn't until I was in my forties, happily married, committed to keeping my baser impulses in check and staying faithful to my wife, that I realized that he wasn't talking about using the car to attract women at all—*He was talking about the sensation* of owning and driving a cool car. In this chapter of the book, I went on to explain how indulging one's car passion in middle age can actually be of value in maintaining a healthy marriage—better to serially swap cars than spouses—and how spouses who understand this and enable rather than block their spouse's car passion actually contribute to the stability of the marriage.

To my surprise, this chapter, rather than proving to the agent that I could deliver on the mind-of-a-car-guy concept and smack it out of the park, rubbed her the wrong way, and not long after, she dropped me. One of the things she said was "I think you need a boy," by which she meant a male agent or editor who understood what I was talking about, because she didn't.

Fortunately, the manuscript got picked up by Bentley Publishers, with whom I'd had a relationship 20 years earlier (long story). I worked on it for almost two more years, then signed a book deal with them. The book was published in June 2013 as *Memoirs of a Hack Mechanic*. Although the content had evolved quite a bit from the draft the agent saw, the chapter on "So why DO men love cars, REALLY?" remained the central pillar of the book, containing observations on men, women, passion, and cars you'll find nowhere else. And what was especially satisfying was the fact that my editor at Bentley was their director of publishing, Janet Barnes, who was definitely not "a boy."

I was still working full-time at my engineering job and giddy with the imminent publishing of the first book. I began thinking about pursuing a Europa. After all, the 1973 BMW 3.0CSi I've owned since 1986 is gorgeous and elegant, like a perfectly-accessorized 40-something-year-old woman you just can't take your eyes off, but most of my other 1970s-era BMWs—cars like 2002s and Bavarias—are, well, a little on the frumpy side in terms of the whole *get sex out of* thing. (I did, though, already own the BMW Z3 M Coupe, a vehicle that I've described in print as "a vixen of a car, a dirty mistress who sticks her tongue in your ear and unbuttons your trousers in full view of polite company and doesn't care." So it wasn't like I wasn't getting any sex out of cars. I guess I just wanted *more* sex. I am, after all, a man.)

Of course, other than the one Europa that gave 13-year-old me a boner, I didn't really know anything about them. However, there's another chapter in *Memoirs* where I talk about stepping outside your comfort zone. I did that 15 years earlier when I scratched a lifelong itch and bought a 1982 Porsche 911SC. I immediately loved that car, and with its exotic shape and good performance, I certainly got my jollies in it. It was definitely German, and yet was completely unlike the BMWs. Learning about the 911SC, looking at several, finally pulling the trigger on one, living with it, and learning how to fix it was an enormously satisfying experience. (And selling it was the biggest automotive mistake I ever made, both in letting go of a passion object and terrible timing in terms of its appreciating value.)

In late 2012, I called about a Europa in Connecticut listed on Craigslist.

But it was too early in the process; I hadn't really even started doing my homework. I blurted out the question I'd ask with any other car—"How bad is the rust on the body?"—and the seller patiently answered "Well, as I'm sure you know, the Europa is a fiberglass-bodied car, so there's *never* any rust on the body; the question is whether there's any on the steel backbone." Oops. Called out by the teacher for trying to wing the book report without ever cracking the cover.

So I began reading. I learned how Lotus founder Colin Chapman and his team designed and built lightweight race cars using innovative techniques such as stress-skinned monocoque construction (designing a fiberglass body so it's not merely cosmetic, and instead is part of the structural rigidity of the car, which in turn allows the car's frame to be lighter), with a Lotus winning its first Formula One championship in 1963 and the Indianapolis 500 in 1965. I learned that Chapman's design philosophy was encapsulated in his oft-quoted maxim "Simplify, then add lightness." I learned that Lotus took those tricks and used them in the Europa, the first affordable mass-produced mid-engine stress-skinned sports car with race-car-like handling. I learned that the reason why a Europa looks a little like a Ford GT40 is that the Europa is based on a design sketch that Lotus did when it was competing for the Ford GT40 racing project (and that the "40" in "GT40" stands for the roof being 40 inches off the ground; on the Europa, it's 42 inches). I learned that the early S1 and S2 cars have Renault engines, but the later ones have a more powerful Lotus-Ford Twin-Cam engine. I learned that the rear windshield is a comically-narrow four inches high, and that the early cars have rear quarter panels (sometimes referred to as "sail panels") that start at the height of the roof and gradually slope down as they near the back, but the Twin Cam cars have cut-down sail panels whose height immediately drops off from the roof line in an attempt to increase rear visibility. I learned that, because of this drop-off, many people feel that, when viewed from the side, these later Twin-Cam cars look a bit like a very low El Camino, and that some folks refer to the car derisively as a "bread van." I learned that the last batch of Europas were "Twin Cam Special" (TCS) cars, most of which were equipped with a highly desirable 5-speed. Lastly, I learned that the only difference between the highly-coveted "John Player Special" (JPS) cars wearing the familiar black-and-gold John Player livery, and the other TCS cars, was the color and nothing else.

Armed with enough knowledge to be dangerous, I began looking, albeit half-heartedly, for a Europa Twin Cam Special I could afford.

Now, the traditional advice with vintage cars is to buy the one in the best condition you can find. You'll sometimes see this phrased as "you can never spend too much money; you can only buy too early." I, of course, ignore that advice. Some of that is a question of budget—it's easy to say "you can never spend too much money" if you *have* too much money. I, on the other hand, simply don't have the coin to buy the best of *anything*. But some of it is also simply my preference. I *like* buying needy cars and bringing them back from the edge. Plus, if I bought something well-maintained, what the hell would I have to write about?

But really, so much of car-buying comes down to crimes of opportunity. You keep your eyes open. You look on Craigslist, Facebook Marketplace, eBay, Bring a Trailer, and down actual physical driveways. You err on the side of cars you can look at with your own eyes, and only consider far-flung cars if someone you know and trust can look at them for you. At least that's what I do.

And you try to find something that speaks to you, and hopefully that's niche-y and weird enough that fifty other people don't want to buy it as well and push the price up. I wasn't willing to sell any of the BMWs to fund a Europa, and that meant bottom-feeding in the $5000 range. That didn't seem completely out of the question; there was no shortage of beat-to-shit Renault-powered S2 Europas in the muddy end of the pond. Finding a Twin Cam Special, though, for that kind of money didn't look easy.

And so it began.

Chapter I:
Shit Gets Real

Other Things Happening in 2013

- My first book, *Memoirs of a Hack Mechanic*, was published.
- Rebuilt the Kugelfischer injection in Kugel, my '72 2002tii.
- Drove my 3.0CSi down to the BMW event "The Vintage" in North Carolina for the third time. This also was the book release event for *Memoirs*.
- Went to BMW CCA Oktoberfest in Monterrey for another *Memoirs* book release event.
- Began a short stint writing for *Road & Track*.
- Sold the hovercraft and trailer (the only project I've ever bailed out of).
- Bought a beater 318ti so my kids had something disposable to drive other than my cars.
- Bought a running $3000 6-cylinder Z3 roadster.
- Bought a 1972 Bavaria, though I didn't take delivery on it until the following spring.

The Teaser

My first Europa-related Facebook post appeared on Feb 24, 2013.

FEB 24, 2013, 8:47 AM

So this is how well Maire Anne knows me. I look at this on eBay, note that the address is Westfield NJ, look it up in Google Maps, see that it's near Elizabeth, and say, subconsciously, UNDER MY BREATH, "that's not that far," and Maire Anne says "looking at a car, are we?"

[link to expired eBay ad for 1972 Lotus Europa Twin Cam]

5 LIKES, 19 COMMENTS

Franklin Alfonso Oh Oh…There is no cure for Lotus…

Carl Witthoft Hey, I'm going on vacation to Manhattan in a couple weeks. Be glad to drive it back to MA for you!

Jenni Morgan A good friend of mine just sold one of these… cool car, when it didn't leak oil. Oh wait, it always leaked oil.

Larry Michelove It only leaks if you keep it filled with oil.

Ed MacVaugh Why don't I believe the supposed mileage :)

Rob Siegel Ed, it's been said that low mileage on British cars is, uh, self-enforced. Having owned a '70 Triumph GT6+ that barely ran half the time I owned it, WHEN THE CAR WAS ONLY SIX YEARS OLD IN 1976, I can believe it!

Ed MacVaugh Well, that interior driver's footwell would suggest then that the driver stayed in the car moving his feet around while the car was stationary. Check out the brake rotor which supports your thoughts.

Ed MacVaugh I still want it and would add John Player stripes even though it isn't black.

Rob Siegel and the 5-speed.

Franklin Alfonso And then there is the Lucas Prince of Darkness electrics to deal with as well.

Rob Siegel Yes. Well. There is that. But at least the electrical gremlins are guaranteed to be less than 42" high.

Oen Kennedy When I lived in Cambridge in the early 70s there were a few Lotuses floating around, and of course, whenever I found one parked, I had to check the speedometer: "160!"

Rob Siegel …which would've been wildly optimistic. But the Europa was fast for its time, posting a 7-second 0-60 time and top speed of, I believe, 120. Now, any Corolla will do 120. With the air conditioning on. The Europa's charm is that it is a 1600-pound mid-engine car.

Franklin Alfonso Well they were known for their handling more than anything else. Well, handling for its time… I had a 60 MGA and it was slow as molasses but with the top down at 40 you felt like you were doing 80 lol. The joys of old British sports cars sigh…

David Weisman I had a 71 'Cuda, highway differential. The speedometer read: "0 - 150", then… the words: "Oh Wow!" +1

Franklin Alfonso Warp 1 lol.

[The eBay auction got picked up by Bring a Trailer and listed there, which caused the number of eyeballs and thus the bids to jump from about $7500 to a closing price of $10,055. I sarcastically commented on BaT, thanking them for putting the car out of my price range.]

John Lenham Saw your BaT post… rats… a term needs to be coined for when that happens.

Rob Siegel "The price got BaT-ted."

Interjection: The purchase

In late May of 2013, *Memoirs of a Hack Mechanic* was published by Bentley Publishers. The kick-off was at the BMW enthusiast event "Vintage at the Vineyard" (later called "The Vintage") in Winston-Salem, NC, that I'd attended for the past few years. I gave a talk before a good-sized crowd, and sold and signed hundreds of books. Shortly after, the book was reviewed in *The New York Times* and *The Boston Globe,* and there were book talks in several east coast locations and a radio campaign in Boston. Coincidentally,

I also began a short stint as a contributing writer at *Road & Track* magazine. I still had my engineering job, but it was increasingly clear that it was ramping down and the automotive writing was ramping up sharply. It was a very busy and exciting time.

As per my contract with Bentley Publishers, there was a modest advance associated with the book release. It made me think about buying myself a present to mark the event, maybe something I wouldn't normally buy.

As I said, most car purchases are crimes of opportunity, and an opportunity presented itself. Of course, I *was* actively looking for trouble.

I'm surprised that I don't appear to have directly recounted the blow-by-blow of the purchase of the Lotus in emails or on Facebook, but here's what happened. For years, I'd been using Searchtempest—a Craigslist multi-city search tool—to look for cars. One night, I set a search radius of a thousand miles and typed in "Lotus Europa." Just barely inside the radius, I got a hit in suburban Chicago for a '74 Lotus Europa Twin Cam Special. The ad said:

```
"1974 Lotus Europa twin cam special. 99% complete!!! Very rare
optional 5 speed transmission, only missing original radio! Interior
in very good condition, Original shop manual included! This one
has only 24,613 miles on it! A true barn find! Last driven in 1979!
Don't miss out on this once in a lifetime opportunity!!! A terrific
project car either restored fully stock or custom. Engine rated at
120 H.P. Car weighs 1600 lbs. Located in Algonquin, IL. PLEASE CALL
ONLY — (630)XXX-XXXX. $7000.00 or best offer / interesting trades!"
```

There were a few pics in the ad of the car outside, but others showed the car in what looked like the lobby of an office.

Photos from the original Craigslist ad

Earlier, I said that, when looking for a car, hopefully you find something that's niche-y and weird enough that fifty other people don't also want to buy it and drive the price up. And so, at this juncture, I'll reveal to you a crucial bit of information about this particular '74 Europa TCS: It was not only dead, it was brown and dead. Now, you have to understand that I *like* brown. For the years I was at UMass, I think I wore nothing *except* brown.

My beloved Porsche 911SC was brown (well, "rosewood"). So I'm dead serious when I say that, for me, the fact that the car was brown was actually a plus.

I can't find any email back and forth with the seller, so it's likely that communications were carried out via phone. I recall him explaining that he owned an independent repair shop that worked mostly on Porsches, and that the Europa was basically a coffee table in the waiting room.

Obviously, with the car located in Chicago, there was no easy way for me to look at it myself. Fortunately, my friend Ben Thongsai, BMW mechanic extraordinaire, not only lives in Chicago, but owns a Lotus Elan, so he knows what's what. There's an email trail with Ben that begins just after the purchase (it follows below), but as far as pre-purchase archived communication with Ben, there's nothing. So, as with the seller, I must've done it over the phone; I probably called Ben and asked him if he'd look at the car for me. If memory serves me correctly, Ben made a trip to check out the car and then called me from the premises. His nutshell was that if I wanted an original unrestored project Europa, I was unlikely to find a better one, though he gave me the expected amount of shit about it being brown.

I authorized him to negotiate for me on my behalf ("Oh, you don't want to do *that!*" he said). He must've fronted me the deposit, because there are references to my paying six grand for the car and I have my receipt for the $5800 bank check that went to the seller and an email correspondence with Ben that mentions my sending him $200. I thought I also had paid him for his services; I'm a bit unclear on how that math worked out.

But regardless of the transactional details, sometime on the morning of Saturday, June 8th, 2013, Ben called me up and uttered these words that will ring in the immortal halls of automotive literature:

"Congratulations. You own a little brown thing."

---

email: Janet Barnes at Bentley Publishers

```
From: Rob Siegel
Sent: Friday, June 07, 2013 2:24 PM
To: Barnes, Janet; Iglesias, Maurice
Subject: Hack Mechanic Review Now Out in Roundel

I just got my copy of Roundel magazine with Paul Wegweiser's review
of Memoirs of a Hack Mechanic in it. So we can add this to the list
of things we're crowing about...!
```

```
From: Iglesias, Maurice
Subject: RE: Hack Mechanic Review Now Out in Roundel
To: "Rob Siegel" "Barnes, Janet"
Date: Friday, June 7, 2013, 4:00 PM

Just added the review to the Bentley website
```

```
From: Rob Siegel
Sent: Friday, June 07, 2013 4:16 PM
To: Barnes, Janet; Iglesias, Maurice
Subject: RE: Hack Mechanic Review Now Out in Roundel
```

Damn, Maurice, where do we go from up?

Have a great weekend you two. Maire Anne just called me and asked me if I wanted her to buy a couple of lobsters for dinner.

Oh, and Ben Thongsai is having a look at this for me in Chicago tomorrow morning:
http://chicago.craigslist.org/nwc/cto/3835852104.html

See you Sunday.

--Rob

```
From: Barnes, Janet
Subject: RE: Hack Mechanic Review Now Out in Roundel
To: "Rob Siegel", "Iglesias, Maurice"
Date: Friday, June 7, 2013, 4:19 PM
```

Whoa! A Lotus Europa??? The book is doing well, but don't get too far ahead of the royalties!!!

I'm really looking forward to Sunday. Can't wait to meet Maire Anne. And I think she just gave me an idea for tonight's dinner . . .

Janet

```
From: Barnes, Janet
Subject: RE: Hack Mechanic Review Now Out in Roundel
To: "Rob Siegel"
Date: Friday, June 7, 2013, 4:29 PM
```

Forgot to mention that that Lotus Europa is sooooo '70s that it hurts. I mean someone really hit that with the ugly stick. Ouch!

Not that I mean to criticize your sense of aesthetic or anything . . .

J

```
From: Rob Siegel
Sent: Friday, June 07, 2013 4:37 PM
To: Barnes, Janet
Subject: RE: Hack Mechanic Review Now Out in Roundel
```

Whoa. You're honestly going to diss my Europa? When I gave you the idea for LOBSTERS? Sheesh. I thought we had a relationship.

```
From: Rob Siegel
Sent: Friday, June 07, 2013 4:47 PM
To: Barnes, Janet
Subject: I'll even take the Rambler over the Europa
```

The Europa is the first time you have diverged so sharply from my own sense of automotive charm, beauty, and/or sex-appeal. I'll even take the Rambler over the Europa.

Just saying.

But I do owe you for the lobsters. David and I have been flailing over what to have for supper all day.

J

```
From: Rob Siegel
Sent: Friday, June 07, 2013 4:47 PM
To: Barnes, Janet
Subject: I'll even take the Rambler over the Europa

You are dead to me now.

From: Barnes, Janet
To: Rob Siegel
Sent: Friday, June 7, 2013, 04:48:25 PM EDT
Subject: RE: I'll even take the Rambler over the Europa

Well, at least we got the book published before this fatal schism
occurred.
```

Pre-TLC: A low-key purchase announcement

Although it wasn't yet labeled "The Lotus Chronicles," the first post on my just-purchased passion object was this. Considering what it kicked off, it's surprisingly low key and matter-of-fact.

JUN 8, 2013, 11:40 AM

I just bought this 1974 Lotus Europa Twin Cam Special. 'CCA guy Ben Thongsai (a.k.a. Zen Ben) looked at it, acted as my agent, and bought it for me. He just called me up and said "Congratulations. You own a little brown thing."
[expired Craigslist ad for 1974 Lotus Europa]

16 LIKES, 24 COMMENTS

Harry Bonkosky Looks very nice. Congrats!

Jeremy Novak I didn't think there was a way to make a Europa less attractive than it already was. I didn't realize they came in brown! Good luck with it.

C.R. Krieger Ben knows damn well that you aren't going to be able to drive (especially in this thing) 1100 miles, one way, to kick his ass... +2

Dale B. Phelps hahahaha

Ali Jon Yetgen There is a Europa parts car at the Hammond PickNPull... maybe pull some spares on your way home!

Franklin Alfonso Love the first 2/3rds of the car. Back 1/3rd, not so much lol.

Delia Wolfe "Brown. It's the 'new silver'." Don't forget to submit your new ride to "The Brown Car Blog." If Maserati can paint a Khasmin brown, will an Aston Martin One-77 be next?

Harry Bonkosky Like this one? [link to photo of brown Aston Martin one-77] +1

Rob Siegel I was brown before brown was cool +2

Delia Wolfe [link to Mothers Of Invention "Brown Shoes Don't Make It"]

Franklin Alfonso But can you get it in beige?
Rob Siegel I believe the paint code on this one is sepia. There's a beige that's burnt sand. +1
Rob Siegel Of course, having "burnt" describing anything in a British car is just asking for trouble.
Delia Wolfe Dammit! Now I'm searching craigslist for a Siennabrun 2002. +1
Franklin Alfonso I believe burnt comes standard with Lucas electrics.
Delia Wolfe Burnt Sand sounds like the name of a British Game Show Host. Announcer: "And now, here's your host, Burnt Sand!" *Applause*
Chris Weikart Rob, I can now see that your holy grail just might be a Studebaker Avanti.
Rob Siegel yeah those are very cool too
Steve Kirkup I once read in a book by a friend, I believe it was in chapter two, "The British Piece of Crap" ..."I never bought another British car (proving that men, in fact, can learn, contrary to direct experimental evidence by Dutch scientists)". Congratulations by the way, nice find!!! +1
Rob Siegel that guy's an asshole +2
Steve Kirkup I used FedEx believe it or not to ship my 912 and was very satisfied. Pricey but safe.
Scott Sislane Cool! I love those. Get some racing decals on it.

Interjection: Shipping

With the Lotus purchased, I had to get it home. I'd never done anything like this before; this was the risk quad-fecta in terms of buying a car sight-unseen, having it be something I was totally unfamiliar with, having it be not only dead but sitting for 34 years and having a seized engine, and needing to contract for shipping.

The handful of times I'd previously shipped a car, it had been a running commodity car (something neither rare nor potentially valuable), so either I as the seller or the other party as the buyer had used inexpensive open-transport shipping. However, when you do this, you are almost never contracting directly with a shipper; you're contracting with a broker who contracts with a shipper, maybe several, and the car may get moved on and off transports several times before it gets to your house. With a valuable car, you typically want not only enclosed shipping, but enclosed *point-to-point* shipping where you're contracting directly with the actual shipper and where the car isn't moved around between different transports without your knowledge.

Further, although the Europa wasn't particularly valuable, it *was* particularly dead, and shipping a non-operational ("no-op") car carries its own risks and expenses. Shippers understandably need to know if a no-op car rolls, steers, and brakes. I verified these things with the seller. He said that the tires were ancient but held air, and that he had the keys so the steering wheel could be turned. The brakes, he said, appeared to be seized but not locked, meaning that the car rolled but the brake pedal appeared do nothing. The handbrake appeared to be functional. I passed this information along to shipping companies from whom I was soliciting quotes.

Since the car was a fragile little fiberglass thing, I didn't want it shaken to death on the trip, so I began looking for point-to-point enclosed *air ride* shipping. The best quote, just about $1200, was from Intercity Lines. In addition to the price, the fact that they're headquartered in Massachusetts, that Jay Leno uses them to ship his cars, and that the drivers work exclusively for Intercity helped sway me.

So, I paid. And I waited.

Phone call: Paul Wegweiser

JUNE 9TH, 2013:

"And if, after that thing arrives and you've rolled it into the driveway, you don't crack a beer and just sit and look at it for half an hour, you bought the wrong car."

email: Ben Thongsai

```
Sun, Jun 9, 2013 8:18 am
From: Rob Siegel
To: Ben Thongsai
```

Hey, Ben, the smile still hasn't worn off my face. I can't thank you enough. A $200 check will be in the mail Monday morning. Send to your B&D Auto address?

I'm occupied today with the barbecue Bentley Publishers is throwing, but my mind is ticking through the possibilities. As you know, despite my wing-nut persona, I actually think things through pretty carefully.

Can you give me a rough order of magnitude estimate for what it would cost for you to take the car, pull the engine, have the machinist you mentioned go through the engine, and put it back in?

No rush.

--Rob

```
Mon, Jun 10, 2013 11:27 am
From: Ben Thongsai
To: Rob Siegel
```

Hey Rob,

First off, before I forget again, have to say I loved the line "magnificent Semitic schnozz" in reference to Yale in your book. Almost done with it. Great read.

Having talked to the machinist briefly the other day before going to look at the car, he explained the engine removal process and after looking at the car, it looks pretty straightforward. Certainly a lot easier than getting the engine and transmission out of an Elan. So I'd guess 3-500 bucks should cover the removal and reinstall of the engine and dropping it off at the shop. Don't know how much he'd charge, I can ask him later today since I need to drop off an M3 cylinder head to get worked on anyway.

There's a chance that the engine may rock free? Cylinders 2,3, and 4 looked good with the borescope. Cylinder 1 has been sitting with some oily fluid in it for at least the last 2 years. Might be worth a shot to pull the valve cover and verify that the cams and cam bearings look good and try to gently rock the motor free.

Ben

Mon, Jun 10, 2013 11:41 am
From: Rob Siegel
To: Ben Thongsai

I'm reading up on the Zetec conversion. Man, Zetec engines are dirt freaking cheap. But of course you pay for it in the adaptation.

My current thinking is that I may just want the car delivered to me, where I can see if I can break the piston loose, see if the engine is rebuildable, and if not, look at other options including Zetec.

Check in the mail.

Thanks again.

--Rob

Mon, Jun 10, 2013 12:03 pm
From: Ben Thongsai
To: Rob Siegel

I really think the engine should either be rebuildable or okay as is if it breaks free. You'll still need to do the water pump on it though...

I would really, really try to give the twink [the Lotus Twin-Cam engine] a chance. It's a great motor, good power for its size, sounds good, surprising mid-range punch too. Plus it has a cool history to it. I mean where else would you find a motor where a crazy cheap Brit genius decides to have a twin-cam head designed to fit a run of the mill Ford block? And have it turn out well? Parts aren't too expensive and you can correct the couple of weirdo flaws with modern parts.

Although if you don't want it, I'd love to build a Seven of some kind with that motor, or an Escort Twin Cam…

Mon, Jun 10, 2013 12:08 pm
From: Rob Siegel
To: Ben Thongsai

Oh, believe me, I intend to give the motor far more than just "a chance." As you know, I am above most things a very practical guy. I'm not a purist to whom anything other than the twink motor would be an affront to man and god, nor do I want to stuff some 300hp supercharged beast into it. But when I look on eBay and see this [expired eBay ad for freshly rebuilt Lotus-Ford Twin-Cam engine for $6800] and then look on CL and see this [expired Craigslist ad for Zetec engine for $200]... like I said, I'm a practical guy.

:^)

--Rob

Mon, Jun 10, 2013 12:27 pm
From: Ben Thongsai
To: Rob Siegel

That's not a very fair comparison now is it? :)

Hell, for 6800 bucks I would expect that you could have this car fully sorted and great short of perhaps a high-level paint job. Which it needs to be a show quality car but it has a great "patina" look to it as is. I figure you'll end up spending in the 2-5K range depending on how much you do yourself and how much you pay others to have a sorted Europa without the fresh paint.

Ben

Mon, Jun 10, 2013 12:32 pm
From: Rob Siegel
To: Ben Thongsai

I will be doing a happy dance if I can get the car sorted and running in the $2k to $5k range (no paint).

Coming into this as a Lotus noob, it was jarring to hear "oh they made thirty thousand of these Lotus-Ford Twin-Cam engines they're not exotic" and then to see the first one I saw advertised have a $6800 price tag.

I spoke at length with Wegweiser. He said "admit it. When this car arrives, you're just going to sit in the garage, crack a beer, and look at it for 90 minutes." He's right. If you DON'T do that, buddy, you bought the wrong enthusiast car.

Mon, Jun 10, 2013 3:46 pm
From: Ben Thongsai
To: Rob Siegel

Not exotic, but 30,000 with the last one made almost 40 years ago they're not exactly common either. And maybe I've been in this business too long but $6800 for a fully done (assuming it was done properly) somewhat rare motor sold outright doesn't seem outrageous. Might not make sense for a $12K Europa, but since when does much of this hobby make sense? :)

Besides, try pricing an M motor sometime... (US E36 excepted)

I'm thinking that getting the brakes sorted is going to be more of an annoyance than the motor. Twin remote servos, fun...

And yeah, I usually end up staring at any cool new acquisition for a while, just because.

Wed, Jun 12, 2013 8:46 pm
From: Ben Thongsai
To: Rob Siegel

Hey Rob,

Just checking if everything went okay? I don't have to go break any kneecaps? :)

Are you transporting it straight home or should I plan on trailering it somewhere?

Doesn't matter either way to me, just want to plan for it if I need to.

Smile still on your face?

Ben

....
Wed, Jun 12, 2013 8:58 pm
From: Rob Siegel
To: Ben Thongsai

Hey, Ben, sorry, been trying to juggle Europa obsession with pretending to have a real job.

No, no kneecap breaking is necessary, though I appreciate the offer, and know that you have the tools.

Yesterday I contracted with Intercity Lines to transport it here to Newton. $1195, enclosed, no moving it between trucks, and I believe their trucks have air ride (having seen geophysical equipment I've towed get beaten to shit, this was quite appealing). They were the best price for enclosed no-broker shipping. Shipping date not yet scheduled, but I'm assuming I'll have it here sometime next week. I'm already trying to do the calculus of garage spots, where the lift goes, imaging stashing the fiberglass body in the 5th garage space beneath the deck, etc. If I could find a second mid-rise lift for short money (since I imagine the Europa will occupy mine through the fall, winter, and spring), I'd buy it.

I've been posting and lurking on several of the Europa forums. The body of advice seems to be to rebuild the twink—that that'll be the fastest way to get the car back on the road. The Zetec route is very appealing, but it's not like SVX engines in Vanagons where there are web sites devoted to it and vendors who sell you everything. It's more like a handful of guys who have done it, and each one is different, and no click-and-buy solutions.

Spoke with Sam Smith today. He said, and I quote, "WHAT THE HELL ARE YOU THINKING actually Ben sent me a pic and it looks pretty nice."

More when I know it. Thanks again.

--Rob

....
Mon, Jun 17, 2013 9:25 am
From: Ben Thongsai
To: Rob Siegel

Was hanging out with Sam at the track yesterday and we both thought about you as we were watching this happen. Welcome to Lotus ownership!

```
From: Rob Siegel
To: Ben Thongsai
Mon, Jun 17, 2013 1:46 pm
```

Yes, Sam was kind enough to send that to me, with the comment "I call this the plight of the Lotus owner: a symphony in one part."

My response was "bite my shiny fiberglass ass."

....
```
Mon, Jun 17, 2013 2:56 pm
From: Ben Thongsai
To: Rob Siegel
```

Hey, at least it's not on fire. Yet. :)

Pre-TLC: Shipping arrangements

JUN 12, 2013, 9:17 PM

I'm getting the Europa shipped from Chicago to Boston. Turns out the firm Jay Leno uses (Intercity Lines) is local to Massachusetts and gave me a great price for door-to-door covered shipping, no brokers, no moving between trucks. The idea of covered shipping with air ride seemed utterly frivolous until I thought about a fragile fiberglass-bodied car that's sat indoors since 1979 being shaken to pieces.

26 LIKES, 35 COMMENTS

Adam Pepper A wise choice...Intercity was (and likely still is) the way to go for damage-free transport.
Chris Weikart Rob are you sure this is a good idea? +1
Rob Siegel oh hell no
Richard Memmel Looks fun
Kenneth Robb The Ford expert in San Diego I mentioned to you is Bill Schlossnagel. It's possible I misspelled his name. Well the "Bill" part was easy.
Roy Richard Don't put the big Bruins sticker on it until it is clear of Chicago. +2
Elizabeth Brackett My husband restored a Europa many years ago. He's currently working on an Elan and a Renault 5 Turbo 2...
Rob Siegel No kidding... I may need to pick his brain.
Elizabeth Brackett Let me know! He says he still has the shop manuals (for the Twin Cam) and a spare

gas tank. He's actually had two, one before we met. Both Specials, a '73 and a'74. ₊₁

Lance Johnson Wayne Carini uses Intercity too and oh by the way, I've always loved the Europa.

Jeremy Novak Exactly. If it's going to rattle itself to pieces, it should do it while you're driving it!

Dale Robert Olson I tried sitting in one once, didn't fit.

Christopher Wootten Tried to use Intercity for a late July MD to TX shipment but they are booked up due to Pebble Beach. Thoughts on others?

Rob Siegel After Intercity, Passport, Applewood, and Enclosed Vehicle Transport had the next most competitive quote in the high-end world. I didn't solicit specific quotes from brokers (as opposed to shippers) and non-enclosed transport; instead I posted it on UBid and saw where they started to come in, then cancelled the auction. If this was a mid-priced BMW (say, for the sake of argument, a $6k M3), I don't think I would've had a problem going with the low-priced spread, but a dead fiberglass car is different. ₊₁

Franklin Alfonso Ah. Soon the madness begins lol. There may be a lot of "youmayaswells" when you start work on it. But I guess you already know that.

Jonathan Bush Drove one once 20 years ago with a hot-rod Twin-Cam Lotus engine and 45 DCOE Webers. It was terrifying, but in a good way. Like when you get away with cheating death, all you can do is grin stupidly.

Rob Siegel cheating death and grinning stupidly are the twin goals we're going for here :^) ₊₄

Charlie Gordy Have always liked Lotuses, but the question is, can they speak German???

Roy Richard I'm looking forward to the "youmayaswells." Bet most of them include Lucas somewhere. I never tire of Prince of Darkness jokes :)

Roy Richard Hey Rob since you are in Lotus mode how about a real classic so your Europa has someone to talk to. [link to eBay auction of Lotus Seven]

Rob Siegel yes, well, by all accounts, the Europa is the gateway drug to the stronger stuff, of which the Seven is crack cocaine ₊₁

Christopher Wootten Rob, thanks for the information on shippers. I made the mistake of going to a broker and am getting hammered by a robodialer. Another guy got a hold of me and tried to close me on the spot asking for a credit card deposit. I pushed back and searched them later and found that they have a horrible reputation for service and sales tactics.

Rob Siegel Wow, Christopher, I've read about that happening. The high-end shippers I mentioned all had good on-line quotation services.

Dave Farnsworth I once went to look at a Europa. Rosemary quickly went out and bought me a 1/20th kit version. After fiddling with it for an hour or so I gave up on both ideas. Smart woman.

Rob Siegel I have no doubt I am wading into waters way over my head. But I decided that, with 24k miles, with it having been stored indoors since 1979, with Ben Thongsai having looked at it, with the cost of $6k, with appreciating values of Europas, there's not that much risk in the purchase. The risk comes when I start taking it apart.

Scott Aaron The rear bumper is in an interesting place vis a vis the lights...just some interesting weirdness back there.

Rob Siegel I posted the question to a Europa message board "how did the '74 Europa escape the 5mph bumper madness?" There appear to be two answers: 1) nearly all of them were actually made in '73, and 2) production numbers were so low that they petitioned for and were granted a waiver, as did other manufacturers. ₊₁

Roy Richard Rob, not much risk in taking it apart, hell most Brit cars do that on their own. The risk comes in trying to put it back together.

Franklin Alfonso Hey, how can you pass on British Racing Brown lol. ₊₂

Franklin Alfonso So when is the new chariot arriving?

Rob Siegel Trying to find that out; shipping order is booked but have not yet heard arrangements.

Franklin Alfonso You got your 7/16ths 9/16th 1/2 inch spanners ready? I would be like a kid just before Christmas going crazy lol. ₊₁

Rob Siegel Actually I own very few non-metric wrenches. Reportedly, since the engine is Ford-Lotus and the transaxle is Renault, the car requires both.

Franklin Alfonso Those three I mentioned took my MGA apart lol.

Pre-TLC: Impending arrival

JUN 20, 2013, 7:25 AM

Anyone else having a Lotus Europa delivered today? No? Just me I guess.

27 LIKES, 21 COMMENTS

Paul Wegweiser Really? All the cool kids got theirs LAST week, Rob. [photo of a Europa in the shop Paul works at] +5

Rob Siegel Damn! Once again, I'm a day late and a dollar short.

Paul Wegweiser I'm seriously looking forward to loads of project updates on yours!

Rob Siegel Paul, tonight, as per your suggestion, I intend to enter into the "sitting in the garage with a beer in my hand and looking at it" phase, because, let's admit it, it's just going to go right into the crapper from there. +11

Paul Wegweiser I expect you'll mutter the following phrase many times during your resurrection of this vehicle: "Gee... whoever designed THIS sure was an asshole." +3

Daniel Senie Hope not, anyway :)

Rob Siegel So... anyone with whom I went to jr high school in Amherst may in fact recall the roots of this Europa obsession. When I was in 8th or 9th grade, I worked at a little stereo store called "Sound Ideas" in the alley, in that same building as The Amherst Bicycle Shop and For The Record before they moved to Faces of Earth (when Faces was still in the old Amherst Candlepin Lane building), and, going back even further than that, "The White Light" bookstore and head shop. This building was razed to make way for Boltwood Walk. Anyway, the stereo shop's owner, Joel Artenstein, had a red Europa. Tom Porter, do you remember this?

Tom Porter Robby I do recall the stereo shop and the alley, I recall the beginning of a generalized obsession, but I don't recall the Europa nor its owner. But I suppose at that moment the die was cast and your "hack" fate settled!

Alexander Wajsfelner Of course you have it running by British car day at Larz Anderson this Sunday...

Rob Siegel Yeah I have a can of official Lucas Electrical Wiring Harness Replacement Smoke that'll fix the whole thing right up +4

Franklin Alfonso Well I hope you will get many hours of enjoyment out of it. But be forewarned she might be a real ball-buster. I heard Tony Soprano had one and look what happen to him...

Matt Pelikan No, but I've got a box full of dead moths.

Rob Siegel they're probably more reliable

Kenneth Robb well we got a new dog a week ago so we are ahead of you.

Nick Macdonald Beware of falling asleep in the garage staring at it—you may never wake up.

Ruth Goldberg WAHHHHHHHHHHH!!!!!! That is amazing, Rob. Congrats!

Delia Wolfe Eat your heart out, Joel Artenstein.

Paul Muskopf So when do you want an M42 in it?

Steve Collins Bad Luck! Sorry.

Interjection: Delivery

On June 20th, 2013, what appeared to be the largest car carrier in the world arrived in West Newton. Due to my street being narrow with overhanging wires, and my house being at the corner of two streets with a sharp right angle turn from one into the other, the big auto transporters can't fit down it, so whenever I have a car shipped, I need to tell the shipper to meet me three left turns away on the other side of the park behind my house.

When I received the call that they were just minutes away, I frantically corralled my next-door neighbor Dave and his son Jake, my wife Maire Anne, and our close friend Kim who lives on the other side of the park. We all walked out onto River Street and watched as the tiny fragile-looking little Lotus was unloaded from the upper level of the enclosed trailer and rolled out onto the street. The five of us then pushed the 1600-pound car the three right turns to get it to the top of my driveway.

Initially I walked beside the car and leaned in with my arm while they pushed, but the car was so light that they did this with no difficulty at all, and they goaded me into getting inside and "driving." I was giddy, smiling like an idiot, making *vroom vroom* noises inside my new purchase.

But then, when we got to the top of my driveway. I realized something. The driveway slopes down into the garage. It's not a steep slope, but it's not something you want to be wrong about either. I remembered about the Lotus' brakes not working. I hit the pedal, and found, as the seller had said, that it felt like the brakes were seized, though whether it was the pedal itself or the hydraulics I had no way of knowing. The handbrake, though, appeared to work. I had my helpers move a wheel chock about six inches at a time, and by engaging and releasing the handbrake, we inched the Lotus down the incline, making sure it wouldn't gather too much speed. We all breathed a sigh of relief when its nose was in the garage and it needed to be pushed again.

I got out, not knowing that this would be the only time I would "drive" the car for almost six years.

Just then, the glass on the left mirror fell off and landed on the concrete floor. Miraculously it didn't break. We all laughed that the car I'd bought was not only dead, but parts were literally falling off it. Six years later, when I got the car running, began driving it, and cared about trivial issues such as rear visibility, I remembered this and miraculously located the mirror glass and reattached it.

My garage is shaped like a shoebox, 31 feet long by 17 feet wide. It's attached to the corner of the back of my house. There is a single roll-up door the width of one car. The garage can hold up to four cars if they're each less than 15 feet long and if they're arrayed in two rows of two, but if you do that, the two cars on the left become sardined in. Since clearly the Lotus wasn't going anywhere anytime soon, we rolled its nose into the least-accessible left rear space, then jacked up the back of the car and shoved its ass end over. That's where the car would stay for almost six years, but I had no way of knowing that either.

Incredibly, I suppose due to the excitement, there are no photographs or video of any of this.

Pre-TLC: Lotus Position #1

JUN 20, 2013, 8:22 PM

Night gathers, and now my watch begins. It shall not end until my death. (Note: Photo inspired by an offhand comment by Paul Wegweiser...)

Blair Robinson Very poetic! You sound like a "Game of Thrones" fan! Lol

Kevin McGrath That's funky. I bet it sounds like a Lotus too. Nice lines.

Elizabeth Marion I thought you swore off the color 'brown' years ago?!

John Whetstone Ha! You did it!

Paul Nerbonne But ... winter is NOT coming

Blair Robinson The time of discontent shall bear upon ones soul as the harsh winds sever the bones! Seriously... Very happy for you Rob. I know how you feel. I often sit out in my garage, drink a cold Yuengling and admire my 745i. It's my Europa. I will even sleep in my garage! It helps that I converted it into a bedroom though, lol

Bernie Sharpe I am the shield that guards the realms of men.

Blair Robinson ...for the realms of Men wIther...

Blair Robinson Rob, I know it's not a 1968 yellow VW Bug, but you should still get a 28IF LMW license plate!

Chris Lordan "...the deeds of a man ensnare him; the cords of his sin hold him fast. He will die for lack of discipline, led astray by his own great folly..." I modified it a bit; it's now from the Adverbs.

Rob Siegel I am the wrench in the darkness

Steven Anthony Smith Wench in the darkness? Hmmmm.

Blair Robinson You are a wretch alright!!...oh sorry, you said wrench... my bad... lol

Harry Bonkosky Another unsuspecting soul lost to Lucas, Prince of Darkness.

Scott Aaron You beautiful freak. :)

Matt McGinn Time to bake the bread. For that van. Rob other than a like that's the best I got! Great car just love it like you've loved all the others! And write. Take parts of that car, or the whole car apart and write about what the builders/previous owners were thinking.

Rob Siegel Matt, yes. I began sussing out how to pull the engine and transaxle, and noticed that the long thin rear wishbone doesn't mount to the frame, but to the transaxle itself. Pull the transaxle and—unless I'm mistaken—it looks like there's nothing holding the rear wheels vertical. The car has to be up on jack stands.

Matt McGinn Stressed member.

Matt McGinn VERY stressed member. [Note: this is the first of MANY stressed member jokes]

Rob Siegel Me? Yes. Oh, the transaxle. Right.

Blair Robinson Stressed member! I broke up with a girl once cause of that! Lol

Steven Anthony Smith [link to The Road Warrior "How's the rig?" scene with long recitation of everything that's wrong with it]

Rob Siegel If I'd bought a Jensen, I could say "last of the V8 interceptors..."

Richard Memmel Its my bday why did you get the present?

Blair Robinson The past, we cannot change! The future, we cannot predict! Here, right now, is a gift. That is why it is called the "present!"

Paul Wegweiser This photo. Is a poem.

Blair Robinson A Greek tragedy?

Paul Wegweiser Yeah... like you needed inspiration. NOT.

Scott Sturdy It's asking Rob "Does this garage make my butt look big?"

Blair Robinson Looks like it would be a good getaway car!

Kenneth Robb Perhaps you should consult with Peter Egan on the best techniques for pondering English Garage Projects.

Rob Siegel He IS the Brit Zen master, yes.

Kenneth Robb And an astute judge of one's mechanical ability. I was visiting him some years ago the day he was about to install the engine in his just-restored XKE. Though I offered at least three times to stick around and help him he sent me off on a scenic drive of Wisconsin. No dummy, that Egan.

Daniel Senie It's so... short!

Kenneth Robb Hurry and dig into this pile to find all the problems. Then I can harass Thongsai when he's here for The Fourth!!

Rob Siegel No fair harassing Ben. I roped him into this, not the other way around.

Kenneth Robb I don't care. It's just fun to harass Ben.

Jim Sillery Jr No you didn't...

Franklin Alfonso I have heard it said..."There is no cure for Lotus" only treatment. Yours begins now...

Richard R Mackey Good practice for your 2nd career—suicide watch at prisons—plus, a new chapter for your 2nd book.

David Weisman Lotus! damn boyeee

Rob Siegel "damn" is not exactly the reaction I'm having; WHAT THE HELL DID I DO is a bit closer.

Charlie Gordy You know, I bet if you concentrate you can get it off the ground without the lift!

C.R. Krieger I see that you've found its highest and best use...

Steve Kirkup I had the "What the hell did I do?" feeling when I got the Model AA delivered to a storage barn that I had to rent. Now I just say "What the hell should I do now, and why haven't I attempted to do it?" when I write the rent check. BTW the watch is much more fun in a recliner.

Rob Siegel Steve, the only reclining allowed in this garage is flat on your back beneath a car.

Steve Kirkup Well I suppose if you put the Lotus up on your lift you could sit with a beer under it. I can't do that.

Rob Siegel Actually, you'll notice the lotus is not up on the lift. It's a fiberglass car with a steel backbone. Jacking it up and securing it is challenging. There are lifting points on the fiberglass but the Loti (Lotia?) say not to use them. I'm not certain I could put it on the lift. At least it answered the question "where in the garage does it go?" (Answer: In the corner. On very carefully-placed jack stands.)

Steve Kirkup So did anything get displaced for winter storage? Or do you have room sine you got rid of the other 02?

Janet Barnes Shouldn't you be in the Lotus position, Rob???

Rob Siegel HAHAHAHAHAHA shit I should've thought of that

Steve Kirkup I think you need to take that photo!!!

Roy Richard Is the Lotus position the one where you pull your head between your knees and kiss your ass goodbye?

TLC Day 2: The drivetrain drop

JUN 21, 2013, 7:17 PM

The Lotus Chronicles, Day 2:
Began the engine pull. Not too bad, actually, just the usual assortment of stuck exhaust nuts you'd expect in any 40-year-old car. But it's going to need a lot. Any thought that a few raps on the stuck piston with a rubber mallet, throw the engine back in, gas and oil, and Bob's your uncle were squelched by looking inside the coolant hoses. 33 years is a long time to sit.

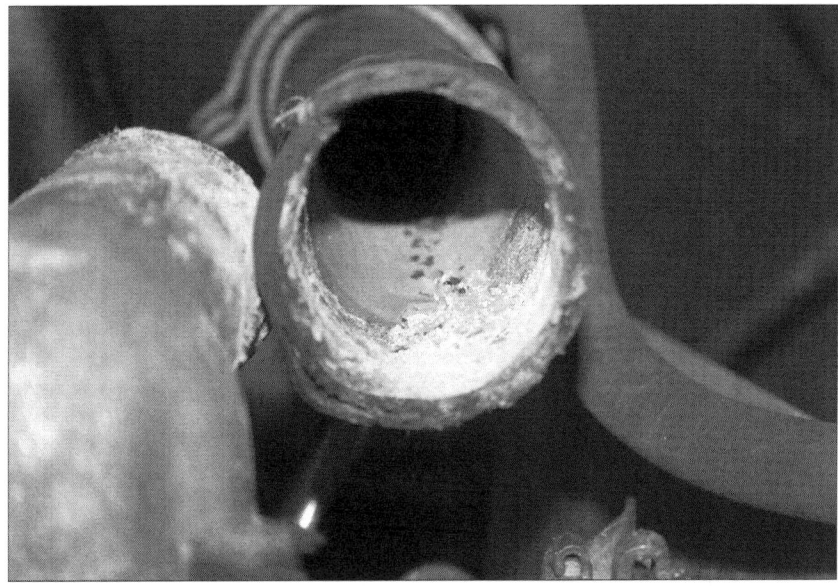

Paul Wegweiser Child's play! Get some Marvel oil and soak the cylinders each night for a week. Then see if it rotates! I'm pullin' for ya!

John Whetstone As per Roger on Chasing Classic Cars!

Ed MacVaugh Marvel Mystery Oil and STP, be running like a champ. My Dad (much older, born in '05) used to say "The only mystery is why people would pay good money for it."

Kenneth Robb It looks like it could benefit from a stent.

Carl Witthoft It's no fun if it's easy.

Rob Siegel FYI the previous owner had poured so much Marvel Mystery Oil into cylinder #1 that, when I pulled the exhaust, a giant red puddle of Marvel Mystery Oil spread out on my garage floor. The engine is still seized. It's not going to be that easy.

Matt Pelikan That's kind of what my arteries look like.

Franklin Alfonso Repeat after me...There is no cure for Lotus, only treatment... Looks like it will be a while till you have sex lol.

Roy Richard Rob are there any strange looking pieces of metal sticking out of the oil pan or block? If not, one word: SILIKROIL

Franklin Alfonso On the bright side you will probably get enough material for a second book, Hack Mechanic Vol 2, The Lotus Project.

TLC Day 3: The Lotus Position #2 and #3

JUN 22, 2013, 8:47 PM

The Lotus Chronicles, Day 3:

The Lotus Position, and the Advanced Lotus Position. (Kudos to Janet Barnes for The Lotus Position. The Advanced Lotus Position is me projecting. Nothing went wrong today; just continuing the engine drop. [Note: As time went on, I would, of course, be completely correct about The Advanced Lotus Position.]

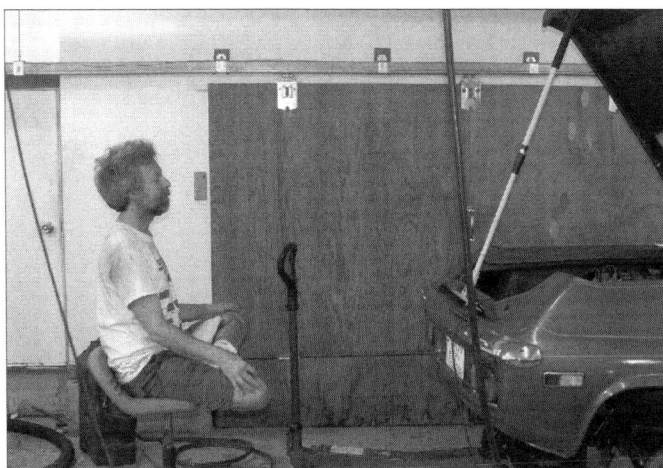

The Lotus Position.

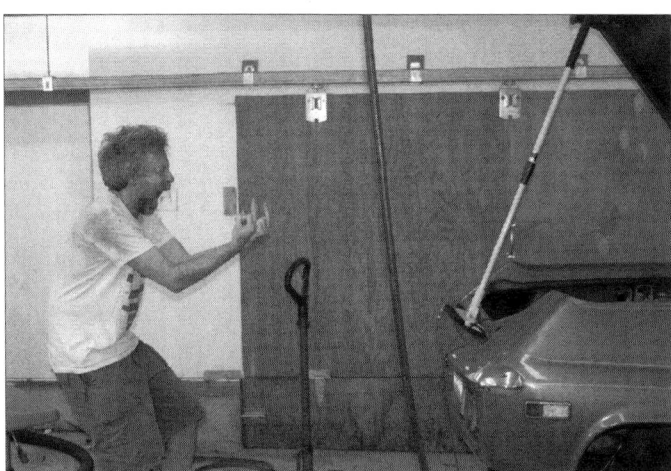

The Advanced Lotus Position.

26 LIKES, 20 COMMENTS

Faith Senie I'm just impressed you made it into and out of the Lotus Position!
Steve Kirkup I love the show of one single digit per hand expressing you're delight

Kenneth Robb Rob is just signaling Colin Chapman that he is still Number One! +1

Harry Krix Having been a long time Brit car fan before getting into 2002s back in the 70s... I often think back on what it would be like to own another of the Queen's tin worm ridden, no sense electrics, totally backward Loti, and then I wake up. +2

Franklin Alfonso Wax in wax out... patience grasshopper will be required on this project for sure. I think I am mixing my movie metaphors.

Jay Neal My 69 Triumph GT6 was enough punishment for me. BMW FTW!

Rob Siegel Jay, I talk about my '70 GT6+ in the book, and how I never bought another Brit, "proving that men CAN be taught, contrary to direct experimental evidence by Dutch scientists." Guess the Dutch were right!

Franklin Alfonso I guess I was lucky to have not kept my MGAs very long. Just long enough for fond memories before passing on the really big problems.

Jay Neal My Triumph Trident motorcycle was the only good Brit thing I ever owned.

Chris Weikart Rob, the Dutch scientists were male. +1

Chris Weikart Wait, I misread that. They were female.

Franklin Alfonso My brother in law Steve is a car guy. When he was young he had a Mini Cooper. Rebuilt the engine a few times. It was a blast to drive. Later he bought a Spitfire and modified it for racing. Now he works part-time during the winter down time from his pool business rebuilding Porsche transmissions. He now drives a GTI as his daily driver and has a Porsche Turbo 2+2 older model around 18 yrs old I think like Tom Cruise had in the movie Risky Business.

Janet Barnes Perfect yin and yang you've got going there, Rob!!!

Nick Macdonald Getting in character for road side antics?

Eric Heinrich Awesome, Only what, 3 days? Four?

Ruth Goldberg Love this photo!

Dan Tizzano Did you do the chant while in the Lotus Position? Ekancar, satnam...

Bo Gray Carmasutra... (3)

Rick Viehdorfer Oh dear... A Lotus. God help you! *smile* BTW: My brother was a factory-trained Lotus tech...

Thomas Bradley How now is that brown cow?

TLC Day 4: The stressed member

JUN 23, 2013, 8:12 AM

The Lotus Chronicles, Day 4:

Everything a BMW person needs to know about acclimating to living in Lotus Land is contained in this photo showing the so-called "stressed member," where the rear wishbone is attached to the transaxle itself, saving the weight of having a rear subframe. Toto, I don't think we're in Bavaria anymore...

CHAPTER I: SHIT GETS REAL   35

Ben Greisler Now that you have gone British, you will switch from being the Hack Mechanic to the "royally pissed off and mad mechanic." In general, the Brit cars were able to make me go from zero to hyperventilating in the least amount of time back when I worked at a foreign car repair shop. Rolls, Jag, TVR, Lotus, didn't matter; they all made me want to cry at some point, no matter how much I loved them as cars.

Rob Siegel Ben, I couldn't agree more. And yet something made me want to engage this project.

Ben Greisler And thus is the lure of British cars! I have fancied an early Elan for ages!

Franklin Alfonso This is exciting. Every morning before work I get to see how your Lotus adventure is proceeding. Thanks for the postings. I am enjoying the book very much too.

Scott Sislane Mapping out the attack? That looks scary under there.

Franklin Alfonso British engineering is different alright. It can be very maddening at times too.

Ernie Peters A good example of Chapman "adding lightness."

Rob Siegel Yes, Ernie, exactly.

Steve Kirkup That's an interesting photo, Rob. Gotta love the Brits. Trying to orient myself… is the exhaust partially removed, or does it normally take the route of large intestines? (My favorite part on my TR6 is the cardboard cover for the transmission. It's the only biodegradable part that doesn't last as long as the frame.)

C.R. Krieger I think you've overlooked the electrical system. BMW folks know that brown is always ground. In a Lotus, every color is a ground...

Rob Siegel Exhaust is partially removed. Of course couldn't get it disassembled (yeah right none of us live in THAT universe) so I dropped it intact.

Steve Kirkup My electrical system was eaten by mice before I got the car, so I assumed that it was edible, although I guess that would make it biodegradable too.

Matt Pelikan "Stressed member." "Dry Prong." What is it with you, Siegel?

Rob Siegel Siegel's third law: given sufficient time, any car discussion will eventually devolve into dick jokes.

Steve Kirkup Guess I took the wrong leg of the fork in the road with the large intestine bit.

Eric Heinrich I can't stomach these jokes anymore.

Carl Witthoft Doggone, Matt Pelikan beat me to it.

Roy Richard That looks properly OCB (Oily Coated British).

Paul Goldfine Damn, someone else beat me to the 'stressed member' comment.

Richard Memmel Good time to start over with the electronics in this car and make it correct while its apart screw originality how about functionality and works even in the rain I would be more than happy to guild you through it if you like not that hard or expensive?? Wouldn't it be nice to have s British car with no electrical problems?? No one has ever said that last sentence ever have they just saying:):)

Jay Neal The parts even smell different. That should have given you the first clue. :P

Rob Siegel At the moment, the interior stinks of brake fluid, as did, oddly enough, my 3.0CSi when I bought it (also a dead car; that one only took me 27 years).

TLC Day 4, cont'd: Roomy engine compartment

JUN 23, 2013, 4:27 PM

The Lotus Chronicles, Day 4 (continued):

The garage is now officially recognized as a territory of Lotus Land. You have to love an engine compartment with so much room you can stand up in it. Of course, there's even more room now that the drivetrain is out.

(You other folks go to the beach or something?)

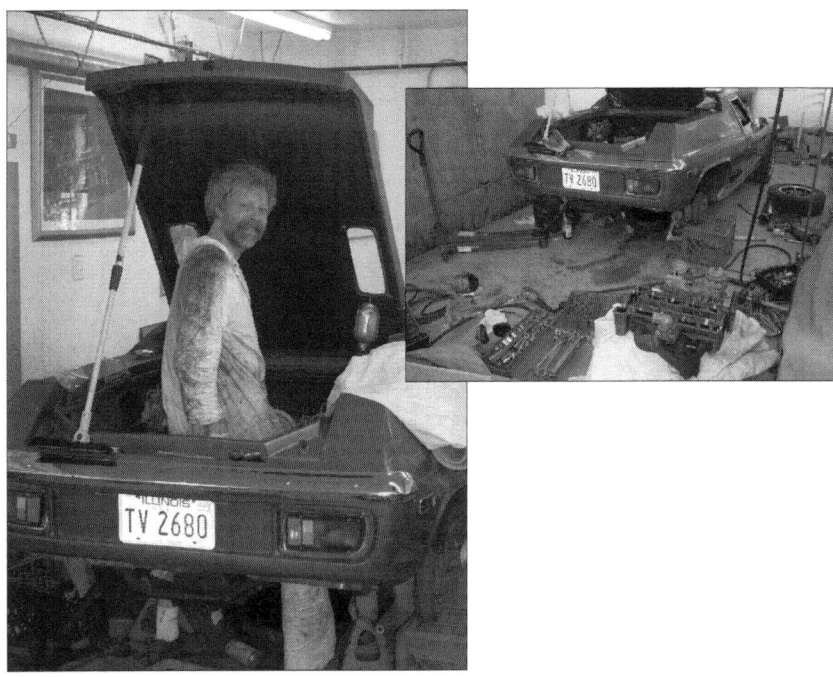

20 LIKES, 22 COMMENTS

Alexander Wajsfelner Yeah, the beach.
Rob Siegel Oh, Alex—THANK YOU for installing the air conditioner in the garage in 2005...!
Carl Witthoft Size isn't everything. And nope, I'm not at the beach. BTW, you do know what happens in the Land of the Lotus (eaters)?
Rob Siegel Uh, the electrical system doesn't work for shit? +1
Barbara Radford Alfonso Frank will be very impressed. He is enjoying your book so much... I am hearing all the stories. +1
Rob Siegel "The Hack Mechanic—living on the edge. So you don't have to." +1
Debra Adams Will this be your next book?

Rob Siegel Let's get this first book selling like gangbusters first, and we'll talk :^)

Franklin Alfonso Lolita would be a good vanity plate for this one Rob lol. Just when you thought you were out of the Brits, she sucked you back in...

Rob Siegel Actually, you're right; that WOULD make a great vanity plate. But the on-line MA registry where you can check such things (where I checked KUGEL prior to applying and getting it for the 2002tii) shows "LOLITA" as unavailable. How about 42INCHES?

Rob Siegel (that's the height of the roofline. GET YOUR MINDS OUT OF THE GUTTER.)

Franklin Alfonso Ha ha. Well it still may be many moons before you have sex with Lolita so you have time to check what's available. Maybe you can get the abbreviation of "Wet Dream." (1)

Rob Siegel Massachusetts lists restrictions on the selection of vanity plates, including "Such combination is obscene in that it refers to a sexual body part, a term for or most closely associated with a sex act, or the availability for sex." So I doubt WETDRM would cut it. I'm surprised if they actually issued LOLITA; it just says "not available."

Franklin Alfonso Oh they are such Prudes lol. Buzz killers.

Rob Siegel Actually, by the time I get it running, FLOTUS might be about right... and it appears to be available!

Franklin Alfonso POTUS1 Not available.

Franklin Alfonso I guess there is no timeline on how long this might take to get it on the road. Every day can bring a surprise.

Rob Siegel My goal is to NOT slide all the way down the slippery slope, to NOT restore it (no paint or body work), to sort out what it needs, and to have it drivable by spring. [Note: HAHAHAHA oh god that's funny]

Franklin Alfonso I get it. The main thing is to have it road worthy and hopefully enjoy driving it a bit.

Roy Richard Spring 20??

Carl Witthoft FLOTUS ok, FLATUS not?

TLC Day 5: Lubricating the seized piston

JUN 24, 2013, 9:26 PM

The Lotus Chronicles, Day 5:

Not much to report. Cleaned up the garage after the weekend Lotus-O-Rama drivetrain drop. Squirted 1/2" of Silikroil into cylinder #1, which appears seized (no visible gap between piston and cylinder wall). Hope to get the engine on the stand tomorrow and drop the pan.

Wouldn't want you to think I'm slacking.

(10 LIKES, 3 COMMENTS)

James S Eubanks Finished watching the part II episode of Overhaulin' Lotus Europa. The owner had a very reserved reaction. He also didn't fit in the cockpit very well. The power to weight ratio looks pretty good, though!

TLC Day 5, cont'd: The mouse nest

JUN 24, 2013, 9:51 PM

The Lotus Chronicles, Day 5 (continued):

After posting, I obviously felt I hadn't done enough, so I separated the transaxle from the engine, and found a mouse nest INSIDE THE BELL HOUSING. Now THAT'S a car that hasn't run in a long time!

Daniel Senie So is cylinder 1 fused due to mouse pee deposits corroding it, or will you find skeletons of mouses past under the piston?
Marc Bridge I never heard of a tranny mouse before. Not that there's anything wrong with that.
David Kerr That nest looks OEM to me.
Brad Day yeah, I think those are Lotus crumbs from the bread van.
Kenneth Robb Colin's idea of sound insulation.
Rob Siegel Is THIS what Colin Chapman meant by "add lightness?"
Craig Thomas Reisser Rob, Lotus Europa = you are a glutton for punishment
Ernie Peters Assuming there are original design drawings, I suspect this is not part of the original plan.
Franklin Alfonso I guess in old KIAs it would have been a hamster's nest.
Carl Witthoft So you found the mouse. Just a keyboard and monitor left to seek out.
Steve Kirkup Otherwise it looks pristine in there!!

Letter: Carl Baumeister

I spoke with the previous owner, Carl Baumeister to find out if he knew anything about what damage the engine had suffered. Carl said he'd write up his recollections. A few days later, I received a nine-page hand-written letter which I am reproducing below.

"Hi Rob. In regards to the history of the Lotus Europa Twin Cam Special, the prior owners told me that the car was originally purchased at the Lotus dealership in Chicago in the fall of 1974. The owner of the dealership was said to have driven the car personally and he ordered it himself (hence the expensive optional 5-speed). The man that I purchased the car from said that he drove it for several years (summers only), that in the fall of 1979 he underwent back surgery and had several of his vertebrae fused together. Although he was unable to get in or out of the car, he remained determined to drive it again and refused to sell it despite his family's pleading. The car sat in his garage for many years and had become a secondary work bench / catch all.

In the 1980s, his son did clean it off and got it running. He stated to have driven it up and back on their street only, since it was not plated or insured. Again the car was placed back in the garage and covered with other things.

His son had taken a job down south, got married, and started his own family. With his mom passed away and his dad struggling to remain independent, it was decided that they would sell his dad's house and move him down by them to live so they could help him when needed.

I met both the father and son by coincidence. I was stuck in construction traffic and was detoured to an area I would not normally be in and sitting in a long line of cars that seemingly never moved. To pass the time I was just looking around and glanced down this old driveway where there was a lot of activity—rented moving truck, a big green dumpster, and a steady stream of people throwing papers, boxes, old furniture, and other random items into the dumpster. After watching this for a while, I noticed the out-of-state places on the two cars in the driveway. I was intrigued since they were obviously not moving in. I continued watching as traffic started to inch forward (I was made aware of this by the blast of the car horn behind me—he was even more impatient than I was). I moved forward five or six feet and again stopped.

Again glancing back at the house, now having an unobstructed view of the open garage because of the dumpster, I saw it! Buried under a variety of items, the dusty, dirty, long-forgotten Europa Twin Cam. I knew what it was but had never really seen one in person. It just called out to me. Being the car guy that I am, I had to know more.

When an opening in oncoming traffic appeared, I pulled a fast U-turn with only a slight amount of traction loss :^). I didn't want to seem too eager but I was! I pulled into the driveway, got out, and approached a man coming out of the house. I introduced myself and inquired about the car. I was directed inside to meet the owner. He was older, average height, and very thin.

I again asked about the car, and he gruffly responded "I'm not selling my damned car!" Followed by similar pleasantries, I was asked to leave. I walked out of the house somewhat dejected.

I then heard footsteps approaching me from behind. I turned around defensively, not knowing what to expect. A much younger and slightly stockier version of the old man was approaching, and profusely apologizing for his dad's attitude. He explained that his dad is being relocated but not of his own free will. It just needed to be done due to his absent-mindedness.

I explained my reason for stopping and he invited me to look at the car up close and personally.

Under all the dirt and debris, I could tell it was a beautiful car and it appeared mostly complete. The son joined me in the garage a short while later and said that they were definitely not going to move the car down south, regardless of what his dad says! He went on to explain that they had lost the title many years ago but had gotten a duplicate a few years back. The rub was that the keys were lost as well. I was asked to stop by over the weekend and they would try to have it dug out from under all that stuff and hopefully he could work on changing his dad's mind about selling the car.

After a couple of sleepless nights, the weekend arrived and I drove back to the house. I was greeted by the son and was told that I had to sit down with his "old man" and explain my interest in his Lotus and my plans for it. After a few very uneasy moments his dad realized I was a real car guy and cared about the future of the car as much as he did and agreed to sell it to me. I picked up the car that afternoon and trailered it to my shop.

I eagerly unloaded my newfound toy and promptly started cleaning it up and writing a list of the missing/broken parts I needed to find. It looked pretty good cleaned up so I started pulling all the locks out to have keys made.

After much searching, I found a locksmith, a real old-timer that had the blanks and the books to look up the proper codes to cut new ones. A day later I had three keys [ignition, doors, and hood/trunk] that would lock and unlock everything on the Lotus. I started hunting for the missing pieces to make it complete. Knowing this was going to take some time, I relocated the car to a spot of prominence in my customer/showroom area

Everyone that enters my shop inquires about it. Most of them have never even seen one! I know that this car needs to be restored back to the greatness that it once had. Start dates on the project came and went, and it really deserves that attention! All I can say is that it is rich in history, but moreover, IT'S JUST COOL!!!

It is now time to pass the torch and start a new chapter of the rebirth of a classic Lotus Europa. I know that you will do right by it.

Please stay in touch during the process as I would love to know about the progress. These classics all need to be found, restored to their former glory and shared with future generations of car guys, gearheads, and all automobile enthusiasts!

Good luck!

—Carl Baumeister

[Note: After reading the letter, I called Carl to confirm whether I had it correct that the car "ran when parked" and the engine had seized from sitting as opposed to the car having suffered some catastrophic failure and then rolled into storage. He concurred that that was his understanding. Four years later, I would write my book, *Ran When Parked*, about a decade-dead 1972 BMW 2002tii that I bought sight-unseen in Louisville, drove a rented car to where it was, resurrected it in situ, returned the rental, and drove the tii home. In the book, I discuss how, the longer a car has been sitting for, the less relevant the fact that it "ran when parked" is, and how you need to address the issues of damage from having sat. So, having written *Ran When Parked*, the title of this book could and perhaps should be *Seized From Sitting*.

More to the point, even if the car had been driven into the garage under its own power at some point in the 1980s, it was now 2013, and in addition to dealing with the seized engine, the car was going to need… everything.]

Interjection: Electronic community

After I bought the Lotus, I posted to several Europa forums, introducing myself and looking for advice on the engine rebuild. I received a very welcoming response. There were multiple folks who knew me from my writing in the BMW community and who also had Lotuses. It appeared that I was not alone in my dark dirty craving.

It even turned out that a few people were local. One fellow, Rob Thomson, was just out in Carlisle, about 25 miles west of me, and had had a fellow who I later will refer to as "The Lotus Engine God" rebuild and tune his engine. I struck up an email correspondence with Rob. Being able to post questions to enthusiast forums is a great thing, but nothing beats an actual friendship, especially when the other party is local and can recommend vendors and services that are actually close to you

I believe the initial back-and-forth occurred on one of the Lotus forums, but much of the rest of it was email captured below.

---

email: Rob Thomson

```
From: Rob Thomson
To: Rob Siegel
Mon, Jun 24, 2013 9:29 am

Hi Rob. How does the head look? Much corrosion or wear? I used
a guy in Harvard (Bill McCurdy at Williams Racing) to rebuild my
engine. He used a machine shop in the same area (Ayer or Harvard). I
do not remember the name but could find out.

The water pump is another area that often needs special attention.
These can fail from just having sat for a long time. Cheapest
```

option is to rebuild with original parts, but if the timing chest components are corroded, this may not be an option. I replaced mine with a Dave Bean cartridge unit. Was actually a pain, and did not fit as well as it should have [Note: this would turn out to be a prophetic statement]. Another option is the Burton unit from the UK. Not cheap, but maybe better quality than the Bean unit (McCurdy had one for another twink he was rebuilding). Also, the Burton unit would require some mods to work with the Europa—the water return is to the side (for an Elan) and not the front. Yet another option is to just to bypass the mechanical WP and use an electrical unit.

—Rob

....
From: Rob Siegel
To: Rob Thomson
Mon, Jun 24, 2013 9:38 am

The main thing I need to deal with is the fact that this engine is seized and has been sitting since, I believe, 1979. The head looks surprisingly good. Over the next week I'll tear down the block and see what's what. Someone had loaded up cylinder #1 with so much Marvel Mystery Oil that a huge amount of it ran out when I detached the exhaust, so it's my assumption that's where it's seized. At least it's had a good long soak.

I've read about the water pump issues. I haven't settled on an approach yet. I find the electric water pump idea interesting, except that it creates a new path for failure if the electric pump fails. If it isn't one thing, it's another.

Thanks again.

-Rob

....
From: Rob Thomson
To: Rob Siegel
Tue, Jun 25, 2013 12:10 pm

Hi Rob:

I thoroughly recommend Bill McCurdy—he knows these engines top to bottom; they are his stock in trade. He has been rebuilding them (mostly for racing applications) for 40 years or more. The heads in particular are quite delicate and can be easily messed up if you have someone work on them that does not know what they are doing.

It may be worth just stopping by with the head and getting a free consultation from the man himself. He is a one-man shop working out of a shed in Harvard. Let me know, would be happy to make the intro.

Cheers,

Rob

....
From: Rob Siegel
To: Rob Thomson
Tue, Jun 25, 2013 10:58 pm

Rob, I talked for 1/2 an hour with Bill. What a great guy. He said I shouldn't make any decisions about what to do until I drive your car [Note: Rob's car has an almost stock de-smogged Federal-spec '74 Twin-Cam engine]. Seriously. That's what the man said. So I'm under orders. :^)

--Rob

TLC Day 6: The Lotus Engine God

JUN 25, 2013, 10:51 PM

The Lotus Chronicles, Day 6:

Talked with the local Lotus Twin-Cam expert on the phone for half an hour. Great guy. Finally I said "Well, clearly, you're the guy."

He said "Oh, don't say that. I only have that title by default."

"What do you mean?" I asked.

"The other guys are dead."

(22 LIKES, 9 COMMENTS)

C.R. Krieger Clearly, someday soon, YOU could be the guy! +5
Paul Skelton LMAO
Willy Wiley Ah yes, the slippery slope of specialization. Or everything I learned about _____ I learned on the internet.
Franklin Alfonso So you mean there is a "Curse of the Lotus" like the Pharaohs for trying to resurrect this dead Lotus?? +1
Charlie Gordy Modesty and a sense of humor. Guy must be good! +1
Scott Aaron Sounds like Rob Siegel best friend material.
Jeremy David Walton The best Lotus specialist is a dead'un? Yup, there's a few of those, including the founder/ creator...
Carl Witthoft This better not end up with your Lotus being named Christine. Just sayin'

email: Rob Thomson

```
From: Rob Thomson
To: Rob Siegel
Wed, Jun 26, 2013 9:06 am
```

Well then, orders is orders! Looks like you have no way out now. So when you are going to be up this way next, we will make it happen.

Yes, Bill is a great guy. And his shop is everything you would imagine. A tin shed in a field in Harvard that is an Aladdin's cave of engine parts. Shelves of disorganized stacks of Weber carbs, twin-cam heads lying around, a couple of engine stands with motors at various states of rebuild. I will warn you that he is a perfectionist (which can get expensive if you are not careful) and will try and convince you to rebuild your engine as a long-stroke 1700cc with a twin Weber head. All good for at least 150HP, but again at a cost.

I had a real battle with him to keep mine close to stock (1600cc with the Strombergs). But glad I did. The 115HP it makes is plenty for such a small and light car.

Welcome to the fun.

Rob.

....
From: Rob Siegel
To: Rob Thomson
Wed, Jun 26, 2013 10:29 am

<<Bill will try and convince you to rebuild your engine as a long-stroke 1700cc with a twin weber head.>>

Yes, in fact, that is exactly what he recommended. He said my '74 with the Stromberg head will run "like seven bags of crap."

If you don't mind me asking, just so I have one of the cost goalposts, what was your cost to Bill for rebuilding your 1600 Stromberg motor close to stock?

Thanks!

--Rob

....
From: Rob Thomson
To: Rob Siegel
Wed, Jun 26, 2013 11:57 am

Seven bags of crap! I think I resent that! Bill tends to be a bit opinionated in case you have not realized. In truth I think he was starting to warm to the ZSs near the end; they are less finicky and temperamental than the Webers, and make for a more drivable car at the cost of some high-end HP. Many moons ago I had a Triumph Spitfire with a lumpy cam and a side-draft Weber. Was a bugger to get tuned, and once you did, was gutless under 3k and burst to life over that.

As for the cost of the rebuild—well this gets a little embarrassing. Although I certainly did not mean to spend as much when I started, it ended up setting me back about $13k. Shocking I know. And for that you get pretty much a stock configuration although we upped the compression a tad (10:1 I think) and balanced the moving components, but that was about all. Needed all new valves, seats, guides, springs, followers, oversized forged pistons, crank re-grind and bearings, plus the Bean water pump, ports polished, and fully dyno tested and tuned. If I were to do it again, and were feeling braver, I would probably just have Bill do the head and do the block myself. These engines are tricky, and the parts are just damn expensive. You have probably noticed the difference between Lotus and BMW. For as over-engineered and Teutonic BMWs are, Lotus are delicate and a bit kit-car like. Hand-built is not necessarily a virtue.

Rob.

....
From: Rob Siegel
To: Rob Thomson
Wed, Jun 26, 2013 12:54 pm

Rob, I am so sorry to have put you on the spot with regard to asking about cost. But please know that that information, won at the cost of embarrassment on your part, is worth its weight in gold to me. I cannot and will not enter a process where the cost to come out the other side is $13k. My car isn't worth that. It needs too many other things. I would go the Zetec route before doing that.

If you wind up reading my book, you'll learn that despite my wing-nut Hack Mechanic persona (and who isn't a wing nut who buys a dead Europa sight unseen?), I'm actually a very deliberate guy who thinks things through and views risk very carefully.

More as I know it. Thanks again. Seriously. I owe you big time for divulging that info to me.

--Rob

....
From: Rob Thomson
To: Rob Siegel
Wed, Jun 26, 2013 4:14 pm

Not a problem at all. Embarrassment was of my own making. When I got the car, the engine was running OK, although was a bit smokey, and leaked oil all over. I drove it for about 2 years like that, fixing minor things on the way (brakes, electricals, wheel bearings, suspension and those bloody door hinge pins etc) but the smell and oil puddles started to get to me. I probably could have just got away with a re-bore and lapping the valves and some freshening up (without doing anything major to the head) and been happy for quite a while. But once I started into the project, shipwright's disease kicked in, and I was $13k in the hole. Started off as a 5-6k project and just kept creeping up. I don't blame Bill; he is a good guy and clearly not out to take advantage and I trust his judgment. I think he just does not like to compromise on his work (an admirable quality for the most part), and that can sometimes make for an expensive rebuild. But I made every decision on the way knowing the costs, so cannot blame anyone but myself. It's like every house renovation project I've been involve in—take a guess at what you think it will cost and then double it.

In any case, the offer [to drive my car] still stands. If you are out this way and want to take the Europa out for a feel free to give me a shout. And do let me know how it all goes with yours.

Cheers,

Rob.

....
From: Rob Siegel
To: Rob Thomson
Wed, Jun 26, 2013 4:21 pm

Thanks for everything, including the vote of confidence on Bill. Is he a drinking guy? Does he do scotch for favors? This is the sort of thing I hate to be wrong about.

--Rob

....
From: Rob Thomson
To: Rob Siegel
Wed, Jun 26, 2013 5:34 pm

Mine was similar—a 15,000-mile engine but corrosion in one bore. I have two theories-either water collected in the valley between the cam covers, and seeped past a spark plug into the bore, or enough moisture/condensation got past an open intake valve while it was parked for an extended period.

You may find the water pump is shot. Apparently these die from just sitting for a long time, but worth a try.

Not sure if Bill is a Scotch man. I do know he has had some health issues so may have been once, but perhaps no more. In any case, pretty sure he would not be offended. Even if he cannot drink it himself, he would find a home for it.

Are you still looking for a set of pistons? I have the stock ones that came out of my engine still. They look okay—probably need cleaning up, but ran fine in the engine before the rebuild. I have no use so you are welcome to them. Gudgeon pins too.

Rob.

TLC Day 7: Dropping the pan

JUN 26, 2013, 10:08 PM

The Lotus Chronicles, Day 7:

Got the engine on the stand. Dropped the pan. No metal. Other than piston #1 being seized, things look clean as a whistle.

Talked with the previous owner. Asked "Was the big bad thing that happened in 1979 that caused this car to be parked for 34 years the fact that it seized?"

"No," he said. "The previous owner had two vertebra fused and couldn't drive the car. I bought it from his son, who'd planned to restore it, but eventually bailed."

"SO IT RAN WHEN PARKED IN 1979, AND IT SEIZED FROM SITTING?"

"Yes, that's my understanding."

I'm feeling lucky.

Matt Pickering Ya, but what about that previous owner and their spine? What about them?
Sam Smith Larry Webster needs to be watching this.
Bill Howard "Other than piston #1 being seized, things look clean as a whistle." Sounds like asking Missus Lincoln how she liked the play otherwise.
Rob Siegel BOOM!
Scott Sislane Very impressive...
Bill Howard Is that an inverted orange drywall compound bucket supporting they brake assembly? Is that a factory-authorized tool?
Scott Sislane I used the same orange bucket to assemble my head, header and side drafts. It's hack authorized.
Rob Siegel Bill, it's a Lotus. Colin Chapman's aphorism to "add lightness" extends to tools, too.
Kenneth Robb I hope you get to drive it before your back gives out Oldtimer.
Ernie Peters Some folks use an adapter to mount the engine to the engine stand using the exhaust side motor mount bolt holes rather than the transmission mount holes. Gives better all-around access. Like so.

Rob Siegel Thanks Ernie!

Bill Howard Is the paint color the same as our parents' Coppertone refrigerators of the '70s? Did you find platform shoes in the trunk (boot?)?

Rob Siegel Yeah... well... I happen to like brown. My 911SC was brown. But my sister, who I adore, looked at the Europa, kind of cocked her head and scrunched an eye, and said with devastating honesty, "why would any manufacturer ever paint a car that color?"

Bill Howard That color is becoming popular as seat leather ("tobacco" or similar) and lord does it look good for a lot longer than tan or parchment or beige leather let alone ivory.

Dale B. Phelps Wasn't "tobacco" sort of a caramel color?

Bill Howard More like cordovan, maybe minus some red.

Faith Senie So it was parked because the previous owner seized, then?

Ben Greisler The side mount is exactly how I hang M10 engines from my stand. I even modified the plate to allow direct bolting.

Barbara Radford Alfonso As easily as the day is seized, fingers crossed the piston can be un.

Jeremy David Walton Think lucky & mebbe the Lotus gods will be with you. After all they usually hoped for lucky best with each debut, back in the (Europa) day.

Franklin Alfonso British car repairs 101. Lolita might just be taunting you with an easy repair before she bitch-slaps you back to reality. But I hope you get lucky.

David Weisman I'd put in a small block V8 while you're at it.

Nathan Smith My guitar is, in fact, a Lotus... best wishes to you Rob. Can't think of anyone I'd rather see on this job.

Adam Pepper Is that the cylinder head? (ha!)

Daniel Diamond Lotus: Lots Of Trouble, Usually Serious... but perhaps not this time!

TLC Day 8: Unsticking the piston

JUN 27, 2013, 10:04 PM

The Lotus Chronicles, Day 8:

Unstuck the piston by tapping around the periphery of the crown with the wooden shaft of a hammer, then banging on it with another hammer. The fact that it'd been filled with Marvel Mystery Oil for years (after being seized for years) couldn't have hurt. Need to clean the bore out with a Dremel tool and a Scotch-Brite wheel before I can draw the piston out the top. Disassembled the rest of the block.

Damn, this is good. I'm beyond feeling lucky. I'm feeling... saucy.

18 LIKES, 18 COMMENTS

Jeremy Novak So the piston bores are minty fresh? MMO is good stuff. Any gouging or pitting evident? It's tough to see in the pictures.

Rob Siegel 3 of the 4 are fine. The one on the left, that still has the (now unstuck) piston in it, has corrosion near the top that's bad enough that I want to take it down with a Scotch Brite wheel so I can draw the piston out the top. Once I do that and get the piston out, it goes into The Lotus Expert who will tell me what the bores need. There's a used standard sized piston set on eBay for short money, but there's no sense buying it if the cylinders need to be bored oversize. +1

Faith Senie Careful, the Lotus gods may have a problem with "saucy"... +1

Rob Siegel oh, I know, I tempt, I tempt... +1

Blair Robinson That thing will be on the road before Autumn!

Charlie Allen You're doing fine. Slow and easy wins the game. Take care and do it right. Plus the Lotus gods are obviously pleased and looking on you as another Lotus Expert... +1

Mark Douglas Hastings Joy: that wonderful moment when you realize you have found and proven a solution to the problem that was dogging you. It must have felt great when that piston finally moved.

Rob Siegel Actually it scared the crap out of me, because I was tapping it DOWN. All of a sudden it went down A LOT. I imagined the rings popping out of the bore at the bottom and not being able to draw the piston back up (there's not space to pull it out the bottom; the crankshaft bearing journals are too close together).

Mark Douglas Hastings Glad it worked out for you. I almost got to drive a Lotus Europa years ago. The drive fell through when I couldn't jam my 6'2" frame into the driver's seat. (Sigh). I did get a nice ride as a passenger though. I remember taking a rotary at 60 and having it feel like 20. Amazing car. +1

Kenneth Robb You'd best play a recording of The Hallelujah Chorus so Jaweh remains on your team.

Richard Memmel how cool is that :)

Jonathan Poole these updates are now a daily ritual like the Dry Prong saga. keep them coming! +2

Franklin Alfonso Glad the gods are smiling on you so far. I watched Overhaulin' last night and saw them pillage a Lotus to the point they may as well have blown it up and just made a car from scratch.

Rob Siegel see my blog post at thehackmechanic.blogspot.com

Franklin Alfonso Agree 100%. I know if somebody gave me back my 1960 MGA I would be thrilled and it would bring back great memories but I would not like to have it anything other than be restored to good original condition not some alien bastardization of something it never was.

Franklin Alfonso I noticed that on Overhaulin' the Lotus had a Renault engine originally and I seem to remember you saying something about Ford engines a while back. What is in your Lotus, the Ford Twin-Cam?

Rob Siegel Yes, mine is a "twink."

Carl Witthoft You used that toolset just to mock all those people talking about "when the only tool you have is a hammer..." didn'tja?

Roy Richard It was the Silikroil :)

TLC Day 9: Engine disassembled!

JUN 28, 2013, 8:37 PM

The Lotus Chronicles, Day 9:

Block and head are fully disassembled, ready to be loaded in the back of the 'Burb to go to The Lotus Engine God. Along with a bottle of scotch. (It's lubricant. You know. For the cylinder walls.)

9 LIKES, 7 COMMENTS

Jason Thomas My Machine shop guy prefers a 30 pack of Chump oil and gets well lubricated before he starts!

Steve Collins Your block head has long since been disassembled, smelted and rendered. Wing nut. :-)

Meron Lavie Isn't the Lotus the car that Patrick McGoohan drove in "The Prisoner"? – Miller

Rob Siegel Meron, that was a Lotus Seven. My car is a Europa, which makes much less sense than a Seven.

Meron Lavie Ooops, my mistake. But it still was a beautiful car.

Franklin Alfonso Let's not spoil this by making any sense of any of it. It's an adventure.

Franklin Alfonso I used to give my German mechanic a bottle of scotch at Christmas 'till I realized my Volkswagens and I were getting to know him too well.

email: Rob Thomson

....
From: Rob Siegel
To: Rob Thomson
Sat, Jun 29, 2013 5:50 pm

Rob, thanks so much for that offer [for the used pistons]. I may take you up on it. I have the block completely disassembled and hope to get everything over to Bill for a look-see next week. The seized piston came out of the block looking surprisingly good once I cleaned it up. The other three pistons look fine and have no stuck rings. The seized piston has the top ring stuck, but no obvious score marks on the piston. We'll see what Bill says.

--Rob

....
From: Rob Thomson
To: Rob Siegel
Sun, Jun 30, 2013 6:41 am

No worries. You have probably figured this out, but always take what McCurdy says with an enormous grain of salt. When Bill sees your block and head I am sure that there will be a great deal of muttering and grumbling and he will tell you that you cannot possibly get away with a rebuild unless you "do it right," and will give you all sorts of doom and gloom stories of what can go wrong if you don't. Given your plan to get the car running first, I am starting to think I might have done you a disservice in introducing you to Mr. McCurdy.

Rob.

TLC Day 10: A slow day

JUN 29, 2013, 9:09 PM

The Lotus Chronicles, Day 10:

To paraphrase the late great Richard Brautigan, nothing Lotus happened today. But I had a fabulous drive in the 2002tii with the Nor'East '02ers, ending in a barbecue at Steve Kirkup's house. I now feel well inside the car guy bell curve. Sure, I bought a dead Lotus Europa sight-unseen, but Steve has a 1931 Model A flatbed (which is mad cool, btw), and a TR6 that's been on jack stands in his immaculate garage for four years. I feel much better.

(21 LIKES, 7 COMMENTS)

Delia Wolfe Rejoice in the moment, Rob! You won't be that far behind on your Lotus project for another three years and 355 days! +2
Bob Shisler Beat me... my '02 only spent two years on stands. I think. :)
Steve Kirkup British cars enjoy jack stands. They are much more at home there. +2
Franklin Alfonso You always feel better after hanging around with someone more crazy than you lol. +12
Paul Wegweiser Rob, you *are* aware that Richard Brautigan ranks as perhaps my all-time favorite writer... and I mean ever, right? We'll probably wear matching outfits on the 19th, too. *sigh*
Rob Siegel Paul, I sneak into your digs when you're not around so I can continue to fake this spooky connection that we seem to have. I thought you would have figured that out by now. +1

Chapter II:
The Project Runs
Into Molasses

Interjection: What *is* the Lotus Twin-Cam engine?

Facebook posts are kind of like CNN broadcasts from soldiers embedded with troops during a military operation—chaotic, rooted in the immediate, looking at things through a soda straw, bereft of larger context. I had no way to know it at the time, but despite the feverish pace with which I'd yanked out the drivetrain and tore down the engine (all of which was done within nine days after The Little Brown Thing rolled into my garage), the project would immediately bog down due to what to do about damage to the seized cylinder.

The issue was that I didn't have a clue what a rare bird the Europa's Lotus Twin-Cam engine (or "Lotus-Ford Twin-Cam Engine," as it is sometimes called to distinguish it from the Lotus 907 16-valve twin-cam engine that succeeded it) was. In my vintage BMW world, if you need an M10 four-cylinder or an M20 or M30 six-cylinder engine, you can almost literally decide what you want to pay and find one for that price. $300 will buy you an unseized motor with zero provenance (sitting for years, condition unknown). $800 will often buy you an engine whose seller you can look in the eye and tell you he drove the parts car from which it was pulled and the engine ran well and didn't leave an oil cloud in its wake. $2500 may purchase a recently-rebuilt engine with documentation. In contrast, for all intents and purposes, *there are no such used Lotus Twin-Cam engines like this for sale.* Anyone lucky enough to have one is keeping it.

The Lotus Twin-Cam, or "twink" engine as it's sometimes called (and why it isn't "twinc," I don't know, but it's not), is a Ford 1558cc block (spoiler: there is no such thing; we'll get to that) with a Lotus 8-valve twin-cam head on it, different pistons, a different front cover and water pump, and other modifications. They were in Lotus Sevens, Elans, Cortinas, Europas, Ford Escort Twin-Cams, and not much else. About 34,000 Twin-Cam engines were built. That's a funny number. It's more than a thousand like the build numbers for some rare homologation specials, but it sure as hell ain't 2.3 million like a real production engine.

Even understanding exactly what the Twin-Cam's Ford block is is confusing. The block is often referred to as a Kent block, a Formula Ford block, or even a Pinto block, and there's some truth to each of them, but none of those descriptions are accurate enough for you to buy a block and get one that'll work. Books have been written about the history of Lotus and the twink engine, so my intent here isn't to capture it all but instead to explain why it's so daunting to a neophyte who just bought a car with a seized engine and simply wants to get it running.

The story I've read is that Lotus founder Colin Chapman was unhappy with the cost and the performance of the Coventry Climax engine in the Lotus Elite, and wanted a new more powerful less expensive mill for the soon-to-be-unveiled Elan. In 1959, Ford released their 997cc three-main-bearing 105E "Kent" engine, so-named because it was produced near Kent, England (think of Ford "Cleveland" and "Windsor" engines). The 105E powered the new Ford Anglia and went on to be an extremely popular motorsport engine, with different versions becoming successful in Formula Junior, Formula 2, and Formula Ford. Chapman put together a team to design a new engine that combined this block with a new Lotus twin-cam head. Thus, during early development, the twink began life using standard Kent 997cc 105E and 1300cc 109E three-bearing blocks. However, Ford soon came out with its higher-displacement 1500cc five-main-bearing 116E block, so Lotus switched to using that. To get the additional displacement, the 116E block had a taller deck height (the distance from the crank centerline to the top of the block) than the original 105E block, so the Twin-Cam design was modified to accommodate it.

Very soon after twink production began, Lotus increased the bore to 1558cc. Note, though, that the twink was and still is commonly referred to as a 1600 engine. Indeed, the Elans it went in were called "Elan 1600s."

As I understand it, discussion with Ford on the purchase of blocks for the Lotus twink engine lead to an arrangement with Ford on their Cortina, where Lotus not only built the twink engines but also installed them, along with Lotus-modified suspensions and other components, in homologation-special Lotus-prepared Cortinas. These were then called Lotus-Ford or Lotus Cortinas. So, beginning in 1963, the new twink engine went into both Elans and Cortinas.

The blocks used for production Twin-Cam engines in '62 through

'67 were reportedly stock Ford 116E and 120E blocks that were "graded" and selected for cylinder wall thickness (remember, Lotus was boring out the 1500cc blocks to 1558cc), but beginning in 1967, this changed to "L" blocks cast by Ford specifically for Lotus to use for the twink engine. I've read differing reports online on whether the cylinder walls were actually thicker or fell within normal statistical variation. There were four different stampings of the "L" block, the final one being the 701M block. Thus, there appear to have been a total of six different casting versions of the block used in production twink engines.

Around 1968, Ford began producing a true 1600 block. This is the "tall" Kent 711M block used in the 1600 Crossflow, the regulation engine used in Formula Ford. The block also was used in the Kent I4 1600 engine in the '71-'73 Pinto.

So, is the twink block a "Kent block?" A "Formula Ford" block? A "Pinto block?" Yes, no, and no. Remember, the twink block is a 1500 block bored out to 1558cc, not a 1600 block.

The easiest way to understand things is to look at deck height. In the family of Ford Kent four-cylinder blocks whose numbers I've mentioned above, there are three commonly-used deck heights: "short" (7.12 inches), "medium" (7.78 inches), and "tall" (8.21 inches), and the twink engine used the block with the "medium" deck height.

Got it? Neither did I, for a long time. Let me present it a different way.

- Twink engines used medium deck-height blocks based on the Ford 1500 Kent pre-Crossflow engine but bored to 1558cc.
- The Kent 1600 Crossflow engine that was the basis for many Formula racers uses a "tall" block.
- The 1971 through 1973 Pinto had a Kent I4 1.6-liter engine, which uses the same tall block.
- In my humble opinion, the confusion of the twink being referred to as having a "Kent block," a "Formula Ford" block, or a "Pinto block" arises from the fact that the 1558cc twink engine is often referred to as a 1600cc engine, but it's not—it's a 1500cc engine bored to 1558cc—and these 1600cc blocks aren't in fact directly compatible because they have a taller deck height.
- A tall Kent block can be made to work with a Lotus head, but it requires a different crank and connecting rods, a longer timing chain, and a different timing cover or a spacer to take up the gap in the front. There's also an adaptation needed for the exhaust side engine mount.

- I've read conflicting reports on whether a "tall" block can be machined down to be dimensionally equivalent to a medium deck block. 7/16-inch is a lot to deck. I've read that if you take that much off, it comes too close to the water jacket around the cylinders.
- I've read that several different 1500cc medium deck-height Ford blocks can be used in the twink engine, provided the cylinder walls are thick enough to bore it to 1558cc, but that there are other important differences, including (if the block is early) the design of the main bearing endcaps.

As you'll read shortly in the book, when I found that the seized cylinder in my block was damaged, I looked into replacing it. For the reasons above, I found that it was non-trivial making sure you're buying a block that will actually fit. I learned that Ford has twice re-issued the 1600cc Formula Ford block, reportedly with an improved bearing cap design, increased cylinder wall thickness, and reinforcement around the clutch housing bolt bosses. And apparently the latest reissue is available in both "tall" Kent and Lotus-appropriate "medium" deck heights. However, I was told there was some concern of how plug-and-play the block is with the Lotus Twin-Cam head and other engine components, or whether additional machining would be required. (Note, though, that, as you'll read near the end of the book, with the level of additional machining required to get my cartridge-style water pump kit installed, I may have been frightened away by something that may not have been that big of a deal.)

But even beyond the block, not all Lotus Twin-Cam engines are the same. The European engines had a "Weber head" with Weber carbs, but the U.S. Federal-spec engines used Stromberg carbs, and since the intake manifold is a cast part of the head, the carbs aren't interchangeable between the two heads. The engine in the Elan Sprint had higher compression and better cams. The last round of twink engines were marketed as "Big Valve Twin Cam" motors, but within those there are Weber, Stromberg, and Sprint variants that cause the output to range from about 105 to about 135 horsepower.

FURTHER, the Twin-Cam engine in the Europa has a different front timing cover than that of the other cars it was in; the cover has the coolant inlet on the front instead of on the side, and replacing it isn't just a 30-minute job like on a normal engine (you'll learn a *lot* more about that when I finally reassemble it). So, in terms of buying a used motor I could drop in, there weren't 34,000 of them; there were only 4,710. World-wide.

It took me years to fully appreciate what all this meant, but in 2013 what it meant was that a) No, you don't find used $300, or $800, or $2500 Lotus Twin-Cam engines you can drop in a Europa; b) Any good a la carte used engine is rare and expensive; c) I didn't have the expertise to buy a new or a used block without the possibility of buying the wrong one, and d) I needed someone familiar with the engine to advise me and do at least some of the machine work or risk it getting screwed up.

Cue The Lotus Engine God. Enter, stage left. Single spotlight. Big music. Maybe *Also sprach Zarathusra*. Yeah, I can hear those timpani banging *DOM dom DOM dom DOM dom DOM dom*…

TLC Day 11: TLEG #2

JUN 30, 2013, 9:54 PM

The Lotus Chronicles, Day 11:

The disassembled block and head are loaded into the truck to meet their Lotus Engine God deity tomorrow, who is going to rip my head off when I say "my strategy is to reassemble it as inexpensively as possible. I just need you to tell me what's already been done to the engine, mic the bores, advise me on cost-effective improvements, and help prevent me from doing anything really stupid."

Clearly, this strategy relies heavily on the aforementioned bottle of scotch.

9 LIKES, 9 COMMENTS

> **Steve Kirkup** You are at his mercy, my friend. Actually from what you described, if he is a true lotus god, he will be honest and just. +1
> **Bruce MacPherson** You really think one bottle is enough? +1
> **Jeremy Novak** If it's the right bottle, one will be sufficient. +1
> **Roy Richard** One for each of you is probably best. +1
> **Rob Theriaque** So... What kind of scotch?
> **Carl Witthoft** Wow, I remember people de-rusting fenders with Coke, but I never heard of cleaning cylinder walls with Scotch. /// Hey, why not play the Tom Sawyer thing. Ya know, "I really love rebuilding Lotus engines, but maybe for the right price I'll let you help."
> **Scott Aaron** Too funny. Good luck.
> **Franklin Alfonso** May the Force be with you my young Lotus resuscitator.
> **David Kerr** Go with something barrel proof, Marty's used to have a pretty good selection of private-label stuff.

TLC Day 12: TLEG #3

JUL 1, 2013, 7:16 PM

The Lotus Chronicles, Day 12:

Took the disassembled engine to The Lotus Engine God. He raised a big question as to why an engine with purportedly 24k on it has a sleeved block, but said the head looks pretty good. My "I want to reduce risk and get out of this at a reasonable cost" approach did not sit well with him. "I build race motors," he said, "that you can beat the living crap out of all day long for thirty thousand miles." It wasn't a boast; it was the flat factual statement of someone who knows his shit. But clearly he was proud of it. I explained I wouldn't be putting thirty thousand miles on this car in 20 years of ownership. It seemed that I was leaving with my disassembled engine.

Then I gave him the scotch.

Then he smiled.

"Bring it all inside," he said...

23 LIKES, 13 COMMENTS

Kenneth Robb The "bend over" was understood? +1

Rob Siegel Him or me? Oh, me, right... I'm paying him, initially, to sus out the block, crank, rods, pistons, head, valve, guides, and cams, and tell me what's reusable and what's not. Might be about five hundred bucks. One step at a time. The tough part will come later when I want to take it back, re-use the pistons, and assemble a sloppy motor. +1

John Herbert Don't be such a cheapskate...go big or go home. :) Rebuild it once and be done with it. I am speaking from experience where I had to rebuild my big block three times in 700 miles. Too much detail for this forum!!

Rob Siegel Sure, I don't want to do that either, but neither do I want to take every recommendation from a guy who builds race motors and wind up with a $15k bill. +4

John Herbert Hopefully he will give you some options.

Rob Koenig Simple. Go racing! +2

Skip McCauley if the part about the scotch is true then this story is, OH so full of WIN. +3

Richard Memmel I'm on the rebuild it once camp myself been there done that with the cheap sloppy rebuild over and over +2

Franklin Alfonso You must have regaled him with your charm....and the scotch didn't hurt.

Carl Witthoft Too bad you couldn't just trade him even up for some engine that was "not race-ready" but in fine running shape.

Carl Witthoft ^^runs out to see if TheBottleOfScotchTrick works with wives, too. <--s//Scotch//Champagne//

Franklin Alfonso Chardonnay. Scotch if she's a biker chick…

---

email: Rob Thomson

```
From: Rob Siegel
To: Rob Thomson
Tue, Jul 2, 2013 8:03 am
```

Rob:

I took the disassembled unseized engine to Bill yesterday and spent several hours there. Your first paragraph below absolutely hit the nail on the head. He looked at my block, which is sleeved, and he said he thought the "car with 24k on it" story probably wasn't true, and told me how bad a sleeve job it was. Then he looked at my head, pulled out two valves and buckets, and said they didn't look too bad.

Then we sat there for at least 1/2 hour with him telling me why my reassemble-it-myself strategy had a very low probability of success, telling me the pitfalls of someone like me trying to buy pistons and size them for a block that needed to be honed, and essentially telling me that he didn't want to be part of something that was driven entirely by cost. I said that that wasn't quite fair, that it wasn't entirely driven by cost, it was driven by RISK—that I've spent money like water on certain of my BMWs, but on this rebuild, I did not want to enter into a process whose cost was bottomless and unknown.

We seemed at an impasse, and I thought I was going to be sent packing with my disassembled engine.

Then the phone rang, and it was his girlfriend, who he'd explained had been out of town, "and she's letting me live with her, so I need to be nice to her." When he got off the phone, I said "well, why don't we start here," and I gave him a bottle of scotch I'd bought him, explaining it was a moderately-priced Isle scotch, with a lot of peat in it, but not the sea-weedy notes you get from, say, a Laphroiag, which can taste like the bottom of a tackle box (an acquired taste, to be sure). He was very quiet for about 10 seconds and had a surprisingly emotive wistful look in his eyes. Then he said "well... why don't we just bring it all in."

So he now has my engine, and the deal is he's going to clean and check out the block and head, make sure they're not cracked, have the crank and rods magnafluxed, and give me an opinion on whether the pistons are re-usable. He estimated maybe $500 of his time, and maybe $150 for the magnafluxing. And then we'll take it from there.

I would like to get over and look at your car sometime. Are you around this weekend?

Thanks for everything, Rob...

--Rob

....
From: Rob Thomson
To: Rob Siegel
Tue, Jul 2, 2013 9:56 am

Hi Rob.

I have not seen Bill for must be three years now, but as you describe the conversation with him, I can just hear his voice, the tone and intonation etc. It's all coming back to me. I hope you agree, but my experience is that Bill is a good guy and will do the right thing for you, although will complain about it every chance you give him. And don't you just love his shed? Look closely and he has some real gems tucked away in there.

Yes, I am around this weekend—no horse shows. So whatever time is fine. I might try and get out for a trail ride so we can figure out an approximate time frame but am flexible. Should be fun.

Cheers,

Rob

TLC Day 13: TLEG #4

JUL 2, 2013, 10:44 PM

The Lotus Chronicles, Day 13:

The Lotus Engine God says that my block, head, cams, crank, and rods all basically check out, but the block needs to be honed to get rid of the rust from the seized piston, and that'll enlarge the bores enough that the standard pistons will be too loose, and the only new stock pistons available are standard-sized pistons, so he's recommending custom pistons. Slippery slope, meet Rob. Rob, slippery slope.

27 LIKES, 33 COMMENTS

Bob Shisler .060 over…?

Charlie Allen In for a penny, in for a pound.

Bob Shisler .090 over, pop-up pistons, wet-lapped for squish (that's a thing, right?)… offset-grind the crank for stroke… piston oilers w hi-pressure oil pump…

Kenneth Robb And while you have the boring bar out why stop at the first oversize? Oh yeah, your stroke looks a little short as well.

Jeremy Novak Would you like some extra surfactant on your already slippery slope? We can surely provide some!

Bob Shisler cross-drill the block for lightness? no, that's one click too far on the insanity ratchet… :p

Bob Shisler maybe a few clicks.

Eric Heinrich While you're in there you might as well convert to dry sump since you're getting new pistons. That way you can get lighter pistons that only need 2 rings instead of three and with dry sump that will increase reliability, the extra power is a nice side benefit and then you might as well… +2

Eric Heinrich …ummm…did you say something about a slope? +1

Eric Heinrich For every person in history that has suggested to E30 M3 diehards to do an S50 conversion, now I get my chance back: "Why not just put a Toyota 2ZZ-GE engine from an Elise in there?" (ducking, running) +1

Richard Memmel what's first oversize available? more or less then custom? either way think of the big picture finishing said engine reinstalling and going for first drive :) dearly beloved we are here today to witness this… you get the point. +1

Eric Heinrich Are you doing this one on your kitchen floor? +2

Bob Shisler I hear Suzuki makes a lightweight bike engine that makes 175hp and might fit. :)

Bob Shisler [link to a Lotus Seven with this Suzuki engine]

Bob Shisler just sayin ;) +1

Toby Racer The bike racers get the bores chromed—that shrinks 'em back down a bit. Me, I run Ross pistons…

Franklin Alfonso Oh Oh,,. I hear the mayaswells coming in… slip slip…

Franklin Alfonso I think Jay Leno has a Lotus engine kicking around that was not used from that Overhaulin' episode. Oh yeah that was a Renault engine. Yours is a Lotus-Ford Twin-Cam?

Barbara Radford Alfonso Mayaswell

Roy Richard Welp if you are going to have to bore it out and buy new pistons go as big as you can, may as well get some cubic inch while you are at it :) +1

Bill Howard Slippery slope? "We have met the enemy and he is us." There are small block Chevy V8s available cheap.

Rob Siegel Bill, a Zetec engine bolts right up to the same transaxle (bolt pattern is identical), but as with any engine swap, you pay for it in the adaptation.

Neil Halin Won't an LS1 just bolt right up?? ;)

Steve Collins Did you consider sleeves dear?

Steve Collins Bore them out. Sleeves. Maybe a few mics of Teflon and a top end lube.

Steve Collins Ps coat the rings with Teflon too

Rob Siegel Steve, the block is already sleeved

Christopher Spargo Plate with nikasil to build up the diameter. Lots of vintage racebikes are getting plated on cast iron now. I'm sure they could do a 4cyl automotive block. Though I suppose with 4 bores, pistons might be cheaper. :)

Rob Koenig Nikasil!

Ed MacVaugh Have you considered removing existing sleeves, and installing a new set to standard spec?

Rob Siegel Sort of. The bores are scored. I'm waiting for the machinist to tell me if they're out of round. If they're not out of round, possibly they can just be honed rather than bored, in which case the machinist would need to tell me what size they are when they're honed, which will tell me whether standard-sized pistons will work or if I need to go to slightly oversized pistons. If they need to be bored, I'd need to go to standard oversized pistons. There are no more COTS cast standard oversized pistons other than the occasional NOS eBay set—they're all forged, so they're not cheap. So I'm starting to get a handle on costs for these things. I have not yet costed the "use new standard pistons but re-sleeve the block" option. +1

Carl Witthoft I decided this morning that, whenever the Lotus is finally rebuilt to your satisfaction, that we'll have some nice pigeon invite you out for dinner or something while a crack team disassembles the whole thing and reassembles it in whatever room in your house is 6» greater in each dimension. +2

Roy Richard How about the ole knurl the pistons trick?

TLC Day 14: Pistons on the brain

JUL 3, 2013, 10:44 PM

The Lotus Chronicles, Day 14:

Nothing. Absolutely nothing. But I can't stop thinking about pistons.

8 LIKES, 6 COMMENTS

Tom Samuelson How about sleeve-ing the cylinders, boring to stock size and reusing the stock pistons?

Rob Siegel cylinders already are sleeved—very curious for a 24k car.

Tom Samuelson Replace the sleeves? Check out this video on YouTube [link to Kent Moore sleeve removal kit]

Dale Robert Olson OK, so you know that the block will need to be bored due to the rust. This means new pistons no matter how you look at it. What are the thoughts of the Lotus community? Any substitutions? I know someone who successfully used VW pistons in a BMW M30 block. Anybody got a set left over from a rebuild that never happened? You are the King of Hack, let's get going! +1

Franklin Alfonso Already has sleeves at 24k that is strange. I agree with Tom. Why not replace the sleeves if that is possible?

Bill Howard Custom pistons could be cheaper than Lotus OEM. [Note: Bill's comment turned out to be prophetic.]

email: Rob Thomson

```
...........................
From: Rob Siegel
To: Rob Thomson
Thu, Jul 4, 2013 11:53 am
```

Rob, I have a radio interview about my book 8-9 Saturday morning, then am going out to breakfast with my wife and a cousin, but am free Saturday afternoon and all day Sunday. Any slice of your time within those windows I can sign up for?

--Rob

```
...........................
From: Rob Thomson
To: Rob Siegel
Thu, Jul 4, 2013 5:57 pm
```

Hi Rob:

Saturday afternoon works. I have to drop off my youngest at a friend's at around 1:00 or 1:30, so we could say 2:00 or really any time after that would be fine.

BTW I got a copy of your book for Father's Day and just started it today. Great read so far. Had to smile as you describe your first car, the GT6, and lambasted Triumph for their under-engineering. And now you go and buy a Lotus!!! Oh we really need to talk!

Coincidently my first car was also a Triumph—a 1959 Herald Coupe purchased for the princely sum of $75. And that was followed by a Mk 1 Spitfire. Both cars older than I was. Thinking back, I too am surprised I survived those early years. Talk about "unsafe at any speed."

Rob.

TLC Day 15: TLEG #5

JUL 4, 2013, 5:27 PM

The Lotus Chronicles, Day 15:

It's the 4th of July for heaven's sake; I got nothing. But I will share a quote from The Lotus Engine God. A bare very early Twin-Cam head, never even been tapped for spark plugs, showed up on eBay. I asked him about it.

"You have to ask yourself," he said, "how a head like that even exists. I'd wager it has a casting or machining mistake, was pulled off the factory line, put in a corner, and is now being sold 50 years later like it's something special. For me to buy it, I'd have to fondle it until I couldn't fondle it no more, and then if it bit me, it would be because I was stupid and not because I let someone take advantage of me."

I could sit and listen to this guy forever.

15 LIKES, 8 COMMENTS

Harry Bonkosky Truly a wise man.
Matt Pelikan What Harry said.
Kevin McGrath Happy 4th!
Stephen Fishman At what point does he instruct you to try and take the pebble from his hand?
Nicholas Bedworth A teacher friend of mine was going to buy a Lotus like yours from a former student. The student was driving it to deliver, and he hit a big dog. The car was essentially scrap, and the dog picked itself up and walked away, wobbling but essentially unhurt. +1
Rob Koenig I've often wondered the exact same thing about these mystical and mythical oddities that appear decades later. How does something like that come to exist and, even better, remain locked away for so long. I'd tend to agree with him that the stuff is probably just blems. Unless of course the parts are OEM Delorean parts.
Franklin Alfonso Keep supplying him the scotch and maybe you can listen forever. +1
Rob Siegel That's literally my plan. +1
Roy Richard Wax On Wax Off [link to scene from The Karate Kid]

TLC Day 16: Parts cleaner

JUL 5, 2013, 10:04 PM

The Lotus Chronicles, Day 16:

I have a Lotus-load of parts to clean. I thought I'd scratch a lifelong itch and buy an industrial parts cleaner, basically a sink sitting on a 30-gallon drum of solvent, but then I'd have to deal with a 30-gallon drum of solvent. Researched it, and wound up buying a $149 drum-less cleaner on wheels from Tractor Supply. Wheels are good. I have a tight garage.

George Thielen Didn't you already have a dishwasher in the house?
Rob Siegel NOT cool.
Dale Robert Olson Can't beat a parts washer for big projects. The stuff will evaporate over time, though.
Rob Siegel The parts? I'll tie 'em down with zip ties and bailing wire.
Dale Robert Olson No, silly, the solvent. :-)
Bernie Sharpe Zip ties and bailing wire just made my chuckle audibly while watching the Sox on the west coast. Thank you.
Erik Wegweiser Are you sure it wasn't our baseball team?
Barbara Radford Alfonso Getting Bernie to chuckle is what I live for.
Carl Witthoft Good for bathing the family pets, too, I imagine.
Martin Meissner I love parts washers, it's such a no brainier alternative to going through 8436 brake clean cans.
Rob Siegel Yeah, that's the motivation. I gave up on the "if I'm going to do it I want something big enough to hold an engine block" and went with the "I'm spraying brake cleaner into disposable roasting pans in my backyard anything would be an improvement on this" paradigm.
Carl Witthoft Hey-- just how old is that 'scope there under the sodapop? 1995?
Richard Memmel Try bio clean.com for a natural degreaser that works wonderful its what I replaced my solvents with and it won't kill animals.
Richard Memmel Bio green clean my bad.
David Weisman I'm so jealous.

TLC Day 17: Sharing the day and the cars

JUL 6, 2013, 7:11 PM

The Lotus Chronicles, Day 17:

In my book, I write about cars as a conduit for human connection. I just had a beauty. Through posting on a Lotus Europa-related site, I met another enthusiast, also named Rob, who invited me over for the day to drive his well-sorted Europa. He also has an M6 and a gorgeous Triumph TR250. I drove over in my Z3M Coupe and couldn't help snapping a bunch of pics of two of the more extreme shapes in motordom.

[Addendum: It was a bit too personal to reveal on Facebook at the time, but as Rob and I talked, I learned that he'd lost his wife to cancer four months prior, and was raising two teen-aged daughters. So here he was, killer property with a horse farm for his daughters, architect-designed garage, massively cool cars, and none of it mattered. Against this backdrop, we swapped cars, and for about an hour, bombed around the back roads of Concord and Carlisle, him in the M Coupe and me in his Europa TCS—the only time to date I'd ever driven one. It was a little slice of heaven for both of us.]

10 LIKES, 8 COMMENTS

Sam Smith So wait. Dude. The suspense! Are you going to tell us whether or not you *liked* driving the thing you're trying to give birth to? +1

Ernie Peters How do you give this post, especially the pictures, 4 thumbs up? BMW, Lotus, MINI, nice garage with a lift... Must have been what Adam experienced before he met Eve. LOL +1

Rob Siegel Sam, I LOVED it. +1

Scott Sturdy So everyone who owns a Europa is named Rob? +3

David Handrick Yikes!!

Franklin Alfonso You loved the garage, or driving the Lotus, or I assume both???

email: Rob Thomson

Sat, Jul 6, 2013 10:23 pm

Rob, thanks so much for sharing your time, cars, and beautiful home with me. That was really very special.

Thanks again.

--Rob

From: Rob Thomson
To: Rob Siegel
Sun, Jul 7, 2013 12:11 pm

Hi Rob.

You are very kind. It was really fun and I am always happy to talk cars. Do share news as things progress on the motor. We will have to organize a Europa drive when you get it back on the road. The pics are quite flattering. It motivated me to give the Europa a scrub this morning.

Cheers,

Rob.

TLC Day 18: Piston data

JUL 8, 2013, 10:25 PM

The Lotus Chronicles, Day 19:

And the answers are in from several vendors. The Lotus Engine God was right (not that I expected anything else). Inexpensive, original, cast oversized pistons are not stocked anywhere (unless you luck out and find a NOS set on eBay). New oversized pistons are all forged. Like race pistons. $700-$800 a set. We shall examine the alternatives and proceed very carefully...

4 LIKES, 26 COMMENTS

Ed MacVaugh If you were to post exactly what you are looking for, collectively your fan club might know someone, somewhere. I do not even really know whose engine Lotus used in your car, or what sized pistons were standard.

Rob Siegel Thanks Ed. Right now I'm still in the stage of accumulating knowledge to gain wisdom and avoid doing anything too stupid. And, actually, I HAVE already posted "anyone have a set of NOS 020 over pistons" on two Lotus forums.

Ed MacVaugh Lotus Twin-Cam Engine By Miles Wilkins says: "The factory only ever kept standard and 0.0015 in. oversize pistons and stocks are now well down. However, Hepolite produce the full range of Powermax pistons from standard through 0.040 in. oversize." +1

Rob Siegel Yup, I have the book. But even Burton Power in England says "Bill is quite right, there are no STD type (cast) pistons available for this engine anymore, they are all forged now."

Bob Shisler check with Ireland Engineering... last I heard they were $550 a set for JE customs. That was a while back tho...

Rob Siegel It becomes an issue of how much risk you want to take, and whose time and expertise you're paying for in order to lower that risk. I can't just call up a custom piston manufacturer, order up a set, and stand a prayer of getting it right on an engine I know nothing about. If I pay TLEG to do that, he'll put all the care he puts into race engines into it, and the hourly cost will eat me alive. This is why the sort-of-off-the-shelf 020 over set makes some sense, even though it ain't cheap either. I stand at least a decent chance of being able to have a machinist tell me "yes you can bore this block 020 over," then order a set of 020 over pistons myself, then take the pistons to the machinist, tell him "bore the block for these pistons," and probably get it right.

Ed MacVaugh My point was the book (presuming an error with the decimal place) suggests factory offered 15 over, not the 20 over that you have posted as seeking. The Hepolite brand has been sold by Federal Mogul to a Taiwanese holding company, so new stock is not what you are looking for from that firm. I cannot tell from the box whether Hepolite is cast or forged or whittled by gnomes.

Bob Shisler [link to JE Pistons custom order form] :)

Ed MacVaugh [link to lotuselan.net post on Hepolite Powermax pistons]

Rob Siegel Your attention to detail is better than mine, Ed; nice sleuthing. I have the original Lotus factory manual. I can certainly check what the oversizes were. It's complicated by the fact that the pistons also had "group" stampings (1 through 4) that, within a bore size, micro-sized them.

Paul Skelton Hehe... "whittled by gnomes."

Rob Siegel I hadn't seen that post, but I've seen similar ones that I find similarly confusing. all of which is why, even if I stay with the sleeves that are in it, or hypothetically, sleeve it back to the "original" size, I'm very hesitant to think that I can get it right. It's the machinist who has to get it right. It's why I'm trying to find a middle ground between paying someone way too much money to rebuild my engine, and just running a ball hone through the cylinders and just slapping it back together myself. I'm a guy who likes to feel comfortable that I'm entering into a process that will generate a good result.

Ed MacVaugh Why can't the standard pistons be reused? Has the scored bore damaged the pistons? I originally suggested the replacement of the sleeves so that sleeves and rings would be all that needed replacing. I've done that with Triumphs long ago and far away.

Eric Heinrich I would get the JE pistons. For only a little more you can add Teflon and ceramic coatings and have the wrist pins DLC coated. All told not a lot more on top of the new pistons and will go a long ways towards reducing wear and providing a little extra power/detonation protection. On a $600 set of pistons the coatings might only be another $150ish. Plus the JE's will be a lot lighter.

Roy Richard Hone the bore, knurl the piston and slap it back together. +1

Eric Heinrich Knurl new pistons?

Roy Richard Yup knurl the pistons, an old trick to fit undersized pistons.

Eric Heinrich That was suggested to me one time for my old Pontiac. For the cost of new ones, I figured I would get far more years out of new before a rebuild.

Roy Richard You knurl below rings of course. Some racers use to claim they actually held more oil and got better lubrication.

Roy Richard [link to article on pbase about knurled pistons]

Roy Richard Ya know when I first posted this a few Chronicles ago it was tongue in cheek but I am starting to think this is the way to go. Three out of the four hole are fine, a light hone and put the pistons back in. Mic the pistons, take the largest and knurl it, hone the bad cylinder as little as possible and stuff it back together. An easy 50K miles with no problems. A little blue smoke, hell it's British! Roy Richard Hey Rob if you hadn't gotten rid of our lathe [Note: Roy and I used to work together at my engineering job], we had a very nice set of knurling tools... D'oh!

Dave Farnsworth Look at the bright side, you could need a set of pistons for an S14. +2

Eric Heinrich Aha! Problem solved with the introduction of M3T S14 pistons.

Eric Heinrich Actually, contact Elephant. They could have a custom set of JE's made for not a lot of money.

TLC Day 20: Cost containment

JUL 9, 2013, 9:53 PM

The Lotus Chronicles, Day 20:

I tried to explain my cost-containment plan to The Lotus Engine God (who has my engine). He said "not until you drive Andy Bodge's car with the stroker crank and 30 extra horsepower."

I'm going to spend all my money, and then I'm going to die.

19 LIKES, 25 COMMENTS

Bob Shisler Order a nice micrometer so we can measure your smile afterwards... ;) +1

Daniel Senie You might have to sell one of the herd?

Rob Koenig And so it begins... The fine print at the bottom of the screen mentions that it's only $1 per HP if you act now! +1

Dale Robert Olson Do it right, you will still be happy with the long-term results long after the short-term $$ pain is forgotten.

Rob Siegel Seriously, I totally disagree. There is no "right," there are only choices. All joking aside, The Lotus Engine God builds race engines that are totally beyond my driving needs and financial resources. +1

Bob Shisler Ask him "what if I gave you a budget of $XXXX.XX, what would you recommend I do?"

Pike Perkins $1 per HP? Probably more like $100?

Rob Siegel Oh, I'm trying, I'm trying... he's not a "budget guy." My plan is to give him only the head, and I won't do that until he gives me a number rounded to the nearest thousand, and to take the block elsewhere. That's when he said "you need to drive Andy's car first." He's crafty, this one... +3

Rob Koenig I missed the "K" after the one. :) In all seriousness, I'd collect the pieces and thank him for the help. The longer the test drive the higher the likelihood they'll be buying the car.

Rob Koenig And really Rob? Giving him the head? Let's not get crazy here.

Tom Kinney Sorry for the dumb question, but can't any competent machine shop sort this out? What's so special about an old Ford motor?

Rob Siegel The block, yes. The head, no, at least not from everything I've read.

Ernie Peters High horsepower in a Europa might mean beefing up the transaxle assembly and rear hubs. I'd go as close to stock as possible.

Rob Siegel It's a bit vexing, Ernie. The Euro spec cars reportedly had as much as 130hp. The "Federal" cars were reportedly down to 113, but actual dyno reports puts them just under 100. So the "extra 30 hp" may just put it where the Euro cars were. That having been said, for cost reasons, I'm inclined to keep pretty close to stock. It's a 1600 lb car. 100 (or 113) hp goes a long way. I drove a '74 like mine, just rebuilt to stock Federal specs, and I thought it had plenty of power for a tool-through-the-country driver. +1

Dale Robert Olson I suspect what I thought I meant by "right" didn't translate. You are so correct on the choices. A race motor is not "right" for you, but you do want it put together in such a way as to be dependable, road-worthy and a reflection of your values. +1

Rob Siegel And sorry if I snapped at you, Dale. I have this pet peeve against the phrase "do it once, do it right." It's almost always said by someone who isn't spending their own money... :^) +3

Dale Robert Olson Truer words are seldom spoken...

Eric Heinrich "If there isn't time or money to do it right, there will be to do it over." Rob, I would keep it simple—spend a little extra $$ on some lightweight coated pistons and a lightweight flywheel so it's a little more rev happy and be done with it. Keep the headwork to what needs to be done to make it serviceable.

Jeff Hillegass I love how you piece together little bits of info about this rare beast. Much different than a Corvette or Mustang of the same era, where you'd have excess, redundant info... and where's the fun in that? It's the same thing that draws me to 1960's Vespas and Lambrettas...much simpler mechanically, but each a puzzle to be sorted. And it's cool imaging them being thrashed by their first owners, some forty years ago. +2

Franklin Alfonso Well Rob as for spending all your money then dying, you can't take it with you unless you have an Asbestos suitcase. +1

Debra Adams Sounds like a sound plan to me. Y'all can't take it with you!

Carl Witthoft Dumb question, but: Can stock pistons be built up via anodization to match the final cylinder diam?

Franklin Alfonso Can a sleeved engine be re sleeved and the stock pistons be used with new rings?

Maire Anne Diamond Okay. Just don't spend my money. [Note: This is my wife's first FB response to all of this. She's pulling my chain. She is incredibly tolerant and supportive of my automotive passions. But she seriously means "Just don't spend my money."]

Dave Farnsworth 195 HP, 420 lbs, 2 wheels, $15k out the door. Do the math and get a S1000RR. After all, you can't live forever but you can go really fast. +1

TLC Day 21: Shhhh

JUL 10, 2013, 10:12 PM

The Lotus Chronicles, Day 21:

Shhhh... the Lotus is sleeping...

6 LIKES, 9 COMMENTS

George Thielen seeping? +2
Rob Siegel Actually, I'm not sure there are any fluids in it to seep at the moment. +1
Jeremy Novak How can it mark its territory if you deprive it of fluids? +1
Rob Siegel It's a Lotus, not a Jaguar. The Lotus doesn't mark its territory like a big cat. It gets... how shall I say this?... pollinated. +1
Jeremy Novak All properly British cars are fiercely territorial and want to mark out *their* turf. MG, Triumph, TVR, Austin ... brand or "catness" doesn't particularly matter. It's British. It's colonial. It marks its territory. It might become despondent if denied this basic urge. +1
Tom Schuch When I had my Jaguar XJ6, I preferred to frame it in a positive way: The Jaguar Automatic Dynamic Underchassis and Driveway Lubrication System. It comes standard on most British cars, no extra charge. ;) +1
Franklin Alfonso British cars...I believe they are referred to as "Garage Queens."
Carl Witthoft Meowlpy Down! Coma Time for Lotus!
Dave Farnsworth But they did perfect the flow through oil change system. On the other hand, my Triumph two wheeler (2009) has been rock solid and dry as a bone—they got that together.

TLC Day 22: Mouse smell

JUL 11, 2013, 10:34 PM

The Lotus Chronicles, Day 22:

Pulled out the seats, vacuumed up the rodent detritus, and soaked the rug in Febreze. The smell of mouse is now no stronger than the smell of brake fluid.

9 LIKES, 18 COMMENTS

Paul Wegweiser My gal's father used to perform high-end upholstery work back in the 60s. When they had a car with immense odor (think: cadaver) he says that a bowl of ammonia in the center of the car, and slices of apple throughout the floor, cured it. (Thinking the ammonia sucked the funk out of the material, and it was absorbed by the apple wedges). No idea if it works... I suggest a Hack Mechanic experiment!
Paul Wegweiser That mouse smell WILL come back next summer if using standard methods of cleaning. Ask me how I know.
Rob Siegel Okay. I'll bite. "How do you know, Paul?"
Paul Wegweiser 7 days of rain and high humidity followed by one 91-degree sunny day is all it takes. I'm leaving the windows down in the F Bomb tonight. +1
Rob Siegel What's the cleaning secret? Do you get different mice to piss in it backwards? +1
Paul Wegweiser Sadly, I have only one "reverted urine"-producing mouse at my disposal. He's named Nigel. I feed him nutmeg and Tylenol cold medicine. I have however trained him to change records on my phonograph in the off season +4
Bob Shisler They have pet-urine stuff in the cleaning aisle? I'd think ammonia would either act as a solvent liberating the urine stink (like a permanent marker erases previous permanent marker marks) or just make it far worse.
Paul Skelton I had to sell a car because of a corned beef and cabbage incident... +1
Rob Siegel BOOM! Paul will be here all week... +2
Nick Macdonald I had a 1969 Cortina that suffered from a quart of milk spilled all over the back seats and carpet a year before I was gifted it in 1980--ineradicable. For mice, try Petastic—accept no substitutes, it really works. Don't buy Nature's Miracle.

Rob Siegel Yeah, the dairy thing will kill you. My son Ethan spilled a yogurt smoothie in the metro. It was all downhill from there. 'Course you had to push it downhill... +1

Paul Skelton Rim shot Rob everyone! Please tip your servers!

Jim Sillery Jr Wife left a 1/2 gallon of milk to ferment in the rear of our brand-new Highlander in the spring. It pressurized and blew on a Saturday and we found it Monday. Two professional "overnight" details and 6 months later, it still catches me every so often.

Rob Koenig Nature's Miracle. Use it liberally on everything they hit. It won't harm the upholstery or carpet. Febreez is just perfume. It won't break down the funk.

Franklin Alfonso Jim that sounds like grounds for divorce...

Elizabeth Marion Beware of the fallacy of Febreeze being phthalate free! It is not true! So if you feel a sudden urge to shave your legs and have a good cry, you will know why!

Carl Witthoft @ Elizabeth Marion—sounds more like a description of Boehner getting ready for a bicycle race. ba-bing! +2

David Kerr Treat Easter eggs gathered by little kids as you would, well, little kids on a field trip—count them on every transfer to/from the vehicle, and don't move until each and every one is accounted for.

TLC Day 24: TLEG #6

Jul 13, 2013, 8:49 PM

The Lotus Chronicles, Day 24:

Met for two hours with The Lotus Engine God. The bores are too scored to clean up with a ball hone (he tried), but the fact that the block is already sleeved means that he doesn't advise boring it .020 over for standard oversized pistons because the sleeves would get too thin. He's recommending taking off, say, 005, which means custom pistons. This shit is making my head spin.

5 LIKES, 17 COMMENTS

Rob Theriaque Let me display my ignorance while looking at this from my mechanical engineering background: Why not bore out the sleeves and replace with customs of the original diameter? Surely custom sleeves would be cheaper than custom prisons? Or would the water jackets not allow it?

Rob Siegel I asked that. This guy knows an incredible amount about these Lotus Ford engines, and says that, in his opinion, re-sleeving back to the original diameter is unlikely to be less expensive than custom pistons due to the machining required to get the 40-year-old sleeves out and the new sleeves in. I will seek a second opinion.

Rob Theriaque Yeah... Given the number of machine shops I deal with on a daily basis, their quality of work, and dearth of business, I'd guess there's a capable shop out there that can help.

Jonathan Poole You know, for what you're about to spend you could probably get a very nice S14 in there. +1

John Whetstone Or an Audi engine. It looked awesome in the 2002. +1

John Whetstone Yuk.

Paul Skelton That Audi 2002 is gonna haunt us for a long time I fear...

Kenneth Robb And you never got sucked in and in and in on a project before? But you don't want to regret cheaping out a year down the road so-----------bend over.

Ben Greisler Put the original engine in storage and put in a new crate from Ford Racing. Something like this: [link to Ecotec 2.4l 190hp engine]

Carl Witthoft Yeah, aside from the authenticity problem, I go with Ben G's suggestion. Yr never gonna get the original running smoothly. So either run w/ sloppy pistons (hey, Brit cars leaking fluids is required, eh?) or wait 'til you win Megabucks.

Rob Siegel With engine swaps, you pay for it in adaptation of intake, exhaust, water rail, wiring, engine mounts, etc. Plus, on a Europa, because it's mid-engine, the crankshaft hub front-faces the "firewall," so there's zero room for accessories on that end. That's why a stock Europa taps the back end of one of the camshafts to drive the alternator. An engine swap only makes sense if the engine is dirt-cheap, like a Zetec engine. I haven't ruled it out, but I'm teetering on the precipice of committing to the Lotus Twin-Cam rebuild at an unknown cost, and that's what I have trouble doing. I could simply throw it back together, but I know the results would be terrible (burning oil, low compression) because the cylinder walls are scored. I could hone it and throw it back together, but I know the results would be terrible (burning oil, pistons slap) because the honed walls would be larger than the recommended clearance. "Bend over" has one meaning on a stock 2002 engine, and quite another one here that I am unwilling to accept. I'll spend three grand, but not six, and certainly not twelve. I'm trying to find an intermediate solution that is more than throwing it back together with an outcome I know will be bad, but less than "it costs what it costs" (which is what TLEG says). Roy Richard mentioned knurling the pistons, which I mentioned to TLEG, and it has more credibility than I thought, so I'm going to look into it more.

John Herbert Rob, call my buddy and get a second opinion. He does a lot of re-sleeving. Just a thought.

Rob Siegel John, can you remind me his name and the name of his shop? Thanks.

John Herbert Gregory's Machine Shop. Curtis Kidd. He is on my Facebook page.

Rob Siegel Got it, thanks John.

Roy Richard Knurling can add up to .008 :)

Franklin Alfonso Another chapter for the next book… slip sliding away…

TLC day 25: Another piston option

JUL 14, 2013, 9:50 PM

The Lotus Chronicles, Day 25:

A professional machinist (Curtis Kidd) who an EOD guy I work with says does all the Chevy engines in his club, had an interesting suggestion: Why not just bore the engine enough over to get rid of the scuff marks, and send an original piston to a place that makes custom pistons for them to duplicate? Gets around the problem of needing an expert to specify the new piston. I love the value of tapping human contacts and other people's experience. Thanks Curtis; I'll follow up on this tomorrow.

(7 LIKES, 9 COMMENTS)

Curtis Kidd Rob I wasn't thinking about the rings when we talked earlier don't forget to ask if they can supply a ring with that custom piston

Rob Siegel Got it, thanks Curtis.

Bob Shisler JE does this 5 days a week…

Jackie Jouret Your Lotus Chronicles have finally managed to convince me that my father was right when he refused to buy me the Lotus Europa I asked for in high school. "You'll have nothing but trouble with it!" he insisted. Yes, I thought, especially with boys! Which was part of the plan, of course…

Carl Witthoft Now Jackie, if ya wanna have troubles with boys, you should have asked for an Econoline or a VistaCruiser. Knowhaddimean?

Jackie Jouret If this van's rockin'…

Jackie Jouret Or the ever popular "Grass, gas or ass…no one rides for free" also seen on vans of that era. Fortunately for my dad, I wasn't looking for THAT much trouble!

Rob Siegel Anyone want to vote which is more trouble: The Econoline with that bumper sticker, or a car whose acronym is "Lots Of Trouble Usually Serious?"

Jackie Jouret Twin pillars of trouble, they are!

TLC Day 26: Parts cleaner again

JUL 15, 2013, 10:25 PM

The Lotus Chronicles, Day 26:

Was about to pour the five gallons of solvent into the parts cleaner when I noticed that the drain plug just basically dumps the solvent on the floor. I think 10 minutes at Home Depot to install a stopcock with a hose fitting is time well spent, no? (Of course my garage floor could use a good solvent cleaning.)

5 LIKES, 5 COMMENTS

> **Martin Meissner** Maybe also a piece of mesh over the drain hole to at least catch the BIG chunks?
> **Kenneth Robb** A lot of kids like to huff it too so maybe it's time for a rave in your garage?
> **Richard Memmel** I run a semi truck air filter in my tank to further filter the solvent +1
> **George Thielen** You just wanted to post the word stopcock. +3
> **Jay Neal** Does it have a drain petcock? +1

TLC Day 27: And again

JUL 16, 2013, 10:19 PM

The Lotus Chronicles, Day 27:

Went down to Home Depot to buy a shut-off valve (won't call it a stopcock anymore; too many minds in the gutter) for the parts cleaner. Took me a while to match the drain plug—looks like it's 1/4 NPT. Thought I'd adapt it to a hose faucet so I can just screw a hose to the damned thing, but HD ain't so great for this sort of stuff. McMaster-Carr (mcmaster.com) to the rescue—the best web site in the entire world.

9 LIKES, 7 COMMENTS

> **Kevin McGrath** Thanks for the tip!
> **Rob Siegel** It's always fun to type something like "aluminum" into mcmaster.com. You want balls, sheets, rings, beams, what? What hardness? What temperature range? What finish? It's truly remarkable. +3
> **Neil Halin** It is a good thing that MMC doesn't have a local store or I'd never leave. I went to the JC Whitney store once and had a brain freeze walking around.
> **Noa More Rensing** I love McMaster, but sometimes with having to daisy-chain plumbing adapters you really need to do it hands-on. Schwartz has a decent selection but I miss the Stanford physics stockroom.
> **Rob Siegel** Yes, both Schwartz and the True Value on Lexington Street in Waltham have excellent fastener / adapter sections. Turned out what was needed was 1/4" NPT to 1/2" NPT adapter and a hose faucet with a 1/2" NPT input. I'm sure that would've been easy pickin' at Stanford!
> **Roy Richard** Rob, one caution if you are using a water valve the washer may not hold up well to degreasing fluid.
> **Rob Siegel** understood

email: TLEG

```
From: Rob Siegel
To: The Lotus Engine God
Sent: Tuesday, July 16, 2013 6:56 PM

Subject: What do you think of this Formula Ford block on eBay?

Bill, can you give me your opinion on this? Would it be less
expensive to buy this, have it bored to 82.55mm, and go with stock
pistons, than do the whole custom pistons thing? Or does the check-
out that the block requires make it a zero-sum game?

[link to eBay listing for Formula Ford 1600 block with the word
```

"Lotus" in the title]

....

```
From: The Lotus Engine God
To: Rob Siegel
Sent: Tuesday, July 16, 2013 6:56 PM
```

Subject: What do you think of this Formula Ford block on eBay?

Rob:

That's a 2737 block. They came to the US in 1600 Pintos as a non-Uprated X-flow. I've seen them in two configurations, early (like this block) and late (like an Uprated block). One of the differences between the two is the type of main [bearing] cap fitted. The caps on uprated motors are much more robust!!!!! The early caps failed often enough that Ford was compelled to redesign them. That's probably why this block is still around. No one wanted to spend the money to put the uprated caps on it. Now, not all the early caps failed but it's universally accepted that they need replacement if you're going to build any kind of a performance engine. Additionally, 2737 cylinder walls may not be thick enough for a TC bore size. Dave Bean suggests a "hydrotest" after boring one to 83.5mm.

[Note: This was one of several times I looked at buying a block and came away realizing that I'd never have enough knowledge to do it and actually procure something that'd be less risky than fixing the original block.

TLC Day 28: A second opinion

JUL 17, 2013, 10:05 PM

The Lotus Chronicles, Day 28:

I wanted a second opinion on the machine work, so I stopped by The Little Foreign Car Garage in Waltham and asked for a recommendation. They recommended Century Machine in Waltham. On the way there I passed another repair shop with a bevy of classic German, British, and Italian marques parked outside and asked them the same question. Same recommendation. Spoke with the owner, who said the other guy who works there used to work on Lotus engines back in the day. Spoke with him [Erik Lindskog]. I talked in loving yet frustrated terms about my relationship with The Lotus Engine God who has my engine. "This guy sounds extremely knowledgeable," he said; "What's the problem?" I explained how he won't give me a price, and how his house is built at the bottom of a very slippery slope. He listened patiently, and finally said "Tell this guy what you want, and if he STILL won't give you a price, I will." This is progress.

10 LIKES, 12 COMMENTS

> **Rob Koenig** Lotus Guru seems to follow the same rule of thumb that car dealers do. If a prospect takes a test drive the odds of "dehorsing" are exponentially greater. Especially if they've taken the car home and/or showed it to friends. In this case you made the commitment of having him do the preliminary review. He's hoping to "decash" you since, you know, he's the guru and all... and he still has your motor.
>
> **Rob Siegel** I do not believe The Lotus Engine God's motivation are to fleece me. Quite the contrary. I think he is a consummate professional who takes enormous pride in the perfectionism necessary to build race motors. I don't need a race motor. Federal-spec Europas are way down on power from the earlier un-smogged cars. I believe his motivation is to get me an engine that will substitute for sex. I just can't afford it. And I LIKE sex. +5

Rob Koenig I didn't mean he's out to fleece you. More so to get you "horny" per se.

Martin Meissner The Little Foreign Car Garage... I think my friend David Masi works there.

Scott Sturdy Go with the guy who will actually do the work and not dick you around.

Bernie Sharpe "28 Days Later" starring...

Thomas Jones I see where the Lotus Engine God would like to take your engine. He does have a point, but maybe finding a middle ground with him can work. Not that I'm advocating a race engine, but who wouldn't want their '02 to have a nicely done motor with a mild cam, properly tuned 40's, and some well-done port work. On the other hand, the new machinist sounds like they can give you what you want, a stock build. I'd be inclined to sit down with the Guru and calmly lay down the rules. If he doesn't want to play, take the motor elsewhere.

Being an '02 guy that loves anything sporty with wheels but whose carburetor experience is limited to "normal" carbs like DCOEs and 32/36s, Strombergs and SUs have eluded me when pressed to attempt tuning them. Based on that I like to stick with what I know and usually recommend Weber conversions... I say go for a mild performance upgrade but only mild. Forged pistons, and Webers are not the end of the world and not the slippery slope they may seem.

Rob Siegel Tom, these Lotus Twin-Cam heads have the intake manifold as a cast part of the head. There are Stromberg heads and Weber heads. To get Webers, you either have to change the head (very expensive) or have the Stromberg manifolds machined off (very expensive).

Rob Koenig You know Rob, maybe it's time to finally just jump off the fence and buy that mill you've always wanted. Imagine all the machining you could do and all of the stuff you could make yourself!

Rob Koenig By the way, welcome to PA!

Thomas Jones Ah, so your Europa IS a Twin Cam. I somehow (embarrassingly) thought that it was Renault engine. So, that would be the same engine as my coworker's '67 Élan Super Safety, except for his Weber head. Did I mention SUs don't like me... (1)

Paul Skelton I've heard rumor of a Lotus w a V12 BMW motor. Could be a lot cheaper. :-P

TLC Day 34: An affordable engine

JUL 24, 2013, 7:47 AM

The Lotus Chronicles, Day 34:

I finally found a Lotus Twin-Cam engine I can afford.

[Note: It's not obvious from the photo, but this is a tiny plastic model.]

15 LIKES, 9 COMMENTS

Franklin Alfonso Now you need to shrink the Lotus to fit +1

Carl Witthoft Nonono, Franklin, you hit the engine with the Big-O-Beam! (see Attack From Mars, Bally 1995) +1

Franklin Alfonso But where can you find a cheap used Big-O-Beam... CRAIGSLIST?

Franklin Alfonso Maybe you could enter the Lotus in a Soap Box Derby...

Steve Collins That would fit in my lawn tractor nicely. +1

Jonathan Poole That might fit nicely onto a creeper. +1

Paul Muskopf Add lightness. +1

Elizabeth Marion This item description is hilarious! Apparently, it is very 'rare' and 'would look mint, in Ford Escort and much more... feel free to use your imagination.' Perhaps if you imagine it to be 18 times larger than it is, than it will work for you!

Nicholas Bedworth Shipping seems a little high. :)

---

email: Rob Thomson

```
From: Rob Thomson
To: Rob Siegel
Thu, Sep 12, 2013 12:02 pm
```

Hi Rob.

Been a while since we caught up and was wondering how things were going on the engine rebuild and Mr. McCurdy. Any updates or still waiting on the final diagnosis?

Also I have been ruminating on my own car ownership status and am thinking of parting with the Europa. Although I do love the car, I do not drive it (has not been out since we went for a spin in July) and given my changed circumstances of late, it does not make much sense. Anyway I have not listed it anywhere yet, and may well wait till the spring, but just thought I would give you a heads-up in case you are reconsidering your own project, and really still want a Europa but without the hassle. Or if there is anyone else you know who might be interested, please put the word out there. I was thinking $15k is probably a reasonable price (about ½ of what the car owes me but c'est la vie and all that).

Cheers,

Rob

```
From: Rob Siegel
To: Rob Thomson
Thu, Sep 12, 2013 4:35 pm
```

Rob! I hope you and your daughters are doing well.

My engine saga has taken many turns. As much as I adored Bill, and could sit at his feet and listen to him for hours, I could not enter into a process whose cost was unknown. About six weeks ago I pulled the engine out and took it to Century Automotive Machine in Waltham (2 miles from me), who is the machine shop The Little Foreign Car Garage and others use. He quoted me $2k in labor to bore the block, spec and procure custom pistons (the block has sleeves that are .060 thick; the first standard oversize is .040; that's too big, thus custom pistons), and fully assemble the head. I supply all parts. I'll assemble the block. This process is now finally moving forward; I wrote them a deposit check and placed the first order for parts with Dave Bean in CA.

I would love to buy your beautiful sorted-out Europa, but it's not brown, you see, which scotches the whole deal.

Seriously... regarding taking the plunge on the engine rebuild, I was standing with my toes in the water for months, but have now begun to wade in and write checks. I guess I'm up to my knees. Maybe my balls. If I look at the whole thing rationally, I should still probably bail and buy yours, but I think I need to keep moving forward. I won't rule it out entirely, but... I just bought an eighth car—a '95 318ti, for $850, rust-free, under 100k, automatic, for my youngest to learn to drive.

Hey—I'm now writing for Road and Track. My first piece will run in October. And, incredibly, I'm talking with an agency about negotiating the film and TV rights for "Memoirs of a Hack Mechanic." They asked ME. We'll see what happens.

I'm actually typing this from a hotel room in Rahway NJ—I'm speaking at a NJ 'CCA chapter event tonight. Drove the Z3M Coupe down. Most I've driven it in years. So I understand the "if it's not used, what's it for?" logic.

Stay in touch.

--Rob Siegel

....

From: Rob Thomson
To: Rob Siegel
Thu, Sep 12, 2013 9:01 pm

Hey Rob. Thanks. The girls are very well, getting back into the swing of the school routine. Hard to believe the summer is over already.

I think you are on the right track with the engine rebuild. I love Bill too, and his work is flawless, but honestly it's hard to justify his level of work and obsessiveness on the Europa. And by doing more of it yourself, think of all you will learn about that fine British engineering! Sometimes you do need to just plunge straight in. I am rarely brave enough for that.

I am still debating the future of my Europa. If I drive it again this weekend I know I will likely fall in love again, as irrational a choice of car it is. But on a whim I am going to look at an E38 BMW 740i tomorrow that is advertised locally. Not very rational either and a long shot at best. More out of curiosity than anything but it just seems like an awful lot of car for the money (assuming it does not become a money pit) and worth a peek. Not sure if you know anything about these beasts but let me know if there is anything I should look out for.

I was wondering when I should start to look out for the R&T articles, and will be sure to get the October edition rather than just wait for it to turn up at my dentist's waiting room (which is how I usually get to see it). And film rights! Do you get any influence who plays your part? You could have some fun there.

Anyway I look forward to hearing how things progress (on all fronts).

Cheers,

Rob.

....

From: Rob Siegel
To: Rob Thomson
Fri, Sep 13, 2013 7:16 pm

I'm trying to think what car I could trade you for your Europa, other than my Europa, that makes sense, but none of it makes sense.

I told the representation agency that I wanted The Hack Mechanic to be played by Danny DeVito. Because no one says "HEY, ASSHOLE!" like DeVito. And I feel strongly that, in any honest telling of my story, there should be a lot of "HEY, ASSHOLE!"

--Rob

....

From: Rob Thomson
To: Rob Siegel
Sun, Sep 15, 2013 8:12 am

I was thinking Danny DeVito but thought you might be offended by the suggestion (and your good lady wife would certainly be if Rhea Perlman was cast opposite him). Or Joe Pesci (of "My Cousin Vinny," and definitely not "Goodfellas"!). And then Marisa Tomei could be the Mrs. Watched that movie again recently with the girls, I had forgotten how funny it was.

Enjoy this wonderful weekend.

Best,

Rob.

....

From: Rob Siegel
To: Rob Thomson
Sun, Sep 15, 2013 8:17 am

LOVE "My Cousin Vinny." I rather like the choice of Joe Pesci and Marisa Tomei as The Hack and His Wife. That way, you could work in the scene where she says that they'd be a team, and the wife would have to help The Hack Mechanic on every repair, and she says "OH MY GOD WHAT A FUCKIN' NIGHTMARE!"

TLC Day 35: Changing machine shops

JUL 24, 2013, 9:35 PM

The Lotus Chronicles, Day 35:

I'd decided a while ago that I don't need The Lotus Engine God's race engine-building skills, and not having a price estimate from him for engine work was a deal breaker, but it took me a while to pull the trigger, drive out there, and take everything back because I really like this guy. But I did. I spent another 1/2 hour with him, going over it one last time, trying to see if I could leave the head with him but not the block, but he still talked passionately about how price estimates prevent high-quality work. I asked him if he'd charged me for the four hours of wisdom-absorbing I'd spent at his feet. He said he didn't. I told him I didn't expect that to be free, I'd learned an enormous amount of value, and handed him a hundred-dollar bill. "Think of it as a tip," I said. Without missing a beat, he said "that's nice. You gave me a tip, now I'm giving you a gift," and handed it right back to me. What a class act. I'll try an Amazon gift certificate next. [Note: he did accept the emailed $50 Amazon gift card, which I sent with a note saying that he could view it that I was being generous, or that I was being thankful, or that I was pre-paying my ticket back into his shop if I needed to crawl back with my tail between my legs.]

I drove everything directly over to Century Automotive Machine, the other machine shop I'd scoped out. "I'll draw you up a detailed estimate for reassembling the head, honing the block enough to get rid of the seize marks, and spec'ing custom pistons," he said. "You want to assemble the block yourself, right?"

"Right."

He took one of the valves and slid it into one of the existing valve guides. "This other guy said that the valve guides were bad?" he asked.

I thought for a moment. "No," I said, "he said they were nearing the outside of their wear specification."

"Are you building a race motor or a street motor?"

"Street."

"I'd reuse these valve guides without hesitation."

I paused, then smiled at him and said "We're going to get along just fine."

But I'm not going to sit at this guy's feet and absorb knowledge.

23 LIKES, 4 COMMENTS

Bernie Sharpe Well, the SME (subject matter expert) did all he could for you, or should I say, all that you would let him. May the spirit of The Lotus Engine God live on forever in story and possibly song. Now, get dirty and have that Lotus ready for a nice ride when we come down next week! =-) +1

Jeremy David Walton You really do do it your way. I would not have had the moral courage to take the bitz back...hope it works out for you as well as the book did.

Carl Witthoft Ya did the right thing all around. TLEG's got his principles, and you've got not only a budget, but a different perspective on cars (street vs. peak racing, more or less).

Franklin Alfonso Slippery slope averted and some valuable knowledge gained while on your quest to the land of Lotus. Well done Grasshopper.

TLC Day 88: Finally, an estimate

SEP 5, 2013, 9:00 PM

The Lotus Chronicles, Day 88 [33 days—more than a month—since the last post]:

Episode IV: A New Hope (or: Today is the First Day of the Rest of Your Engine)

Machinist #2—Erik Lindskog, The Machinist Who Is Not The Lotus Engine God (TMWINTLEG?) called me, saying that, after having had the engine a month (I'd been in and out of town four times), he had an estimate. I immediately went down and he talked me through it. Hot-tank everything. Reassemble the head with new valves and bearings. R&R the rods. Deck the block. Spec and procure custom pistons .010 over, as .010 over is the smallest set of commercially-available rings. Bore the thin (.060) sleeves in the block to fit the pistons. Return to me the fully-assembled head, and the block and pistons that have been checked and test-fit. I will assemble the engine. No painting the block, no glass-beading the head; I'm not going to detail the rest of the engine compartment. Cost estimate: $2k, including rings and custom pistons. I buy all other parts.

Houston, we have lift-off.

31 LIKES, 15 COMMENTS

Rob Koenig Damn, that's a deal. Especially WITH pistons.
Rob Siegel Yes, in fact, so much so that I need to recheck the math... +1
Steve Kirkup Congrats! Progress that will result in... power!
Ernie Peters Make sure the valve guides and seats are up to spec. Otherwise, quite a deal. +1
Carl Witthoft Hey, remind us again how much you paid to get into this mess? :-)
Lindsey Brown Lotus, eh? That's how they got Jochen Rindt.
Roy Richard Sounds great only comment would be I agree with Ernie.
Marc Bridge Not bad!
James Laray That's a good price.

Rob Siegel But I'm wrong. that's labor only, doesn't include the custom pistons (about $550). Still, it's forward motion.

James Laray Still a good price.

James Laray 2002 M10 engine rebuilt with love, care, and high end pistons (from a noted shop) is upwards of $4,500.00 is it not?

David Weisman That kind of talk makes me hot.

Christopher Wootten Wow! The labor costs are really great. Picking an average cost of $100/hour that's 20 hours and totally amazing. I have paid that for labor on less lofty and sexy projects.

Ursula Bauer Wow, Rob, very exciting!

TLC Day 101: New engine parts

SEP 18, 2013, 9:15 PM

The Lotus Chronicles (yeah, we're back baby), day who the fuck knows [it was actually day 101]:

A box full 'o engine parts (the first of many) arrived from Dave Bean Engineering. It goes straight to the machine shop tomorrow.

And, in a nightmarish development, the guy who let me drive his Europa with the engine rebuilt by The Lotus Engine God offered to sell it to me for a price I can't afford but will certainly approach if I thoroughly sort out mine.

Damn.

9 LIKES, 4 COMMENTS

Lindsey Brown Does the Europa come with the requisite ballet slippers? +1

Kenneth Robb I am so glad that my ample girth and size 12 shoes preclude my fantasizing too much about owning a Lotus as I can't even qualify as a hack mechanic. I do still have a dwell meter in my tool box. Youngsters can look THAT up on-line.

Franklin Alfonso Focus Grasshopper... One Lotus at a time lol. +1

Rob Knochenhauer Stay strong Rob! You will master the Europa... whatever the cost may be!

Not technically TLC but Lotus-related: The call of the Vixen

Doug Jacobs to Rob Siegel, Sept 28th, 2013: You can thank me later...

[Craigslist ad for 1970 TVR Vixen S2]

2 LIKES, 18 COMMENTS

Rob Siegel I have, in fact, always been very attracted to these.

Doug Jacobs Yes, and I've always been attracted to waif thin, heroin-addict-looking strippers with absurdly large, round, silicone breasts. Attraction isn't really the question, is it? +3

Rob Siegel (best retort to an obvious car attraction statement EVER) +3

Doug Jacobs The question is: "Do I stand up, face into the hot wind and jump into the chasm, knowing I'll never be the same person again, or do I sit back on my beige couch and castrate myself with a plastic spork?" +1

Liz Shapiro Hallsworth Do models like you to kick their tires? Cause this guy ain't havin' none of that.

Rob Siegel Liz, Road & Track recently did a highly politically incorrect one-pager of "Supercar vs Supermodel." They appear to have pulled it from their online content. But some of the quips were:

"Cost of Major Breakdown."

"Supercar: 1/3 of the purchase price."

"Supermodel: All the glassware in your house."
And:
"Reliability of Used Models: You know, we're just not going to go there." ₊4

Ed MacVaugh Folks who tease like this make my blood boil. Seller wants you to set the price for what he is offering.

Ed MacVaugh Rob Siegel remember me when you want to clear out the Lotus. Seriously. ₊1

Rob Siegel You may come to rue that, Ed, but, will do...! ₊1

Ed MacVaugh Well, it's that or a Ferrari :) or a Morgan ;)

Carl Witthoft I believe the reason Rob S. only does BMWs (and a certain Lotus gathering grime in a corner) is that, while he likes to work on persnickety cars, he also likes them to function for a few hundred miles afterwards. That rules out all English-made cars, right?

John Morris My Lucas and Amal-equipped British motorcycle will easily go 1000 miles before it needs something major. Yeesh!

Ed MacVaugh You have Amal Alamuddin on your motorcycle? Jono Morris

John Morris Any comment I can think of in response would get me in trouble, I'm sure!

Doug Jacobs Yeah, well, everyone misses the point here. It's not about the practicality of something you know and can handle. The point is the challenge of either taming or being tamed by a dangerous mistress. With something like a TVR, you can end up at the foot of the bed, bleeding from your ears, and still be entitled to claim a win. Just showing up is a victory. ₊2

Rob Siegel Doug, you think you might have a book in you? :^)

Doug Jacobs Rob, I used to write a column for a couple of motorcycle magazines, but I made everyone really angry, so I quit (was fired).

Franklin Alfonso Well at least you wound up not have to deal with body rust... The frame, well, a welding torch and plumbers pipes should do...

TLC Day 163: Checking out the hone

NOV 19, 2013, 10:12 PM

The Lotus Chronicles, Day 163 (I know, you're shocked):

[75 days—nearly 2 ½ months—since the last post]

TMWINTLEG called me up and asked me to come down and look at the honed cylinders. Remember, these are sleeved with thin sleeves and can't be bored to take an oversized piston, so the approach is to hone them just enough to take the scoring off them, then fit custom pistons. He cleaned them up ten thousandths, which is as much as he recommends. Three cylinders look fine, but the cylinder that held the seized piston still has one small corrosion blotch up high, above where the rings will touch, but has one band of corrosion about 2/3 of the way down the bore. I can see them but I can't feel either of them with a finger or a fingernail. The alternatives are re-sleeving or replacing the block, neither of which, obviously, would be cheap.

"What do you think?" I asked.

"This is just a driver, not a vintage racer, right?"

"Right."

"If it was my car," he said, "I'd use it."

Oh, and it looks like a Lotus bomb went off in my garage.

As Oat Willie said, onward through the fog...

13 LIKES, 28 COMMENTS

Rob Koenig The rings will eventually clean the rest of that up. Nicely done! +1

Rob Siegel I like you, Rob... +1

Jeff Keener Looks my parents garage up in NH after Scott and Matt helped me pull my engine. Picking it up from machine shop this Saturday! +1

Jeff Hillegass Close enough for Parliament work. +4

Jeff Hillegass (You can use that line, if you want. I have no application for it. Early holiday gift.) +1

Rob Phelan If you can't feel it, seriously, let the rings, oil and combustion gasses resurface it for you. +1

Rob Siegel I'm seriously loving this feedback. Thanks guys. +2

Richard Memmel driver. enough said.

Bruce MacPherson I'm sure there are running British engines with cylinders worse than that, and their owners just don't know about it, right? +1

Lindsey Brown Don't worry about the one cylinder; concentrate on the hyperdrive instead. +3

Jeff Keener Rob- I'm planning on assembling the M10 with Joe Gibbs assembly grease and breaking in with their BR 30 5w30 oil for first 500. What are your thoughts on this?

Rob Siegel Jeff, I'm really out of my depth on those sorts of questions. I can follow a manual and take an unfamiliar engine apart and put it back together, but that break-in stuff, I don't know. I stopped in on Rob Torres at 2002Haus a few years ago and spent part of a Saturday as an extra set of hands helping him reassemble an engine. He coated the bearings with some stuff from a squirt oil can. "What is it?" I asked. "Lubriplate cut with something?" "No," Rob said, "it's just a mixture of what was in the bottoms of quarts of oil, 5-30, 20-50, whatever." For the Lotus, I'll follow what's in the manual. +1

Dale Robert Olson Looks good to me, the fingernail test is the key, if it doesn't pick up anything, neither will the rings.

Tom Samuelson Weld the area, and re-hone?

Ernie Peters It would drive me crazy knowing the block wasn't "perfect," but I agree it should be fine as is. Now start getting it back together already!

Paul Wegweiser Rob, for assembly lube, I swear by gray Moly mixed with motor oil... if you can still find Moly grease, that is! Applied with a flux brush / acid brush. For the 300SL motors, I use Torco break-in oil, but change it after about 60 minutes of run time. It's an expensive 12 qt (!) process... but it makes me sleep better at night. +2

Harry Krix Does not look critical, I agree with Dale Robert Olson, if it passes the fingernail test; street engine, rebuild it and forget it.

Franklin Alfonso Like we used to say in the Band when tuning up: Close enough for rock&roll...

Rob Siegel Tom Samuelson--> When someone in a position to make more money off me (a.k.a. the machinist) says "I think it's fine," I tend to take their advice.

Tom Samuelson Rob Siegel, "...but while you're already in there..." To quote a famous Zen car book. +1

Steven Anthony Smith Sometimes the opposite of "good" is "better."

Roy Richard Given the patented "British Continuous Oil Change," adding a quart a week to make up for leaks, this should be fine.

Justin Gerry +1 for the Joe Gibbs stuff. They know what they are doing. I'm amazed at how much better this stuff deals with track duty in my tii. I destroyed some other well-known synthetics and this stuff had no issues.

Chris Lordan Hey Rob, cylinder barrel scuffing like you describe is very common when rebuilding the piston engines that are used in light aircraft. The rebuilders, who are heavily regulated, have a lot of leeway to use professional judgement when deciding what to use and what to scrap. Because of the expense of the components, re-use of scuffed barrels is common, especially when the operation and safety of the engine is really not in question. So I agree—this is a good call.

Rob Siegel Thanks Chris. I don't rebuild a lot of engines, and don't have a body of experience and feel to go through. I expected everyone to jump down my throat and get all "do it once do it right" on my ass. I'm delighted that the cognoscenti are saying the opposite! +1

Chris Lordan Wow! I've never thought of myself as cognoscenti-like!

Rob Siegel I decree it—you are cognizant!

Other things happening in 2014

- Rebuilt the engine in Kugel, the '72 2002tii.
- Broke my foot.
- Drove the tii with the broken foot 3000 miles round-trip to MidAmerica 02Fest.
- Went to Southeast Sharkfest.
- Drove the hastily-resurrected Bavaria to The Vintage.
- Went to the Manchester Auto Show in VT and saw my first '59 Lotus Elite.
- Bought a ton of parts from a restoration shop that had closed. I still have many of them.
- Bought a Euro 1979 635CSi.
- Bought an '87 E30 325is.
- Went on the last Nantucket vacation on which any of the kids came with us.

- Went on a geophysical survey to Mare Island CA, looking for unexploded shells in San Francisco Bay with a remote-controlled crawler.
- Began working part-time for Bentley Publishers.

TLC Day 211: Custom pistons ordered

JAN 16, 2014, 7:01 PM

The Lotus Chronicles, Day 211 (or, "I'm not dead!"):

[48 days since the last post]

Just spoke with TMWINTLEG. The custom pistons, and rings to fit them, have been ordered from Diamond Piston. An original piston was sent to Diamond. They'll replicate it, with minor tweaks, and sized slightly larger than the cleaned bore size (which is I believe 0.006 over). When the pistons arrive, he'll measure them and do a final hone to accept the pistons. Forward motion!

21 LIKES, 9 COMMENTS

Bob Shisler Sweet!
Rob Koenig Sounds like you're on track for a spring revival.
Rob Siegel I'm not setting any schedules. Plus, I hope to be driving the tii to MidAmerica 02 Fest in Eureka Springs. Plus, I bought this '72 Bavaria with 49k I don't even have here at the house yet. +3
Bob Shisler Pics of the Bavaria?
Richard Shouse Custom pistons? Ouch... there go the book royalties...;)
Rob Siegel Actually about the same as new stock -- $600 +1
Richard Shouse But, but, but... there's only four of 'em!... cough, ackk!!!!!
Rob Siegel No, that's total.
Richard Shouse That's what I mean, $150 each! Gasp... I don't think I paid that much for a set of 8 for the 302 Ford I rebuild years back... I know, I know, that was then, this is now... darnit...

TLC Day 222: Valve seats

JAN 17, 2014, 8:16 PM

The Lotus Chronicles, Day 212:

TMWINTLEG calls me up, asking me if I want to replace the valve seats. "I'm about to start the 3-angle valve job," he said. "It's now or never if you think the car needs hardened valve seats to run on unleaded gas." I'd researched this six months ago but forgot the answer. 30 minutes online later... it seems the oldest Lotus Twin-Cam engines had cast-iron seats, which should be replaced, but anything '74 or later should already have alloy seats sufficiently hard for unleaded fuel. If you're vintage racing the car, you might want to go to something harder still, but for a lightly-used road car, the stock alloy seats should be fine. So this car is a '74, driven 'till '79, then put away. The engine apparently was gone into once before me. Exactly what seats does it have? "Put a magnet on the seats," I ask him. "Does it stick?" "Nope," he said. Alloy. "Cut 'em." "I'll do it tonight," he says.

Onward.

16 LIKES, 14 COMMENTS

Kevin McGrath I love how you can write something about a type of car that I could care less about and make it interesting enough to make me want to read more. :) +1

Rob Siegel yessir... turning pain into art, the function of writers/artists for thousands of years, the Hack Mechanic is carrying on a VERY long tradition, just with cars... +3

Steve Kirby Unless the heads are known for dropping valve seats you're pretty much fine w/the tri angle valve job (how are the guides btw)

Rob Siegel usable +1

Kevin McGrath "Turning pain into art." Now there's a tattoo!

Steve Kirkup I was on a business trip with a foundryman from our business partner. He is getting a tattoo on his back of an induction furnace for melting alloys for castings. That is turning a black art into pain back into back art. +2

Daniel Senie I was curious if your song for the next show at the Serendipity Cafe would be about the Lotus... +2

Jackie Jouret That leaded fuel thing was a crock—read Jamie Kitman's piece in The Nation, archived online.

Daniel Senie We had a 1975 Peugeot 504 wagon (OK, stop laughing). If you put unleaded gas in it, the head gasket would blow. Happened more than once. Also had a filler neck that didn't hold a US nozzle, and resulted in gas geysers more than once. Super-comfortable seats, though. +2

Eric Heinrich You could just run leaded race gas... :-)

Rob Siegel Jackie, I don't have the expertise to know if it's a crock or not. There seems to be conflicting information and gray area. The forward of Miles Wilkins' "Lotus Twin-Cam Engine" book, considered the bible, talks about newly-rebuilt engines experiencing valve seat recession in as little as 500 miles of high-RPM driving. It scared the crap out of me when I read it. But as I read more online, that seemed to only apply to the earlier twink engines in Cortinas and Elans with iron seats, and only if they'd never been run with leaded gas which should've impregnated a certain amount of lead into the seats. To me, the fact that this will be a road car, not a vintage racer put it into the "don't worry about it" category. And, as Eric says, there's always race gas, or additives. +3

Tom Kinney just do it, if nothing else it's a good selling point and the project has your name on it

Tom Kinney oops, didn't see the rest of the post, onward then, probably will be fine...

David Weisman all this talk about sufficiently hard, etc... +1

TLC Day 277: We have pistons!

MAR 13, 2014, 8:39 PM

The Lotus Chronicles, Day 277:

[55 days—almost two months—since the last post]

A big day in the Hack Mechanic world. I picked up the head for the 2002tii from the machine shop. AND... while I was there, TMWINTLEG asked "Wanna see the custom pistons for the Lotus? They just came in this morning." It goes to prove the saying: British cars can sense infidelity.

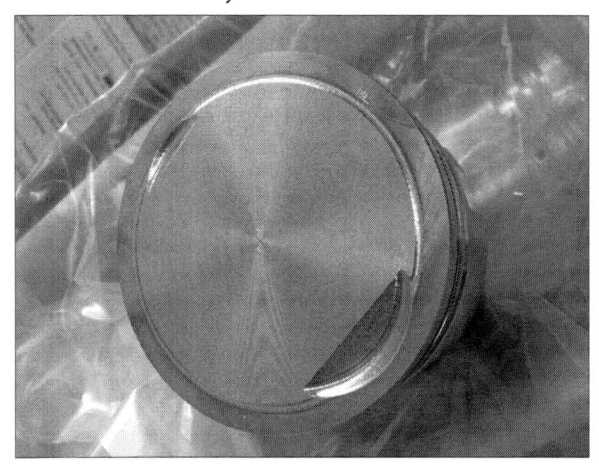

28 LIKES, 11 COMMENTS

Karel Jennings *ALL* cars can sense infidelity. Spend thousands on one, start the other one and it'll throw a rod in protest +4

Rob Jones Nice!

Sam Pellegrino I prefer the refreshed head, myself.

Richard Shouse All British cars are sluts... they don't care who they screw as long as lots of money is involved... +9

Mark Gascoigne Trying to think of something profound to say but Richard has it nailed.

Rob Siegel Hey... how about we put this pair in a dark engine compartment... let them get busy... see what they produce. What could possibly go wrong? +2

Dave Bentz It looks almost too purdy to use...

Stéphane Grabina Brits always start out pretty but after so many years of drinking and smoking their teeth get yellow and then they start to smell. Wait what was I talking about? +1

Greg Calvimontes And don't forget Richard, they spread their fluids everywhere... ;) +1

Richard Shouse Except where it's supposed to be... +1

Trent Weable Freshly machined engine bits. Ahhhhhhh.

Chapter III:
The Doldrums

Interjection: Punctuated equilibrium

Paleontologist and evolutionary biologist Stephen Jay Gould was a major proponent of "punctuated equilibrium," the theory that most biological systems—animals and plants—are stable for long periods of time, then undergo sudden rapid periods of transformation in response to external stimulus.

Boy, does this ever describe the rhythm of work on the Lotus.

In the previous post on March 13th 2014 (Day 277), I went into the machine shop and saw the newly-delivered custom pistons. It would be—spoiler alert—four years before I would pick up the machined block. In the meantime, the car would sit in my garage and get covered with crap.

What happened?

Well I could say "never tell a machine shop that you're not in a hurry." And I actually *did* say that to them. Bad idea. But it wasn't their fault.

Although I can't find it documented anywhere, I recall TMWINTLEG (The Machinist Who Is Not The Lotus Engine God, Erik Lindskog) telling me that he was having problems getting the jackshaft to turn smoothly in the new bearings he installed in the block. He used a bearing scraper to lightly clearance them, but then found that the root cause was that the jackshaft appeared to be slightly bent. It fell to me to determine whether the jackshaft was the same part as a standard Ford in-block camshaft or a unique Lotus part. I read conflicting information on Lotus forums, and didn't want to bother TLEG with the question since, after all, I'd pulled my engine out of his shop.

The years ahead held job change, then unemployment. The Lotus engine naturally slid to the bottom of a very long list of priorities. And so it sat in a back room at the machine shop. And the car sat in the back of my garage in plain sight, the very lightweight elephant in the room.

To barely paraphrase *The Lord Of The Rings*, some things that should not have been forgotten were lost. History became legend. Legend became myth. And for four years, the Lotus passed out of all knowledge.

TLC Day 504: We have head!

DEC 6, 2014, 3:29 PM

The Lotus Chronicles, Day 504:

[227 days—7 ½ months—since the last post]

Are you ready for something absolutely shocking?

We have head! I picked it up from the machine shop yesterday. It had been ready since last spring, but I was hoping that by leaving it at the machine shop and saying I wanted to pick everything up together, it would spur them to completion. New strategy: By taking the head, I create a space on their shelves. Nature abhors a vacuum, right? (Answer: maybe, but nature REALLY abhors a dead Lotus.)

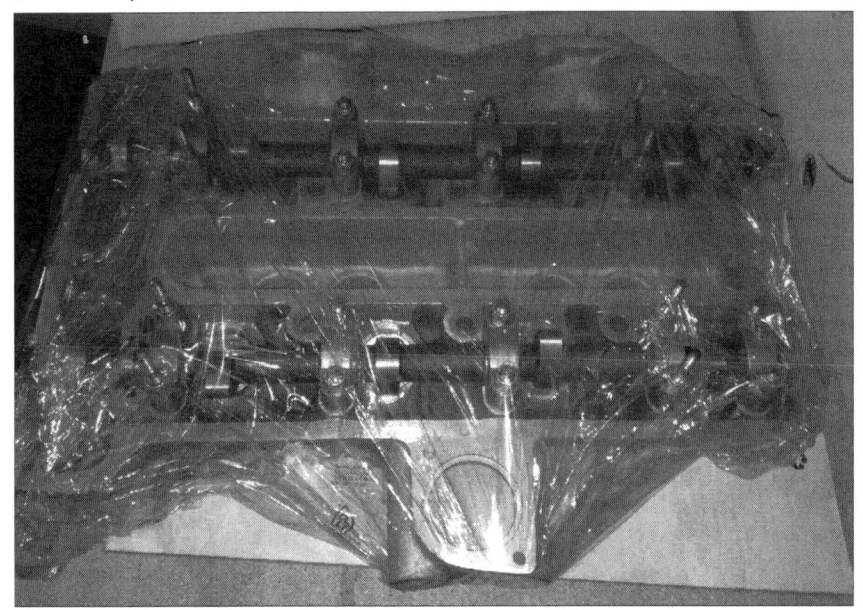

27 LIKES, 26 COMMENTS

Larry Johnson sure is purdy
Trent Cole Remanufactured head. Kind of like Viagra, right?
Andrew Wilson Two things you never get at home, the other is eggs benedict. +2
Franklin Alfonso Well I hope you get the other parts and finally get a happy ending to the Lotus project. +2
Daniel Senie Progress! We might see the car move before the decade runs out? +1
Doug Jacobs Nature abhors a dead lotus? And here I thought 'dead' was their 'natural' condition. +4
Franklin Alfonso At least they don't rust as bad as most British cars...
John Graham Does it leak yet? +5
Richard Memmel Looks like you have even more work to do ya you lol +1
Kenneth Robb "Dead Lotus" may be redundant. Sorry. +1
Maire Anne Diamond I think we should rename the band Dead Lotus. +5
Rob Siegel and you can...
send me a dead Lotus in the morning
send me a dead Lotus in the mail
send me a dead Lotus on your wedding
and I won't forget to park a Rover on your grave... +6

Lindsey Brown I'd like to come over and play sometime soon. Let me know when mommy says it's okay.

Nicholas Bedworth Lotus? [photo of a lotus blossom]

Jeff d'Avanzo Wait—I thought I read from some wise auto-sage that all British roadsters were crap – it's just the different degrees of crapiness that you get to sort through...

Rob Siegel well... yes. but a) it's not a roadster, and b) it's way cool crap.

Rob Siegel I'm a little guy, so I do fit in the Europa. In addition to height and weight, the main factor for fitting and driving is, in fact, shoe size—the pedals are so close together you almost need to be wearing ballet slippers.

Jeff d'Avanzo Right, right, I was thinking Elan... shows you how much attention I pay to British cars. I was interested until I sat behind the wheel of an XK 120 and realized at 6'2" I was 6 inches too tall for the car. MGs surprisingly have good leg room. One thing on Brit cars—they do corner well. Good luck with the rear visibility!! +1

Richard Shouse How does one get into an out of that thing without making old man noises?

Rob Siegel I'll find out

Chris Lordan "I believe that this nation should commit itself to achieving the goal, before this decade is out, of landing a man on the moon and returning him safely to the earth. No single space project in this period will be more impressive to mankind, or more important for the long-range exploration of space."—President Kennedy, Address to Congress on Urgent National Needs, May 25, 1961

"I believe that the Hack Mechanic should commit himself to achieving the goal, before this decade is out, of driving a Lotus Europa around the block and returning it safely to the garage. No single repair project in this period will be more impressive to mankind, or more important for the long-range exploration of garage space."—Chris Lordan, Address to Facebook Comments Section on Urgent Hackish Priorities, December 7, 2014 +1

Rob Siegel Chris, There's a great story that, at that State of the Union address, NASA representatives watched and said "wait... THIS decade?" +1

Scott Aaron The front of the car looks a little like a fish. Ironically, it's the owners that get the hook. +1

[There will be a few other Lotus-related posts, but there will not be another update to "The Lotus Chronicles" with specific news of progress on the project for 1591 days, almost exactly three years, because there will not *be* any real progress.]

Other things happening in 2015

- Left my 33-year engineering job and began working full-time at Bentley Publishers writing repair manuals.
- Bought and sold "The Turkey," a running '72 2002 parts car.
- Bought a salvaged-repaired 2013 Honda Fit Sport as Maire Anne's daily driver.
- Bought and sold a '76 2002.
- Bought "Old Blue," a 'slightly rusty heavily patina'd 73 2002tii that, once resurrected, absolutely screamed.
- Replaced the front axles in the 2003 325XiT wagon, a repair so awful that soon after I sold the car to my friend Alex.

- Drove the Bavaria back to The Vintage, where it was subjected to the now-legendary "feathering episode." (search YouTube for "Rob Siegel Bavaria feathers")
- "L'affaire Brian Ach" (telling a guy I'd just met in a parking lot in Virginia, whose tii began running horribly on the drive down to The Vintage, that if he towed it to my house in Massachusetts, I'd fix it for him for free—well, for parts. Turned out to be a broken spring inside a fuel injector, something that no one I know has ever seen. Brian and I became good friends.)
- Bought a '74 tii ("Otto") and daily-drove it while I was working at Bentley.
- The last Nantucket vacation.
- The ridiculous job replacing all the metal brake lines on the last Suburban.

Other things happening in 2016

- Bought a 2003 530i stick sport which became my daily driver.
- Began recording my first album since 2004.
- Bentley Publishers published my second book, *The Hack Mechanic™ Guide to European Electrical Systems*.
- Drove my '79 Euro 635CSi with no a/c to Southeast Sharkfest in Chattanooga. Got roasted alive by the heat. Resolved to retrofit a/c into it.
- Drove to The Vintage in my tii Kugel, running with Brian Ach in his tii which I'd repaired, making sure that this time it made it.
- Lost my job at Bentley Publishers and went back briefly to doing geophysics.
- Did my final geophysical survey, a two-month stint in Denver.
- Began writing a weekly online piece for BMW CCA.
- Sold "Otto" the '74 2002tii, but not before finding and replacing a cracked head.
- Looked at a Vixen (a rare BMW turbodiesel-powered motorhome) in Albuquerque (google "Rob Siegel BMWCCA

The Vixen That Wasn't Part I II III")
- A psychopath was elected President of the United States.

Not technically TLC but Lotus-related: Europa in Denver

NOV 19, 2016, 11:06 AM

[Note: I was on a two-month geophysical survey in Denver and looking for trouble.]

Yes I knew it was local (Boulder). Yes I knew it was both on CL and on eBay. Yes I am aware it was also on BaT. Yes I think that a Europa with an "aluminum Buick V8" (which could mean anything) coupled to a 911 transaxle is way cool. No I didn't go see it. No I didn't bid. Yes I have "pangs" when I see it sold for $7500. Yes I am aware that, with the Europa's 42" roof line, I could stack three of them in one space in my garage. Yes this proves once and for all that I do have some modicum of restraint and common sense. And yes, as Kor says to Kirk in "Errand of Mercy," it would've been glorious. Now BACK OFF.

73 LIKES, 48 COMMENTS

Trent Cole Shit, did they intend to wrap the motor in headers? Worst looking header job ever. Otherwise "A" for effort on this one... almost an attractive car. Still looks like someone stepped on a Vega. +2

Andrew Wilson "I told them we've already got one." +12

Josh Wyte Don't worry, it's still an ugly car.

Marc Bridge That thing must be a rocket with a V8 in it! What's going on with the Brown Hornet?

    Trent Cole Hornet... LOL! There are so many other things that are brown that would fit better. +2

    Marc Bridge I was trying to be nice! +2

    Trent Cole Figured! +2

Daniel Senie Now you have an alternate drive train plan for yours when the rebuilt engine proves a bust.

Lee Highsmith Your restraint makes the world suck a little more. Stop being selfish and bring more stray dogs home! That thing needed you! +3

    Rob Siegel Yes, but I did not need IT. +1

    Mike Crane Too many projects means you don't get to play in them.

Lee Highsmith Need? NEED???? I thought this was Rob-freaking-Siegel +1

Rob Siegel Oh, rest assured -- I have my eyes on something far stupider than this. +4

Mike Crane There's no helping you…enjoy. LMAO

Lee Highsmith Color me intrigued…

John Whetstone Atta boy!

Franklin Alfonso Yellow meh…you already have brown…

Jenni Morgan

Chris Lordan How about this, instead? [BaT add for 1997 Lotus Esprit V8]

David Kerr "budgeting properly for running costs is critical"

Chris Lordan Gotta love BaT. I can't tell if some of the seller comments are tongue in cheek or not.

Mark Minkler I owned a 1970 Europa. I'll have to tell you about it. +2

David Rose That sold for $7500? Good god. +1

Doug Jacobs If you scratch that, it's never gonna heal…

Ricky Marrero I had a 914 with the same engine. When I sold it, the new owner put in a 3.8 turbo from a GN. It worked much better.

Dayton Burlarley Aluminum Buick v8 should be an early '60s thing. GM sold the design to Range Rover.

Rob Siegel Yes, I know. The question is: what is actually in the car? It's not photographed or well-documented. +1

Jeremy David Walton Europa + Rover nee Buick V8 high profile UK via GKN giant auto components special widely shown & driven to prestige race meets in late 1960s. Good combo as motor so light & comparatively torquey compared 4- pot production units. Thanks for sharing, looks neat… +1

Frank Mummolo Let's just say you don't have room for it…

Scott Aaron If you are giving yourself big restraint points for not buying this, you are worse off than I thought. +3

Scott Aaron I mean that in the best possible Internet way.

Rob Siegel right… +1

Rob Koenig You know 9 out of 10 cars that sell on eBay end up relisted. You can catch it the second time around. +1

Greg Aplin Is it just me, or do the fender flares on this thing throw a baby Pantera vibe? +1

Maire Anne Diamond That is seriously ugly. [Note: this is only the second post from my darling wife :^)] +2

Rob Siegel V8s and fender flares not your thing, baby?

Maire Anne Diamond It was the color and the wheels that made me cringe. +4

Rob Koenig Is cringe another word for vomit? +1

Paul Wegweiser Maire Anne Diamond the only way to make a brown car look worse: park it next to a yellow one. Reminds me of some 70s era funky burger joint color scheme. +2

Jonathan Poole Paul Wegweiser my first car was a topaz Braun 320i, it hid the rust nicely. That's the best thing I can say about it.

Joe Bruckner Glad you got that off your chest. +1

Doug Park Wow, this Europa looks horrible.

Jack Frederick I remember the article in Car & Driver on the Europa when it first came out. The punch line on it was that it was a great little car. After all it is a Lotus. It only problem was when you wanted to get out of it you had to crawl up onto the curb.

Other things happening in 2017

- A psychopath was inaugurated as President of the United States.
- Attended the BMW CCA Foundation for the induction of Yale Rachlin's '74 2002tii into their permanent collection.
- Did the epic re-dyeing of the seats in the 1987 E30 325is, then sold it.
- Bought Louie, the decade-dead '72 2002tii in Louisville, drove down there in a rented SUV with tools and parts, resurrected it by the good graces of Jake and Liz Metz and others, drove it the thousand miles back home, then drove it to The Vintage and back.
- Wrote and released my third book, *Ran When Parked* (the story of Louie).
- By coincidence, Bentley released *Mechanical Ignition Handbook: The Hack Mechanic™ Guide to Vintage Ignition Systems*, the other book I'd written for them when I worked there, two weeks later, making my 3rd and 4th books come out in rapid succession.
- Sold "Old Blue."
- Bought a 1996 Winnebago Rialta (a Volkswagen Eurovan with a Winnebago camper body).
- Began doing Q&A in the column "Wrenching Thoughts" for Hagerty magazine.
- Got a 1987 635CSi for free from the father of my son's girlfriend, who was tired of it dying and his Mercedes mechanic not being able to fix it.

Digression: They're taunting me

FEB 1, 2017, 2:55 PM

"Get your Lotus ready for Spring." They are taunting me.

[advertisement image: r.d. enterprises ltd. — Lotus Parts Specialists for over 40 years — rdent.com — 215-538-9323 — "Get your Lotus ready for Spring" circled]

38 LIKES, 29 COMMENTS

Ed MacVaugh I got that, too. That is why I was researching shipping costs :) +1

Matt Pelikan They don't say which spring. +10

> **Rob Siegel** HAHAHAHA!
> There's a story, possibly apocryphal, where, when JFK gave his "we will put a man on the moon by the end of the decade" speech one NASA guy in the audience turned to another and asked "which decade?" +3
>
> **Scott Linn** And yet they did it anyway! +1

Wade Brickhouse Do it!!!

Sam Smith RD is dangerous. And he knows his shit—will tell you what's good, what's not worth paying for, what's made of repro crap rubber overseas, and which UK stuff is necessary. He got a lot of my money. Then I sold the Elan, and I kind of missed talking to him. Lotuses break you, Rob. Run. Also as long as you're running here I'll help you out sell me the car because something something Stockholm Syndrome. +4

> **Rob Siegel** Sam, of the many conversations we could have about many things, the one about the Lotus would be the least productive. +1
>
> **Scott Linn** I'm a fan of RD too, he has always done right by me. +1
>
> **Ernie Peters** Ray is a great guy. Spent many $$ at his business getting my Elan on the road. +1

Daniel Senie EVERYONE is taunting you about the Lotus... +2

Doug Jacobs You could have brought back some of that unexploded ordnance from CO and got your Lotus ready for spring as well. +2

Nathan Smith I'm gettin' mine tuned up right now! [photo shows Nathan tuning up his guitar] +2

Alan Hunter Johansson Hahahaha!

Luther K Brefo Jeff, hehehehe

> **Jeff F Hollen** I do not own a Lotus. I own two Lotus powered vehicles.
>
> **Luther K Brefo** That's even worse.
>
> **Jeff F Hollen** ...and I've never seen one of them +1
>
> **Luther K Brefo** I bought a car sight-unseen once or twice. Like intercourse without protection.
>
> **Frank Mummolo** Worse. It's not over in 15 minutes. +2
>
> **Luther K Brefo** Lol

Tom Schuch Getting my Lotus ready for spring. [photo shows man in Lotus position] +3

Franklin Alfonso Did you ever check with the place you left the engine as to the status of repair schedule?

Doug Jacobs You might consider having another child and training him/her in Lotus engineering. You might get the motor back sooner. +1

Interjection: The Europa in the wild

In May of 2017, I was driving Louie, the 2002tii I'd just bought and resurrected, as part of a caravan with some friends down to "The Vintage," the BMW event I attend annually down in Asheville. We were in bumper-to-bumper traffic on a section of I-81 when one of my traveling companions called on the cell and said the last nine words I ever expected to hear: "Hey Rob, there's a red Lotus Europa up ahead." Sure enough, when the traffic cleared, I could see the unmistakable silhouette of a Europa Twin Cam like mine. At about half the height of any other car, it was pretty hard to mistake for anything else. I goosed the tii up to nearly 90 mph to catch it

There it was. A Europa in the wild. What are the odds? Tiny and impossibly low and angular, such a light little bit of a thing, looking like a gust of wind could cause it to swing beneath a tractor-trailer and shift lanes, it was thrilling to see in its natural habitat. And the guy driving it was into it; he was ripping along just like I was in the tii. I thought, you know, the next person who calls the car a "bread van" who has never actually seen one tearing up asphalt, I'm just going to punch right in the mouth.

I video'd The Europa In The Wild on my phone as I excitedly yelled "I have one too!" to the driver. As I did, I momentarily swerved and probably scared the shit out of him. The video can be seen by searching YouTube for "The Lotus Chronicles Europa in the wild." You can hear me say "I'm *totally* taking this as an omen." And I did.

That evening, when I arrived at The Vintage, two other friends showed me videos they'd shot of other Europas on their drive down. It turned out there was a Lotus event somewhere on the Eastern seaboard that same weekend. So, okay, the Europa wasn't quite as randomly "in the wild" as I thought. But still, it and its owner were going to an enthusiast event just like I was. I still totally took it as an omen.

However, omens themselves don't necessarily translate into forward motion. When I got back home, there were other projects ahead of poor long stalled Lolita (including a time-critical a/c retrofit into my '79 Euro 635CSi), and she remained ignominiously abandoned in the back of the garage. The only real immediate change was that I felt worse about it.

Not technically TLC but Lotus-related: Footprints

Sep 16, 2017, 6:53 PM

I'm pretty lax about locking the garage and closing all the windows, but I see that that will need to end. These are raccoon prints on the windshield of the Lotus. I have to admit that love the native American pictographic feel, but I don't like the idea of the big varmints crawling around on or in the E9. At least, when I get the Lotus running and it has "gremlins," I'll know who to blame.

96 LIKES, 38 COMMENTS

Dorothy Isenberg This is from the opening credits of the horror film, "Racoon Zombie Apocalypse"! +2

Pam Loeb Looked for you at the British Invasion in Stowe today... +1

Scott Linn Who to blame for gremlins? Isn't that always Lucas?
I will say that I've been driving my MG Midget for 42 years, and MG TD for 12, and my Europa for 6, and electrical problems in all have been pretty much non-existent. Every issue was just due to old parts wearing out. I've had more problems with my Chevy Blazer... +1

Alan Hunter Johansson Chevy Blazer. Way to set the bar! 😀

Scott Linn Actually, our Blazer is 23 years old and 223k miles...

Paul Garnier At least you didn't find this [insert pic: volvo with suggestive body print.jpg] +13

Alan Hunter Johansson Don't kid yourself. You could have hermetically sealed that Lotus when you parked it and it would STILL have gremlins. +4

Jamie Eno Kind of creepy... like they were trying to get out.

Garrett Briggs Are you sure it wasn't one of them there free range children? +1

David Weisman Yes!!!!

Jake Metz Just put the raccoon in the Lotus and send it down to Kentucky. Problem solved. +4

David Kiehna They must be scoping out a winter hibernation spot!

David Kiehna I'm still kicking myself for selling mine. I'll take it off your hands! [photo of Europa and MG] +1

Mike Walker Don't know about raccoon prints, but cat prints can be acid enough to etch the paint.

Jean Laffite Garbage pandas (1)

Vince Strazzabosco Sure it wasn't aliens instead of 'coons? That IS a Lotus... +1

Paul Wegweiser Rob Siegel, he has a point. I'm not sure I could ever drive a car, with the knowledge that it was once "probed."

Harvey Nuttall Who cares... as long as it doesn't get on the BMW +2

Sean Curry The picture with all the prints is creepy. It's like the souls of all the hack mechanics that have gone before you are trying to escape! +6

    Kevin Morrin I LOL'd.

Dave Gerwig That was Rocky Racoon visiting from Florida. At least the roof didn't blow off and it was a dry space to stay. +2

John Morris I have an anti-raccoon device. [photo of a pack of wolves] +1

Mark Thompson "Gremlins under the hood" I remember that line from one of your Roundel articles 25 years ago...

Doug Jacobs It could be the Lotus spirit animal, pleading for attention.

Franklin Alfonso I guess the raccoons couldn't get it started either... they didn't know you pulled the engine.

Nathan W. Naetzker I think it has a certain Chauvet charm- perhaps a motif for the exterior of the car?!

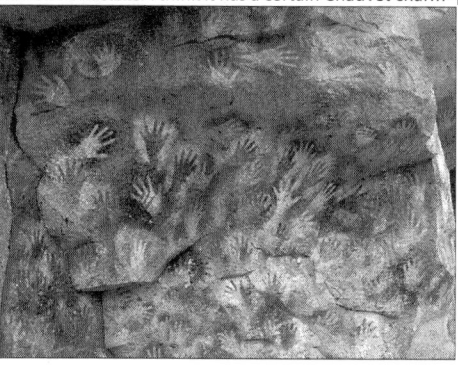

    Franklin Alfonso caveman patina... +1

Jeff Hillegass Based on that dust, I've figured out your plans for the Europa. You're going to stick some cardboard boxes on it, seal the garage with plywood and 2x4s, then tell Wayne Carini you know of a barn-find Lotus in Massachusetts that he should take to auction on CHASING CLASSIC CARS. +3

Brian Ach Those are worth an additional $12k in "barn find" parlance +1

Hugh Forrest Mason Europa!!!!!

Susan Levine Looks like a horror movie

    Kim Niemi Davidson Right? I was thinking walkers trying to get in. LOL +1

Maury Walsh Give them enough time. They'll come up with something more.

+4

    Rob Siegel THIS is what I am saying!

Karin Vander Schaaf That's actually a really cool photo...

John Graham Wow!

John Graham I woke up to this one morning... Talk about scary. [photoshop of cow on hood of 3 Series BMW]

Layne Wylie Raccoons have 5 digits. That's just a cat that couldn't get traction.

Not technically TLC but Lotus-related: That way lies madness

SEP 30, 2017, 3:29 PM

[Craigslist ad for ultra-ratty Lotus Elan 2+2 project]

Oh the poor thing. Still, there's probably no cheaper way to get a parts motor for my Europa's Lotus/Ford Twin-Cam. If it had the Weber head, Weber carbs that weren't corroded to handfuls of powder, and Sprint specs, it might actually make sense. Whew! Dodged that bullet!

13 LIKES, 12 COMMENTS

> **Jeffery W Dennis** Always liked these 130's.
> **Patrick Hayes** Just do it.
> **Samantha Lewis** Just breaks your heart. That was someone's very special car once. +2
> **Michael Trusty** Miata engine in the Europa? +3
> **Dan Kaufman** I think he should power his Lotus with a BMW engine.
> **Daniel Senie** Dodged a bullet... so the Lotus can continue to be a garage ornament for some time to come.
> **Carl Witthoft** Buy it -- the two carcasses will keep each other company. +1
> **Jerry McNeil** Looks almost like an Opel GT. +1
> **Lindsey Brown** I'll swing by your place with a Chevy small block, and a big shoehorn, and solve your Lotus problem. +1
> **Rob Sass** Ouch. Had a nightmare last night that this tuned into that. [photo of Rob Sass' beautiful Elan 2+2] +2
> **Rob Capiccioni** Check out (Spydercars Ltd). They do some amazing stuff with late-model Ford conversions into earlier Lotuses.
> **Wink Cleary** I have a couple Webers sitting around... 😀 +1

TLC Day 1591: The turning of the tide

OCT 18, 2017, 5:15 PM

The Lotus Chronicles, Day 1591 (I know, you're shocked):

[that's 1087 days—almost exactly three years—since the last post titled "The Lotus Chronicles" that detailed any specific progress]

A number of events caused me to perform an automotive archaeological dig in my garage and unearth the Lotus from beneath the layers of crap which included a black 2002 rear bench seat, four steering wheels, and a sheepskin seat cover I have no recollection of owning. These are:

1) I've begun writing about the sad story of the Lotus (the car that may drive me to hobby desertion) for Hagerty Online. The first piece should run on Tuesday.

2) In addition to my Hagerty online pieces, Hagerty has asked me to write for their magazine (I'll need elbow patches if I actually begin to rub elbows with "Jay" or "Wayne"). As part of establishing the first column, they're sending a photographer to, uh, shoot me in the garage tomorrow. Removing all the crap that had been piled in top of the Lotus to make it visible seemed to be the professional thing to do.

With these two events, something Lotus-like seemed to be in the air, so on this gorgeous fall day, I drove the Z3 the two miles over to the machine shop that has had the Lotus' engine for the last four years. I explained to the shop that I wasn't angry or impatient, no harm no foul, that a good part of the inactivity has been due to me not pressing them and me back-burnering the project during my employment ups and downs (there are few things that seem more frivolous to me than throwing money at the Lotus), but said, hey, let's front-burner it so I can get the motor re-assembled this winter.

When I got back from the machine shop, a ray of light was shining through the garage window and hitting the long-neglected Lotus. Another omen?

So, after I finish the air conditioning in the Euro 635CSi, after I finish my long-delayed 3rd CD, but before I crack the hood on the free dead 635CSi that's now behind the RV... mark my words: This shall be the winter of The Lotus (or, as Franklin Alfonso said, "the Lotush. Imagine Sean Connery saying it." :^)

Have faith, pretty Lolita. I'm still captivated by you.

153 LIKES, 86 COMMENTS

Andrew Wilson Mark my words...year of the lowtush is almost upon us. +5

Corey Dalba I'm too tall to ever drive a Europa... I tried

Rob Siegel and I'm too short and stubborn not to +7

Bob Maricle I don't know. I always thought that Lotus looked like a Chevy El Camino with a chopped top and funny design line for the top of the side windows. But, I have never driven one, which I understand, is the excitement. +4

Franklin Alfonso Yes ! I too find the top of the door windows strange how they bow. But handling and light weight are its forte...

Jeff F Hollen This makes me so happy. ^) +1

Christopher Kohler Excited to hear about future updates.

Doug Jacobs The Winter of Your Discontinence. +5

Franklin Alfonso And so the Lowtush will rise like the Phoenix from the ashes and be resurrected at last! By the Hack Mechanic! In all her British racing brown glory! +1

Rob Siegel AND, according to the Massachusetts RMV vanity plate web page, "LOTUSH" is still available :^) (though even money says they'll bounce it if I actually apply for it) +10

    Andrew Wilson Go for it!

CHAPTER III: THE DOLDRUMS

> **Chris Ryan** Yes well that all depends on how you read it... but really, is Low-Tush really something they would kick out? LOL
>
> **Paul Nerbonne** I may have told you this before—most creative vanity plate I've seen was 3M TA3 because it needs a rearview mirror to be legible +4
>
> **Frank Mummolo** Don't be so sure. An Orthodox friend of mine in NJ had DRECK approved done years ago. When I asked him how he pulled this off, he just shrugged his shoulders and said, "There's no Jews in Trenton." +1
>
> **Mike Lapic** Mn did not mind. [photo of Europa with MN "Lowtush" licence plate] +3

Franklin Alfonso Try anyway!

Skip Wareham Must be the warm weather.

Scott Linn Ooh, looking forward to seeing how you fix things you run into on the Lotus. +1

Robert Heitz Well we can thank Hagerty for that nice kick in the pants. Hopefully hack faithful will feel the kick themselves and get that someday project to why not today. +1

> **Alan Hunter Johansson** As we speak... [photo of a vintage truck with interior ripped out] +2
>
> **Robert Heitz** Well I turned the light on and looked at it. That's something... [photo of garage almost as bad as mine] +2
>
> **Robert Heitz** I may have some 907 parts in this mess... if I get around to it. +1
>
> **Doug Jacobs** Bullshit! I've been riding him merciless for two years now and he's finally cracked under the strain! +1
>
> **Alan Hunter Johansson** Nah, he saw my posts and it got him all excited again. [photo of a long-dead Lotus abandoned in a field] +1

Lindsey Brown Paint it Fjord blue, and sell it to an "Avengers" fan who preferred Tara King over Emma Peel. +1

> **Rennie Bryant** Emma Peel drove an Elan
>
> **Franklin Alfonso** She could have driven me anywhere lol. [gif of ground-up pan of Diana Rig as Emma Peel clad head-to-toe in leather] +1

Bill Snider Exciting times to come! Looking forward to your reports on the adventure!

Dohn Roush Inspiration is like a Lotus. You never know when it will start, and how long you can keep it running. Good luck! +3

Scott Aaron Are you flunkin' serious? I thought I was going to be at your funeral with a bunch of guys going "thank God he didn't put it in his will and leave it to ME." +8

> **Rob Siegel** you're a cruel man. +1
>
> **Scott Aaron** I know, I know. I'm in a bad phase. Probably hormones.

John Whetstone OHBOYOHBOYOHBOYOHBOY! I hope you do realize that Bruce's offer on Saturday [to come to my house; Bruce is a retired machinist] was sincere, he would love to help. And he'd definitely keep you on task.

> **Rob Siegel** John, THAT is one of the other things that was in the wind +1
>
> **John Whetstone** One thing he's good at is being your conscience. Not that you need such a thing! But seriously: No. Corner. Will. Be. Cut. +1

Alan Hunter Johansson YAY!!!

Jeffery W Dennis Go for it Rob she will reward you.

Dohn Roush Rob, only you would have a breadvan the color of crust... +2

Rennie Bryant I know some good Lotus sources. If you need them.

Rennie Bryant At least you didn't get this far. [photo of utter basket case Lotus on a trailer] +1

Frank Mummolo This is world-class news. The Hack Takes On The Dream Breaker In A Steel Cage Death Match! +3

Meron Lavie Love your bedroom decor!

Chris Lordan #GoRobGo +1

> **Rob Siegel** Hey, Chris... you know what else I found sitting on top of the Lotus? And I have no recollection of how it got there; it's filthy and been kicking around for years but I never had the heart to throw it out. look at the location and the date on it... NHIS 10/16/1992. [photo of disgusting old sweatshirt from BMW CCA track event] +2
>
> **Chris Lordan** I've gotta say—wow. Would only be spookier if unearthed on Monday (10/16/2017)!

Dave Gerwig That's a Lotus Eloped. Been engaged for 4+ Years, time to make the relationship official. +5

George Whiteley Oh boy, this is looking like a serial of Lotus articles come to life. +1

Chris Roberts Go Lolita! <3

John McFadden Sorry, I couldn't resist.

> I'LL TAKE SHIT THAT DOESN'T WORK FOR $500, ALEX

+6

Chris Lordan Chippy, chippy, Mr. McFadden. Very chippy. +1

John McFadden I didn't even have to use a meme generator, it was just out there on the interwebz! :-)

Chris Lordan Few things are perfect. And here we have the perfect thing. dingDingDING! Trebeck: "It's the Daily Double! What will you wager?" Contestant: "My soul, Alex. My very soul!"

Faith Senie Woo hoo, it's gonna be an interesting winter! +1

Rob Siegel Wait 'till you see the solar panel :^) +1

Marshall Garrison And so, boys and girls, the rest of the story is a mystery. After getting his newly-rebuilt engine delivered, Mr. Siegel was last seen stepping into his garage on a cold day in March of, ohh.. what was it? 2018, I think... with his Bentley Lotus Europa manual in hand, and hasn't been seen since, to this very day! There were some who said maybe the Lotus ate him after the perhaps unfortunate coincidence of drawing upon the model's plant-based name for inspiration, naming it 'Audrey II', after the insatiable people-eating plant in "Little Shop of Horrors." +2

Rob Siegel or maybe something something flux capacitor something 88mph no wait different car never mind

Marshall Garrison Rob Siegel Once you get it running, maybe stay on the safe side & stick to no higher than 87mph just in case! :D +1

Carl Witthoft Or naming it Christine. +1

George Saylor Europas are easy. It's the wiring you have to figure out. Running changes to electrical system. Instead of 'Model Year' the designation was 'Chassis xxxxx to Chassis xxxxxxx'. +1

Rob Capiccioni I eagerly await your Lotus Chronicles! 🙂

Franklin Alfonso I once had a metallic brown 78 Corolla station wagon. It didn't help it go faster though…

Tom McCarthy When the sunlight hit the car was there sounds of an angelic choir? Or wiring harnesses cracking? (1)

Scott Crater Bring a Fire Extinguisher… +3

Wade Brickhouse YAY!!! At Last! Can't wait. +1

Maire Anne Diamond Well, I guess that means I won't see much of Rob this winter, unless I want to visit him in the garage. +2

Justin Lippert I had the pleasure of friend status with one Harry Applebee, renowned Lotus grand master. Just a tip of the cap, RIP Harry. Now Rob, isn't it just a tad disingenuous of you to alter the crime scene? Those that have yet to know you should be afforded the opportunity to clearly see what's in store down the spark plug hole with you. +1

Brad Purvis She is an evil mistress and will crush your soul. +2

Justin Lippert "Lotush" a zz top hit played quietly. Or A person in possession both an amply junked trunk and unfortunately a diminutive height. +1

Brad Purvis You got a problem with "diminutive height?" +1

Daniel Senie So will you do a GoFundMe called "Get The Lotus Out Of The Garage" fund?

Rob Siegel Don't get me started on the GoFundMe thing… +1

Larry Johnson Can Richard Rawlings help ?

Jack Landreth great cars but fragile and trying to get an arm into the center tunnel can be frustrating and painful! +1

> **Rob Capiccioni** That's what she said! 😄😄😄 +2

Scott Aaron Hopefully your car turns out better than this song did… [video of REM's "Lotus"]

Lindsey Brown Two words: Tii motor. +2

Rob Siegel | DO have a spare one under the porch... +1

Interjection: Dinner with the pros

I have a friend / colleague who's a big vintage Chevy guy. He was active in organizing the 2017 Coastal Virginia Auto Show and was gently twisting my arm to come down and give a talk. It's not a German or a Euro-centric car show, mostly American iron (not that there's anything wrong with that). Plus it was right before Thanksgiving which we host for 25 people, and weather-wise, November is always iffy. But despite all this I decided to drive my beloved BMW 3.0CSi down. The weather held, and it was worth it just for the back-and-forth trip over the 18-mile-long Chesapeake Bay Bridge-Tunnel.

The show *was* a lot of fun. I really like mixed shows. I mean I love BMW 2002s but there are only so many of them I can look at (oh crap I think I said that out loud).

After the event, I wound up having dinner with two pros, one of whom has a cable show and the other of whom owns a nationwide chain of transmission shops. Of course, we talked cars. They were both very interested in the Europa. They were totally aware of the whole first-mid-engine-stress-skinned-road-car thing. I spun out the whole long sad story about the stalled engine rebuild, but ended on an upbeat note, explaining about the omen in seeing The Europa In The Wild, and my recent trip to the machine shop who has the engine to look them in the eye about restarting the dormant project.

I then said rhetorically "So I think it's going to be all right" as I took a sip of beer.

"How do you know?" one of them asked.

"*I'm sorry?*" I said as I practically spit out my beer. It was the last thing I expected someone to say.

"Did you actually *see* your engine?" he asked.

"He's got a point," the other man chimed in.

I paused, composed myself, thought about it, and stammered out "Well... no. But it's in a room in the back. I'm sure it is."

"Did... you... *see*... your... engine?" the first man repeated.

I momentarily froze.

"AAAAA-*HAHAHAHA!* I'm just fuckin' with ya," the first man said. "Still, leaving an engine at a machine shop for years is a really bad idea. But you sound like you know what you're doing."

I think they were just giving me some good-natured ribbing, mixed with a little friendly jab-in-the-ribs to the Yankee who was a little bit of an outsider at this show.

But, during the drive home, it did make me think. This was my fault. All of it. I'm the one who had taken a wrench to the spigot of spending on the Lotus. Of course, I had good reason. My employment situation had gone from uncertain to chaos to a stabilized fraction of my former income. Even in the best of times, with all these cars, I can never give every one of them everything they need, and in the worst of times, throwing money at the dead Lotus seemed the absolute height of frivolity (really, is there anything more frivolous than simply *owning* a dead Lotus?). So I couldn't really blame The Machinist Who Is Not The Lotus Engine God. Most of the glacial pace could be laid right at my own feet. The only surprise was that the machine shop let me get away with it instead of calling me and telling me to get my Brit junk out-a there.

While, from an economic standpoint, shutting down spending on the Lotus made sense, my dinner partner was of course absolutely correct. Even if it's at your own house, leaving a motor apart *for four years* is a terrible idea. Parts get lost, shiny exposed surfaces deteriorate, and your momentum dwindles to negative values. If it's somewhere else, like at a machine shop, it's worse. Businesses get sold. People die. One day you go back there to find the building ripped down to build condos, you scream and sob about your jackshaft, your custom pistons, and your bores honed seven thousandths over, and they shoot Thorazine into your neck and stick you in a padded cell.

But if the engine had been back-burnered at the machine shop, at least it was out of sight and out of mind. The rest of the car, though, sat in the back corner of my garage, occupying precious space, an elephant-in-the-room monument to my failure. During the ensuing four years, the "bread van" back deck and roof of the Europa had accumulated layer upon layer of garage junk. It was like watching one of those time-lapse animations of dinosaurs becoming fossilized. Granted, I'd recently cleared off the crap that had accumulated on the dead car, but that felt like ripping off a scab. Inexplicably, I'd never put a cover on the car, so what had been a somewhat shiny paint job when I'd rolled it into the garage had gotten badly dulled by dirt, grime, and particulates. Further, during a brake-bleeding job on one of the other cars, a hose I'd forgotten to fully secure caused fluid to shoot sideways out the wheel well. At the time I hadn't noticed it, but brake fluid spattered all over the Lotus' nose, freckling and blemishing the fragile painted fiberglass surface.

When I bought it, the Lotus had been an intact survivor, so full of promise. Now, it looked like a meth-addicted train-wreck. I knew where the block and head were, but other parts were scattered between the garage, basement, and under the deck. And due to the lack of the "stressed member" drivetrain, I couldn't even roll it outside and leave in the backyard under a tarp like an abandoned project should be. The car had gone from being a source of excitement to a source of humor to a source of shame. People would ask me "So, what's new with the Lotus?" and I'd visibly wince. All I could do was look with regret at the car's disassembled degraded condition and admit that it was the consequence of my irresponsibility and overreach.

Sigh.

In *Memoirs*, I wrote that a dormant project car is "the physical manifestation of possibility. It's your buy-in. It's your passion simmering on low. You might not be able to get the car pretty and shiny and running today, but perhaps you will tomorrow. When you reach the end of the line and put it on the block, it is admission of the end of possibility." Now, this was *not* the end of the line. I was *not* throwing in the towel. To the contrary, I'd had the come-to-Jesus-git-'er-done meeting with the machine shop the previous month. I liked them. I trusted them. I continued to work with them on my BMW projects. In terms of them completing the machine work, I *did* think it was going to be all right. But all I could do was wait. And it's a little unnerving when you realize that that even if you think you've had a sea change in your outlook, that you've internally reset your priorities, put your conceptual shoulder to the metaphorical wheel, your virtual nose to the tropological grindstone, it's out of your hands.

Plus, momentum is a double-edged sword. When you've got it, the physics of it is that it literally keeps things moving. But when things grind to a halt, it's no longer momentum—it's inertia. If things have been stopped for four years, it's going to take a lot of energy to get them moving again. I'd had two omens and had set my jaw into the wind, but I honestly didn't know whether I had the energy.

(Former physics majors: Cut me a little slack. I'm trying to be dramatic as fuck.)

Part IV: The Block Gets Reassembled

Other things happening in 2018

- Ran out of garage space and sold my Z3 to our neighbor and friend Kimberly.
- Drove Louie down to the BMW CCA Foundation Museum, where he spent nearly a year as part of their "2002 Icon" exhibit.
- Bought "the Lama," an '87 E28 535i sight-unseen on Craigslist in Tampa. Had it shipped up. Turned out it had a broken rocker arm. I had to rebuild the head.
- Released my first new CD in nearly 15 years
- Wrote and released my fifth book, *Just Needs a Recharge: The Hack Mechanic™ Guide to Vintage Air Conditioning*.
- Resurrected and sold the black '87 635CSi I'd gotten for free from my son's girlfriend's father. Wound up caravanning with the kid who bought it down to The Vintage and back, driving my now-air-conditioned '79 Euro 635CSi.
- Bought back Bertha, the '75 2002 I'd bought in Austin in 1984, moved up to Boston, sold to my friend Alex in 1990, then sat for 26 years deteriorating in his neighbor's garage.
- Tried to ready Bertha to drive to BMW CCA Oktoberfest in Pittsburgh in Kugel, but had to take Kugel instead.

Interjection: The water pump

I've alluded often enough to the central role of the water pump in the rebuild that I thought it might help if I gave it its own contextual interjection. You're going to need it when the shit hits the fan, which it's about to.

As I described in the "what the Lotus Twin-Cam engine is" chapter, this engine is a Ford Kent 1500 medium deck-height block bored to 1558cc with an eight-valve Lotus twin-cam head on it. However, in that section, I was concentrating on identifying the block and explaining why it's challenging to source a replacement. I didn't describe the unique design issues necessary to make the head work on the block, and why those changes make rebuilding this engine so challenging.

Like any other pushrod engine, the Ford Kent engine has the camshaft inside the block. A gear on the front of the cam runs off a chain driven by the crankshaft gear. A small front cover encloses the chain and gears. Above this cover, there's a hole in the front of the block that's the inlet for coolant pushed in by the water pump. The water pump has a flange that attaches to the front face of the block using three bolts, with the pump's impeller sitting in the hole.

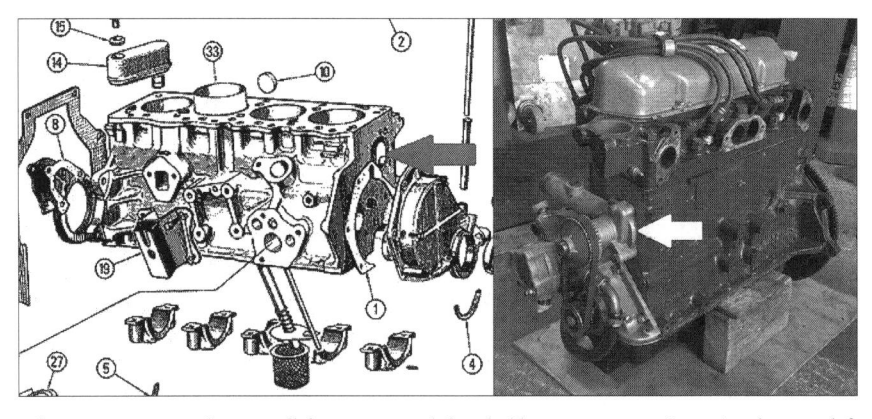

The Kent engine with its small front cover and detachable water pump. Arrow in photo on left shows location of water passage. Water pump in photo on right is mounted over that hole.

Now, if you think about it, if an engine isn't a pushrod engine but is instead a single or double overhead-cam engine, there has to be a timing chain or belt connecting the crank in the block to the cam gear(s) in the head. That creates two complications. The first is that the belt or chain must be shielded from the elements. If it's a rubber belt, the shielding can be a minimal cover, in which case it's possible for the water pump to still bolt directly to the front of the block with the cover fitting around it, as it does on the M20 engine in BMW E30 3 Series cars. But if the engine has a timing chain, the chain requires lubrication with engine oil, which means that a metal front timing cover (generally a lower cover attached to the front of the block and an upper cover attached to the head) must completely surround and seal the timing chain and gears.

But the second complication is that, if the front of the block is now completely enclosed by a timing cover, how does the water pump attach to the block and pump coolant into the water passage? The answer is: The engine is designed so it does. Generally the water pump bolts to a flange on the lower front cover, and there's a water channel in the cover that leads to a water passage in the front of the block. Note that this means that the front timing cover now not only has to provide an oil seal for the timing chain, but a coolant seal as well. On almost any OHC engine built in the last 50 years, though, that's not a problem. BMW 2002 folks, go look at an M10 engine. That's how it works. The water inlet is on the right front of the block, the timing chain runs to its left, the lower timing cover has a water passage sitting directly over the water inlet, the water pump bolts to cover, the pump impeller sits in a separate recess but swirls water into the passage, and thin paper gaskets between the block and cover and cover and pump seal everything. Easy peasy.

Water pump flange and water passage in BMW 2002 M10 engine (photo courtesy "Tom1" on bmw2002faq)

Got all that?

So now, imagine that you're the Lotus engine team trying to drop a newly-designed twin-cam head on a Ford Kent block. There's already an existing water passage in the middle of the front of the block to which a water pump directly bolts, but that configuration won't work anymore because now you need to hang a timing chain in front of the engine and enclose it inside a front cover. How do you graft on a water pump?

The answer isn't great, and proves to be one of the banes of the Twin-Cam engine. The design includes a "timing chest" with an inner and an outer cover where the water pump *is integrated into the outer cover*—that is, it's not a separate bolt-on part like nearly every other water pump in the automotive kingdom. This all requires some explanation.

Because the Lotus head is a twin-cam head, and because the cams each need to have a gear on the front for the timing chain to turn them, the front of the head is substantially wider than the block. Thus, since the cam gears and their timing chain have to be oil-lubricated and this all needs to be sealed, the timing chest, like the front of the head, is substantially wider than the front of the block. It's configured and attached in an odd way.

The timing chest with its integrated water pump about to be removed from my twink engine. Note how much wider it is than the block.

The inner cover, also called the backplate, has a large hole that fits directly over the water passage hole in the front of the block, and attaches to the block via a single bolt. There's an adhesive-coated paper gasket between the backplate and the block. Once the timing chain and tensioners are installed, the front cover is attached with three bolts that thread into the three holes in the block used by the original water pump, securing both covers against the block and sealing the hole around the water passage. There are additional nuts and bolts around the periphery of the cover/backplate pair, squeezing them together. Adhesive is used between the backplate and the cover, but there is no gasket. Again, this is all to provide an oil-tight "chest" for the timing chain and cam gears on a block that was never designed for them.

The backplate. Note the hole that fits around the water passage into the block, and the three holes for the bolts that held on the original Ford water pump. The arrow points to the single bolt securing it to the block. The gear shown is on the front of the "jackshaft," the vestigial in-block camshaft that's still present and used to run the distributor, oil pump, and fuel pump.

Next, the water pump itself. It uses an integrated spindle and bearing which, along with a seal, are press-fit into a small hole in the front cover. An impeller is attached to the back, and a hub to which the pulley attaches is pressed onto the front. When the front cover is mounted on the backplate, the impeller sits in the circular water passage into the block, and the timing chain runs around it. There's an aluminum spacer with two rubber o-rings that seals the hole in the backplate. Yeah, it's pretty strange.

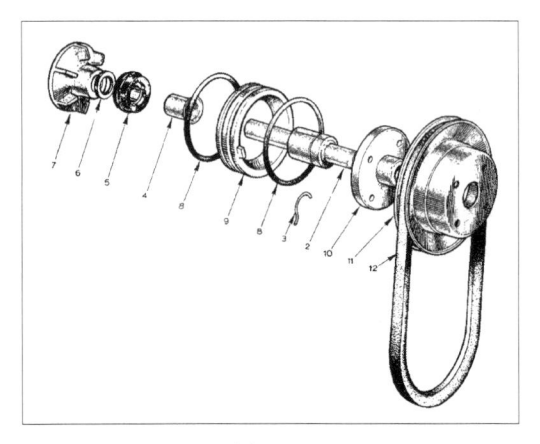

The individual pieces of the integrated water pump.

Front cover and backplate removed from engine showing chain path around impeller.

There are a number of problems with all this. First and most obvious, if the water pump fails—if the bearing develops play, or the pump leaks because after a long sit, the seal sticks and then tears when you start the engine, or, as was the case with mine, the whole thing just becomes a seized corroded mess after sitting for over 30 years—the front cover must be removed to replace it. And in order to do *that*, according to the workshop manual, the head and the oil pan must come off (although there are posts on Lotus forums that say you can do it with the head in place, though it is challenging to get the timing chest re-sealed). But remember that the twink engine was originally developed for the front-engine Elan, where front cover removal is possible with the engine in situ, but in a mid-engine Europa where the front of the engine is right up against the firewall and the water pump is at the small of your back, it's reportedly next to impossible to do this with the engine in the car. So, in a Europa, if the water pump goes south, you need to pull the engine. It's for these reasons that you read posts where, if the pump is seized *but not leaking*, people get by until the next engine rebuild by disconnecting the belt and splicing in a Craig Davies electric water pump.

Second, owing to the peculiar design of both the water pump and the timing chest, if the gaskets or o-rings fail and you find coolant in the oil or vice versa, you have no choice but to re-seal the case, which also requires pulling the engine. Plus, at assembly time, it's very easy to screw up the sealing. It's not like a German motor where there are dowel pins that snap precisely into holes. There's a lot of play. Again, there's *one bolt* holding the backplate to the block. You need to set up the adhesive, mount the backplate with that one bolt finger-tight, get the holes for the water passage concentric, test-fit the front cover, get the upper block and lower pan surfaces aligned, remove the front cover, tighten the single bolt, apply the adhesive between the covers, align everything again, tighten it all down, and pray that you got it right. This is why The Lotus Engine God's advice was to pressure-test the cooling system so if it was leaking externally or into the timing chest, you could catch it before dropping the engine in.

And if all this isn't bad enough, it turns out that, of the 34,000 twink engines built, most of them—all the ones for Elans, Sevens, and Cortinas—had the water passage entering on the right side of the front cover. However, with the Europa, that wouldn't work because that part of the engine is right up against the inside of the Y-part of the car's steel backbone. So the Twin-Cam engines on the roughly 4700 Europa Twin Cam cars are different in that they have a modified front cover where the side water inlet is blocked and a new inlet is welded on at a right angle to the cover. You can see this in the photos of my front cover above.

So, given all, this, and given that I had a seized water pump but the engine was already removed from the car and disassembled because it needed to be rebuilt anyway, when I heard that there was a removable cartridge-style water pump solution from two different vendors *that solves the problem* of needing to a) pull the front cover, b) possibly remove the head, and c) in a Europa, pull the engine from the car, I think you can see why I thought I'd be nuts not to make this be part of my engine rebuild.

But I had no idea that it would introduce nearly a year's delay into the process and create a whole other layer of problems by introducing a whole new set of parts that were never originally in the engine.

I eventually learned that, on Elans and the other front-engine cars the twink motor is in, the failure rate on water pump bearings is higher because, like on most front-engine cars, the water pump belt also runs the alternator, so keeping the belt tight puts a lateral load on the bearing, but since the Europa has the alternator relocated to the back of the engine, that stress isn't present and thus the water pumps tend to be less troublesome on Europas. So my whole "I'd be an idiot not to install a cartridge pump at rebuild time" thing wasn't even the slam-dunk that I thought.

As Kenny Rogers, Sarah Vaughan, Lady Antebellum, and Backstreet Boys all sang, if I knew then what I know now…

TLC Day 1688: The block is ready

JAN 25, 2018, 12:14 PM

The Lotus Chronicles, Day 1688: The guy at the machine shop just called me saying that the machine work I'd contracted them for four years ago on the Lotus engine (boring the block to receive the custom pistons, assembling the pistons on the rods, sizing the rings, getting the jackshaft to fit) is done, he's prepared the bill, and I can come pick up the pieces. Pretty serious milestone right there.

However, I need to decide what to do about the water pump. The original water pumps in these motors are leak-prone and impossible to change with the engine in the car. There are new cartridge-style pumps, but they require an entire front housing that probably needs to be machined to mate perfectly with the upper timing cover, a decision I should've made and an action I should've taken when the block was decked. The entire pump/housing kit is about $800, which is real money, at least it is to me.

Decisions to make, but forward motion at long last.

57 LIKES, 58 COMMENTS

> **Luther K Brefo** Which engine? 907?
>> **Ed MacVaugh** Ford-based, Twin-Cam.
> **Matt Pelikan** Skip the water pump altogether and just use it for very, very short trips? +5
> **Patrick Hayes** Electric pump?
>> **Matthew Zisk** It's British—probably a squirrel running on a wheel. Damn tricky to feed it on long trips, I am guessing. +1
>> **Rob Siegel** Patrick, People DO that, but there still needs to be a mechanical water pump in place in the housing or it'll leak.
>> **Patrick Hayes** Can you pull the impeller and weld up the shaft and weep holes?
>> **Rob Siegel** Patrick Hayes, Interesting idea, but I think I would be loathe to do that. +1
> **Robert Myles** Sounds like an expensive no brainer to me. Unless you're a masochist.
>> **David Kemether** He purchased a non-running Lotus Europa. Clearly he is a masochist. +4
>> **Rob Siegel** Right, I was about to say that the jury has already weighted in on the charge of self-inflicted pain. +2
>> **Martin Meissner** Put it back together as is, drive it like you stole it and get it out of your system, then forward it on to the next care taker to worry about +2
> **Jake Metz** Are you going to keep it for the long haul? If so, bite the bullet and do the upgrade. If you're going to sell it before the water pump starts leaking, go the cheap route. 800 is nothing to shake a stick at, but how much will it save you in the long run? +4
> **George Saylor** We never really had a problem with the water pumps. The factory did come out with a procedure to r&r the pump with the engine still in the car. Barbie hands I think. +1
>> **Rob Siegel** George, did you have an Elan or a Europa? On a Europa, the pump is basically up

against the firewall at the small of your back.

George Saylor I worked on both. I do not recall the factory procedure. If I had to replace one I would have still removed the engine and trans.

Russell Baden Musta My Dad did mine in car on creeper in our home garage in about 75… it can be done. (signed… the red Europa spied on I-81 by author!)

Doug Jacobs Prone to fucking leak? That's what you're worrying about? IT'S A FUCKING LOTUS! A little leaking water pump is the very least of what the eventual owner of this car is going to have to get used to. Quit gift-wrapping a dead beaver and make the fucker run. +8

Doug Jacobs Seriously, no matter how hard you try, you're not going to make this car into a BMW. If you wanted to nct leak, you should have put a Subaru engine in it. Plus, you're losing sight of the hack mechanic canon —you're talking big dollar resto here —and while that might engorge your ego a bit, it's really not what it's all about.

George Zinycz Rob Siegel, your answer is right here

Franklin Alfonso Now is the Lotus of your discontent… and so it begins. +2

Dave Gerwig Hooray! Happy for you and the Lotus garage mate. +1

Marc Bridge With the rebuild you are doing, what's the likelihood of that pump going bad? +1

Frank Mummolo If he doesn't do the pump repair/upgrade, I'd say 100%. That's just how it is with these projects. +2

Brian Ach do it right once rather than wrong twice +1

Russell Baden Musta I just want to make my Lotus Twin-cam oil tight!!! I keep having to buy more products that come in cardboard boxes just to get the box to protect the floor from oil!! +3

Benjamin Shahrabani Depends if you're keeping the car long term…once it's running that is.

Karel Jennings M10! M10! Etc. +3

George Zinycz He's ALREADY owned the car long term. +3

Travis Lehman Just stuff an LS in it. All the cool kids are doing it, we meet out back behind the 7/11, smoke cigarettes and talk about stuffing LS motors in various cars. You know you want to! +3

Ken Sparks Just do it. Leave it at the machine shop and get ur done. +1

Ernie Peters I replaced the water pump on my Elan with replacement parts, basically the shaft with bearing and impeller, using the old pulley. It's not difficult, but there is a clearance measurement between the impeller and the front cover that needs to be adhered to. Not hard to do, but you need to be OCD about it. I chose to not go with the cartridge pump. When I worked at the Lotus dealer, original-style pumps seemed to last around 45,000 miles. We had a mechanic who could remove the front cover on a Europa with the engine in place, so it can be done without removing the engine assembly. +3

Rob Siegel Thanks Ernie. I need to suss out what state my original front cover and impeller are in. First I need to find it. It's been almost five years at this point. They may not even be in the state of Massachusetts :^)

Ernie Peters Rob Siegel PM heading your way in a few…

Chris Lordan Wow, five years? This surprises me just a bit. I thought it was, like, two. +1

Rob Siegel Nope, it was a present to myself just after the first book came out. Memoirs was released at The Vintage 2013, which would've been Memorial Day or thereabouts. I bought the Lotus sight-unseen in June. +1

Wade Brickhouse Rob Siegel Do you WISH it were still sight unseen? :-) +1

Rob Siegel At times yes.

Brad Purvis I don't care if it's an MG, Lotus or Triumph. An English car will make you cry. +1

Bill Howard You can "come pick up the pieces"?

Rob Siegel Poor choice of words. I paid them to do some machine work so I can then assemble the engine. So I am not picking up an assembled engine. +1

Larry Webster This is all after the supercharger, right? +3

Sam Smith Do the cartridge pump. All the available setups are nicely made and worth it. Actually adds real value to the thing when you sell it. On top of the fact that the pump-in-cover is terrrrrrrrrrible. +4

Rob Siegel Thanks Sam. I'm likely to do the one from Burton Power. +1

Scott Linn Mine came with one installed and I haven't had a problem in 8 years of driving it. Of course, now I'm doomed… +1

George Zinycz Is the whole R&T masthead going to chime in? +1

Chris Lordan Not the whole masthead—only those sufficiently Hackish (TM) +2

John Graham Pretty soon his emails will start with 'Hacks Log, Stardate 1118844'. +2

Emery DeWitt This is exactly the same decision as using the stock BMW water pump versus the expensive one that lasts forever on an E36. Have you ever bought the expensive water pump on a BMW? Having to pull the engine to replace it on the Lotus shouldn't be a problem. You're the Hack Mechanic after all.

Daniel Senie Crowdfunded water pump. Premium: a copy of the book that will be the Lotus Chronicles. +1

Nicholas Bedworth Look at the Lotus Chronometer. Divide by 365. Tempus fugit and all that. Close to five years. When will it ever end, and will it end well? +1

TLC Day 1692: The water pump decision

JAN 29, 2018, 7:37 PM

The Lotus Chronicles, Day 1692:

I resolved to, uh, resolve which of the two cartridge-style water pump systems I'd buy—the one from Burton Power in Great Britain or the one from Dave Bean Engineering in California. For a number of reasons, I'd given the nod to the Burton Power one. I was about to order it when I noticed an alarming statement on their website: "Not for Europa." I sent them a message asking why, and heard back this morning. "As you probably know," it said, "the Lotus-Ford Twin-Cam engine was in Cortinas, Elans, and Europas. There are subtle differences between the motors. This pump kit is for the first two but not the Europa."

So I called Dave Bean Engineering, from whom I'd ordered some parts four years ago before I ceased spending while the engine sat, and asked them about the applicability of their pump to the Europa. An enormously helpful gentleman [Ken Gray] explained that, although their website lists only one version of the kit, there are, in fact, two. The one for the Europa is out for fabrication; they expect to have it in stock by the end of February. [Note: I got this wrong. There is now one version of the kit that can be made to work on both the Elan and the Europa.] I asked him if I needed to machine the new front cover flush with my newly decked block. He said that he didn't think so, that the gasket should take care of up to about ten thousandths difference, and advised that I could start to assemble the bottom end while I waited for the pump kit.

I LOVE it when options collapse and you're left with the correct one.

45 LIKES, 21 COMMENTS

Martin Bullen "He didn't think so." So, hopefully the correct one... (buzzkill)

Kevin McLaughlin Cannot wait until you get her back on the road Rob! +3

Robert Myles Excellent. Now if you were only so clear minded about the lights in the workshop... 🙂 +1

Brett Kay I love all your BMW stuff, but I absolutely love watching you bring the old Europa back to life!!! +1

Rob Koenig If you loathe leaking timing covers as much as I do, use this: (I still use Elring 461.680, but it's not as readily available) [link to Permatex anaerobic gasket maker]

Steven Bauer A dab of RTV where the front cover, block and head come together can't hurt.

Ernie Peters I bought bunches of parts from Ken at Bean (and Ray at RD Enterprises). Both good companies/good people. +2

    Rob Siegel Yup, that's who I spoke with.

    Russell Baden Musta Bought majority of parts from RD over the years, and a few at Bean. Never disappointed from either!

    Rob Siegel Thanks Russell. I bought from RD as well before I stopped spending four years back while the engine was marooned at the machine shop. They didn't sell either removable water pump kit. +1

    Russell Baden Musta Rob Siegel I'm planning on pulling my TC out for rebuild most likely next winter... has to be better crank seals out there. There's another place in UK that has TONS of parts very reasonable. I got double the amount of clips to hold those chrome strips and Fiberglass side panels on... soon as I remember the name I'll blurt it out to you!!

    Rob Siegel Burton Engineering, I assume you mean. +1

    Russell Baden Musta Rob Siegel Paul Matty in UK!! Check with them too. +1

    Rob Siegel Ah, I do not know them, thanks! +1

    Russell Baden Musta Rob Siegel I'll have to check Burton too then! Lots of stuff, reasonable!

Scott Linn Haven't yet bought anything from Bean but have heard great things about them. And their catalog rocks. RD Enterprises is good too.

Ed MacVaugh That looks suspiciously similar to an M44 water pump :)

Ed MacVaugh The water outlet comes out at a different direction on your Europa from the Elan and Cortina and Lotus 7. The timing case cover you show has the pipe going off to the left of the image. I think if you look at the cover on your Europa, the pipe goes off at a different direction, maybe towards the viewer? [Note: Ed was absolutely correct and this turned out to be a crucial issue, but I didn't yet understand its significance.] +1

Rob Siegel Ed, that's just the one they have the photo of on their website. As I said above, when I spoke with them, they said they sell a Europa-specific kit [to repeat, I got this wrong; they now sell one kit that works for both configurations]. I need to wait about a month for it to be in stock.

    Ed MacVaugh Understand. My point was to illustrate that the "subtle difference" in the engines is really a different casting :) [Again, Ed was absolutely correct]

TLC Day 1734: Picking up the pieces

MAR 12, 2018, 1:30 PM

The Lotus Chronicles, Day 1734: Holy crap, there are freshly-machined Lotus Twin-Cam motor lower end parts in my garage. Because, between the soon-to-be-released new CD and the soon-to-be-released new air conditioning book, I just didn't think I had enough to do.

(That's Eric Lindskog, the machinist at Century Automotive Machine, posing with the block he's had for nearly five years—more my fault than his.)

(I'm going to do this very slowly. It'll probably be six months before I have a fully-assembled engine, nine months before it and the transaxle are in the car, and a year before it moves under its own power. But still, a major milestone.)

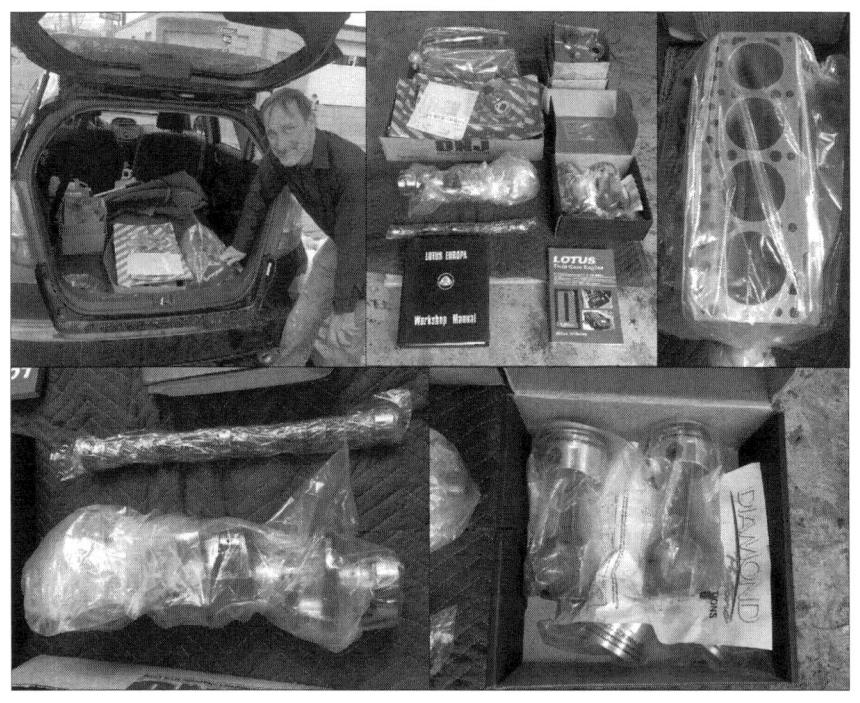

93 LIKES, 69 COMMENTS

Doug Jacobs A year? Sounds like someone wants to stretch out the Lotus's literary possibilities.

Rob Siegel Yes. I admit it. I have made a career out of having a dead Lotus. A running one isn't as funny. And is, I'm sure, a lot more expensive. +7

    Doug Jacobs I meant providing you with 52 weekly rehab stories...

    Rob Siegel Well, I DO have that Hagerty weekly bucket to fill... +1

    Daniel Senie And a book about resurrection (of a Lotus) to write later.

    Andrew Wilson Running one...now that's funny +5

    Rob Siegel Andrew Wilson, you're a cruel man :^) +1

    Francois Doremieux The good news is, it may not run for an extended period of time, giving you opportunities to revert to one of the things you do best, hack :P

    Andrew Wilson Rob, Next book: Didn't Run When Parked, How I Resurrected the Twin Cam Lotus & Other Tales of Woe.

    Rob Siegel Actually, the Lotus supposedly ran when parked in 1979, but whether that's really true, I don't know. I bought it from a guy who got it from the son of the original owner. He said his father developed back problems and parked it.

Wade Brickhouse Andrew Wilson "Didn't Run When Parked: Still Doesn't"

Francois Doremieux Wade Brickhouse "Might run briefly until parked again" +2

Jonathan Poole Well, you've also banked at least part of a career on NOT buying a Vixen, which is considerably less expensive than a non-running Lotus. +3

Ed MacVaugh I spy the old-style timing cover and water pump.

Rob Siegel You are correct. The Dave Bean cartridge-style one is out of stock and on order.

Marc Schatell Bring the parts to the Vintage and we'll assemble it in the parking lot. +7

Andrew Wilson Well, it is coming together. Said it would be a Winter project...2013, 2018, 2019 etc. +2

George Murnaghan "One year at a time, and I know it's mine..." +1

Josh Wyte Jesus and I was pissed at my machine shop for taking over 2 months to work on my E36 M3 head... +1

David Shatzer If we buy copies of all the books, do we get a ride in it? Assuming we're still respirating with a pulse and above room temperature? +1

Rob Siegel You forgot "above the ground." Even if you're dead, that's really the gating function to a ride in any car :^) +1

David Shatzer Rob Siegel Well played, BUT, would you install a roof rack? +1

Shaun Doherty Or you could hide 5 golden tickets in the books as part of a worldwide competition. Invite the winners to the garage for a ride, and then make a movie! 🌍 +3

Dave Gerwig David Shatzer A roof rack on a Lotus is commonly called an Elf Shelf. +2

David Shatzer I'm just envious because a) I only rode in a non-Twin Cam Europa in College, and 2) I always wanted one. 2)a) But I don't fit anymore.

Rob Siegel Fortunately, I'm still a little guy with size 10 feet, which is about the absolute foot size limit a Europa can handle. +1

Brian Ach Rob Siegel no one wears shoes when driving a Europa, or a Countach +2

Martin Meissner Ooooohhhh blingy!!! +1

Ernie Peters Very cool! +1

Alan Hunter Johansson 🖤🖤🖤🖤🖤 +1

Adam Pepper Lindskog still in business? They balanced my first Volvo B18 rebuild way back in the 1980's...

Rob Siegel Erik is the son of the founder of Lindskog balancing. I don't believe the balancing shop is still in business.

Adam Pepper Well, congrats on getting "parts in-house"... +1

Glenn Koerner Getting this back together should get you another book and at least one killer Lotus Blues song maybe more. Looking forward to reading about the process. 🙂

Rob Capiccioni Oooooooh! Fresh, clean parts! Let the fun commence!

Steve Woodard Brings back memories of an MGB I had completely reworked in 1969 —block bored over .040, balanced, Crower cam. I loved that car. Good luck with the Lotus. +1

Maire Anne Diamond Hey, Rob, which car was used to transport those freshly machined Lotus parts to your garage? +1

Rob Siegel Uh, I called an Uber. +1

Alan Hunter Johansson Why am I suspicious...?

Maire Anne Diamond Alan Hunter Johansson The Honda has to work hard to get Rob's attention. Like having a Lotus engine inside of it. +3

Scott Linn Ooh, shiny bits. +1

Rob Siegel yeah, not anyone can put a block in a bag like that :^) +1

Russell Baden Musta Get the book open, get to work. Weekend project!! +1

Rob Siegel I WILL get the books open. Both books. And read on the forums. Very slowly and carefully. Over a period of months. +2

Russell Baden Musta Rob Siegel Don't make me come down there!!!! A week max! +1

Dave Gerwig It Fits right in. +1

Jake Metz You've got to be careful, you don't want the reputation of those lotus parts rubbing off on that Honda. +1

Rob Siegel though the other way around would be nice :^) +1

Jake Metz Maybe it's like one of those Harry Potter / Star Wars battles—what side will overcome the other. You should write a screen play. I'd invest 11 bucks and 95 minutes in a theatre on that storyline. ₊₁

Carl Witthoft Lindskog, huh? I remember when his (Dad's) business was a bunch of machine tools in their basement. Good to see the family is still cranking out quality stuff. ₊₁

Lindsey Brown Please extend my warmest regards to Eric. His father once told me something that has stuck with me ever since: "I never speak ill to a customer of a competitor; it looks unprofessional, and they can never be sure of your motives." ₊₂

Lindsey Brown Obviously, the spaceship should be named The United Planets Cruiser C-57D. ₊₁

Rob Siegel Lindsey Brown, had to look that one up :^)

Phil Morgan You're awesome for not quitting on this. I think of that line from your first book that you said would be written on your tombstone. Something about building a spaceship but sadly lacked the room to finish it. I know I butchered it but I do admire & appreciate your perseverance. ₊₁

Rob Siegel Thanks Phil. I really appreciate it. For the record, what I said in the book was (giving it the proper context):

"I met someone recently who spent twenty grand restoring a 2002. These days, that's very easy to do. I mentioned this to Maire Anne as part of my consistent attempts to establish that I'm not as far out on the right edge of the bell curve as it might appear. "I'd never spend twenty grand rejuvenating a car," I said, confident that this would help place matters in their proper context.

"No," she said, "You'd spend two grand on each of ten cars."

Ouch.

But she's right.

Why do I do that? Is it Siegel's Seven Car Rule run amok? Is it my peculiar attempt to reduce risk by spreading around my bets? Or is it a simple inability to commit to any one car?

I don't know.

I think I need to lie back down on the couch one last time and work this out.

There's no doubt that, in fact, I spent well over $20k rejuvenating the 3.0CSi, in chunks of several thousand dollars at a time, over a period of 25 years. But I fit it into my temporal and financial schedule, doing things when I had the money, performing all of the work except paint and bodywork myself, in a way that rendered the car undriveable only for short amounts of time. This is who I am. I'm an implementation guy. I'm a guy who gets things done. I'm a guy who's gotten very good at seeing a path to the possible. When I die, just chisel on my tombstone "he was a very practical guy with a really good track record of achieving short-term goals. He thought he could build a rocket and go to Mars in bite-sized chunks, but, sadly, he lacked the garage space." It's not Rimbaud, but neither am I." ₊₅

Chris Lordan Rob Siegel, the first book is now a healthy, happy five years old. Congrats! ₊₁

Daniel Senie Short term goals... except the Lotus. ₊₁

Brian Ach Rob Siegel No, It'll read "He saw a path to the possible, and led many others to see it too. And they found it, and it was good." ₊₂

Neil M. Fennessey I used to bring my stuff to Lindskog Balancing back in the 70s and 80s. My t-shirt of the seal balancing a ball was proudly worn by me for years. My sister HATED IT and so I wore it specially for her amusement. I might still have one tucked away now that I think about it. ₊₃

Franklin Alfonso Meh... I predict you have it running by this summer. ₊₁

Mike Savage I dropped off an M54 head at Century yesterday 10am. Must have just missed you. You left their dog out of the picture...I bet he barked at you for taking the lotus parts...

Rob Siegel Mike Savage, Oh that's too damned funny. Yes, I was there just after that. I've been going there for years. That was the first time I ever met the dog. What a sweetie.

Carl Witthoft Fun fact about the Lotus: there's a line of high-end driver's chairs produced on Djerba (ancient Meninx), off the coast of Tunisia. That's why that place is known as... the Land of the Lotus-Seaters. ₊₂

> **Rob Siegel** *groans*
>
> **David Holzman** That story is odd, I see. ₊₁

John Lenham There is something special about parts returning from the machine shop, all clean and wrapped in plastic and smelling of their magical oils! ₊₁

Scott Chamberlain Good luck! One of my law school autocross buddies drove a Europa Twin Cam back then as a daily driver. He had to overhaul the engine THREE TIMES before he had a runner. Always thought it was a sexy, sexy car... but in an Anna Nicole Smith kinda way, like "Is ANY car worth that much aggravation?"

> **Franklin Alfonso** Only to die on you at some point...

Larry Chew Lotus Notes? +1

TLC Day 1740: One thing per day

MAR 18, 2018, 2:37 PM

The Lotus Chronicles, Day 1740:

One of my secrets to productivity is doing one thing per day, even if that's reading up on a web forum on one thing, or ordering one part, or installing one bolt. Today I installed four. They happened to be the four that hold the engine stand's rotating flange to the back of the engine block. I can now begin the engine reassembly. Well, actually I can't—I'm waiting on delivery of the Graphogen bearing lube. But ordering that was the one thing I did yesterday :^)

65 LIKES, 57 COMMENTS

Mike Miller Yesterday I put a black Craftsman roller cabinet and toolbox in my kitchen. Today I filled it with tools and kitchen stuff. +2

Tony Pascarella Four bolts today. Does that mean you can take the next three days off? +3

Rob Siegel Yes. It does. +2

Michael Cari I also do the same with my 2002tii project.

Jeff Baker Where do you get the theater poster from? It's quite awesome. +2

> **Rob Siegel** Thanks. All three of my kids were theater kids in high school, mostly backstage scenic design and construction. My middle one was in a production of Grease. They made these giant paintings that were reproductions of classic movie posters. The kids did them. At the end of the show, I asked Kyle what they were going to do with them. "Break them down and throw them out," he said. I asked if I could have Forbidden Planet. I showed up at the high school with my Suburban, but it was too big to fit. I had to borrow a pickup truck, strap it lightly across the bed, and drive home very slowly. +14
>
> **Charles Gould** This poster is incredible. I can't believe that they made this. I remember the original film and I loved it. +2
>
> **Rob Siegel** And, as you can see, it barely fits in my garage.
>
> **Charles Gould** Looks fabulous +1
>
> **Gene Kulyk** Warren Stevens should have won an Oscar for his portrayal of "Doc."
>
> **Eric Chasalow** I knew the composer, Bebe Barron [YouTube link to Bebe Barron]

Carl Witthoft What you don't have any other lube around? Did you check in the bedroom? +1

Tony Pascarella Could you elaborate on that grease? I'm gonna be assembling an engine soon. Never heard of that grease, I usually use assembly lube.

> **Rob Siegel** It's British, and what's recommended in the Lotus Twin-Cam Engine book I have. I usually use assembly lube too, but when in doubt, follow the directions, right? +1
>
> **Franklin Alfonso** I thought the British recommended Epoxy glue to keep the engine all together... +1
>
> **Tony Pascarella** Franklin Alfonso tough crowd here +1
>
> **Tony Pascarella** Rob Siegel right oh mate. By the book definitely. +1
>
> **Franklin Alfonso** Meh. It's like working out, eating well and taking care of yourself and then you still die... anyway. It's British so you only prolong the agony...
>
> **Eric Heinrich** Krytox or go home

Charles Morris I built a Mopar 383 big-block on my kitchen table once... why do you need an engine stand for such a tiny block? Just roll it over on the bench when you want to work underneath :) +2

> **Rob Siegel** Charles, I literally don't have a bench!
>
> **Charles Morris** In 1983, neither did I. No wife either, which is how I could build it in the kitchen :D +5

Chris Roberts That's some serious discipline, Rob. ;-)

> **Rob Siegel** Says the guy who strips, welds, and paints cars :^)

Andrew Wilson Good rhythm of repair, consistency is key. +1

Bailey Taylor Patrick and I met Anne Francis at Wonderfest many years ago. She autographed a Honey West publicity shot for my Ann (it was her favorite show when she was a kid). +1

David Kemether Not to rain on your parade, but I had a lengthy conversation with the gentleman who owns this Europa at Cars and Coffee this morning. He said that he had spent fewer hours building a 39' boat used to cross the Atlantic than he had spent restoring the Lotus. [photo of Europa at C&C] +3

> **Rob Siegel** I appreciate your asking and relaying that back, but remember that I don't restore ANYTHING :^) +1
>
> **Dave Gerwig** David Kemether what is the light blue car in the front of the picture? Looks inviting
>
> **David Kemether** https://en.m.wikipedia.org/wiki/Marcos_Engineering
>
> **Dave Gerwig** Looked like a cross between a Cheetah and a TVR. Never saw one before. Thx!
>
> **David Kemether** The first time I saw it at C&C I thought TVR until I looked closer.
>
> **Dave Gerwig** Ad a V8 and you would have a Griffith. Neat looking ride, for sure.
>
> **David Geisinger** Here's a great photo of a friends early Marcos. Wood frame!

Rob Siegel Does it come with thousands of pairs of shoes? Wait... wrong Marcos... +2

David Kemether Rob Siegel, you are thinking of the ill-fated Philippine auto manufacturer. Their first and only model was the Imelda which failed due to a propensity to get stuck on metal bridge grates.

+4

Rob Siegel HAHAHAHAHAHAHA!

Rob Siegel that, and the rear visibility sucked

Rob Siegel but it looked great with a hundred-foot-tall cocktail dress +1

Dave Gerwig Rob Siegel I heard it got the Boot +1

Lindsey Brown If Robbie the Robot can synthesize hooch, surely he can do the same with Graphogen bearing lube. +4

David Weisman (Homer voice) mmmmmMMMM Graphogen +1

Rob Siegel Regarding MGB V8s, coincidentally, Rob Capiccioni just posted this: https://silodrome.com/mgb-v8-buying-guide/ +2

David Weisman Kinda front heavy?

Rob Capiccioni The all-aluminum V8 is apparently lighter than the all-iron 4 cylinder from what I understand.

Rob Siegel I would love, at some point, to own any of the little Brits with the little Oldsmobile/Rover V8. +1

Brian Ach for the last time, sell me that poster

Rob Siegel Not a poster—actually a painting on a very large frame. it's enormous. and not for sale :^)

Brian Ach Rob Siegel I know, I am undeterred

David Kerr What's your crazy, I don't want to sell it price? /AmericanPickers

Rob Siegel Throw me another softball, why don't you? [Dr. Evil gif "ONE BILLION DOLLARS"]

TLC Day 1743: Laying in the crank matey

MAR 21, 2018, 9:00 PM

The Lotus Chronicles, Day 1743:

I'm not REALLY starting to assemble the engine, just test-fitting the main bearings, yessir, that's all I'm doing... we can stop anytime... unless that feels really good and I can, you know, keep going and lay that crankshaft in... oh you dirty little Lotus...

90 LIKES, 52 COMMENTS

Thomas Jones Go, go, go!
Erik Wegweiser Just watch... we'll see a black bar obscuring the crank shaft any day now.
    Rob Siegel HAHAHAHAHA!
Eric Heinrich Send the bearings and rings out to WPC in CA.
Josh Wyte They didn't assemble the short block for you?
    Rob Siegel I didn't pay them to, so, no +2
Doug Jacobs Is the face _supposed_ to have those gouges along the right side?
    Rob Siegel Probably not, but they're highly exaggerated by the lighting, and only the front-most part of it is a sealing surface. +1
Chris Roberts Sweet photo. I really just like looking at my fresh block too, so I'm savoring the assembly for now +1
Andrew Wilson You need to name the Lotus, Zoot. [link to Monty Python and the Holy Grail "And then, the oral sex" scene] +3
    Rob Siegel "and then spank ME!" +1
    Rob Siegel (and no, the Lotus is already named Lolita) +2
    Bill Snider Sung to the Sting tune...
Chris Mahalick Hate to bum you out Mr. Siegel, but I am a week or two away from firing mine off. 10k original miles. [photo of Chris' engine compartment] +3
    Rob Siegel oh baby!
    Francois Doremieux Third engine?
    Rob Siegel oh, snap!
    Chris Mahalick Francois Doremieux LOL! I think it is still the first. Hmmm, now I have to check. +1
    Rob Siegel mine has 24k, but I also need to check if it's the original. as you can see from the pic, it's sleeved, and it's not supposed to be, so I think something big and bad already happened to it once in the distant past. +3
    Chris Mahalick Rob Siegel My friend is a total Lotus expert, and he told me that these cannot be over-revved at all. I am so used to just pounding on my BMW and 911 that this Lotus may be too

weak to endure the usual violent and prolonged beatings I administer. +1

Christopher P. Koch Sometimes when a new engine is first produced they are sleeved only for the manufacturer to later discover it is much cheaper to coat the raw walls with something like Nikasil... +1

Tom Kinney Usually the bearing in the center has a saddle, to control endplay.

Lou Millinghausen That's OK. When you turn it right side up it won't be noticeable.

Charles Gould They certainly are seductive little sluts, aren't they? +2

Francois Doremieux It's all about preliminaries... And lube!

Kevin McGrath 'Mmm... lookin' purdy'. [photo of creepy looking guy with teeth missing] +1

Rob Siegel hey, let's not get weird about it or nothing +1

Kevin McGrath Rob, too much? Sorry, Goober got overly excited with clean parts. +1

Lindsey Brown When it's fully operational, I challenge you to a run down Rte. 202 to Amherst. Loser pays for drinks at the Blue Wall. +1

Rob Siegel When it's "fully operational," we no longer will be. +2

Francois Doremieux It's both a transient state and an ideal that can never be reached...

Franklin Alfonso You're not gonna risk actually driving it without a spare Lotus available in the garage for parts are you? +1

Andrew McGowan Damnit...you got your Lotus stuff back from the machine shop while my Alfa stuff is still there! +1

Robert Myles "50 Shades of Hack Mechanicing" +1

Andrew McGowan I always find it funny to look back at my years of dealing with a Triumph TR4 as a daily driver. One day, when driving home from college, there was a terrible "lunge" or hiccup from the motor, very violent. At that, it ran like complete shit, but still got me home. I knew something terrible had happened mechanical in the motor, so I wasted no time with the usual trouble shooting, and just pulled the motor. After I had it on the engine stand, pulled the oil pan, and then main bearing caps, I saw the crankshaft had broken in two pieces at the #2 main bearing. Here is. 2l motor with a 92mm stroke with 3 main bearings. An M10 is bombproof in comparison... except when we get to the cylinder head. +1

Scott Lagrotteria Basic difference of British vs. German. No one has ever accused the British of over engineering their cars.

Russell Baden Musta I remember that! +1

Frank Mummolo "I can quit anytime." +1

Chris Roberts This photo fascinates me. There's something fragile and beautiful in a British/Italian way about the in-block cam configuration and it's easy to see why these would be hard to find now. I tried looking for some Wiki info about the S3 motor but didn't find much, if any, details about this aspect. When you have time (ha!), can you post a link to more info for inquiring minds?

Rob Siegel Chris, it's commonly called the Lotus-Ford Twin-Cam engine. Posts on Wikipedia and Hemmings sum things up. [en.wikipedia.org/wiki/Lotus-Ford_Twin_Cam]

Rob Siegel Block is very similar to the one used in Formula Ford, Lotus Twin Cam head. I've never spent the time to delve into it to understand the similarities and difference between this block and other Ford 1600 blocks, which is one of the reasons why I felt it necessary to fix this one as opposed to buying another one, and why it took so long to get this one fixed.

Chris Roberts This is going to be great watching it come together... and for Lolita the Lotus to make her debut.

Rob Siegel Chris, you need a big bucket of popcorn, and it will get stale. This is going to take a looooong time.

Chris Roberts I get that, and understand why also... but that's all part of the fun. It would almost be a disappointment if you threw it all together in a week. I will stay tuned for the duration.

Brad Purvis Be careful of an English temptress. I should know. Not only do I drive one {okay, two (that is when they actually are working)}, but I married one as well. +3

Chris Roberts I do say.

Rob Siegel It's said that British cars can sense infidelity. True of British women? +1

Brad Purvis Rob Siegel, I dunno. I never tried.

Rob Siegel smart man.

Franklin Alfonso Why take unnecessary chances with your own parts...

Brian Ach awesome

TLC Day 1747: Crank bearings in

MAR 25, 2018, 9:44 AM

The Lotus Chronicles, Day 1747:

Got the jackshaft [the 3rd camshaft left over from when it was a cam-in-block engine] installed, then lubricated the in-block bearing shells with Graphogen and laid in the crankshaft. Placed the Plastigauge and torqued down the bearing caps. All five spec out at .002, right in the .0015 to .003 clearance range. I still need to deal with the thrust bearings, but not too god damned bad for before 9:30 am on a Sunday morning.

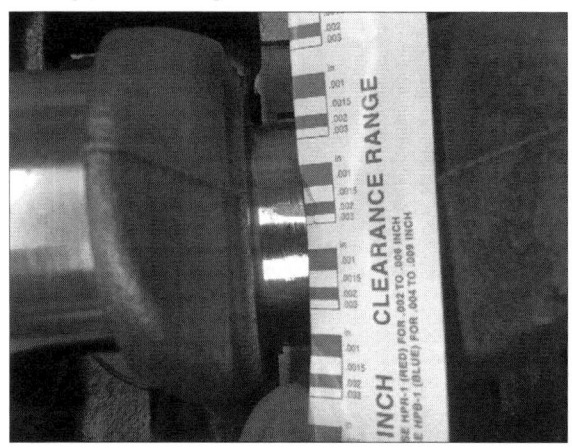

63 LIKES, 37 COMMENTS

Frank Mummolo Yep. You're bit! 😊 +1

Franklin Alfonso You do good work before church... or should I say temple... +2

> **Paul Wegweiser** I think he refers to this portion of the day as his "Pre Fruit Loops Nutritious Breakfast" time. +3
>
> **Franklin Alfonso** Well we all know he's a bit loopy...

Charles Morris I would trust telescoping gauges and micrometer before Plastigage, but it's good that they are all identical +1

Paul Wegweiser I'm unfamiliar with British assembly techniques, but that looks like a HUGE amount of lubricant on those first couple bearing shells. I'm a "barely visible thin sheen of moly/oil" guy, myself. As an added note: I've been told that you want to use reasonably fresh plasti-gauge when performing this task... not sure how "old" is "too old" but that it has a shelf life that can affect its accuracy. +5

> **Rob Siegel** thanks Paul. I don't pretend that I know what I'm doing :^) I'll go lighter on the shells in the bearing caps. +1
>
> **Paul Wegweiser** Doug Moore Luckily my chest cavity is pre-lubed from the factory. +3
>
> **Ed MacVaugh** Rolls Royce slathers it on from a 5 lb can in their engine assembly line (at least back in the 1970s when this car was first made) with the thought that it would preserve the metal joints until first running, which could be years in the case of an aircraft or tank engine. Maybe the same, here. I wondered then, and still wonder, why this product never found its way overseas, and what the Yanks and the Germans used instead.
>
> **Rob Siegel** The Wilkins book says to "smear with Graphogen." Other web forums say that moly products are better, but when in doubt...
>
> **Ed MacVaugh** Rob Siegel I've been doing some back channel investigation since I saw your post. Most everyone else, beside the British, used motor oil in engine assembly for the first hundred years, based upon the presumption that engine oil is what each morning startup will see for the remainder of the engine's life. Then folks became more interested in materials and research and high pressure loads on cams, and cranks and piston pins, and started making and selling "assembly lubes" with all matter of ingredients. Then, folks started noticing that oil filters were clogging immediately after first startup, and all matter of ingredients have now been reduced to

those ingredients that will dissolve in motor oil and not clog the filter system. +4

Paul Wegweiser Ed MacVaugh Today eggs are bad for you... next Wednesday they will be healthy for you again.

Ed MacVaugh Paul Wegweiser Now you tell me. Remind me, is it bad cholesterol or good cholesterol? +1

Frank Mummolo Important to change oil filter pretty soon after startup on a rebuilt engine using assembly lube. I have been researching the same topic for a BMW R90S motor I'm rebuilding. That seems to be the consensus in Beemer land and for the very same reasons you state, Ed. +1

Rob Siegel Got it, thanks. FYI the Wilkins book was written in 1988, so not ancient, but not contemporary.

Ed MacVaugh Rob Siegel From what I can see, it is ancient in so far as assembly lube. Graphogen might easily have been the only game in town. A review of their web site shows that they have only been selling to consumers for the last couple of years. Before that, it was quantity to the trade. Red Line issued their first MSDS safety sheet in 1995 for assembly lube.

Rob Siegel It's still sold domestically by RD Enterprises and Dave Bean, and in Britain by Burton Power.

Ed MacVaugh Rob Siegel They just started marketing to consumers in 2016, they are looking for a US distributor :) [link to Graphogen website] +1

Marshall Garrison Y'know Rob, if you need a petition to send in to be a Graphogen stockist, you could probably garner some signatures here; I'm just trying to figure out how you'd sound saying "stockist" and "hack mechanic" in an English accent... +2

Lindsey Brown Don't worry about excess lube in a British car. Most will leak out before it ever even runs for the first time. +3

Lindsey Brown I worked with a Russian for several years, and he told me something he'd heard from one of the country's top chemists: "Of all the motor oil additives on the market, there are only a few that won't actually damage the internal friction surfaces. Those that won't are all made up of 100% motor oil." +1

Tom McCarthy So you will be driving the car on Monday then? +4

David Weisman That's hot man +1

Kevin Pennell Looking really good Rob +1

Ezra Haines [link to Fletch "it's all ball bearings, Fletch"] +12

Rob Siegel Fletch—the movie where Chevy Chase was only 15% the bumbling idiot he was in most of his other films. Okay, 20%.

Ezra Haines Rob Siegel Burt Reynolds and Mick Jagger were up for the roll. That would have been a different movie, with less bumbling.

David Weisman Rob Siegel ahh but he played drums with the Ultimate Spinach and early steely Dan! +1

Cameron Parkhurst I got up and had waffles. Also not bad for before 9:30 on a Sunday morning. +1

Robert Myles I changed the air filter on the Econobeast, bought bagels and went grocery shopping by 0930. You win. +1

Brandon Fetch You're a jackshaft! (Am I helping?) +4

Martin Meissner Jackshaft... that sounds like an insulting name to call someone...ie That guy Trump is a total jackshaft +3

Russell Baden Musta Well done!! +1

Mel Green Lots of newer Beers in South Africa, and many right-hand drive Cobras! +1

Sal Mack Life is a circle. At 16 you are working alone with the jackshaft and again at retirement. Onward and upward to the thrust bearing. Some assembly required. +2

TLC Day 1747: Thrust washers

MAR 25, 2018, 2:13 PM

The Lotus Chronicles, Day 1747 (continued):

And the crank is in. Thrust washers too. Properly torqued and lubed. Man does that feel good. Oh baby. I need a cigarette, then a shower. Hope it was good for you too. Seek medical attention for a post-crankshaft-installation high that lasts more than four hours. I am such a juvenile. But I'm not the one who named the goddamn things "thrust washers."

80 LIKES, 23 COMMENTS

Doug Jacobs I can feel a shift in the force… +1

Marc Schatell Nice work! +1

Marc Touesnard The reflectory period can feel pretty damn good when you stand back and look at what you accomplished! +1

David Shatzer Thrust washers? Are you the thruster, or the thrustee? Interesting that little sewing machine motor has separate thrust surface… +1

Franklin Alfonso With Lotus you are the thruster at the begging but in the end become the thrustee… +4

Jonathan Tinker Hopefully there was a shaft collar involved… +1

    Rob Siegel THAT's the spirit! And a ball hone! +2

Luther K Brefo My favorite part is that the bearing caps when put in proper sequence, spell FRONT.

    Rob Siegel you mean NRTFO isn't a word? crap! +2

David Shatzer At a minimum a crank protection sleeve for protecting the main seals… +1

Russell Baden Musta Good job Rob! +1

Henry Noble And coincidentally there in the background is Robbie the Robot carrying a damsel in distress. +1

Robert Myles I assume you were wearing protection. +1

Jeffery W Dennis Soon the garage art will be mobile.

    Robert Myles … or banned in Boston. +2

Tony Pascarella Now time for nap.

Franklin Alfonso The motor will never be this clean again. At least till the next rebuild… +1

Joshua Weinstein Assembling a motor is one of the satisfying thing a car geek can do. And the action and satisfaction last much longer than sex. At least the missus tells me so. +1

    Franklin Alfonso I'm sure Lolita feels the same way…

Brian Stauss Nice work Rob! Can't wait for the next chapter in the Lotus chronicles +1

Brad Purvis Yes. I remember you talking about "jackshaft lubricant" among other things. BTW, the proper terminology when working on an English car is "fag" not "cigarette." You needed a "fag" after all those thrust washers and lubricants. I hope it was a "Players." +1

Nicholas Bedworth Instead of Players, one can also enjoy Senior Service. [pic of cigarettes]

Frank Mummolo Still in foreplay mode. Just wait until you turn that key for the first time! +1

TLC Day 1748: Rod bearings

MAR 26, 2018, 1:59 PM

The Lotus Chronicles, Day 1748:

Engine builders: Rod bearing clearance on my Lotus engine is, as measured with Plastigauge, from 1 to 4: .002 (a little tighter, actually), .0015, .001, and .001. Spec is .0005 to .0022. Rod side of bearings smeared with Graphogen, cap side dry, torqued in two stages, 25 and 45 ft-lbs as per the book. Crank and rods were inspected by two machinists. Rods were reconditioned, though I don't think anything was done to the big ends.

So, I'm good, right?

21 LIKES, 50 COMMENTS

Ed MacVaugh You are within the specs (and within the generally recognized limits of "one to three thousandths", but whether you are good or not, is between you and your judge. :) +7

    Rob Siegel Thanks Ed. The latter is, of course, a much bigger question...

C.R. Krieger SURE! What's the worst that could happen?

Ed MacVaugh I am curious how you «recondition a connecting rod."

    Eric Heinrich Machine the big end, balance, stress relieve, treat with oil shedding coatings +3

    Ben Greisler Make sure it is straight too.

    Eric Heinrich Ben Greisler and in plane +2

Rob Siegel plus they were re-bushed

Gene Kulyk Like the country was in 2001? +4

Ed MacVaugh Customarily, the shop checks for twist between the big and little ends, and roundness for the big end. +2

Sal Mack Wait did I miss the news about a Brazilian? What got rebushed? Ed has people checking big and little ends. This build is ripe for my immaturity. +2

Matthew Zisk And mine, as well. +1

David Shatzer Double entendre ahoy!!

Tom Kinney You are Good to Go, IMHO

Rob Siegel thanks Tom!

Ben Greisler Send it. And next time get a set of inside and outside micrometers to do it right. ₊₁

Rob Siegel thanks Ben.

Ben Greisler Actual mics will be more accurate and allow you to check in multiple places pretty easily. Useful for other things too. You just need to check that your inside mic agrees with your outside.

Rob Siegel I think that the machine shop who had the bottom end and did the machine work actually did that, but I still like to lay on the Plastigauge. It makes me feel involved with the process, and it's a final QC check. ₊₅

Bill Howard Rob Siegel "I still like to lay on the Plastigauge"... before you lay on the plastic

Neil M. Fennessey Although you can get decent bore-gages pretty inexpensively, The circular bore "standards" you need to calibrate the things are really $$$. Starrett pin-gages are a decent substitute if you have a clamp to hold your mic while you measure the pin-gage.

Ben Greisler Actually, as long as you calibrate the inside mic to the outside mic and they agree on the number, that is all you need for clearance. I bought all my base stuff decades ago used. Starrett and B&S mostly. Recently I have been buying more specialized stuff, also used. Can get great deals used. Pennies on the dollar.

Doug Jacobs Since it was put together first time with a yardstick, yeah, you're good. ₊₅

Rob Siegel that's a Whitworth yardstick ₊₃

Doug Jacobs Well, yeah...

Bret Luter More like a Cubitstick

Franklin Alfonso Not to worry. Will be just as good as Lucas Electrics. ₊₁

Sal Mack I am just happy to see that you have moved on from fiddling with your jackshaft. I guess the thrust bearing is also behind us. ₊₁

Rob Siegel Well, you have to grow up sometime. I'm fiddling with my pistons now. I've got FOUR of those.

Marshall Garrison Confucius say: better to be pissed-off than piston. ₊₂

Franklin Alfonso The piss-ons have been off long enough...

Charles Morris I'm in my late 50's and have not grown up. So far it's working for me :D ₊₁

Robert Myles You will note with all the sexy talk that Rob is, in fact, wearing protection.

Rob Siegel This was nearly five years ago, but I was wearing protection even then!

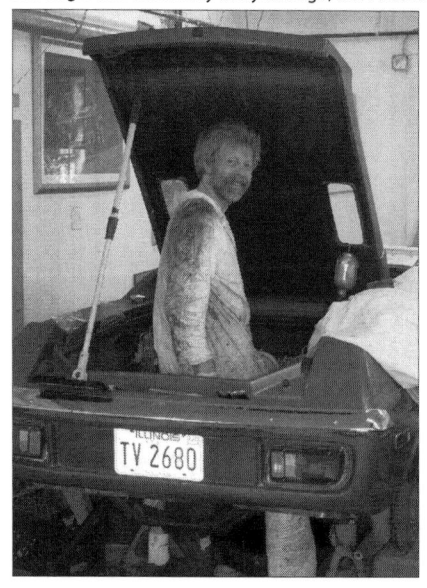

Robert Myles You gotta hang with a better class of car if you need to wear a HazMat suit, Rob. ₊₁

Rob Siegel Well, it IS an Ermenegildo Zegna... +1
Scott Aaron You are really making some progress, there. +1
Lindsey Brown You must be good, because nothing could ever go wrong with torqueing bearing caps. +1
Duane Sword My mechanical skills are on par with Jeremy Clarkson such that my question is "what size hammer do you need to hit stuff into place?" [link to video of Clarkson hitting BMW engine with hammer] +1
Paul Goldfine Or, as my experience repairing military aircraft taught me, "If it don't fit, force it." +1
Pete Winegardner If it fits... it ships.
Robert Myles As we used to say in the electronics lab, "If it beeps, ship it." +1
Rob Siegel As my friend Lindsey says, "As we sometimes say in the shop, PUT THE GODDAMNED PART IN THE GODDAMNED CAR!" +2
Russell Baden Musta Rob Siegel, so I went and did this last week... the one on the left, one of your inspirations will be getting the same engine treatment next year!!... again!!! [photo of Europa and Elise] +2
Rob Siegel damn! +1
Russell Baden Musta Rob Siegel it's staying, just for the record! It has a grandchild now finally at age 45!
Russell Baden Musta Rob Siegel why are we up so early? Or are you still torqueing bolts?
Scott Aaron I have multi level garage envy all of a sudden +2

TLC Day 1749: Fully-stuffed block

MAR 27, 2018, 10:46 PM

The Lotus Chronicles, Day 1749:

Block is fully stuffed with pistons, rods, and crank. All clearances within spec, all bearing caps torqued down. It'll be a while before there's more progress like this, as the Dave Bean cartridge-style water pump I need is on backorder, and along with it comes a different lower timing cover, but this is huge. Freaking BOOYA!

116 LIKES, 43 COMMENTS

Luther K Brefo [gif of "DAMMMNNNN"]
Dave Gerwig Do you gap your piston rings or have the shop attend to that task? +1
    Luther K Brefo Same question.
Rob Siegel Dave, The machine shop did that. He checked all ring gaps in the bores (very important as they're custom non-standard-sized bores and pistons), then put the rings on the pistons and the pistons on the rods. He said they used spiral lock clips and they were a little tricky and recommended

I let him do it. +2

Steven Bauer Progress! +1

George Saylor Rob, now that you're getting closer—have you driven a Europa before?

> **Rob Siegel** Yes, just once. a local guy I met on lotuseuropa.org invited me over to drive his ‹74 TCS (same car as mine). The guy's wife had passed away of cancer three months prior and left him with two teen-aged daughters. I brought my Z3 M Coupe. we swapped cars and bombed around the woody suburbs between 95 and 495. It was a slice of heaven for the both of us. +2

Francois Doremieux I remember what my gambling friend taught me: Quit while you're ahead! +2

> **Rob Siegel** Ha! if it only worked like that! It's more like in for a penny, in for a pound (or a pounding :^) +1

JP Hermes Oil pump? +1

Rob Siegel At a minimum, can find it in my garage, clean it, and check it

Charles Gould It's really finally coming together. Beautiful work, congrats! Will it be done in time for our Microcar Classic event in July? We'd love to have it here to participate. www.bubbledrome.com (1)

> **Rob Siegel** Thanks Charles, but zero possibility of it being done/
>
> **Charles Gould** Ok. Perhaps 2019!
>
> **Charles Gould** Well, please consider attending anyway. I can even lend you some bizarre and interesting vehicle to drive on all of the tours and activities. +1

Chris Roberts Whoah—super cool 😎 +1

Lance Johnson Is there a head waiting in the wings somewhere?

Rob Siegel The head has been fully assembled and sitting in my basement for several years now. +2

Ken Doolittle L.O.T.U.S: Lots of the unusual snafus. +1

Thomas R. Wilson It's a thing of beauty! +1

Robert Boynton What's the compression in one of these?

Kevin Pennell So, Rob, did have a purpose in placing the work in progress (good progress I may add) under the "Forbidden Planet" poster? Just wondering ;) Looking great BTW! +1

Russell Baden Musta Freshly painted too!! Looks awesome Rob!

Rob Siegel Russell, BTW, I usually don't write about the Lotus in my online column for BMW CCA (it is, after all, a Lotus :^), but I made an exception last week, summarizing the story thus far. Your closeup didn't make it in, but the photo of your car from behind did.

> **Russell Baden Musta** Cool Rob! Very nice article Rob! Still cracks me up... Europa in the wild!

Daniel Senie What, no video showing the pistons going up and down? 😀

Rob Siegel no, that's porn +1

Daniel Senie Rob Siegel accompanied by sex music?

Rob Siegel HAHAHAHAHA! You know how hard I try not to have these words collide, right? I can imagine nothing more disjoint than a video of the pistons in the Lotus engine going up and down, accompanied by a sound from "A Landscape of Ghosts" [my new CD] It's such a conceptual disconnect that it's actually funny. +1

Daniel Senie Rob Siegel you know I keep encouraging you to cross the streams 😀

Rob Siegel yeah. no. +1

Brian Ach wow

Adrian Sultana Looks a lot like the M10 I just built. [photo of stuffed M10 engine on stand] +2

Jon Beales Here's mine, at the 35 year mark for the engine refresh. #kadunza [photo of another stuffed M10 engine on stand] +1

TLC Day 1750: A leaky stag

MAR 28, 2018, 7:43 PM

The Lotus Chronicles, Day 1750:

I hate a leaky stag. Very messy.

15 LIKES, 21 COMMENTS

Charles Morris You need a 55 gal. drum of sealer in which to submerge any British engine. Then some of the oil might have a chance at staying inside it... +1

Fred Larimer Easily portable though...and, "non-hardening"... ok, try not to laugh... repeat after me; Lotush... Lotush... Lotush... +1

    Rob Siegel I know, right? +1

    Robert Myles You beat me to it.

    Matthew Zisk Me too

    Drew Lagravinese Me too

    Rob Siegel "...and me!" "And then... the oral sex!" +8

Marshall Garrison Stag Party @ Rob's! +2

Tony Pascarella Can you say pipe dope? Who you calling a dope?

Dave Klink In my experience, nothing seals a Stag well...

Clyde Gates and Triumph Stags did tend to leak...

Dave Klink Oh yeah, just a little...

Kieran Gobey No, this is a leaky Stag [photo of a Triumph Stag] +2

Lindsey Brown I've worked on way too many Stags, and they all leaked.

Steven Bauer Get him some Tupperware

Wade Brickhouse Your deer is incontinent?

Russell Baden Musta I'd recommend welding the TC together once torqued!!! That should keep it all in!! (1)

Steven Anthony Smith Not as messy as blowing a seal, though. +3

Rob Siegel Few things are. +1

Robert Myles "I swear, we just held hands!" +1

Franklin Alfonso Maybe some penicillin to follow that up with might be good... +1

Kent Carlos Everett So... spread this stuff on my shoulders or what? +1

TLC Day 1881: That way lies madness too

AUG 3, 2018, 2:11 PM

The Lotus Chronicles, Day 1881:

I've done stupider things. But not recently.

Plus, I've got one dead Lotus in my garage. Put another one in close proximity, and it's like particle physics. The universe could literally end. I simply can't take that responsibility. I'm passing. For the universe.

[Craigslist ad for very rough 1974 Lotus Elite "This hard to find car was purchased for a restoration project, but…"]

76 LIKES, 87 COMMENTS

Chris Roberts Holy crap. Such a rare find. It could probably sell for a lot more to overseas buyers.
Ed MacVaugh Carburetors in the footwell! What is not to like. Paging Wegweiser to the white courtesy phone . . . +1
    Paul Wegweiser no. +1
    Paul Wegweiser hell. no. :D +1
    Rob Siegel But you had to think about that for FIVE MINUTES. That means yes. +1
    Paul Wegweiser Rob Siegel Only because I brain-bleached myself after the first viewing, and forgot. +3
Scott Linn Had a chance to buy a running one a number of years ago, but ended up buying a Europa instead. Glad I did. +1
Rob Siegel the woodgrain makes the interior look like some mid-70s GM product +2
Adam G. Fisher I'm a few miles away if you want me to go look. ;-)
    Rob Siegel Thanks Adam. Really, I'm good. One more project and my house will eject me like an immune system attacking a virus. +3
Noam Levine …as much as I like quirky hatchbacks, my e24 project has the garage space… wonder if the seller would be open to a trade.
Zoé Brady As soon as I saw the picture I knew it wasn't your garage. Too clean. +1
    Doug Jacobs In other words, you can _see_ the garage. +4
Jeff F Hollen Buuuuuht it's got a 907! We could be close family! +1
David Layton Yes, the 'Mid-Seventies Fake Wood Grain' epidemic v
Jeff Dorgay Didn't we used to say back in the day "Lots Of Trouble, Usually Serious?" :) +1
    Rob Siegel We still do! +1
Robert Boynton Misery loves company 😊 +1
Bill Gau Rob, You're either a man or a mouse. Tell him you'll take it if he throws in the wheel dollies, because something tells me you're going to need them. Indeed, let's see what yer bones are truly made of … come along now … pull the trigger :^) +2
Frank Mummolo As Paul Newman once commented (on another matter): "Why should I got out and have hamburger when I can go home and have steak?" +1
David Kiehna I vomit every time I see one.
Russell Baden Musta Finish it Rob!!! It will finance the Europa! +1
Greg Aplin I thought it was an Eclat. What's the difference in that and an Elite?
    David Kiehna About 1 point on the ugly scale. +2
    Donald Sheldon Same car, different name per market
    Greg Aplin Donald Sheldon so we got the Elite and Europe got the Eclat or vice versa?
    Harry Bonkosky The Eclat was the fastback version. [photo of yellow Eclat]
    David Kiehna Donald Sheldon They have the same chassis/drivetrain for the most part, but the roof line and rear bodywork is different. +1
    Greg Aplin Then I have been mistaking Elites for Eclats all this time, as I thought the Eclat was the ugly one. That almost has a passing similarity to a Bitter, just less trunk.
    David Kiehna Actually looks decent in this green! [photo of green Eclat]
    Donald Sheldon David Kiehna great clarification. My dad had an Eclat.

Larry Schwarz The Universe thanks you for a wise decision! +1

Jamie Eno If the answer will always be "No" then why is it still in your Craigslist search???? +2

Brian Hart Love that new wiring job under the hood. Used up the whole roll of red. +1

Steven Hage I saw a Lotus Europa yesterday if that means anything +1

Donald Sheldon Rob, I gotta send you a message in this one. It's a multi-generational affair in my family that involves multiple purchases spaced across decades and a car fire on Father's Day. I'll send later.

Rally Mono I like British cars but I have my limits. Lotus...mmmm no thanks. +1

Rally Mono I remember a friend having one of these new... after about 9 months of ownership he gave up and went German +1

Cameron Vanderhorst Did you notice a sign in the front of my garage that said "Dead Lotus Storage?" [pic of Quentin Tarantino] +3

Bob Sawtelle The best part of the ad is the picture of the sticker "EVS protected by Electronic Vehicle Security" pretty sure the security is the fact that it has British electrical aka Lucas. Hell anyone who tried to steal this would probably get an assist from the owner taking it away. +3

Franklin Alfonso I kind of like the look of this one better. Reminds me from some angles of an AMC Gremlin lol. +1

    Greg Aplin Nah, Gremlins were way nicer looking. 😀

    Greg Aplin See? 😀 [photo of Gremlin]

    Franklin Alfonso LOL I had 2 back in the 70s...

    Greg Aplin Franklin, I had a Pacer, and I was proud to call it mine. My college car. It hauled frat pledges, tubas and tympani to and from marching band practice, any number of kegs, and the occasional shapely coed. I get prickly when I see articles slamming the Pacer! +1

Franklin Alfonso And that fancy steering wheel...

Franklin Alfonso Meh you should hold out for a Jensen Interceptor... May as well go all in...

Brad Purvis Total plutonic reversal? +1

    Rob Siegel "I thought you said crossing the streams was BAD" +1

    Brad Purvis Rob Siegel—Crossing the Lotus' too.

Franklin Alfonso Well at least Rob's Lotus is British racing brown... You know it's not like he's ever gonna paint it...

Wade Brickhouse Do it. You need 2 excuses. :-) +1

    Franklin Alfonso Only one car not running now. But with limited garage space does he really need two?

Francois Doremieux If you are posting this, it's because you are tempted... +1

Patrick Hayes That red one I sent you awhile back is still for sale here in Charlotte +1

Franklin Alfonso Oh red!... his only weakness...

Tank Şimşek That car will not last...

Scott Linn Lots of Trouble Usually Sitting +3

George Saylor If I bought any 60's or 70's Lotus I would budget for a new frame. Especially if there was rust on any other metal surface.

    Franklin Alfonso You need at least 3 so 2 can be parts cars... +2

Dirk Rasmussen Daddy? [photo of Gremlin]

Robert Boynton Just a thought. Maybe the "$3,500 or OBO" is what they're offering to pay to have it taken away. +1

JT Burkard The cheapest part of this car will be the purchase price. I see many wallets emptying with this restoration +1

    Franklin Alfonso Better off with the Gremlin lol +3

    Franklin Alfonso 4 on the floor and a V-8 Gremlin X I test drove one and it was fun. Just slightly nose heavy...

Brad Purvis Lotus used to make some truly ugly cars, not the latest this Elite and the Eclat, but God I love them. God help me I do love them so. Kind of like Patton and war. +1

    Rob Siegel And you love the smell of burning electricals first thing in the morning, right? +1

    JT Burkard On a quiet night you can smell the Lucas roasting +1

    Brad Purvis Rob Siegel—Smells like Victory! +1

Robert Myles

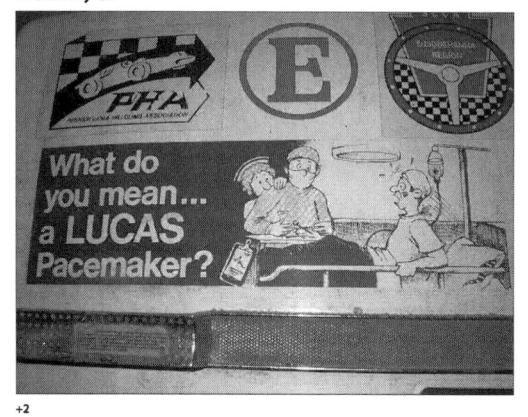

+2

Andrew Wilson Rob, do not F with time continuums, they always end badly.

Erik Wegweiser Looks like it's powered by Mr. Fusion! +1

Brian Ach Sorry, I love Lotus, but the Elite and Eclat, you can burn them

    Franklin Alfonso Lucas electrics they burn themselves… +1

Scott Aaron It will be a weird day in the world when you post up your first Lotus drive. That will be bizarre.

    Rob Siegel I'd bet on the spring of 2019. [Note: I was correct :^)]

    Franklin Alfonso Still waiting for part eh…

David Geisinger I'd go with this one [eBay ad for 1976 Lotus Eclat]

Franklin Alfonso Eclat= Every Common Lousy Automobile Trouble…

Shane Kleinpeter That car is the confluence of ugly and unreliable, yet somebody will still buy it because Lotus. +1

Rob Siegel And if they always wanted one and love the look of it, why not? +2

Bill Howard "one dead Lotus" <<< Point of grammar, Rob: If it says "Lotus," then "dead" is implied. +2

TLC Day 1883: Talk amongst yourselves

AUG 5, 2018, 7:30 PM

The Lotus Chronicles (sort of), day 1883:

To complete the trifecta of this automotive-intensive weekend, I swapped Kugel for WARP9. When I pulled it into the garage in front of Bertha, I realized that these two cars have not seen each other in almost 30 years. When I close the garage and shut out the lights, oh the stories they'll probably tell.

Bertha: "He sold me, and then the bastard he sold me to locked me up for 26 years! It's fucked me up for decades! Talk about abandonment issues! Damn am I glad to be back home!"

WARP9: "You have been usurped so many ways, I don't even know where to start. Don't look to your right. There is a monkey-shit brown British thing. Pretend it's just not there. That's what HE does."

The Lotush: "My time WILL come. He said so. I believe him."

The M Coupe: "He likes my ass the best."

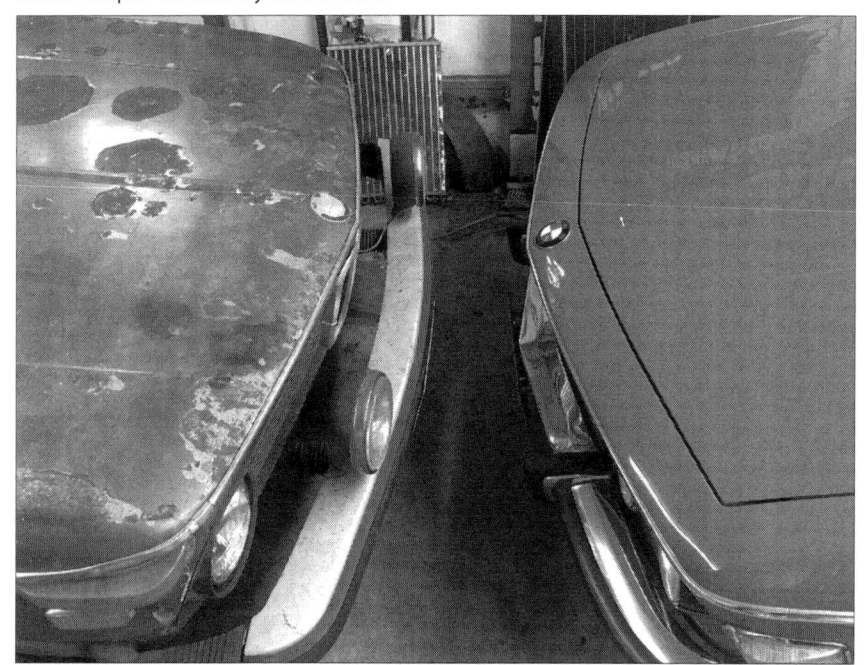

160 LIKES, 38 COMMENTS

Andrew Wilson Damn funny, Rob! +1

Paul Wegweiser I initially read this as "...this automotive insensitive weekend" and thought "well yeah! How cruel!" *reaches for another Vicodin* +5

Aashish Dalal Cars talk. Mine are all still mad at me for making poor financial choices like buying a house and food to eat, instead of car parts. They are completely jealous of my job, which consumes most of the time that these should consume. Here's my greeting card from mine, shortly after my back surgery. It's clear that they care. [photo of 5 cars in warehouse with two-level lift] +9

    Rob Siegel The tii: "I am king of the mountain! Anyone want to knock me off my pedestal? Hit me with your best shot!" +1

> **Aashish Dalal** Haha the e30 thankfully says "and the first to catch the dust!" +1
>
> **Clyde Gates** Is that a white Honda 800 next to the Microbus? +1
>
> **Ethan Diamond** You know that VW definitely didn't buy the greeting card but still got to sign his name. +2
>
> **Charles Gould** Clyde Gates White Honda AN600. Two-cylinder air cooled economy car, predecessor to the Civic. The earlier (1964-1967) S800 was a four-cylinder water cooled with a double overhead cam and a redline of 9,750 RPM! +1
>
> **Aashish Dalal** Yep! The AN600 was directly competing with the mini. It's amazing to see the build and casting quality of the early Hondas in comparison

Lindsey Brown I still say you should paint the Europa Fjord Blau, and put Tara King behind the wheel. +2

> **Lindsey Brown** Of course, you would then require a Bentley, and an umbrella. +3
>
> **Rob Siegel** and leather.

Lance Johnson LOL... literally!!

David Holzman But is it love? +1

> **Rob Siegel** Actually, I just submitted a two-part piece to Hagerty about my 32-year-long love affair with the 3.0CSi. That IS love. The rest of it is just tawdry infatuation, cheap motels, and lots of lubricant. +9
>
> **David Holzman** Hilarious! (But I was referring to the way Bertha and WARP9 were looking at each other.) +1
>
> **David Holzman** Send me a link to the two-part piece when it comes out, or post it on here and tag me. Would love to read.
>
> **Rob Siegel** Oh. That's different. No, not love. Wife and mistress #5 understanding. +1
>
> **George Zinycz** You mean USED lubricant Rob, right? +1

T. Ladson Webb Very fun to stay connected with your AutoLife +1

> **Rob Siegel** Usually you have to PAY for this level of entertainment! +1
>
> **Kristin Dillon Webb** Rob, dearly! ;) But worth EVERY penny! Laughter in life is the BEST Rx for long healthy life! :-D Your vintage car enthusiasm takes me back to my dear Momma who could tune her Camaro engine and do all sorts of other 'non-feminine' talents back in the day! I do wish I had some of her gifts! BUT these dang computer codes intimidate me!! 😊 But I am THE BEST at detailing and keeping them looking their best! Just ask Goldilocks!! ;) 😊 +1

Eric King WARP9 is thinking, "Wow, Bertha is /younger/ than me by a few years... and led a hard life!" (1)

> **Rob Siegel** Speed is an awful drug... +2
>
> **Franklin Alfonso** Bertha has hood envy... +1

Daniel Senie The Lotus is an optimist. +1

Jonathan Poole This is the title of the song I think. +2

Daniel Senie Would make a fun song, the cars all talking about you. +2

Larry Filippelli this is hysterical... and SO TRUE in a Pixar films sort of way!! +2

Steven Bauer I see a series in the vein of Thomas the Tank engine +1

John LoVerde Too funny!

Brian Ach That's hilarious. A new column for you: Car Talk: The secret lives of Siegel's cars. You heard it here first. +2

Michael McSweeny thanks for the chuckle...

Elizabeth Marion "And he wears white socks! " +1

Ethan Diamond I am so very grateful they can't actually tell you stories. Oh the joy rides we've never had ;) [to be clear, this is my oldest son admitting something I've long suspected :^)] +4

Vicki Wegweiser Amazing that all of your other cars didn't turn tail. Bertha looks like she has leprosy. 😊 +1

> **Rob Siegel** Yes, but not in a bad way 😊 +3

Scott Aaron So you're doing a thing where you're making the cars talk? This is becoming some sort of My Mother the Car/ Mr. Ed mashup. I knew this Bertha thing was going to be a trial for me. +1

TLC Day 1968: Meeting a troublemaker

OCT 29, 2018, 10:28 AM

The Lotus Chronicles, Day 1968:

Some of you may remember my telling the story of how, on the way to The Vintage in 2017, while caravanning with Jose Rosario and Brian Ach, we hit traffic on I-81, and Jose rang me on the walkie talkie and said "Hey Rob, there's a red Lotus Europa about ten cars ahead." When the traffic started to flow, I sped up to nearly 90 to catch it. Sure enough, there it was, impossibly low, angular, and gossamer-like. I mean, when was the last time you saw one of these in the wild? (It turned out there was a Lotus event in Asheville, but I didn't know that at the time). I regarded it as an omen to re-engage the engine rebuild on my long-dead Europa in the garage. I snapped the photos below, and wrote up the story for Hagerty. A week later, I got an email from the Lotus driver's wife. Turns out a Lotus friend of theirs in England had read the Hagerty article and seen the photos. The driver was Russell Baden Musta, a pilot for NetJets. We became Facebook friends.

Well, last night, after Maire Anne and I watched Dr. Who (an episode with giant spiders; how could we not?), I got a FB message from Russell saying that he'd flown into Boston, was unexpectedly here for the evening, and was staying overnight in a hotel not five miles from me. I ran over, and we hung out in the bar for nearly three hours talking cars. The Sox game was on, but neither of us paid any attention to it. Turns out Russell has owned his Europa for 44 years, so there was a lot to talk about. I told him the trials and tribulations of trying to complete my engine rebuild. I also recounted the "a car like that, you can get SEX out of" story.

We said our goodbyes at midnight. It wasn't until I was driving home that I realized I'd paid no attention to the game. I'm happy for the Red Sox, but I am very comfortable with my priorities.

167 LIKES, 100 COMMENTS

Chris Mahalick I feel like I am in a slow race with you as to which of us will complete our Europa first.
Rob Siegel you will +1
Chris Mahalick Rob Siegel I am sitting here in front of the computer trying to rationalize why I should not go out to the garage. :) +1
Joey Hertzberg Hahaha..
Russell Baden Musta Great to meet You in person Rob!!! This may be the first time in 44 years now with this car, that I've come to know someone after a friendly exuberant wave being passed!! Now to keep the progress going on Your Europa! Will get you the info on the 2019 LOG (Lotus Owners Gathering) which is near you this coming year! (8) [Note: This turned out to be a prophetic comment.]
    Daniel Senie If it's not running, Rob could tow the dead Lotus to the event behind Bertha. :)
    Scott Linn Europa can't roll, engine/transaxle removed so no rear wheels.
Greg Lewis You have your priorities straight in my book. Has anyone ever tracked you down to talk baseball? No offense but I'd wager not. +2
Franklin Alfonso Anything new to report on Lolita? Nothing back from the machine shop yet?

> **Rob Siegel** Still waiting on the custom lower timing cover and cartridge-style water pump. Without it, the water pump can't be changed without pulling the engine, and the pumps tend to leak from sitting. +1
>
> **Franklin Alfonso** What the hell is the matter with them? Don't they know keeping a Hack mechanic celebrity writer is a bad business practice??? +3

Steven Hage I sat in a lowered Europa once, I still have no idea how I got out of it haha +3

> **Franklin Alfonso** You don't sit in those cars. You wear them.
>
> **Brian Ach** Can they even be... lowered? +1
>
> **Russell Baden Musta** Brian Ach actually yes! The model I have had higher ride height in front because of headlight height regs in USA. When I restored it in 85, I put 2" lower springs on front only, Spax gas adjustable shocks w ride height adjusters... plenty of clearance still... proper caster/camber/tow to spec... drives great!!
>
> **Steven Hage** It was so low that the tire rubbed through the fiberglass in the rear [photo of a very low S2 Europa] +1
>
> **Russell Baden Musta** Steven Hage wrong offset does that...

Kai Marx I started reading your articles after researching about Europas! I always wanted to have one but I was always weary of the stories etc. Of course I would end up doing a 13b swap if I could. Hahaha +1

Sean T. Going I really need to take you to a game. But I appreciate your passions. +1

> **Rob Siegel** Sean, as I think you saw from my FB comments, I DID watch the others. I was planning on watching the last one. It was just how the evening spun out. +2
>
> **Wade Brickhouse** Rob Siegel Didn't pay any attention either, I was at a Bob Dylan concert. :-) +2

Kevin Pennell Great story Rob, as always. BTW, Dr. Who was really cool last night, wasn't it? +1

> **Rob Siegel** I laughed at the title: Arachnids in the UK. It was a play on the title of the classic Sex Pistols song Anarchy in the UK. +3
>
> **Kevin Pennell** Nice catch! Yes it was!

Doug Jacobs Wait a fucking minute. "gossamer like?" Give. Me. A. Fucking. Break. +3

> **Rob Siegel** tough crowd :^) +2
>
> **Russell Baden Musta** Hey, that was me driving that car... +2

Thomas Jones That's the game where they dribble the orange ball right?

Thomas Jones Across the street from a childhood friend of mine, there was a guy in his mid-twenties or maybe early thirties who was stripping and repainting his Europa in his mom's garage. I would be riding my bike by, stop and chat with him for hours at a time. He was painting it red. There was another, dark blue, Europa in the area that was around after I had gotten my driver's license and my red 1600-2. While I didn't get to play around on the local backroads with him, my friend with a '66 Mustang did. Then, nearby, in the north Berkeley hills, there's a Super 7 under tarps in a driveway that's been there for at least 35 years. I still dream of knocking on the owner's door with the story of dreaming of owning, restoring, and enjoying their Super 7 since I was eight or nine years old when I started riding my bike all over those hills. +3

Brian Ach An excellent day and a good omen.

Bill Howard But this wasn't one of the hot tub conversions? +1

> **Russell Baden Musta** I have a separate hot tub, but driving in a pouring rain, my Europa does take on water!!! +1
>
> **Bill Howard** Russell Baden Musta Route it over the exhaust manifold. +1

Mark Jeffery Hello Rob, I do remember the write up on the sighting of the red Europa, great to read that you met up with Russell +3

Tim Warner I still think a motor from a 2002 would be perfect in the Europa.

Brian Hart My last sighting was this sorry example. Had a closer look at it this summer and found a VW transaxle where the Renault unit should be—thorough mess but there's always the possibility... [photo of Europa buried by snow]

Andre Brown I love to hear about stories like this.

Matthew Rogers Your priorities are on point.

Neil M. Fennessey I seem to remember a Europa running in the NESCCA back in ‹74 of ‹75.

George Whiteley There's nothing like following an old guy in a Lotus Europa and waiting until he stops to see him get out of it, LOL! +4

> **Russell Baden Musta** Okay, you asked for it... I believe I'm old...65 and 9 months... I qualify... and I have a Lotus Europa obviously... stay tuned for an old guy getting out of a Lotus Europa

movie clip... kinda reminds me of Mike Myers' Sprockets Character Dieter showing Fat Man in a Sprinkler... gimme a day or two.. +2

Scott Linn I find head first the easiest way. +1

Al Mancuso About 3 weeks ago in Pennsylvania. :-P [photo of red Europa at a Cars and Coffee] +4

Tim Warner A friend of mine had his Europa in his living room- for drive in Movie night! +1

Susan Rand Great story! +1

Scott Linn A bit over 3 weeks ago in Oregon. [burgundy Europa in a parking lot]

Melanie Pray How wonderful for you both. Cars have a way of bringing good people together. +3

Brad Purvis Let's see... NetJets Pilot. So I assume you got stuck with the bill? +3

>**Russell Baden Musta** Who, Me or Rob?
>
>**Brad Purvis** Russell Baden Musta , you'll be ostracized by your brother pilots if you didn't stick him with the bill. I trust you got your FREE USA Today and breakfast at the hotel as well. 😊 +1
>
>**Rob Siegel** Brad Purvis, Russell graciously bought my Jack Daniels +2
>
>**Russell Baden Musta** Brad Purvis I forgot my damn paper... no sudoku to do on flight home dammit... +2
>
>**Brad Purvis** Rob Siegel Wait until the word gets out. He'll be the laughing stock of the pilot community. Pilots are notoriously cheap bastards who stick others with the check, or scrupulously dissect it so they don't pay more than their "fair share." We are all about freebies; free breakfast, free newspapers, and "crew discounts" on meals, rooms, flights, cruises, vacations, etc. It's a running industry joke for those of you uninitiated to our industry "perks." That said, if I ever get a layover in BOS I just my buy you a whiskey as well, but don't expect any of the high-falutin Jack Daniels, top shelf stuff. It's Old Granddad for you Mister. +2
>
>**Russell Baden Musta** Brad Purvis I'm the freakin envy!!

David Holzman Sounds like a great evening! You should'a called me!

JohnElder Robison We have a blue Europa at J E Robison Service Co right now.

Chapter V: Full Engine Reassembly

TLC Day 2013: Water pump ordered

DEC 14, 2018, 7:00 PM

The Lotus Chronicles, Day 2013 (seriously!):

I made great strides on the Lotus' engine rebuild this spring (I got the block assembled), but then it was back-burnered—again—due to my deciding that I needed the aftermarket front timing cover and cartridge-style water pump from Dave Bean Engineering that allows you to change the water pump without pulling the engine. Others sell this kit for the Elan and other cars that use the Lotus Twin-Cam engine, but the water routing on the Europa is different. Only Bean offers the kit for the Europa, but it's been out of stock for 18 months. I'd call Bean every month and ask about the kit. Ken, the manager, would patiently say "Yeah, we're dealing with a new machine shop to make these for us... they're close, but we don't have them yet... and I see that you're on the list, so you actually don't need to keep calling every month." So I stopped calling.

Having put Bertha away in Fitchburg last weekend, and having gotten The Lama (the E28 535i I bought that turned out to have a broken rocker arm) to the point where I can simply get in it and twist the key and drive it around, I found myself wondering what my next project should be. With winter settling in, that next project needs to be in my garage, like, NOW. I decided that, unless something unexpectedly tasty came on my radar screen, my attention should be bent back toward the Lotus that's not only already IN my garage, it's been a bloody beached whale in said garage (perhaps a beached albatross? at least it's very light) for five and a half freaking years.

So, today I called Bean. I spoke with Ken, asked about the Europa cartridge-style pump and lower timing cover kit, and was stunned when he said "We've had these in stock for a couple of months now." After the shock wore off, I laughed and reminded him of what the history was. He paused, then said "I remember... I'm REALLY sorry; I have no idea how you slid off the list."

So, at long last, the removable water pump kit I need to continue with the engine rebuild will be shipped to me on Monday.

I know I've said this every winter, but this WILL be the winter of Lolita. If I can find her and the pieces of the motor under all the crap in the garage.

69 LIKES, 53 COMMENTS

Bob Sawtelle And Ken wondered why you called every month... Rule 64: "Always count on the laziness of others"... +7

Doug Hitchcock Oh HELL YES!!! +1

Marc Touesnard Ken's list must have a terrible memory +4

Dave Gerwig With the short block reassembled: let's call that first base. Now with the long-awaited water pump assembly "dislodged" from Bean, you are close to stealing second base. Time to inventory all gaskets, nuts and bolts: could be a Holiday Engine Assembly marathon! Great news Rob Siegel! +7

Dale Robert Olson Wow! Now that is what I call a motivating factor to spend some time in the garage. +1

David Shatzer She's old enough now that you can possibly re-name her to just "Lola"... she will be a showgirl... +3

Scott Aaron Let me just cast my ballot that this IS the thing you should be working on this winter. This is so perfect. I'm popping my popcorn... +6

Paul Fini [gif from "Thriller" of Michael Jackson eating popcorn] +3

Robert Myles That's a big "if." Remember, we've seen pictures of your shop. +2

Brett Kay Anxiously watching for progress reports with pictures Rob Siegel. Bring her back to life. +1

Brad Purvis Yes, yes. You definitely, definitely need that Jag. +1

    Rob Siegel for, you know, company

    Brad Purvis Rob Siegel—Company's coming +1

    B.a. Miller Brad Purvis everyone needs a Jag. [photo of XJ6]

    Brad Purvis B.a. Miller, been there. Done that.

    Tim Warner B.a. Miller I prefer my all aluminum 2005 XJ8 L

    B.a. Miller Tim Warner, nice choice. My DS21 is one of my Favorites [photo of DS21] +1

    Tim Warner B.a. Miller my neighbor just got a Turbo Citroen! Loves it! +1

Bob Shisler Happy winter! +2

Brian Stauss I've got the cash ready to buy your next book "Lost and Found in My Garage: Is That a Europa Under All That Crap in the Corner?" +6

Bob Sawtelle Wait 2013 days, and you picked up Lolita in... wait for it... 2013! So I figure in another 5

and a half years when it's running you will have had Lolita for 11 years, so 2013 minus 11 is... yup 2002. Thank you I'll be here all week try the veal. +3

> **David Shatzer** Bob Sawtelle Groan emoji...lol

Paul Wegweiser Hey! Just a reminder: Those little aluminum "C" shaped things that go on the crankshaft and connecting rods and stuff... don't forget to install those! +4

Paul Wegweiser "In other news: Prices of kitty litter in the Boston Metro area are expected to rise, as sudden demand outpaces supply! Adult film at eleven." +4

> **Rob Siegel** that's mean. you're mean. +3

Scott Lagrotteria It could have been worst; he could have said we produced and shipped out the last one this morning. +6

> **Rob Siegel** Yes, that would've sucked +1

> **Franklin Alfonso** Did they advise you they are now made of plastic so as to have the same longevity as the engines? +1

Chris Lordan Way, way in the background w/this. The car life, distilled. [link to Hagerty piece by Aaron Robinson "I Know a Guy"] +3

Steve Park As someone who has a garage full of long stalled projects, I hope this gives me an incentive to get one of them going. I look forward to your words of wisdom. +1

Tom Egan 'bout time there was a little love for the Lotus! I suppose it's a little late to point out the water pump issues that plague the Elan don't apply to the Europa because the alternator loads are removed. +1

> **Rob Siegel** Tom, I didn't really know what to believe. My car had a seized water pump, so it needed to be done anyway as part of the engine rebuild. If Bean didn't come through with a water pump, I was prepared to go the conventional route, but the reputation for pumps leaking through disuse, and the fact that this is one of 12 cars and will likely sit for long periods, spooked me.

> **Tom Egan** There will probably come a day when my water pump craps out and I curse you and your foresight. +1

> **Scott Linn** I believe the one in mine is a Bean pump. I should check the rebuild details... +1

Steve Kirkup Time to shift the contents of your garage to the BMW side to expose where the Lotus was last seen. Are you sure it's still there?

Rob Siegel uh, no.

+4

Rennie Bryant In reality, Europas are not that rare. A RUNNING Europa on the other hand... +4

> **Rob Siegel** Insert meme showing Alec Guinness saying "a running Europa... that's a phrase I haven't heard in a long time" here 😊 +2

Andrew McGowan It looks like I have a shot at getting my Alfa on the road at the same-ish time...I thought for sure you would beat me, but... this happened yesterday [photo of gorgeous Alfa engine dropped in engine bay] +3

> **Rob Siegel** dude, you have an assembled motor and it's in the car... you're light-years ahead of me!

> **Chris Raia** *pun intended
>
> **Andrew McGowan** Rob Siegel yeah but this is the rear of the car... [photo of back of car missing rear subframe] +2
>
> **Josh Wyte** Get that paper weight moving! +1
>
> **Wink Cleary** These Europa stories actually kind of make me glad I didn't win one in a contest I entered at 16. I actually won but the organizers cheated so I never got the Europa—there was a class action suit filed—and won!-$0.36 was my share! (never a word about the $10 entry fee) —all good lessons for a 16 year old to learn!) +2
>
>> **Scott Fisher** I made sure my kids learned that on those claw machines where you put in 50 cents, move the claw for nine seconds, and it shuts off before you even position it over the toy you want. Same lesson, MUCH cheaper...
>
> **Joey Hertzberg** Rob, I don't know if you've read this thread, but it blew my mind. [link to article on grassrootsmotorsports on 1968 Lotus Europa Restomod]

[email: Dave Bean Engineering]

```
From: Ken Gray, DBE
To: Rob Siegel
Mon, Dec 17, 2018 5:42 pm
```

Hi Rob. Thanks for the order. Yes, our cassette pump can be fitted
with the hose nipple and threaded plug to outfit the Europa front
entry or the Elan/Lotus Cortina side entry Twin cam engines. The
kit contains: new front cover with the cassette pocket, water pump
cartridge unit, new back plate and water pump fasteners. The water
pump cartridge features a larger bearing/shaft and seal from the
Turbo Esprit. Advantages: more reliable/durable, easier service
(no removal of the head/oil pump once installed). Yes, an extra
cartridge is possible. We can provide the cassette water pump
rebuild kit too (shaft/bearing, seal, impeller, hub, o rings and
instructions). Please let me know if you have any further questions.

Ken Gray
Dave Bean Engineering Inc.

TLC Day 2023: Water pump arrived!

DEC 24, 2018, 5:18 PM

The Lotus Chronicles, Day 2023:

My Christmas present to myself, paid for by holiday book sales, arrived on Christmas Eve. Damn, I have a good life. #WinterOfTheLotus

154 LIKES, 50 COMMENTS

Kent Carlos Everett Nice new and shinny.

Paul Wegweiser So... uh... an M42 water pump couldn't be made to fit? Where is the Hack I used to know and love? +5

    Luther K Brefo Send it to Paul and I bet it could. What's the diameter?

    Paul Wegweiser Rob Siegel The last M42 water pump I removed made me work for it, but I won. I always win. It came out in at least 5 pieces. +3

    Paul Muskopf It does look remarkably similar/ +1

Kevin McGrath Shiny, new parts require a good luck lick before installation. +3

Roger Scilley For the Lotus? +1

Rob Siegel yup +2

Andrew Wilson It's a Saturnalia miracle! +2

    Andrew Wilson May nary a cuss word be uttered at the installation. +3

    Rob Siegel I'll save the cuss words for if the fucking thing leaks +9

    Andrew Wilson That's the spirit!

    David Shatzer Rob Siegel Noooo. Casting flaw if it wasn't x-rayed, eddy current tested or other non-destructive examination... it's the part, not the installer... he said full of eggnog and cheer... +1

    Kevin McLaughlin It's British. That's guaranteed 😀 +1

Scott Linn So what you're saying is, is that I helped buy those... You're welcome! Those are shinier than anything on *my* Europa... Just sayin. +4

    Rob Siegel Mine too. Scott, did you see the white Europa TCS needing a water pump that just closed on BaT?

    Steven Bauer Those last two bidders must have really wanted it.

    Scott Linn Rob Siegel yeah, saw that one. Prices seem to be rising up a bit. Does your TCS have a 5-speed? I am really glad mine has one.

    Rob Siegel Scott, yes

    Chris Mahalick Is a 5sp worth more?

    Scott Linn Yes, a little.

    Scott Aaron Rob Siegel It kind of reads like it was a project that lasted 43 years. +2

    Rob Siegel When I first saw the auction, I thought "boy, it's risky selling this thing in need of a water pump. Everyone will jump all over it and chime in that it can't be done with the engine in the car." I was very surprised when it went to $15.5k. But the car presented itself well.

    Paul Forbush That seems like a lot of money for something with so much "Patina". Wonder what a really clean Europa might fetch. Or is that one is nice as they get? I'll stick with BMWs.

    Scott Linn I've seen people asking $36k for 5sp TCS' in great condition, but I don't know if they eventually sold for that or not. Hagerty rates a 73 in #3 condition at $17.3k. #1 condition $38k.

Bob Sawtelle Europa!! +1

Mark Thompson Wow! Stud! Holiday Book Sale lucre bought parts... awesome 🍸! +2

Daniel Senie Let's hope this doesn't result in 3 years of winter. 😊 +3

Rob Capiccioni I love car parts as gifts!! +1

Samantha Lewis Ah the water pump, don't get me started, I had almost forgotten... ah, and the heater valve, such a quick way to lose all coolant, in a jiffy. Memories. +2

Jim Denker And the Europa will live again! +1

> **Daniel Senie** Jim Denker I wonder if Rob will have a "coming out" party when the Lotus finally clears the garage door?
>
> **Jim Denker** It will be orgasmic. +4

Brian Stauss Is that, is that, is that the famous Lotus water pump? Good holiday project for you! +1

Brian Stauss I treated myself to two new run flat tires, this Christmas Eve, since a nail decided to take up residence in the sidewall of one. +1

Franklin Alfonso Lolita will dance again next summer! +1

Brad Purvis Proud to serve the resurrection.

Scott Aaron Where's the Lotus version of Vintage? Because we're all showing up in our 02s. +1

> **Rob Siegel** When I saw "the Europa in the wild" on the way to The Vintage a few years back, I only learned later that there was a Lotus event in Asheville at the same time.
>
> **Brad Purvis** It's called LOG (Lotus Owner's Gathering) [link to lotusltd.org. It was in this post that I learned that LOG was scheduled to be in Sturbridge MA.]
>
> **Rob Siegel** Wow, my neck of the woods end of next summer. Not impossible. +2
>
> **Dohn Roush** No connection between the acronym and the current state of most Loti, I assume... +2
>
> **Andrew Wilson** British Invasion in Stowe Vermont. http://britishinvasion.com +1
>
> **Tom Egan** Expecting to see you in Sturbridge next summer, Rob. Not sure about Vermont, I try not to drive the Europa further than I am willing to push it home. +3

Josh Wyte No excuses for not getting that enormous paper weight moving now! +1

Phil Morgan This is awesome. Can't wait to read about it. +1

TLC Day 2030: Oil strainer

DEC 30, 2018, 10:38 AM

The Lotus Chronicles, Day 2030:

Yesterday, with my new water pump and timing cover in hand, I finally began to re-engage the assembly of the bottom end of the Twin-Cam engine. I thought I'd do two trivial things. The first was re-attaching the oil strainer that hangs down into the pan. It looks so much like a tea strainer that there are posts on Lotus forums of people USING an actual tea strainer instead. The problem is that it's a crimp-fit onto a plate, and even after crimping it, it slides around. I have posts to several forums asking if that's right, but have not yet received an answer.

The second thing was that I thought I'd bang in the oil seal on the front timing cover. Almost immediately I put a small gouge in the side of the seal. Then I read the manual and learned that you don't "bang" it in; you coat it in WellSeal (the sealant used for most of the bottom end) and gently press it in. I tried that, and unlike any other seal I've ever pressed in, instead of seating, it wants to cock and twist, and any attempt at straightening it just cocks it more. I read on the forums and learned that there's a "wide seal" you can buy that does a better job staying in one place, but as with most of the other parts for this motor, it's not click-and-buy from anywhere in the United States.

CHAPTER V: FULL ENGINE REASSEMBLY **145**

51 LIKES, 34 COMMENTS

Dave Gerwig Such fertile territory for your writing! Every piece and part has a personality. +3

    Rob Siegel and an attitude :^) +4

    Franklin Alfonso Well just because you finally got the part after all the delays you really were not expecting things to go easy now were you? This is British engineering remember… +2

    Peter Potthoff Franklin Alfonso There is no British engineering, just British manufacturing… +3

Ed MacVaugh Any seal like that, which lies on a removable piece, I squeeze in the seal using a vise with a couple small pieces of wood for protection. Seeing as it is a Lotus, I would use a vice instead. +7

David Kemether [Simpson's "HAHA" gif] +2

Jeff Hecox Typical British seal in fractional size… I believe that the orange color of the seal would have it made of Silicone. Possibly look for one made of FKM or Viton, typically brown or black in color.

Doug Jacobs Can you drill a couple of small holes nearby and safety wire the tea strainer in place? +1

    Rob Siegel maybe

    Tim Warner Rob Siegel strainer may just rest on the bottom of the sump [Note: this turned out to be true]

Matthew Rogers I sense a new book in the works, something along the lines of "Ran At Some Point In The Indeterminate Past." +5

Steven Bauer Weren't you looking for a tight seal a year or two ago? Looks like you found it… +1

Matthew Rogers And ironically, it will doubtless leak no matter how tight it is, because it's British and has a longstanding tradition to uphold. +7

Andrew Wilson Hoping everything from here on out, to quote your urologist, "Is smooth and unremarkable." +5

    Roger Scilley Andrew Wilson that's called self-lubrication.

Kent Carlos Everett Hummm? [photo of chamber of speakers] +1

    Joey Hertzberg Kent Carlos Everett hey, I've seen this mini-truck in my neighborhood!! +2

Samantha Lewis Having experienced the quality and longevity of original parts back in days of yesteryear far too well, I'd highly recommend sending off to Blighty for all the improve design parts you can find. Baffled oil sump being among the really needed. +1

Russell Baden Musta I seem to remember it may be spring loaded and moved… don't know why…

    Rob Siegel Russell, yes, it IS spring-loaded, but in addition to that there's side-to-side play. There's a notch in the plate that fits around the other oil tube. Between the up-and-down and the side-to-side play in the plate, and then the play in the screen on the plate, it gives me the willies. +1

    Russell Baden Musta Rob Siegel I understand… wish I had mine open to help…

    Rob Siegel I'm glad SOMEONE's is closed, running, and happy! +2

    Russell Baden Musta Rob Siegel but mine is soaked in oil!!! That why I want to remove and tear down again… maybe water pump conversion like yours…

Skip Wareham How would Edd China solve the problems?

    Chris Lordan Edd China would solve this problem the same way he solved all the other problems. He'd tower over the problem and force the thing to be fixed. And at 6' 8" or so, he towered over almost everything! +2

    Dohn Roush First he'd put on the rubber gloves…

Roger Scilley Day 1997 is like 5.47123 years, time sure goes fast when having fun. +1

Rob Siegel Bought it June 10th 2013, so that's correct. The "having fun" part, not so much... :^)

Rob Siegel Ed MacVaugh sent me a PM saying "On regular Fords, it is spring loaded downward. The fine scratches on the bottom plate suggests the same for your engine." I cleaned the oil pan, and look at what I found. It's MacVaugh for the win! Thanks Ed!

+4

Scott Aaron Rob Siegel yay! Something got solved! +1

Ed MacVaugh Up until 1974, Pinto cars came with the 1.6 L "Kent" engine. The fantasy in many young car-fond teenagers was upgrading the engine to the Twin-Cam. +1

Rob Siegel Oh, and Ed MacVaugh, regarding what "regular Fords do," look at the FoMoCo stamp on the piece :^)

+2

John McFadden Ed MacVaugh reminded me that Chevy "Cosworthed" the Vega.

TLC Day 2031: Wintah beatah europah

DEC 31, 2018, 11:04 AM

The Lotus Chronicles, Day 2031:

Who knew that there was a winter beater Lotus Europa right up the road in Maine? (Note to novice Craigslist posters: Don't title your post "sports car," even if it IS black with yellow stripes. I've been doing this a long time, and I don't think I've EVER searched for "sports car.")

[expired Craigslist ad, "Sports Car: Early Europa 1971 S2 S/N 54/1824, unrestored, needs work in and out"]

33 LIKES, 62 COMMENTS

    Wade Brickhouse Maybe "exotic?" 😀
    Joseph Oneil Excellent tires. +3
    Doug Hitchcock I wish I had the long green.
    Wil Birch "Must sell before winter." What year? +5
    David Kiehna I owned #1808 😎
    Skip Wareham Complete with cherry bomb muffler +3
        Rob Siegel Ch-ch-ch-CHERRY BOMB! [link to The Runaways video of "Cherry Bomb"] +4
        Philip D. Sinner fanTASTIC !!
    Andrew Wilson I could join the Lowtush Club! [Andrew lives in Maine] +4
        Scott Aaron Andrew Wilson oh I had the same thought!!! But then I thought "yeah, right!" One of us can buy it from Rob for $12k after he's put a year and $39k into it 😀 +1
    Russell Baden Musta Wow...
        Rob Siegel yeah, but that way lies madness ^)
        Russell Baden Musta Rob Siegel 4 large... almost tempting! Little clean up... S2... tight fit fer me, perfect for you!
        Rob Siegel Russell Baden Musta, I have a rule: Only one indefensible British car in the garage at any one time. +4
        Russell Baden Musta Rob Siegel fully understand! You know, I'm thinking you could be ready for LOG 39 at Sturbridge in August! Would LOVE to hang with you kids for that weekend!! +1
        Rob Siegel Russell, we shall see. Primary goal is getting the motor fully assembled this winter. Secondary goal is getting it and the transaxle back in the car. Then there's the rest of the sort-out, and the car, dead since '79, will need EVERYTHING. Of course, I could simply go to Sturbridge without the car... +1
        Russell Baden Musta Rob Siegel yes you can with pictures!!! Or fly to PIT and drive MY Europa up!! +1
        Daniel Senie Rob Siegel if you have the transaxle back in, you could roll the car out, put it on a trailer, and take it to Sturbridge. +1
        George Murnaghan Rob Siegel proudly violating the Siegel Rule since 1989! +1
        Christopher Kohler park one outside.
    Wil Birch Gotta admit though; if I thought there was *any* chance of my 6' 7" self being able to get into (and drive) one of these I'd be tempted. +2
    Bailey Bishop Jr. I need to do some winter work as well. First project is a leaking fuel tank. [photo of Bailey's TCS] +3
    David Kiehna 5'11,175 lbs, 9.5 shoe is a perfect fit. +1
        Bailey Bishop Jr. I'm 5'7», 175 and an 8.5 shoe and it's tight. Anything other than a driving shoe is too big for me.
        David Kiehna Bailey Bishop Jr. I wore Puma speed cats when I drove it. +1
        Russell Baden Musta David Kiehna I'm 6'1", 200 or less..8.5 shoe... boxing shoes cheaper alternative to nomex... Normally wear Clark's Shoes "Senna" loafers... perfect fit!! In my TC... S2 a little tighter...
    JT Burkard My Alfa has BWA's [photo] +2
    Frank Mummolo "Sub-compact" takes "Euphemism" to a whole different level! +1
    Clay Weiland How sacrilegious would it be to motor swap one of these?
    Steven Bauer Put an LS in there...
    Clay Weiland M42 all the things +2
        Wil Birch Clay Weiland ... M42, definitely!
        Rob Siegel Zetecs and Toyota motors appear to be the most common swaps for Europas. The later Lotii have Toyota motors. +1
        Clay Weiland [Dr. Who gif "You're better than this"]
        Clay Weiland Lotii?

Rob Siegel When I bought the Lotus, I knew it had a seized engine. I looked at the Zetec conversion, figuring it would be cheap. The motors certainly are, but I couldn't really find any kit for the conversion, just individual websites or build threads on forums from people who had rolled their own. So the motor is cheap, but you pay for it everywhere else, not unlike other things we know. At that point I thought that if the car's original Twin-Cam engine was rebuildable, rebuilding it would be the quickest (ha!) least-expensive (HA!) least-painful (OH GOD THAT'S FUNNY) path to getting a running car. I'm not sure what I did in this life to have had the universe perform this colossal bait and switch on me. All it had to was not allow me to unseize that stuck #1 piston, but NNNNNNnnnnnnnoooooo.. +2

    Clay Weiland Rob Siegel I bet we could fabricate some M42 mounts. Just sayin' +2

Steven Bauer Ran when new.., +5

    George Murnaghan Steven Bauer never safe to assume! +2

    Lee Highsmith ...for dozens of days!

Rob Morelli If it has been on the market for a long time because of a poor listing the asking price may be low.

Skip Wareham Posted 11 days ago.

Rob Siegel It's been an on off CL for quite a while. Mid-coastal Maine isn't the most convenient location from which to sell this kind of car. +2

Alan Hunter Johansson "Sports Car"... that's burying the lead. +2

Samantha Lewis All I'll say is that I'm delighted that it isn't within 5K miles of me, as things go, it ain't half bad for that price, and if it was in Santa Cruz, I would be thinking crazy thoughts best left unexplored. (1)

Rob Siegel "because suicide is painless..." :^) +2

Samantha Lewis Absolutely! +1

Rob Siegel On a recent CL search, there was actually an inexpensive Twin-Cam car on the west coast. Let me find it. +1

Rob Siegel You're in luck. Looks like it's been sold. Unless you want this one in Myrtle Beach :^) [Craigslist ad for highly-messed-up Europa]

Dave Gerwig Rob Siegel the Myrtle Beach 2x Cam Lotus has an interesting rear flank body model? Are those Lotithrusters or Loticoolers?

Samantha Lewis Way too messed with. I'm safe. Whew!

Rob Siegel Wou wanted this one that sold on BaT last week that sold for $15.5k even though it had a seized water pump: https://bringatrailer.com/listing/1972-lotus-europa/

    Samantha Lewis Oh hell yes to the nines, that would be the one, original except for those cross over hot air pipes between the exhaust manifold and the carb intakes that everyone removed immediately. Glad it's sold, so glad it's sold, and for not much, for all the exquisite torture, oh that perfect memory of a summer drive, where everything worked perfectly, what a beauty those cars, I masochistically love them still, that perfect day, when everything worked, worth everything, definitely. Get yours running! +2

    Rob Siegel workin' on it :^] +3

Paul Wegweiser I've been watching these crime detective Medical Examiner documentaries lately. All the murder episodes begin with; "He seemed like just an average guy, but after he bought that second Lotus, we started to smell something bad under his back porch." +10

    Alan Hunter Johansson Paul Wegweiser I can't imagine anyone referring to Rob as an average guy. I'm sure he's been called a lot of things, mind you, but I doubt that! +1

    Andrew Wilson What? Behind the Lotus? No it IS the Lotus!—Monty Python +2

    David Kerr She's filing her nails while they're dragging the pool of coolant/oil +2

Jeff Hillegass Or... [Facebook post in "Obscure cars for sale" for "Anyone interested in an LHD 1971 Lotus Europa S2 recommissioned after 40 years in dry storage? $17,000"]

Charles Gould And so it begins. The helpless accumulation of all parts, parts cars and restorable projects to match the one example that you just almost finished without any of those parts or spare projects. +1

Scott Lagrotteria YES! Another annoying overpriced ad for a really beat up car, and he can't even write a post.

Other Things Happening in 2019

- Drove down to the BMW CCA Foundation museum with Andrew Wilson for the closing of the "2002 Icon" exhibit. Intended to pick up Louie, but a blizzard moved into the Northeast. Lance White took pity on me and trucked Louie, along with his own cars, back to his warehouse in Cincinnati.
- Sold "The Lama" to Jim and Susan Strickland.
- Was treated for prostate cancer.
- Drove Bertha to and from The Vintage.
- My beloved 89-year-old mother passed away.
- Wrote and released my 6th book, *Resurrecting Bertha*.
- Got a free "faux 2002tii," resurrected it as a running parts car, and sold it.
- Bought Hampton, the 48,000-mile '73 2002, from its original owner on Long Island.
- Sold Kugel to Jim and Susan Strickland and drove it to the 50th BMW CCA Oktoberfest to deliver it to them.
- Picked up Louie, which had been sitting in Lance White's warehouse in Cincinnati, found that it was leaking oil through a crack in the upper rear part of the head, fixed it with J-B Weld, and drove it home.
- Bought a 270,000-mile triple-unicorn X5 (6-speed, sport package, tow package).

TLC Day 2031: Engaging the pump

JAN 1, 2019, 6:05 PM

The Lotus Chronicles, Day 2031 (and yes, I calculated it incorrectly last time. this one is correct. It's too bad; I was all set for it to be day 2002 so I could say "screw this Twin-Cam motor, I'm installing the spare tii engine that's been sitting under the back porch"):

Making forward progress on a long back-burnered project like this is all about doing one or two things a day. Today I thought I'd simply replace the rear crankshaft seal. But nothing is simple; I ran into almost as much trouble as I did with the front seal. The rear seal at least seats against a step (unlike the front one), but it's big enough that I couldn't use the trick of using a socket whose diameter matches that of the seal to push it in. Plus, even after it's in, these damned things are unlike any BMW oil seals I've ever used. They naturally work themselves up and out, like dumplings or tortellini that you're trying to keep submerged while boiling them. The manual says you need to coat the outside of the seal and the inside of the bore it sits in with WellSeal. It takes a while for the adhesive to set, so you have to just keep pushing the damned thing back in until the adhesive holds it. What a pain. But at least this one is done. For the front seal, first thing tomorrow morning I need to order the "wide" version.

Then I thought I'd go into extra innings and begin test-fitting my $900 Dave Bean timing cover and water pump. As per the photo, there's an outer and an inner cover. They came wrapped together, with four bolts passing through them. When I prepared to test-fit them, I had them on a table in the garage. I separated them and turned the outer cover over. I didn't realize the bolts weren't secured to anything, and three of the four of them fell on the garage floor, which, as you know, is an instant passport into another dimension. I searched for nearly an hour and still can't find one bolt. I looked in Dave Bean's parts list, and the other three bolts appear to be off-the-shelf items, but not the one that's missing. It, unfortunately, is labeled "5/16-18 2.25 socket head cap screw with modified head." So tomorrow, it looks like I need to call Bean in CA and order A BOLT.

Nothing goes quickly here in Lotus Land…

33 LIKES, 68 COMMENTS

Jeff F Hollen This just makes me want to do the valve replacement on my 907. +1

Chris Lordan This project reminds me of that U2 song, "Kite"… Who's to say where the wind will take you? Who's to say what it is will break you?… +3

Chris Lordan And look around fercrissakes. That bolt didn't make its way to the International Space Station! +2

    Rob Siegel you say that, and yet… +5

    Chris Lordan … and yet. Mine's just as bad.

    Buck Hiltebeitel That bolt will be in plain sight the day after you install the new one… BTDT

Chris Mahalick I am so sick of mine at this point. Two years and counting. I just want it done and sold already. I'll stick with Porsches and BMWs from here on out. +1

    Chris Lordan Chris Mahalick—you are a wise man. A wise man indeed. +1

Jim Gerock Rob Siegel You should get some short pieces of PVC pipe for pressing those seals in place. +1

Bill Snider An exercise in Zen… soon as you order and it's shipped. the bolt will miraculously reappear. ;-)

    Rob Siegel true dat

Rick Roudebush Will your ring compressor go that small? Those bolts still exist in this dimension somewhere.

Josh Wyte I think your New Years resolution should be to clean up your garage to the point where if you drop something, you at least have a fighting chance of finding it. +1

    Rob Siegel yeah, that's not going to happen +3

    Josh Wyte Rob Siegel my profession has ingrained a certain level of OCD cleanliness regarding my work spaces. In your garage I'd just bring a clean stool and a 6 pack of nice beer and sit there and regard everything with complete dismay. Hoping that as I empty the 6 pack I reach a degree of contentment. +3

Rob Siegel I try cleaning, and invariably stop after an hour or so because it's gotten me over some hump to the point where I can get shit done. Cleaning or organizing past that isn't focusing on the task at hand; it's focusing on cleaning or organizing. It's just how I'm wired. +1

Karel Jennings Rob Siegel Hear Hear! :) (Mine is such a mess) +1

Rick Roudebush Any flat surface becomes a catchall, but I have to have a clear floor for this very reason.

John Jones Rob Siegel Maybe you could spend an amount of time equal to what you spent looking for the missing bolt.

Rob Siegel John Jones, that's dumb. You're dumb. Shut up. +5

Josh Wyte Rob Siegel just don't allow messes to happen period. If you start from a clean and organized spot, cleaning up to get back to that point after a job is easy.

Rob Siegel Josh Wyte...

Josh Wyte Rob Siegel you've got work to do. There's at least 4-6 columns worth of material in how you cleaned up and organized your shop +3

Rob Siegel again, I'd rather work on the cars +1

Josh Wyte Rob Siegel how much time and money have you wasted looking for and repurchasing parts you've lost in your garage? That's time and money you could've spent working on cars.

Rob Siegel Josh, agree to disagree on this one and move on. It's just personal preference and work habits, not a moral issue. +3

Scott Aaron I think I see Jimmy Hoffa's beanie in one of those blue things +1

Joey Hertzberg Rob, you need an unpaid intern or young family member or a smart local kid you can pay for a day a week. This is exactly the type of job (clean and organize my shop) that 1) Young people freaking love and 2) will be so impressive to you that you will swear the kid is a genius. +2

Joey Hertzberg I'm like you... kinda messy, but I am equally if not more at ease and impressed by orderliness. +1

Luther K Brefo Rob Siegel, it's like we're brothers. You saw my shop, but I knew exactly where everything was. ;) +2

Steve Park Rob Siegel I know what you mean. My garage gets crazy at times. However I find the time spent cleaning & organizing to a certain extent reaps long term benefits. It's never something you'll see in a magazine however.

Greg Lewis Nothing like a little English 'engineering' to make early BMW's seem simple and easy to work on. You get a prize when this thing is in the road, which might be a sale and funding another BMW. Could happen. +2

John Lenham I was recently given a "free" 2003 Volvo s60 (yes, I accepted the terms...haul it off and it is yours...) Free because the motor was spread throughout the passenger seat, rear seat and trunk... All that is to say I feel your pain. Just ordered another tube ($40) of the proprietary Volvo sealant used for the oil pan and the cam shaft cover... Also spoke to the previous owner today... "wouldn't you have come across an exhaust manifold in your basement would you...???" +4

Bob Sawtelle Almost an hour and no post from Paul Wegweiser about your messy garage floor, we need to send out a search party, he must be lost or held hostage somewhere. +4

    Paul Wegweiser "Cleanliness saves money!" 😀 +2

    Josh Wyte Paul Wegweiser I agree! +1

    Franklin Alfonso Lolita is gonna fight you all the way Rob. +1

    James Pollaci Read what you wrote but it left me wondering... what kind of tortellini? +1

    Rob Siegel uh, the kind that floats? +1

    James Pollaci Rob Siegel—cheese? Meat ragu? Packaged? Homemade?

    Paul Forbush James Pollaci Had to laugh out loud, I am boiling a pot of tortellini as I read this thread. +2

    Paul Forbush I share your disdain for cleaning, Rob. I cringe when we are slow at work and the boss calls for a cleanup day. +1

John Whetstone Tough crowd. CLEAN crowd! But one of the few perks of a floor that's covered with debris is the fact that the bolt can't travel far. Start dead center where it probably dropped, and work your way out. I know you know this, you're super logical. But you will find it. And of course, check any open header pipes. +2

John Whetstone Is there a Jewish equivalent of Saint Anthony? +1

    Andrew Wilson I'm thinking Saint Rita of the Impossible.

Brian Ach I adjusted my valves, replaced the spark plugs, valve cover gasket, spark plug tube lower and upper seals, PCV valve, and valve cover grommets. Granted that was on a Honda Element, but still, ringing in the new year with an accomplishment. +2

Faith Senie Is that bolt magnetic? If so, run a magnet around the neighborhood where it fell and sort through the random stuff it picks up.

Zack Ricketts Professional garage organizer and reorganizer here... when you have only a one car garage with an open drain to who-the-F knows where, the car, and every tool and piece of lawn and garden equipment jammed into... plus car parts... yeah. No room for error. Literally.

Keith Roth Clean garage or not the activity is supposed to bring a sense of accomplishment versus moving backwards. Get that where-is-the-next-rabbit-hole feeling on that motor.

Paul Wegweiser Weird shit happens when you Google images of Mr Paul Wegweiser. "Here we see a professional technician effortlessly pressing in a front timing cover crankshaft seal ON A BMW which wasn't designed or built by sadists." https://www.shutterstock.com/editorial/search/wegweiser +6

    Mike Kovacs That's too good, Paul. I hope you're getting your piece of the $199 licensing fee!

Andrew Wilson Magnets, yeah, magnets. +1

Bob Sawtelle [weird magnet boy gif] +3

George Zinycz Cars that sit get angry and resentful. Ask me how I know. This is but a page in the story of your European penance... +1

    George Zinycz How the phrase "your European penance" works and is pronounceable is beyond me... +1

Scott Lagrotteria You know the workaround to not having a socket big enough—make right size one of hardwood on a lathe. I've seen your garage & there wasn't a lathe in there. If you give me the dimensions I'll make you one. If interested I can make you what you need within a day. Let me know Rob. All I need are the dimensions. +2

Rob Siegel Thanks Scott. This one is pressed in, but I'll holler if another need arises.

Scott Lagrotteria You misunderstood me Rob, I would make a wood piece to be used to press in the seal you were trying to put in. I saw the pic and your socket was too small; making a piece to do the same task of providing even tension with either a hydraulic press or not (why you need hardwood strength).

Joey Hertzberg Man, as time goes by, I keep getting hung up on you putting the tii motor in there. That would be so cool.

TLC Day 2035: Small parts

JAN 5, 2019, 5:08 PM

The Lotus Chronicles, Day 2035:

The "wide seal" arrived from Dave Bean Engineering. This one bedded in without the twisting problem I had with the narrow one.

The package also included a replacement for the bolt that I'd dropped and lost in the entropy of my garage. But as soon as I looked at it, I realized that they'd sent the wrong one. I was composing a polite but you-screwed-up-suggestive email when I looked at the exploded parts diagram and realized that they'd sent exactly what I'd requested, which was not in fact the one that I'd lost. I'd read them the wrong part off the diagram. D'oh! Fortunately, the lost one was a hardware-store-available part, which I then immediately ran out and procured.

This means that I can start the process of test-fitting, sealing, and installing the outer and inner front timing covers, perhaps as soon as tomorrow. Getting those on will be a major milestone.

Small steps get big jobs done.

65 LIKES, 47 COMMENTS

Karel Jennings Are you going to assemble a British engine in a way that it will not leak oil? +2

    Scott Lagrotteria Karel Jennings Is that possible??

    Karel Jennings :)

Rob Siegel Karel, it's no joke. I could do it the way the factory manual says (dry paper gasket between the block and inner cover, no gasket but just Hylomar between the inner and outer cover), the way the Miles Wilkins Twin-Cam book says (WellSeal on the paper gasket, silicone between the inner and outer cover), a different way that there's some consensus on in the forums, or using Permatex The Right Stuff (which Lindsey Brown turned me onto and I've had great luck with sealing problem leaks) on everything. The problem is that thick beads of sealant are really frowned on, as they squeeze out when the covers are tightened and wind up in the pan. It's not just an academic issue. I pulled caterpillars of them out of the oil screen when I cleaned it. +5

    John Whetstone Rob Siegel this whole thing gives me the heeby-jeebs, but I assume that's your intent with theses projects? Lol

    Brett Kay Might I suggest for sealant you try Yamabond4, available at any Yamaha motorcycle shop or on Amazon. My experience is it far exceeds Right Stuff Rob Siegel +1

    Karel Jennings Rob Siegel I love Right Stuff. Works well and just a thin film will do. Scott Fisher Rob Siegel—if it's the Permatex 2H (I *think*), the stuff that comes in a can with a brush in the cap and looks like half-congealed blood, it is proof against the Forces of Darkness. I used that on my MGB race motor with the cut-to-fit cork STRIPS between the main bearings and the pan gasket. Not a drop. Used the same styon the TD last year, same results.

    About half the problem with British engines and leaks are due to using the wrong stuff in the wrong way: light coat, let it get tacky, then assemble. Most of the other half is due to overtorquing bolts on sheet metal bits. (I had an enthusiastic youngster tighten my valve cover bolts so tight they deformed the seal and oiled down the block.)

The rest… well, the TD has no rear main seal, because the engine was designed in 1914. Yes, First World War era. To keep the oil in the block, there's a reverse thread tapped into the backplate. In other words, the oil stays in because of the Archimedes Screw. +2

Brett Kay Yamabond was recommended to me when I was playing with old British motorcycle engines for sealing case joints. We all know how old British motorcycles leaked oil.

Rob Siegel There are so many opinions on this with the Lotus. I spoke with Ken at Dave Bean who sold me the pump and covers. He recommended WellSeal on everything. It just seems so… thin as compared to laying down something in a bead. +1

Brett Kay Rob Siegel not familiar with wellseal. Do they sell it? I've used yamabond on Volvo 850 cam covers that like to leak and never had one leak afterward for years.

Rob Siegel I already have WellSeal. The Yamabond IS available on Amazon, but through a third party, with several day shipping. Still, I may just put it on order and not rush this. I WOULD prefer to use something that has some history on the Lotus forums, so I'd like to check there. +4

Brett Kay Rob Siegel or if you have a local Yamaha motorcycle dealer they usually stock it

Rob Siegel looks like there are a few +1

Brett Kay I have used on Ford timing covers as well with no leaks. It's about $13 a tube with tax at Yamaha shops. I can tell you it's great on BMW engines too.

Brett Kay Sounds like Wellseal is very similar to Yamabond, non hardening, etc but it does get "stiff" after it sets up.

John Whetstone What's the difference in avg. temperature between the Lotus and a Yamaha?

Brett Kay John Whetstone on a Triumph air-cooled twin I would bet higher than the Lotus.

John Whetstone Really? Wasn't what I was expecting.

Brett Kay John Whetstone the rocker covers on an air cooled Triumph Bonneville get pretty hot. As do the engine cases. Water cooled stuff typically runs cooler than air cooled

Charles Morris I'm partial to Hylomar Blue myself…

Ben Greisler Hylomar is the shiznit. Thin coat on the gasket and let it dry for 5-10 minutes.

Rob Siegel You see the problem… +4

Brett Kay Rob Siegel crystal clear! Lol. I've used almost all of them other than Wellseal. I'll stick to the Yamabond. Wellseal is more expensive from what I found on the net also.

John Whetstone Brett Kay learn something new every day. Way more than that around these parts! +1

Brett Kay John Whetstone We all do. If you ever have need give the Yamabond a try. It's all I ever use anymore. I buy and fix up old beaters usually European stuff. Thank you sir.

Paul Forbush Brett Kay interesting that you use Yamabond on Volvo motors in place of the Volvo "pink stuff" AKA Loctite 518. It must be fairly thin. I have never found anything that performs as well as 518 on Volvo cam covers. Right Stuff is WAY too thick. I wonder how 518 would work on the lotus motor. 518 is meant for applications with very tight tolerances. The Volvo cam covers are machined with and "married" to individual heads.

John Whetstone Brett Kay At almost 62, I leak way more than my vehicles!! +4

Paul Forbush John Whetstone I hear that!

Brett Kay Paul Forbush And they leak in a relatively short amount of time. Is that not the same product used on 911 cases? I've done 3 Volvo's all with YB, and in 4 plus years driving over 100,000 miles after the Yamabond, bone dry. I swear by it. 2 850s and a 960. All high mileage cars in excess of 200,000 miles. Rightstuff I'm not a fan of, its ok for old American cars but I don't like it on European stuff. Yamabond has a different consistency, I use a very small bead and smooth across the surface usually with a disposable glove on using my finger. It's just proven to work for me. I detest oil leaks.

Luther K Brefo This is like the "oil question" on the various forms/groups.

Paul Forbush Brett Kay I have no experience with 911 motors. It sounds like your application technique on the Volvo covers is similar to mine. I have never had a leakage problem using 518, and I have probably done 50 of them over the last 10 years. Yamabond could be my plan b if I can't get the Volvo stuff. I bet I can get it locally at one of the many outboard Marine shops here. +1

Brett Kay Must just be my luck, although my 960 had Permatex on it when I resealed it due to leakage. Somebody else had obviously been in it before. I use the YB due to my experience with

> Old British and Japanese motorcycles. The British bikes are notorious for leaking. YB has worked flawlessly for me. 518 must be a good product as well then. I'll keep that in mind. If they sell Yamaha boat engines they more than likely have it or Amazon sellers have it.
>
> **Brett Kay** Luther K Brefo Many more experts and myths out there about oil. I don't even normally get in that battle, I worked in the automotive oil industry for 9 years in the late 90s to mid-2000s and have been to many R&D labs, Mobil, Penzoil, Quaker State, Etc. But you are right. +1

Paul Wegweiser Ringo Sez: [one of Paul's weird goat-related memes saying "Hates oil in his fur / But would still hit it"] +4

Paul Wegweiser Ringo Also Sez: [another of Paul's weird goat-related memes saying "Shiny things / Put Ringo in the mood to breed." *Author rolls eyes*] +3

Franklin Alfonso Well, I'm glad you are making some progress. Can't wait to read the next piece on the next SNAFU challenge!

Glenn Koerner "Small Steps Get Big Things Done" sounds like a book title. Good luck on the rebuild.

> **Rob Siegel** Thanks Glenn. One of my Hagerty articles on trying to restart the long-dormant Lotus project highlighted the value of doing one thing per day, even if that one thing is just reading about something or ordering one part. But installing one part is better :^) +1

Doug Jacobs Yeah, yeah, yeah—THIS my friend is the sealant you want—and the only one you should accept. [link to Wurth Engine and Housing Sealing Compound DP 300]

Lindsey Brown Once you finish up with Tara King's Lotus, I expect you to start on John Steed's Bentley. +2

> **Bob Sawtelle** Lindsey Brown ooh I was very young but after Emma Peel, Tara King was like her hot young sister. Very nice.

Collin Blackburn In for the electrical fire on the first start. Fucking Lucas electronics. +1

Russell Baden Musta Wow... well... document what you ended up using Rob! Would like to update my TC to stop the progressive oil change! +1

Philip D. Sinner Speaking of the infamous Lotus again, I saw a turbo Esprit at Daytona yesterday. In white, and quite an attention grabber. I didn't get the stoopid fone of my hip in time to snap one, but I really did see it, and it really was moving under its own power.

Carl Witthoft Somehow, a "wide seal" sounds like something Morton Thiokol used even though it wasn't on the assembly diagram. +3

TLC Day 2038: Chain clearance

JAN 7, 2019, 7:00 PM

The Lotus Chronicles, Day 2038:

Not a good day in Lotus Land.

My prior post had to do with trying to select the proper sealant to use between the block and the inner and outer timing covers. Finding very little consensus after asking the manufacturer of the pump kit and reading and posting on several Lotus forums, I asked The Lotus Engine God (a.k.a., TLEG), the guru who I paid to crack-check my engine components 5 1/2 years ago. He gave me yet another opinion (what he kept calling "Silastic," which turns out to be a type of RTV, preceded by a specific and expensive pre-treating primer (Dow 1200RTV) but then said that the choice of sealant was by far the lesser issue, and that the bigger issue was that every one of these aftermarket timing covers he's used has needed some amount of, as he put it, "fettling" to make it fit. He strongly advised that I test-fit EVERYTHING carefully before bolting it up with sealant.

So I did. And almost immediately I discovered that the timing chain hit the inside of the outer cover. Fortunately the problem is completely apparent with the cover off (photo).

I sent the photo to Ken at Dave Bean, from whom I'd bought the kit, and to his credit, he called me almost immediately. We had a lengthy chat, during which he said that these are the first of a new batch of sand-cast covers (part of the 18-month wait), and that he'd stand by them and could test-fit one back at the shop and send it to me if that's what I wanted.

I'm going to try to get out to TLEG's shop tomorrow and get his expert opinion before I do anything. I'm not above taking a Dremel tool to it, if that's part of the standard "fettling," but I don't want to do anything stupid. Or I should say, I don't want to do anything ELSE stupid.

Coincidentally, I just looked on the forum where I'd posed the sealant question, and saw that there's a new response:

"I have done this too many times to count on one hand. I believe I had the first one that Bean made. Or close to it. I had to do mine like 4 times till we figured out a casting issue. This is not any negative thing on Dave Bean's Front Cover. I have bought 2 or 3 since. Anyways it took 4 times of my putting it on to realize that we had a casting hole. Dave replaced it and even gave me some money back. What a great guy. He is missed so much by us all [Dave passed away last year]. What I learned the first time is that you cannot use "the right stuff" on the back plate to block unless you are really fast and put it together really quick. It will set too fast. Then on the second time, I used Permatex RTV grey and in the middle of putting it on, got called to a computer problem call and left it. When I got back about 2 hours later, I should have taken off the rear cover and redid it. I did not. Sure enough when I started the car, it was leaking from behind the rear front cover. Off again. On the third time, I used RTV, got it together but it was leaking around the center bolts. On the 4th time, I put RTV on the back, the front, around the bolts, on mating surfaces where the impeller is, basically everywhere. But upon starting the motor I was still getting antifreeze in my oil. It was then that we realized we had a casting leak. We pressurized the system saw it leaking from a casting hole. Hey, it happens and it was one of the first. Dave sent a new cover to me, with what I seem to remember was a full or close to full refund and I put it all back together the same way. Grey RTV all around. On the back cover, the front edges, the bolts, the seam between the pump impeller flange and backing plate. All went together fine, no leaks and it has been that way for years. Best investment I have made on the car... Gosh, I hated to change that water pump the old way. I am sure I will be buying one for my Europa that I am redoing right now."

(cue Monty Python "But the fourth one stayed up!")

When I announced that my automotive goal for this winter was to get the Lotus engine back together, several people said "Oh, you'll have the car up and running by spring." They were surprised when I said that the odds of that happening are indistinguishable from zero.

I'm as much a prisoner of the familiar as the next guy, but resurrecting Louie, Bertha, and The Lama were child's play next to Lolita. And I'm still on the goddamned front cover.

As I said in my first book, a guy I worked for when I was 13 had a red Lotus Europa. A forty-something guy who also worked for him said "A car like that, you can get SEX out of." It took me decades to understand that he wasn't talking about women digging men who have cool cars; he was talking about the sensation of owning and driving a cool car. That may still be in store for me and Lolita, but this is going to be the longest most delayed Tantric orgasm a car guy's ever had. It had better be good when it, uh, comes.

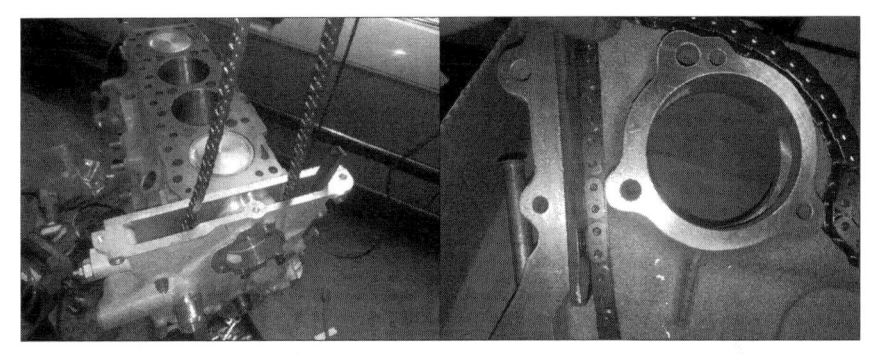

60 LIKES, 62 COMMENTS

Joe Eaton FYI—There are never good days in Lotus Land. Hours maybe. Not entire days. Now stop bothering Ken so he can get to shipping my parts. 😊 +1

John Whetstone I found a typo!! 5 ½! +1

Joey Hertzberg That was cute there at the end

Sean Curry I suspect the "get sex out of" comment has something more to do with being screwed regularly. ;) :D +10

John Whetstone Tell me I'm right?!

> **Rob Siegel** John, 5 1/2 years should be right. I bought the car June 10th, 2013.
>
> **John Whetstone** Not a commentary on a stalled project, but how fast my years are flying by!
>
> **Rob Siegel** I know… which is why I want the engine assembled.

Chuck Vossler You must get pleasure out of torturing your soul with these things, no? +1

Steven Bauer Sounds like you also need a way to pressurize the oil and cooling systems before you take the engine off the stand.. +1

> **Rob Siegel** at least the cooling system +1

Jeff F Hollen This is why I dread doing the valve adjustment on the 907 in the Jensen… those cam trays never seal.

> **Rob Siegel** at least that's at the top and is just flung oil
>
> **Jeff F Hollen** I have to re-valve the original engine to the car eventually. I feel I'll have a similar story to yours when I do 😂😂😂
>
> **Rob Siegel** good. misery enjoys company :^) +3

George Saylor Back in the day about all we had was the dark green Permatex stuff for a sealant. Once the blue RTV came out it was at least a hell of a lot easier dealing with the cork cam cover gaskets. +1

Scott Aaron This whole thing sounds really sucky, man. You might have to throw that cover in the neighbor's pool, grab a guitar, flip on the amp and wail.

> **Rob Siegel** looks like I picked the wrong week to quit sniffing WellSeal… +9
>
> **Scott Aaron** Rob Siegel totally.

Daniel Senie Book title is morphing before our eyes. "Lolita, the Lotus that Broke My Soul."

Brad Purvis And here you thought Alfas were a PITA. 😊 +1

> **Rob Siegel** ah, those were the days +2

Doug Jacobs Or he was yet another stone-cold nuts British car driver and he was actually fucking the car.

David Kemether

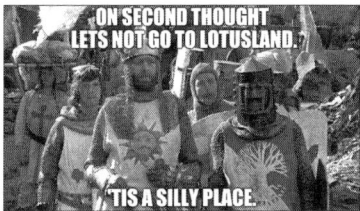

+12

Paul Wegweiser It's amazing how appropriate this meme is. I've posted it at least 4 times in the last few days. Now I realize I had been wasting it, when I should have just saved it for you alone. I'm sorry to have shared something that was truly meant for you, you sick twisted self-flagellating fuck. -Hugs and Kisses, Me.

> Your fetishes are nothing to be ashamed about. Unless your fetish is being humiliated, then you should be very ashamed you nasty little pervert.

+10

Jeffrey Miner The Monty Python line is priceless! And I never realized how universally true and applicable it is... I pulled the exhaust and the driveshaft on my son's son's E30 three times this summer before the rear transmission seal finally took ... a new output shaft finally did the trick. Not the fourth but damn ... how do real mechanics make a living doing this??!!

Rob Siegel wait... real mechanics make a living? +4

Ryan Klug Next time you're looking to seal something up, check out what all the motorcycle/moped guys use, Yamabond... so far, everything I've used it on I've seen awesome results that hold up great...

Rob Siegel ordered a tube this morning

Doug Jacobs Ryan I'm telling you, Wurth Dp-300 is much better than Yamabond.

Paul Wegweiser It takes a lot for me to say this, especially since I just had to perform the ceremonial "annual caliper replacement ritual" (and this time the bonus brake hose replacement) at 248,000 miles, but here it is: I WOULD RATHER WORK ON MY 20 YEAR OLD SUBARU STATION WAGON (OUTSIDE. ON GRAVEL. IN JANUARY) THAN ENDURE THE TORMENT OF LOTUS WORK. Wow. That felt cleansing. Thanks for helping me put things in perspective. [photo of incredibly worn-out brake pads] +14

Rob Siegel As I've said for years, I provide this service free of charge. The Hack Mechanic. Living on the edge. So you don't have to. +3

Rob Siegel But seriously folks... If this careful assembly unearths problems and results in a decent-running, largely leak-free engine, I'm good. TLEG warned me about all this years ago, that there are a lot of small things that can get you in Twin-Cam engine assembly, which is why he wouldn't quote me a price. I really didn't expect that, in going with the cartridge-style water pump, which necessitates these custom timing covers, I was going to trade off an easily swappable water pump for creating even more problems with assembly. But who knows? The original covers were so corroded in place that I had to pull them and the pump off with a fucking slidehammer, and worked so hard at it that I literally thought my right arm needed physical therapy for repetitive motion damage.

Totally not kidding about the fucking slidehammer.

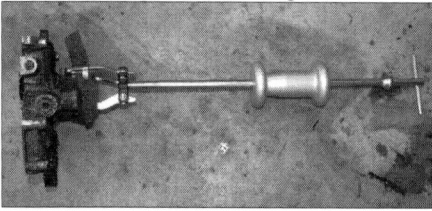

1

Paul Wegweiser Rob Siegel ...too easy. +3

Rob Siegel oh I know

Rob Siegel but this IS a family show :^) +2

Jeremy Novak Says the guy making orgasm jokes. +3

Kevin Taylor It's the sitting in the Lotus position while working on the car that sucks.

Steve Woodard I spent most of my professional life working with suppliers of aerospace castings. It's not too difficult to make an object that looks just like the casting you want, but it's much more of an issue to make one that will pass FPI, x-ray and pressure testing. One of the more entertaining jobs I had during the last few years of my tenure was working with an automotive casting supplier that was trying to do aerospace stuff. You had to be there. +4

Jeffrey Miner Steve Woodard I guess that explains Porsche's first gen Boxer engine casting failures in the late 90's after adopting Kaizan manufacturing practices: pour the oil in the top and watch it leak right through. Everything is harder than it looks. 🙂 +1

Steve Woodard It's significant that doing all those sorts of inspections, with the attendant weld repair and/or scrap, adds lots (and lots) of expense. In the automotive world, it usually doesn't matter (present company excepted), but when the casting has to fly, it does.

Ricky Marrero That will self-clearance itself in a few hundred miles. +4

Rob Siegel and not in a good way +2

Ricky Marrero Rob, self-clearance never ends well... +1

Frank Mummolo And let me get this straight... Lotus actually presented itself to the world as a "company"?? As opposed to a "in my basement hobby shop"?? F$&king amazing. I am thanking the sweet baby Jesus that with all the wacky cars I've owned (and I'm talkin' Morgan, E-Jag, Sunbeam... that sh$t) I somehow had the presence of mind to turn a hard right when I heard "Lotus". And I heard it quite a bit in my youth. There IS a God. Thank you! +1

John Graham As my Elise owner friend would say 'Lotus... Lots Of Trouble Usually Serious'... +1

Daniel Senie When the Lotus is finally done, it needs to be equipped with a set of tubes that blow yellow feathers out behind when you mash the throttle. +4

Franklin Alfonso It's a LOTUS! It will NEVER BE DONE!!! LOL. It may eventually be resuscitated for a time after which Rob will be done... +4

Samantha Lewis What can one say about machines designed by men in brown lab coats, in sheds, and built by sons of turnip farmers from Norwich? They are the antithesis of everything a German engineer would do, they have no right to be so damn seductive, but that's the truth of it. I hope that you have that perfect day, when everything works, and then all this torture won't matter one little bit. Onward. +3

Robert Boynton That "Perfect Day" might also be a good day to consider selling it — before the stress of "What could possibly go wrong now?" has a chance to consume Rob or is answered by the dark sound of a 10-cent broken herkomdizer deep inside its core. Besides, after a brief moment of innocent bliss, what would be the point of owning a car that's all DONE?

Rob Siegel I don't want "done." Right now I don't even want "running." I'll settle for "engine internals not exposed to the atmosphere inside a trash bag." +3

Scott Lagrotteria These problems make converting to electric power a little less blasphemous. +1

Franklin Alfonso I believe this car may be more than a book. You may get a novel out of it. "War and Pieces." +4

John McFadden More like "Warring Pieces" 😀 +2

Joey Hertzberg A poem: Yet, like a siren, she sings to you, a Teutonic lullaby from under the deck. +1

Jim Gerock My Lotus Land. She sometimes leaks past the litter box when ancy. [photo of cat. Author note: unusual for Jim. He's usually so linear... :^)]

Andrew Wilson So this is very similar to what I call "Fuzzy Carpentry" nothing is square, plumb or level. It's not easy to do it unless you know how to and it's a wonderful feeling when it all comes together. 😀 +2

Derek Soltes It's an astrophysical attraction for me: I'd relish a project like that, and the time and space with which to attempt to complete it.

TLC Day 2040: A replacement cover

JAN 9, 2019, 10:06 PM

The Lotus Chronicles, Day 2040:

Two days ago, during test-fitting, I found that the timing chain hit part of the custom front timing cover I'd been waiting nearly a year for. I called the supplier, Dave Bean Engineering, and they were very contrite and offered to send me another cover. But instead, yesterday I took the pump and covers over to the gentleman I refer to as The Lotus Engine God (TLEG) to get his read on the whole thing. He didn't think that the lack of clearance of the timing chain was a big deal—that any aftermarket front cover would require a bit of, as he called it, "fettling." He was more concerned that the o-rings on the cartridge-style water pump had nowhere to compress, and that the edges on the cover in front of where they slid in needed to be chamfered or beveled so they wouldn't rip up the o-rings. I took all that to Bean yesterday with the gentle question "am I the first one to install this new kit?" The answer was "no, we've sold a bunch of them and haven't had complaints," but they agreed to have a look at both issues.

We revisited all this on the phone today. The chamfering may be a minor issue (they said they had no problem fitting the o-rings but would be glad to chamfer the edges of the new cover they're sending me), but regarding the lack of clearance for the timing chain, what yesterday had been "No, we've sold a bunch of these without complaint," today was (and I paraphrase) "Yes, we see the timing chain clearance problem, we're machining more clearance into all the ones we have in stock, including the new one we're sending you… and we wonder why we haven't had complaints about this."

So, big thanks to Dave Bean Engineering for stepping up and supporting their product, and to The Lotus Engine God (Bill McCurdy at Williams Racing in Harvard MA) for his advice to test-fit EVERYTHING before installing it, and spending time with me yesterday.

One step forward, two steps back, one step forward. In the same place, but smarter. That's totally still progress.

63 LIKES, 38 COMMENTS

Daniel Senie They haven't had any complaints because everyone else who ordered the part has their Lotus stacked under everything else in their garages too? +16

> **Rob Siegel** is possible, yes +1

> **Scott Aaron** Daniel Senie EXACTLY

Bill Snider Pretty sure, Smarter merits its own step forward. Glad you have great resources and that DBE is stepping well up. Remember to breathe ;-) +3

Daniel Neal Probably has more to do with the fact that the vast majority of people would find nothing wrong with that because "if it's not hitting then it must be designed that way," something about assumptions in there as well. +1

Luther K Brefo Even BMW has engines that self-clearance. Somewhere I had an M60 lower timing cover that had chain marks well on its way to a bolt… 👀

Luther K Brefo That said glad it's being resolved. +1

Marshall Garrison This is almost like watching stop-motion of a clay-mation changing of your "Lotus Position"! +5

Joshua Weinstein That is progress. And progress is the name of the game!

Scott Linn Great. Just remember this isn't a standard part so problems might be expected, however in the end it will be much better than the original. VERY good advice on trial fitting. I have to remind myself to do that all the time. +2

Mel Green God bless the experts we know… +2

Franklin Alfonso I really hope the sex you get out of Lolita after all this is done and she's good to go is, "worth it" lol 😊 +2

Alan Hunter Johansson "Fettling." Clearly, Lotus is his first language! +1

> **Gene Kulyk** Fettling required a quick trip to Urban Dictionary: giving something a good clean. (yorkshire/northern english), not to be confused with furtling, which is something else altogether. +1

> **Rob Siegel** yes, I found that!

> **Gene Kulyk** Rob Siegel so as you fettle the timing chain cover you can furtle a picture of it?

> **Alan Hunter Johansson** Rob Siegel You kids and your Urban Dictionary. Get a real dictionary! I had only heard fettle used as a verb to mean something like "mess with to get to work right," which seems to coincide well with the OED definition above. I first heard it, I think, and appropriately enough, when my brother's college roommate was talking about working on the SUs on his MkII Jaguar. I'm embarrassed to say that while I was familiar with the noun, from hearing of things "in fine fettle," I'd never seen the connection with the verb. And yes, I have a full set of the OED in my home library. [photos of rack of OED on wall and page from OED] +2

> **Gene Kulyk** Alan Hunter Johansson I have the miniature version however, I was hoping for something off-color from UD. More in character with this Facebook account. +2

Alan Hunter Johansson Gene Kulyk True...

Alan Hunter Johansson Gene Kulyk And I grew up with the miniature version (with its little magnifying glass in the drawer), but when I mentioned to my not-yet-wife Debbie that I wanted a copy of my own, this was a present to me our first Christmas. I married her in short order. You don't let a catch like that get away! +2

Lee Highsmith Project Binky has used the word "fettle" a number of times. I inferred the meaning, but amused it can be found in the OED. +2

Steve Park I think I'm grateful I don't fit in a Europa. +3

Greg Lewis Reminds me of the Disney song "Someday my Prince will come." No one can ever say that you don't work for it. +1

Jay Ford Finagling is yet another term...

Tom Kinney Well fwiw, this is a reminder that the German stuff is really quite good, I was doing some engine repairs on a Norton Manx several years ago and to say that fettling was required was an understatement. I think Carroll Shelby was quoted as saying "assembled by winos, under a bridge." +2

Rob Siegel Yes. Totally. When I put on a 2002 lower or upper timing cover, I never worry whether it will fit in the sense of holes lining up or shit hitting it.

Scott Chamberlain After having pulled out most of my remaining hair, I realized that my remaining life was too short to spend on British cars.

Rob Siegel In my first book, I refer to buying my Triumph GT6+, then never having another British car (this was pre-Lotus) as "proof that men can, in fact, be taught."

Scott Chamberlain My wife, it seems is a hopeless Anglophile, having had a Sunbeam Alpine as a first car. I was cured by my Triumph Spitfire. She's been through two MINIs and a Jaguar, until we final put her into British car rehab. She now drives a new Miata, and is none the worse for her experience, other than 25 years of never-ending repair bills. +1

Scott Chamberlain Even as an impressionable, romantic youth, I knew that Lotus meant trouble. A rich law school classmate had a Europa Twin Cam. Took three rebuilds of his 20k engine to get it running again. Back then, there were quite a few Loti running around Baton Rouge, a college town. In their day, they would run away and hide from everything on an autocross course—as long as you keep feeding them a constant diet of attention and money. +1

Rob Siegel Yeah... when I bought mine, 24k miles, seized engine, I thought that I'd have one of the lowest-mileage most original Europas in the country. I soon learned that there are no shortage of low-mileage Europas, that between the water pumps and the valve guides, engine trouble at 20k is pretty common. Then, when I opened my engine up, I found that it had already been rebuilt once. Yeesh. +2

Scott Chamberlain Long after GM car were computerized, and built by robots, Loti were pretty much hand-built. Not in an exoticar way, but more like 'arts and crafts.' There was a series II Elan that lived in my father in paw's barn for a decade. It was, essentially, a pre-assembled kit car. They were pretty much race car engineering, with cottage industry cars attached to it. At the time we thought of them as hand grenades; explosive, but short lived. +3

Marc Schatell Rob Siegel my ownership of a '66 Land Rover in the mid-eighties taught me this lesson +1

Jeffrey Miner Scott Chamberlain my mother had one of those! (Sunbeam) I still recall her routine pulling it in the garage, turning off the key, and the car continuing to run. Then, blowing the horn, to shut off the engine. I was four years old and knew that was very, very wrong. 🙂 +1

Franklin Alfonso But entertaining lol.

Scott Chamberlain I feel deprived! All we had was a Chevy Nomad, with a stop sign for a passenger floorboard.

Steven Bauer It occurred to me that you may be the only customer who called with an issue because you may be the only one who actually tried installing the part.

TLC Day 2044: The first test fit

JAN 13, 2019, 9:09 AM

The Lotus Chronicles, Day 2044:

The test-assembly of the engine continued. I test-fit the head gasket and valve cover gasket, as well as the Bean water pump in the bore of the new timing cover. Then, continuing to follow Miles Wilkins' "The Lotus Twin-Cam Engine" book," I bought some 7/16-14" threaded rod and used it to make two head studs (the engine doesn't have dowels to hold the head in place like a BMW motor). I was excited at the prospect of letting the head and block kiss for the first time in 5 1/2 years. I unwrapped the fully-assembled head that's been sitting in my basement for nearly three years and tried to lift it onto the studs. The idea was to test-fit EVERYTHING in the timing chain path, including the cam gears, idler gear, and tensioning plunger, and make sure there are no other clearance problems with the chain other than the one I unearthed last week. But the head was heavier than I expected, and I couldn't directly verify by sight that the tops of the studs were going into the bolt holes. I had it lined up, but then while lowering it down, it cocked and wouldn't go down further. With a lot of effort, I pulled it up, to find little aluminum shavings all over the top of the head gasket from where the threaded rods had chafed the edges of the bolt holes.

So THAT was exciting.

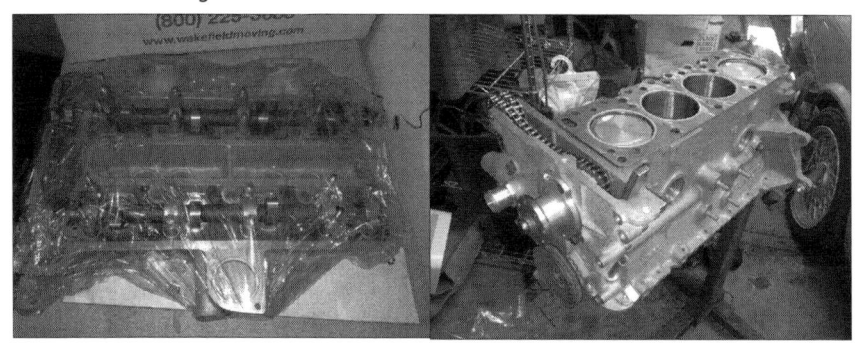

(49 LIKES, 37 COMMENTS)

Marc Schatell So, next step takes two people?
  Rob Siegel Either that, or use my Warn PulzAll electric winch suspended from the ceiling like I do for in-car BMW head pulls and installs +1
Nicholas Mav 😊 +1
Dale Robert Olson Take the rods and grind the threads down a bit, more clearance and no shavings. +2
  John Whetstone Dale Robert Olson or hardwood dowels? +2
Russell Baden Musta So THATS what a TC looks like without all the oil on the outside! +6
Paul Wegweiser "Writes ten thousand books in 4 years. Owns 300 cars. Manages to store ALL of them indoors. Still can't lift a 4 cylinder head." :P (By the way, I did exactly that yesterday with carbs and manifolds attached on "The Fruit Bat".) It was my way of saying :F^&**k YOU, Hernia!!!! +16
Ian Sights I'm a big fan of mechanical assist vs helping hands if it's feasible. The ability to hold things still in close proximity while allowing full hands-free inspection is a big plus. +2
William Jeffers Cut the heads off a couple 7/16-14 bolts so you only have an unthreaded part above the block +2
Michael Ho Channel the power of "The Lotus Engine God." +1
Bob Sawtelle Hehh hehhh "head" hehh "studs" hehh +3
Franklin Alfonso Who knew Lolita didn't like getting head? She is gonna fight you all the way lol.
Scott Lagrotteria Just wrap the threaded studs with Teflon tape. It's cheap, won't affect clearance, and

protects the head from the threads. +4

Kent Carlos Everett Beauuutiful.

Brett Kay Is it possible to set the head on then put the studs down through? Of course you would have to be sure all is lined up. Just a thought. Studs can be a major pita. Lol

> **Rob Siegel** The advantage of the studs is that they keep the gasket in place while you're lining up the head. For test-fitting, I may simply cut some very short studs and that may solve the problem. But for the real fit, if they're short, you can't unscrew them and pull them out. +1

Brett Kay Rob Siegel I get that. Or use a couple pieces of unthreaded round stock that is close to the diameter of the threaded. One on each opposite corner would be easy to pull out once you have a couple studs in and started. It works. Just a thought. +1

Rob Siegel If it's magnetic, I should be able to pull it up with a magnet.

Brett Kay Rob Siegel yes ! Or cut the unthreaded stock long enough to simply grab and slide out.

Rob Siegel The length itself was actually a separate problem I didn't go into. The timing chain guide at the right side of the front of the engine in the photo is at an angle, meaning you have to bring the head in at an angle, which was a contributing factor to it getting stuck on the threaded studs. Not sure how short they need to be to not interfere. +1

Brett Kay Rob Siegel Oh ok. Gotcha. Lotus ! Lol Good luck.

Drew Lagravinese I think you have these misadventures just so you have something to write about. 😊 +2

Rob Siegel What, you just figured this out NOW? +5

> **Drew Lagravinese** Rob Siegel I'm on the west coast. A bit slow out here. +3

Scott Aaron That sounds sucky my friend. +1

Jake Metz If you had a tv show this whole process would have been complete in 30 minutes... sounds like you need to switch mediums. But in seriousness...I'm sorry to hear about these lotus shenanigans. +1

> **Rob Siegel** Ah yes, the "solve it by having a cable show" method. Throw lots of money at it, get whatever experts you need, spend weeks filming it, and condense it into 20 minutes. I'll try that next. +3
>
> **Jake Metz** Rob Siegel I'm here to help. +1
>
> **Rob Siegel** "Tune in again for 'Rob Siegel—Real Agony in Real Time'" +2

Scott Winfield Oy! Good luck and wow, what a nice project. Something about the sound of the word, Lotus! +1

> **Rob Siegel** Scott, this is true. What is also true is how much, uh, mileage I've gotten as a writer out of having a dead Lotus. +2

Scott Winfield That's funny Rob. A toast your someday actually driving it! +1

Clifford Kelly Gotta be a rush when it's finished and you start it for the first time! +2

Dave Gerwig A try out for Wheeler Dealers is in your future! +1

> **Rob Siegel** NEVER!

Dave Gerwig Rob Siegel maybe for one guest resto?

TLC Day 2078: Squishing the cork

FEB 16, 2019, 12:04 PM

The Lotus Chronicles, Day 2078:

Can I bounce something off my engine-building friends? I'm trying to decide how much I need to machine off the top of my Lotus engine's front timing covers to fit the cork gasket that sits between the head and block. It's not like a BMW head gasket where the head gasket extends around the timing chain. They're two separate gaskets. (See photo. Copper gasket shown in place. Cork gasket would go in front of it over the timing case.)

– I've measured the thickness of the layered copper head gasket I'll be using, and it's about 0.04".

– I've test-fit the head onto the block and head gasket, resting un-tightened, and the gap at the front of the engine (pictured) is about 0.046". This is the space that the cork gasket has to fit into. The gap will obviously get smaller as the head is torqued down.

– The thickness of the cork gasket is 0.095". If I assume the cork gasket will compress by 50%, there's essentially zero clearance. This is likely because the block surface was decked during machining to clean it up, but that was years ago; I didn't have the timing covers yet to have them machined at the same time. So it looks like I DO need to take a little off the top of the covers.

– If I start to make some assumptions about how much the layered copper head gasket will compress, it gets complicated enough that I put it in a spreadsheet. If I assume that the copper head gasket will compress, just to pick a number, 30%, the numbers look like this:

head gasket thickness (uncompressed) 0.04
compression (percent) 0.3
predicted compressed thickness 0.028
delta 0.012

measured clearance from head to chest 0.046
predicted compressed clearance 0.034

cork gasket thickness 0.095
squeeze (percent) 0.5
resulting thickness 0.0475

recommended cut 0.0135

So this would say that if the head gasket compresses by 30% and the cork by 50%, I'd need to machine off about 13 thousandths off the timing covers. If the head gasket DOESN'T compress by that much, that's fine because then the cork doesn't need to compress by 50%.

Am I doing this correctly?

17 LIKES, 60 COMMENTS

Alan Hunter Johansson My brain hurts. +3

Rob Blake Like I say to the barber… "A little off da top." +1

Tom Kinney Hi Rob, imho, I don't think you need to worry about it, the cork will easily accommodate another 10-13 thou, without a milling machine you could well make a mess of it, or with a mill for that matter. If you must mess with it, try getting different thickness cork material and make a gasket. +4

  George Zinycz Rob Siegel ^^^ this approach makes waaaay more sense to me, FWIW (not much).

Move Mohammed, not the mountain.

Rob Siegel Yeah, but I'm not sure that the numbers bear it out. It's not that the cork gasket can't accommodate 13 thou compression, it's that that would be ANOTHER 13 thou after it's already compressed 50%, and with zero room for error on any of the other measurements. This cork gasket has a reputation for squishing out of its gap if it's too tight.

Rob Siegel oh, you meant finding a thinner one... the only vendor I can find who sells them is in Australia.

Rob Siegel I wouldn't file it or machine it; it'd take it to a machine shop.

Tom Kinney Rob Siegel Autozone has plenty of cork material. Or wally world. +1

John Rettie Tony Ingram might be able to help. He lives around the corner from me and builds Twin-Cam engines. http://lotus7.com/Home.html +1

John McFadden I once rebuilt a Revell "Invisible V8" model that I found on the floor of my brother's closet in the early 70s to use as a sixth grade science project. Does that count? I'll show myself out... +3

Rob Siegel but did you put the visible V8 inside the visible head? (which I had when I was a kid) https://www.scalemates.com/.../renwal-805-visible-head... +4

Jeff F Hollen omg I had a visible rotary kit as a kid. It was great coz you could see the apex seals fail +6

Robert Myles Jeff F Hollen, of course the visible rotary made the same horrible noises the real one made, and had to be started in the paddock spot next you at 0600 and revved endlessly... #rotardflashbacks +2

Doug Jacobs I'd go with about 4 inches. +2

David Ibsen You need to account for thermal expansion too. If you're worried, get some Plastigauge, and get a base measurement from there. Predicting cork thickness? No... I'll show myself out too...

Ernie Peters Rob, I used the stock cork gasket when reassembling my Elan. I remember using Wellseal to hold it in place. I followed the instructions in the Buckland book to use 5/16" rods to prevent the head from fully seating whilst aligning the breather tube and to help keep that cork gasket in place. And I took my time and carefully installed the three bolts in the front of the timing case to help keep the gasket in place. Of course my block had no machine work done to it, so YMMV. I am tagging Mark Doubet, perhaps he has some thoughts.

+3

Rob Siegel Hey, Ernie, do you have a source for that side-mount engine stand mount? I currently have mine mounted through the bell housing bolt holes, and that's going to make it difficult to mount the flywheel.

Dale Robert Olson Rob Siegel, didn't you get a welder a while back? That looks like the perfect first time fabrication project to take on, an adapter for your current stand. And of course once you have it and are done with it, you guarantee never having to use it again until you lose it and can't find it. +2

Ernie Peters Rob Siegel Rob, I made it. Went to a local metal supply place and got the flat steel and round tube steel (to fit my engine stand). My neighbor migged it for me after I drilled the holes in the plate to match up with the motor mount holes. +2

Franklin Alfonso You will have a better understanding of all this once you have rebuilt the Lotus a couple more times... [gif of piles of hundred dollar bills being thrown into empty engine compartment] +6

George Zinycz Franklin Alfonso But will any of it make more sense?!?

Robert Myles Franklin Alfonso how did you get that footage of me working on my racecar?

Carl Witthoft Some Google-surfing suggests that cork can compress to 50% without losing any elasticity. Presumably it can be compressed even a bit more so long as you don't need full rebound. // I suppose it would be worse than Franklin's suggestion to point out that if you over-grind, you just stick TWO gaskets in there.

JohnElder Robison I don't think you need to machine it at all. The head gasket wont compress much at all, and the cork in front will. There is no engine I know with a step up or down at the front cover. +1

> **Rob Siegel** Thanks John. But most head gaskets I'm familiar with wrap around the cover. This is the only one I've ever worked on where there's a separate head gasket and timing case gasket.

> **JohnElder Robison** Rob Siegel We see that pretty often. It facilitates front cover repair without removing the head

> **Rob Siegel** My mistake. This is the danger of not being a professional and drawing inferences from my small statistical samples.

> **JohnElder Robison** BMW does that now, as does Mercedes. Head gaskets really don't compress and you have to assemble the front cover to the block, and then bolt the head down. But with a separate rubber gasket in front, the cover can be removed and replaced on its own. That is why it's a different compressible material.

> **Neil M. Fennessey** Rob Siegel through the 60s and into the 70s Porsche used a cork valve cover gasket. I think that the problem with them leaking was the the DIYers over torqued the 8x1.25 mm nuts. 5 ft-lbs is the spec but probably didn't "feel" tight enough.

John Oglesby When I rebuilt my Twin-Cam it had a decked block and the stock front cover. I just put in the cork and didn't worry about it. Everything was fine. The whole cork gasket thing is so ill-defined in this engine you are way over analyzing this gasket. The whole idea of the cork was to allow for the ill-defined tolerances. You are working with way too many significant digits in your analysis. +3

> **Rob Siegel** Over-analyze things? MOI??

> **John Oglesby** Vous. +1

> **Alan Hunter Johansson** Rob Siegel Toi +1

> **George Zinycz** John Oglesby Moi aussi... It's in one's nature.

Scott Winfield Good luck, Rob! +1

Nicholas Bedworth See, everyone is overlooking the necessity of consulting with a competent astrologer regarding the exact date and appropriate incantations for the affixing of the cork gasket. Must be some shamans in the area as well. Venerating Lotuses goes millennia. Just saying... +4

Andy Veedub Rob Siegel problem solved. Ultra grey "high torque" silicone. Place a bead around the flange and just bolt the head down. Make sure surfaces perfectly oil free. Will never leak like a cork gasket. [photo of Permatex Ultra Gray Rigid High Torque RTV Silicone Gasket Maker] +1

> **Brad Purvis** Andy Veedub But then it would lose its Britishnessness.

> **Andy Veedub** Brad Purvis you'd never notice it in place. I don't miss that type of Britishness. It can still overheat as per the British norm... that will make him feel better. 😄 +1

Rick Roudebush I would make a cork gasket from slightly thinner cork material and place thin coat of orange high heat silicone on both sides of the thinner cork material and fly with it.

Paul Wegweiser It's British. Assemble it. Watch it leak profusely. Sell it 6 days later. +6

Thomas Jones I like the above idea of using good bead of Loctite Ultra Grey instead of the cork gasket. I'd otherwise measure the thickness of the block and head, compare them to the lower and upper covers, and have the covers cut to match the block and head thickness. But... That's likely me thinking Germanicly, the Brit way of doing it likely makes no sense to me.

Scott Aaron Oh goody. Another episode of "Super Precise Anal Retentive Worrier Mechanic." 😄 +1

Ed MacVaugh The 1970s called. They want their cork back. I'd go with any of the modern "form a gasket" materials. Saves mathematical calculations, too. No machine shop bill, either.

Blair Easthom Unlike an M10 engine, the Lotus front cover is not dowelled to the block. It will move around somewhat if stressed, such as during gasket compression or heat/cool cycles. We used the stock cork with copious amounts of Hylomar. Worked fine for twenty more years. No machining required.

David Weisman Or double the copper gasket?

Kevin McGrath "The Lotus Chronicles, Day 2078:" Rob. So, when are you just going to punt and go Star Date? Day 2078 sounds like Papillon. +1

Scott Lagrotteria Wouldn't this be a non-issue if you use a more modern silicon sealant instead of cork? You wouldn't have to worry about compression as the silicon is much more versatile, doesn't heat age, and overall seals better. I assume it's only an oil barrier.

> **Rob Siegel** I'm very hesitant to do anything that people who know more than I do haven't done. I don't see posts on the Lotus forums about people deleting any of the cork gaskets in favor of

sealant. Sealant is used to coat the cork gasket on one or both sides depending which gasket it is.

Paul Garnier Rob Siegel remember you're also submerged in a world of tradition. I bet most of the Classic Lotus-heads favor original vs contemporary solutions (hence the cork) Cork? really? You know screw cap wine is better.

Bill Mann I've been surprised how sealant is used today. My VW has its oil pan solely sealant gasketed as well as the differential cover on my Ram truck.

John Thomas I'd measure the thickness of the metal rings. It's not gonna compress much, if any, more than that.

Scott Linn Going with the experts on something like this would be prudent. Twin-Cams don't grow on trees. +1

TLC Day 2079: Tight sprockets

FEB 17, 2019, 2:16 PM

The Lotus Chronicles, Day 2079:

Spent hours in the garage doing a test-fit of the head (unsolved gasket issues notwithstanding) with the timing chain wrapped around both cam sprockets and the chain tensioner for the first time to check for clearance and any other fitment issues in the entire chain run. Even the issue of the cam timing notwithstanding (the marks don't line up nearly as well as I'd like), I simply can't get both sprockets on their cams when the chain is on them. There's not enough slack in the chain. It's possible that it's simply a combination of the chain being new and there not being enough play until the head is torqued down (and the too-thick front cover cork gasket, but boy I don't want to be wrong about it because the chain is captured by the front covers and the head; once they're installed, there's no getting the chain out. Think I'll take a break, then pull it all back apart, compare the new chain with the original one, make sure they're the same number of links, and maybe try test-fitting it with the original chain and without the cork gasket.

Lotus: Lots Of Trouble Usually SERIOUSLY GET ME OUT OF THIS FUCKING THING I REMOVED REBUILT AND REINSTALLED THE LAMA'S ENTIRE HEAD IN THE TIME IT'S TAKING ME TO DO A FUCKING TEST FIT

This will be the car that drives me to hobby desertion. Sigh.

Maybe I'll just look at wagons for the next several hours. Happy place. Happy place. Happy place.

73 LIKES, 78 COMMENTS

Chris Mahalick I feel your pain. I just want mine out of the garage and done.

Paul Wegweiser "...all the parts that fall off that car will be of the finest British craftsmanship." +10

Kai Marx No timing chains with a Mazda 13B!! +2

Chris Raia Could you fashion a wood "dummy sprocket" just a smidge smaller than the real thing, and use it on the "other" cam, then test fit each cam individually?

Bob Coffey Relaaaaxx [photo of a kayak on top of a BMW wagon] +1

Carl Witthoft Ya sure the tensioner is fully relaxed? +2

> **Rob Siegel** That's funny, but now I'm totally going to have to check that. +3

Carl Witthoft Wait, I know—you've got the two cams swapped in their mounts.

Daniel Neal This further solidifies why I won't own something Old and British.

> **Rob Siegel** I didn't, For 38 years, Then I fell off the wagon
>
> **Roy Richard** You didn't fall off the wagon Rob Siegel. The wagon fell on you. +2

Harry Bonkosky

[image: meme of a man yelling "SERENITY NOW!!!"]

+6

> **Chris Lordan** Harry—you beat me to the punch! +2

Doug Jacobs Sell the fucking thing to someone who cares. +1

> **Doug Jacobs** When it's not fun, it's no fun.
>
> **Rob Siegel** Doug, I'd take a huge bath on it if the engines' out. With the stressed-member transaxle, without the drivetrain in it, it's not a "roller." My goal is to get the drivetrain in and then make some decisions. +3
>
> **Russell Baden Musta** Rob Siegel and don't make any rash decisions until consulting me personally. You will be fine. Move forward slowly. Be the engine... I'm not talking now... shhhh... not talking... +4
>
> **Frank Mummolo** Be the ball... +2
>
> **David Ibsen** Nananananananah... Bububububububuh... A donut without a hole is a Danish, a Lotus without an engine is very quiet.
>
> **Doug Jacobs** And more useful.
>
> **David Holzman** I'm rooting for you Rob!

Gary Beck Similar to BMW S38. You have to "load" chain in proper direction and check timing marks then put tensioner in. Sometimes the marks aren't perfect because of Head and block milling. You can always make cam sprockets adjustable like I did. Get all the power. Using a degree wheel and set the separation. +1

> **Rob Siegel** Gary, yes, but there's something else going on. +1
>
> **George Zinycz** Rob Siegel you understood that unintelligible mess?
>
> **George Zinycz** Gary Beck I've just learned how much I do not know about engines. It's clearly a good thing that I stay out of them (as well as transmissions).
>
> **Gary Beck** George Zinycz, I also stay out of transmissions. To many special tools for BMW and other modern cars but give me an old 4 speed from a Chevy or Ford and easy, easy. +1

Steve Woodard It's the most fun for me when I'm doing stuff for the second and subsequent times. Put another way, working on cars that you know well is a piece of cake, relative to doing new stuff. My old MGB (as well of those of my roommates) taught me to get the engine/transmission out and back in in very short order. Then my Datsun 510/240Z era involved stuff that I could (or had to) do over and over again. It was fun. That was then, and this is now. Anyway, good luck. +1

> **Rob Siegel** It's a trade-off. New stuff is fun; you can feel the neurons making new connections. But this is over the line.
>
> **Frank Mummolo** Rob Siegel hey, thanks for helping me decide! I'm getting you a hair shirt for Christmas! Maybe throw in a self-flagellating whip, too!
>
> **George Zinycz** Steve Woodard someone please meme Rob into that shot of Ethan Hawks with the barbed wire corset in First Reformed!

Scott Winfield I know it's easy for me to say and when I'm in the middle of trouble I don't feel so good either. You'll get it right. Hell, there's something about the name, Lotus, in the car world that will make it all worth it when you're flying down the road in it. I am assuming you have a car to put that engine in, Rob, lol. We are all sending you British good luck. +1

Tim Warner Have you considered having a gasket made for the top of timing cover? Or making one?

Russell Baden Musta Is that lighting or did you paint it gold?

> **Rob Siegel** lighting. covers are unfinished aluminum, block is painted gray. +1

> **Russell Baden Musta** Rob Siegel looks good either way! I seem to remember step down from the head to the front cover... I've also had major leakage there on cam cover gasket. But it was my fault, didn't use Hylomar to secure it...

Daniel Senie You'll need to write an album's worth of songs as therapy after the Lotus. Even if they're not about the Lotus.

> **Rob Siegel** As well you know, writing ANY song about ANYTHING is both therapy as well as its own exquisite torture +2

George Saylor I put the tensioner in upside down (or backwards) once. Caught it before I buttoned everything up. I understand the new chain being tighter but I'd never seen them that difficult to install. As mentioned make sure the tensioner is backed out.

Craig Lovold Don't give up. After this long you just need to persevere!

Greg Lewis Lotus desertion is allowed in the manner of justifiable homicide. Plus the statute of limitations starts when you buy it, not when you start working on it. +3

Steven Bauer Maybe it's like the pushrod v8 timing chains where you first have to place the sprocket in the chain then bolt the sprocket onto the cam.

> **George Saylor** Steven Bauer yup, that's how it's done on these engines. +1

> **Rob Siegel** that's correct

Joe Eaton Rob. Did you loosen the tensioner all the way? From the photo, it looks like the adjuster is turned in all the way.

> **Rob Siegel** There is no tension on the tensioner at all. The piston and spring haven't even been installed in the cover yet.

Steve Kirkup My TR6 nearly broke my will. British cars... then I sold it fully disassembled for a massive dollar loss. Best...day...ever... +3

Scott Linn I've had bad days (years, actually) with the TD I inherited but things eventually got better. Similar with my Europa. Hope it all works out soon and easily. +1

Seth Connelly Sometimes you're like the movie where the kid is walking down the hallway and the scary music starts... I'm glad I'm not in the movie, but I "won't" stop watching! Thanks for letting us live vicariously thru you! +1

Steve Park The learning curve on a car we don't know can be quite steep. It's the reason many never venture beyond what they are familiar with. As I get older I seem to less willing to adventuresome in this way. I've wanted to get a Lotus for years, but your trials and tribulations are lessening such desire. +1

> **Rob Siegel** Steve, yeah, this has been rough. Buying the 911SC was a much easier step outside the box. It was German, newer, running, and sold in much greater numbers.

> **Scott Linn** Yeah, but you rarely see another Europa coming down the road, vs. the 911. When it has been a while since I've driven the Europa, I'm always surprised how well it drives and handles. A joy to drive for me. But I wouldn't mind a 911SC either. I was surprised you sold that one.

> **Steve Park** You're not one to be easily discouraged. I hope you get through it and are able to enjoy it when up and running.

Hugh Forrest Mason That's a weird-looking M10

Russell Shigeta Hmm. Coincidentally a friend of mine had some serious misadventures with cam timing (also on an orphan car) turned out, even though they had the same # of links, one chain put his cam 6 degrees off. But the other was OK. +1

Fred Bersot Talking about the Lama,.. is that it in the background. Sure looks like it would be..... +1

Wade Brickhouse Have you tried using spanners instead of wrenches? 🙂 +3

Steve Young I hear you. I'm wasting hours and hours trying to get a set of e30 rear hubs out of the trailing arms...

Greg Calvimontes I can't believe I'm going to say this, as I hate people who do this to BMWs: LS this bitch.

Robert Alan Scalla I've been thinking about a Jensen Healey, but it has a Lotus engine. Maybe I should rethink that.

Russell Dejulio Rob I wish that I had your skills!

Rob Siegel Thanks. At this moment I don't feel particularly skilled. I feel like I'm trying to dig myself out from a hole I fell in five and a half years ago. +2

Russell Dejulio Rob Siegel We KNOW the Hack Mechanic will Triumph !! +1

Russell Baden Musta Rob Siegel this will be you... move forward! [photo of Russell's red Europa] +2

Bill Howard Day 2079 of what, the Hebrew calendar? +1

Dohn Roush Is the bronze color a nod to the era of the technology? +1

Karel Jennings Hey, You're the one who fell in love with the awkward-looking British car with the big weird ass. If that didn't tip you off about strange engineering... ;) That said, this is your first lotush engine build, and this kind of stuff inevitable. Do you have an old lotush expert you can call?

Rob Siegel yes; he's been invaluable. +2

John Graham Lotus in British, right? Have you considered using something like this to solve gasket/fit issues? [photo of Nelson wooden contractor shims] +4

TLC Day 2079: Second test fit

FEB 17, 2019, 5:02 PM

The Lotus Chronicles, Day 2079 (part deux):

Okay, your Hack Mechanic has backed off the ledge, and anyone who told me not to worry about that cork gasket can suck it.

At the test-fitting level (head not torqued down, gasket not squashed), the thick cork gasket WAS lifting the front of the head up too much and interfering with the ability to seat the cam gears. Took the gasket out, and got the gears seated. What's more, with everything in the timing chain path installed, I could twist the engine once around the block and verify that it rotates without binding, no valves hitting pistons. Still any number of issues before actual adhesive-and-torque installation, but this is a major milestone. Whew!

I'd hoist one in celebration, but I'm trying to be less, uh, lubed. Looks like I picked the wrong year to quick drinking...

56 LIKES, 34 COMMENTS

Alan Hunter Johansson [Airplane gif]

Joshua Weinstein You also picked the wrong year to stop proofreading before hitting "post." Or is "quick drinking" a cute, ironic turn of a phrase? +4
> **Scott Aaron** Joshua Weinstein I think "quick drinker" is probably accurate +1
> **Rob Siegel** oops! +4
> **Alan Hunter Johansson** Joshua Weinstein Freudian sip. +4
> **Marshall Garrison** Rob Siegel On the plus side, autocorrect didn't replace the 'r' in 'cork' with 'c', despite the "suck it" declarative. Even your phone doesn't want you to blow a gasket. +1
> **Daren Stone** Alan Hunter Johansson Very punny +1
> **Alan Hunter Johansson** Daren Stone We do our best...

Pike Perkins [Craigslist ad for $4000 basket case 1970 Europa]
> **Rob Siegel** As the French said to King Arthur in The Holy Grail, "Ah, no thanks, I already got one." +6
> **Pike Perkins** Rob Siegel It's just in Beverly, I mean should at least go look at it! +1
> **Rob Siegel** You're a troublemaker, you are +3
> **Rob Siegel** Plus that's a Series 2 car, not a Twin Cam Special, so even for parts, nah +1
> **Frank Mummolo** I see a new book/career, Rob: Hack Mechanic Meets Colin Chapman. Becomes Sisyphus! +7
> **Rob Siegel** African or European? +5
> **Roy Richard** I spit in the general direction of your head gasket. +4
> **Alan Hunter Johansson** Pike Perkins I unclog my nose at it +3
> **Robert Myles** Your mother was a hamster, and your father designed British car parts! +5
> **Scott Linn** $4k for THAT? I want what that guy is drinking.
> **Tom Egan** "6" cylinders? Title: None? $4k is too cheap! Here, take my money!!!

Rob Blake That looks like something out of a Rube Goldberg cartoon. +2
> **Rob Siegel** That part of it is fairly conventional. Any twin-cam engine has to have something driving the cams. It's either one chain or belt and two gears, or two chains or belts and two gears. +1
> **Rob Blake** Rob Siegel no, no, no...I'm a rube YES... BUT Cork and Copper gaskets? Must be British Steam Punk type construction. +1
> **Bill Snider** Rob and since you're a belts and suspenders sort of hack, it may be time for whisky, dinner, whisky and sleep. +1

Scott Aaron I like to use more plastic wrap than that in my engine rebuilds, normally.
> **Rob Siegel** I got back the fully-assembled head from the machine shop at least three years ago. It was fully wrapped in plastic. It sat in my basement undisturbed until I cut the wrapping off the bottom to begin test-fitting it last week. I should just rip it all off, but I sort of had the pretext of the remaining plastic protecting the valve train. I throw a clean trash bag over the whole engine when I'm not down there, and bag the entire thing and close it with desiccant inside if I think I won't be down there for a few days. +2
> **Scott Aaron** It's a pretty thing, I'll say that about it. It looks beautiful. +2

Robert Boynton Is this like covering a new living room sofa in plastic? +1
Robert Boynton Seems like, if you ever start it, the plastic might melt? ...Or did I just answer my own question
David Holzman I prefer slow drinking to quick drinking.
Daren Stone Glad you backed off the ledge and sorted this out, as patience and perseverance are just as important to enjoying an old Lotus as a factory workshop manual. +2
John Lenham Not as interesting of a specimen, but I can relate to the relief of going through the first turns of the crank... [photo of a twin-cam engine with a rubber timing belt]

TLC Day 2080: Shaving the covers

FEB 18, 2019, 6:51 PM

The Lotus Chronicles, Day 2080:

Went out through the snow to Williams Racing to have The Lotus Engine God (Bill McCurdy) take 25 thousands off the top of the inner and outer covers. This solved three problems. First, the tops of the covers weren't quite flush with each other. Second, the top of the outer cover was at a slight angle to the inner cover. Third, the cutting provided the little bit of extra clearance that the cork gasket needs. With this done, hopefully I can finally get the covers fitted on the block during the next several days.

I am so grateful for all of Bill's advice and help the past couple of weeks. It is especially remarkable considering that he is the person who first had the motor 5 1/2 years ago, and I pulled it out of his shop because he wouldn't quote me a price on rebuilding it because he said there were so many little things that could run up the cost. Now, being in the middle of it, I understand this in a way that I didn't before, and rather than rubbing my nose in it and soaking me, he is being remarkably supportive. What a great guy.

[search YouTube for "The Lotus Chronicles Timing Covers Being Shaved"]

100 LIKES, 24 COMMENTS

Doug Jacobs Yeah, I'd have bent you over. +1

Dohn Roush Being right is another dish best served cold... +1

Alan Hunter Johansson God, that's fun! +2

George Murnaghan So good that great craftsmen still populate this "parts replacement" world of ours! And that ethical behavior and truth-telling have survived in ways small and large ways in our modern world. +5

Russell Baden Musta Nothin' like a friend with a Bridgeport! Love it!! +3

Frank Mummolo You are fortunate to know him, Rob! +2

> **Rob Siegel** yes. yes I am.

Bob Sawtelle I think I watched that pass like 5 times.. very cool, but I admit I still have a little trouble visualizing how it all fits together and the whole front cork gasket thing.

> **Rob Siegel** Bob, those are the outer and inner front covers, lying as if the front of the engine was underneath them. The part that is being machined is the top surfaces of both the inner and outer covers at the very top and front of the engine where the front of the head would sit. The timing chain reaches through that slot and up to the gears on the cams. This is where the cork gasket sits.

> **Bob Sawtelle** Rob Siegel ok thanks I went back to Saturday's post and now I see the pictures of the test fit. Wow very complicated!

Kevin McCurdy Rob, so this is the other McCurdy you were telling me about when we spoke down in Greer last month . . . glad he's helping solve the gasket and fit issue. +1

> **Rob Siegel** Yes that's correct. Good memory!

Bill Ecker Sooooooo... in other words, he was right all along, payment has just been deferred? +3

> **Rob Siegel** oh he was definitely right all along

Brian Ach I love mechanical machines. they are so machine-like. good to see problem hopefully solved. +1

Chris Lordan I'm betting that after all of this is done, and the engine is fired up and the car is running, that it will seem like it was all worth it. A light, responsive, totally analog car, with an engine built (& improved) with your own hands... Non-car people will never get it. To hell with 'em. You're simplifying and adding lightness 40+ years after the fact. #goRobgo +5

Chris Lordan "Perfection is achieved, not when there is nothing more to add, but when there is nothing left to take away."—Antoine de Saint-Exupery. 'Ol St.-X was right about everything! +1

Rennie Bryant This makes you wonder how this was a "mass production" engine. Every one requires extensive hand fettling. And this isn't the only part of the car like this. The gearbox, the bodywork etc. all done by a cottage industry manufacturer. +3

David Ibsen Also, imagine the Porsche 4 cam motors... all gear driven valve train, Biral cylinders (no cross hatching, but "Dimpled" walls), a roller bearing crank with 1 piece connecting rods (crank had to be disassembled)... whew! The cam timing was easy, it was getting the timing gears to mesh properly, just like a ring & pinion) getting the shims correct etc. it took between 15-25 hours at the factory by a mechanic who had done it many times before. It's no wonder there aren't many cars like that around! +3

Thomas B. Flavin Thank goodness that cork gasket dilemma has been solved! Love, Rudi

Scott Aaron Sounds like a great guy. You're fortunate to have that kind of a resource nearby. +1

[email: Dave Bean Engineering]

```
From: Rob Siegel
To: Ken Gray, DBE
Mon, Feb 18, 2019 8:23 pm
```

Ken, FYI, I discovered another issue with the covers. When the covers were off the block and the cartridge pump was installed through them to force the bores to be concentric, the top and bottom edges didn't line up. The outer cover was below the inner one by 0.016" (sixteen thousandths) at both the top and bottom surfaces. In addition, the top of the outer cover wasn't in plane with the top of the inner cover; it sloped slightly upward. Because my block has been decked, the inner cover was nearly at the level of the deck (not your fault), so in order to solve all three problems, I had Bill McCurdy shave the tops of the covers on a milling machine by about .027", so the tops are now flush. The same 0.016" step, though, still exists at the bottom. That is, the bottom of the inner cover is just about flush with the oil pan mating surface, but the bottom of the outer cover is slightly below it. Because the semicircular relief for the front oil seal is on the outer cover, this surface can't easily be machined. Bill thinks this is a livable step, that the cork gasket and sealant should together be able to accommodate that step, even though it's greater than the 0.010" spec in the manual.

While I was at Bill's, I also had him cut a small chamfer into the bore of the inner cover to help the o-ring slide into it a little easier. I paid Bill $100 for both of these machining tasks.

Again, all of this is just FYI. I'm not asking for or expecting anything. But you might want to do what I did and test-fit the cartridge pump through the bore of the two covers on the bench and check the alignment of the top and bottom surfaces.

--Rob

TLC Day 2081: Time flies when you're fucked

FEB 19, 2019, 8:01 AM

The Lotus Chronicles, Day 2081:

It was exactly two years ago today that I hopped in the rented SUV loaded with tools and parts and headed south to Louisville to begin the adventure resurrecting Louie. First stop was at Paul Wegweiser's in Harmony PA to have dinner with him and Wendy Burtner. Time flies when you're having an adventure. (And I cannot help but note that I did the entire Louie adventure—the purchase of the car, the drive down, the resurrection in Jake's pole barn, and the drive back—IN LESS TIME THAN IT'S TAKING ME TO GET THE FRONT TIMING COVER ON THIS FUCKING LOTUS!)

117 LIKES, 37 COMMENTS

Jerry McNeil Where does the time go?

Harry Bonkosky The Lotus... What were you thinking?! It is like the time I bought a Mercedes 190E. At the moment I closed the deal I automatically aged 10 years. +7

Pete Lazier ...you do realize that no one has the slightest bit of sympathy for you... +2

John Pantel Pole barn sounds kinda kinky. Just sayin... +2

Ian Sights If your theory of being imprinted by your first car experience were universally true, I'd be restoring an old Mini or MGB instead of my E9. My father, in spite of his Germanic background, had many English cars. XK120, XKE, MG Magnette, Minor & Midget, Allard J2X, Morgan Super Sport and a Lotus 7. My first car was a VW Beetle, first of many and I've had several with the three-pointed star before ending up with my 2800cs. Replacing the engine in the Midget and then in my aunt's '71 Toyota showed me that the English way isn't usually the obvious or better one. A joy to drive even with their quirks, I've left owning English cars to others. Notwithstanding the soon to be coming thrill of blasting down the road in your Europa, the happiest moment of your ownership may be when you hand the keys to someone else and watch her drive away. +1

Wade Brickhouse I think it's time for an attitude adjustment. Place a roundel over the Lotus emblem and smile again. +1

Duncan Irving Rob Siegel With all those tools in the suv I have to know: Which one did you forget that screwed up the smooth well being and satisfaction of automotive repair? Like, you know, having to run to Autozone or something disruptive. Hate that.

> **Rob Siegel** Oh, there were several. I came with a set of snap-ring (circlip) pliers, but the snap-rings holding on the slave cylinder wouldn't come off without a fight, and thus dulled the little flared ends on the tips of the ones I had, which are replaceable but not available around the corner, so I had to buy a new set. Then when the clutch master cylinder failed, the upper nut holding it in is maddeningly difficult to reach, so I had to by a set of crows foot wrenches. I already had a set, but they were back in Boston. Fortunately, Harbor Freight is cheap, and they only needed to work once :^)
>
> **Duncan Irving** A credit to your ingenuity and expertise that you made it back as quickly as you

did. Please keep entertaining and schooling us. Thank you Rob.

Jake Metz Rob Siegel Tap and die set too

Rob Siegel Jake, sure, YOU remember the ones YOU ran out and bought... how selfish of you :^)

Harry Krix Louie = Great Adventure. Lotus = Not So Much. +2

email: Dave Bean Engineering

```
From: Rob Siegel
To: Ken Gray, DBE
Tue, Feb 19, 2019 6:14 pm
```

Ken, FYI, I found another minor clearance issue: The top of the gear on the jackshaft, at about the 1:00 position, rubs just barely on the inside of the outer cover (photos). It does this with only the covers installed, no timing chain, and just spinning the jackshaft by hand. This is the inside surface of a water passage. Do you know how thick the casting is at this point? Do I have your go-ahead to give it the lightest touch-up with a Dremel tool?

--Rob

```
From: Ken Gray, DBE
To: Rob Siegel
Tue, Feb 19, 2019 7:37 pm
```

Hi Rob, I understand. Yes please dress the area with your Dremel. I like to use one of the barrel sanding attachments. The cover is plenty thick, .250". I thinking that .050" should be the clearance. Yes, you have my go-ahead.

TLC Day 2081: Jackshaft gear clearance

FEB 19, 2019, 7:31 PM

The Lotus Chronicles, Day 2081 (continued):

Thought I might actually get the timing covers glued on, but the paper gasket that goes between the inner cover and the block was part of a gasket set I bought 5 1/2 years ago, and it's shriveled and dried up like a leaf in November. Soaked it in water and it flattened out, but as soon as it dried, it curled up again. Ordered another one. It'll be here in a few days.

So I did more test-fitting (rehearsal for assembly; need to do things with verve and confidence when the sealant is applied). Found and diffused two more clearance issues.

1) Saw a rub mark on the inside of the front cover. Assumed it was from the timing chain. Reassembled without chain, could still hear scraping. Turns out it's coming from the gear on the front of the jackshaft (the Ford cam-in-block that spins the oil pump and the distributor). It probably would self-clearance without issue, but the lightest touch of a Dremel tool should take care of it. Sanity-checked it with Ken at DBE that touching it up wasn't a problem in terms of casting thickness.

2) Nearly had a cow when it looked like the timing chain coming off the exhaust cam sprocket actually hit the front inside surface of the head. Found that, with the head slid

forward as far as it'll go on the head bolts (no dowels in this baby), it no longer hits, but there's only about enough room for a coat hanger. Man, clearances in this thing are tight. Onward.

42 LIKES, 37 COMMENTS

Paul Garnier "verve" +1

Brandon Fetch He added lightness by removing all but the smallest, tiniest of tolerances. +6

Ernie Peters Rehearsal is definitely a must with everything Lotus. +1

Skip McCauley I hope that when you finally get this thing put back together that this smiles per hour are worth it! +1

    Rob Siegel I bloody hope so!

John Morris "British precision" +1

Josh Wyte Was this motor ever together and running? I'm surprised at the amount of clearancing necessary if it was +1

    Stuart Moulton Came to ask same question

    Rob Siegel It was together when I bought the car, but seized in one cylinder, reportedly from sitting (it literally was sold as "ran when parked"). Upon opening it up, it was clear that something had been ingested before and the engine had been put back together before. But all of that having been said, most of this clearancing is needed because I'm installing a cartridge-style water pump which comes with a pair of custom front timing covers for it to slide in, and those never have been on this engine before. +7

David Ibsen What type of sealant do you plan on using?

    Rob Siegel Probably Yamabond4 on top of an RTV primer (Dow 1200RTV) +1

    Brett Kay Rob Siegel Great choice! +1

Thomas Jones You're reminding me why I've pretty much stuck with BMWs, mainly German stuff at least, for most of my career. As cool as I think Lotus's are, and as much as I want a Caterham Seven… You're cooler much than I am… Here's where this belongs Shad Essex. 🌐 +1

    Rob Siegel no, just more foolish and more frustrated

    Thomas Jones A short commute will do that to a fella.

    Matt Schwartz Thomas Jones you should build an M10 powered Seven

    Thomas Jones I've actually worked on one. Someone brought a Rotus 7 with an M10 in it for me to tune one time. I only got to test drive it a little bit though. +1

Steve Woodard While my MGB engine ('65, before the bumpers were screwed up, among other things) came from a British tractor, I never had to deal with this sort of stuff. With my Datsun motors with the OHC chain and stuff, it all just worked. Oh, and my two Miatas were more-or-less trouble-free, save a Ford-sourced transmission thing. Either way, it's clear that you're a glutton for punishment. Are you going to keep this POS or send it on down the line?

    Rob Siegel Don't know yet. I'm going to get the motor assembled, then get it and the transaxle in the car, and then make some decisions. It IS a very cool car, and I DID buy it because I was attracted to it. I just had no idea that resurrecting it would be so much more involved than doing the same on a long-dead BMW. +1

    Steve Woodard Well, as I said before, working on stuff you know is a much happier place. Good luck!

Christopher Kohler I'm starting to figure out how we won the revolutionary war. +4

    Rob Siegel right, this, and poison ivy +3

Jeff Hecox After sliding the head forward for the cam chain clearance, that is going to change the valves lining up with the valve pockets on the pistons and the squish band between the piston and head

    Rob Siegel I'll keep that in mind. Thanks Jeff.

Luther K Brefo Note to self: Lotus' own engines, great to look at others deal with their… nuances…

Tom Egan Ah, yes. The Jack Shaft. Only the Brits would build a "Twin Cam" with 3 cam shafts. Stiff upper lip, old chap, it'll be worth it when it all comes together! +1

Steven Bauer Subconsciously you knew why you were avoiding getting this done… +2

Alessandro Botta That's the statement why the British motoring is only made from strangers or heritage. They didn't deserve a car industry

Franklin Alfonso The joys of British engineering… [gif of Jeremy Clarkson giving the thumbs up]

Brian Ach Do people swap the 4A-GE engine into these?

Rob Siegel yes https://bringatrailer.com/listing/1971-lotus-europa-3/

Brian Ach Glad I didn't see that

Jim Gerock The Lotus Chronicles. Kitchen table location. tii Goetz piston ring examination. [Jim's cat "Lotus" playing with his parts] +5

Lee Levey After 10 years of Lotus ownership, I learned that Lotus parts suck and will require constant mothering to work properly. +1

David Weisman Heh heh he said jackshaft. Heh heh. Sorry. +1

TLC Day 2084: Covers mounted

FEB 22, 2019, 7:28 PM

The Lotus Chronicles, Day 2084:

Booya! The front covers, the cartridge water pump, and the head are finally installed on the Europa Twin-Cam motor. Next step: Pressure-testing the cooling system. Because if it leaks through an o-ring or a gasketed seal or a metal-on-metal-with sealant seal or a casting flaw, you really want to know before the engine's in the car. If it passes that step, it gets buttoned up (oil pan isn't on yet), and I begin looking at the carbs, exhaust, clutch, and my bank balance. This is big. (Having the motor largely assembled, not my bank balance :^)

94 LIKES, 24 COMMENTS

Bill Snider Perseverance, Patience and Perspicacity! Well done Rob. +1
Doug Jacobs Set that fucking thing on fire!

David Ibsen It would be nice to test it (long block sealed) with near boiling water. It could expose some gremlins before installing it, Ahhh, sleep... perchance to dream!

    Rob Siegel I'll settle for air pressure :^) +2

    David Ibsen In any event, wish you good karma! +2

Jim Strickland Maybe it's the camera angle, but that looks like a big, strong engine...almost industrial. Good looking car in background 😊 +1

Bailey Bishop Jr. I wish my TC looked like that.

    Rob Siegel With everything you've gone through, no you don't +1

    Dave Gerwig 😊

Derek Barnes Congrats! Can I get a ride if u r near Philly?!!!!!!

Carl Witthoft Not sure I should ask, but... once the engine's running, what about the gearbox?

George Zinycz Very happy for you making good progress Rob! Just keep in mind that it's not done screwing with you yet... it wants to taste your tears and hear your cry. +2

Rob Siegel You mean it wants to continue to do those things 😊 +1

Steve Park Congratulations. It's great to get to a milestone, if you will. Fingers crossed the rest goes smoothly. +1

Rob Koenig So, there's gonna be a Europa book at the other end of this, right? The title needs to be The Lotus Chronicles: Day 2xxx with the actual day number being the day on which you first drove it around the block. +5

Brad Purvis 😊

Karel Jennings That is looking fine. There's nothing more satisfying.

John Jones Huzzah! Out of an abundance of caution, I will withhold any congrats pending the pressure test. I am not in the business of breaking up no hitters +1

Franklin Alfonso Well...who knew it would all be so easy? +1

Christopher P. Koch I toured the GM Tonawanda engine plant a few years ago and much to my surprise they don't test each engine on gas/natural gas any more—they run them on compressed air!

Daren Stone Well done!

Mark St Clair How did they ever win the Battle of Britain?

Daren Stone Can I assume the cassette water pump went in w/o too much grief?

    Rob Siegel Yes, pump fit fine; check my latest post for grief (probably unrelated to DBE parts) +2

Ed MacVaugh I like the locating peg, since that engine doesn't use dowels. Any trouble getting it out afterward?

    Rob Siegel Nope, I unscrewed it and lifted it out with the magnet. That part of things went perfectly. +2

Zacherle Ketring What chemical is that on the water pump o-rings?

    Rob Siegel That's just blue RTV being used to lubricate the o-rings before inserting the pump. the sealant I used on the covers (Yamabond4) seemed a bit meaty to use when the o-rings themselves are providing the sealing and the manual says to «smear them with a little RTV.»

    Zacherle Ketring Rob Siegel, you can try Syl-Glide, Napa has it, it's silicone lube. It'll keep thing supple and not messy when it comes apart 😊

    Al Larson Always used Sil-Glide, or good old Vaseline on those. I wouldn't recommend an RTV sealant.

    Rob Siegel Zacherle, I actually have Sil-Glyde, but the Lotus Twin-Cam engine book I'm following specifically recommended a light coating of RTV. When in doubt, follow the instructions 😊

    George Zinycz Rob Siegel English RTV though, right? Theirs is probably somehow different from ours. Or maybe you just have to apply it with your left hand. +3

    Ben Greisler Odd they suggest that.

    Bob Shisler "Brake Grease" is great on rubber parts

    Tom Egan In "British" RTV means Really Thick Vaseline! +3

    Neil M. Fennessey Hmmm... RTV didn't come along until the late 1970s. What year is your engine manual?

    Rob Siegel Neil, it's Myles Wilkins' "Lotus Twin-Cam Engine" book. Google Books shows the original publication date as being 1965. Amazon shows it as being 2013. It may have gone through several revisions. Both the book and the maker of the cartridge-style water pump (Dave Bean Enterprises) specifically recommend smearing RTV on the O-rings. +2

    Neil M. Fennessey Rob sounds good. No doubt the book's been revised and since the guy who sold you the pump has scads of experience, I would follow his recommendation too. Glad to see/hear that you're making solid forward motion on your engine. 🏆🏆🏆🏆 +1

Ben Greisler "It was good enough for Colin Chapman. It is good enough for me!" +1

Christopher P. Koch Make sure to not use the common acetoxy cure RTV silicone as it is acidic as it cures, damaging aluminum. Use oxime cure RTV silicone as it is chemically neutral on aluminum.

John Goddard RTV would not be my first choice here, but then again it's British, so my instinct is probably wrong. [Note: My use of RTV here WOULD in fact come back to bite me, but not for the reasons raised.]

Steve Nelson Where does the pasta fit? 🍝

TLC Day 2085: Chain clearance again

FEB 23, 2019, 9:08 AM

The Lotus Chronicles, Day 2085:

It's not good. It looks like the timing chain is touching the head as it comes down off the exhaust cam gear. It certainly is close enough that it scrapes against the tiny bead of RTV that squeezed out and gets it on the chain. I checked this during test-fitting and determined I could make it go away if I shoved the head all the way forward on the head bolts, and did that during assembly, but with everything sealed up and tighten down, this is where it came to rest. Not at all good. Not sure what to do next. And please, hold your "light on fire" and "LS swap it" comments. They're not helping.

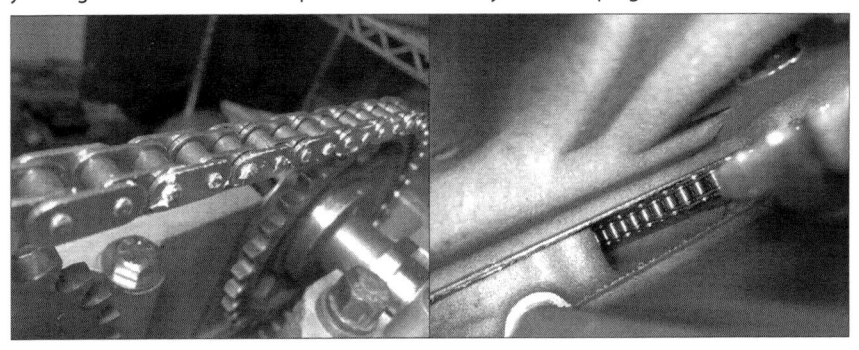

53 LIKES, 135 COMMENTS

Grice Mulligan I take it machining the cam gear is out of the question?

Corey Dalba Send it, chains have a unique ability to carve their own paths. :-) +2

Corey Dalba On a serious note. Is this a common issue? Cam gears backwards? Does the chain seem to align properly by the crank hub, is it def the cams? Are there no guides?

Rob Siegel I haven't found anything about it on the forums. The cam gears are on correctly. The crank gear hasn't been removed from the crank. There is a chain guide on the inside surface of the outer timing cover. It's basically just a rubber-coated piece of metal. The chain is not centered on it, but I looked at the wear grooves on the original one and it wasn't centered on it either. When I get more coffee in me and my wits about me, I may remove the thrust washers from the center crank bearing and try shoving the whole crank forward slightly, but obviously that's not a permanent solution. +2

    Karel Jennings Are there thrust washers on the cams? And is the cam gear new? Could be that is what's wrong. Out of spec part. Or... are there shims available for the cam gears. I'm frustrated right along with you. +1

    Thomas Jones I'm with Karel, could there possibly be spacers behind the cam gears and or crank gear? This can't be right, because there aren't any witness marks on the head from this happening previously. There has to be a reason... +4

Vince Strazzabosco Smart guys, Karel and Tom. Too bad Ben isn't on FB too. +2

Rob Siegel Karel, cam gears are not new. Found nothing on Lotus forums about shims. +2

JohnElder Robison This is the kind of thing where hobbyists need Bridgeports. You might need pins and holes to properly locate the head. Maybe the engine had those and they were lost. You should not be able to move the head AT ALL prior to tightening, this is why. Once tightened down, you might need shims on crank, idler, or cam to position the chain.

Rob Siegel JohnElder Robison, I don't believe that this engine uses pins or dowels to fix the head relative to the block. I've verified this by looking at the exploded parts diagram for the block in the factory manual. The only dowel pin listed is for the block to the clutch housing, and nothing is shown at the block deck. I agree that it's very surprising. The Wilkins book says "[Temporarily] fit two modified head studs (saw off the bolt head and cut a screwdriver slot in the top) at the corners to act as location guides. These must be used to locate the gasket, otherwise damage may result due to the head and gasket sliding around." That having been said, I may have misstated the amount of play that the head has on the bolts. Once all the bolts are in, including the ones in the front that fix the head to the front timing cover, it doesn't have much play to move around.

And yes, I'd like a milling machine :^) +3

JohnElder Robison As for the pins . . . just because an engine was not built with locating pins does not mean it wouldn't benefit from having them. If there is a chain gear on the head and one on the block, there should be some pin to hold the two in precise alignment. Relying on the studs for that is cheap and sloppy. It's like what Land Rover did with their pushrod V8, where they pressed in tubular liners, and the engines are all failing because the liners are not held tight by flanges at the top. A thing that would have added $100 in production forces a $14,000 overhaul 20 years down the road. +1

Dave Borasky [Airplane gif "I just want to tell you all good luck… we're all counting on you"] +11

Dave Borasky I can offer no technical advice. I don't have the gumption to do the kind of work you do on cars and therefore do a little vicarious hack living through your books, articles and posts. Your frustration is palpable, and since reading your posts usually bring me joy I hope you find your way through this part of the Lotus wilderness! +6

Steve Nelson Any chance of adding shims behind the cam gears to move them forward? +6

Skip Wareham You'll come up with a solution, you always do. This is why we enjoy your mechanical adventures. +1

Doug Jacobs Sorry for the "light it" comments—I see now you're a tragic, not comedic character. Sort of like Nixon on the beach with his metal detector. I guess my comment is—could that actually be the clearance? Is the chain going to draw away from the case when hot and under load? I don't know that I see this as an issue, hitting the bead isn't hitting the case. +5

Rob Siegel I need to find out.

> **Doug Jacobs** Rob is that a non-load-bearing cover its close to hitting? Can you file some of that casing down? +1
>
> **Rob Siegel** Doug, that's actually not a bad idea… +1
>
> **Doug Jacobs** Rob—if it's going to happen, better you do it before the motor does it on his own. +3

Corey Dalba I seem to recall having a very similar issue with an M10 few years ago. I think what I found was the "new" chain was slightly thicker than the 35-year-old chain. At the same time I had replaced the oil pump gear and chain. Due to BMW P/N discrepancies I believe I was forced to buy a oil pump gear and chain from s14 which resulted in slight contact with a piece of casting around that area. I filed it down, turbo charged it and sent it. Its now done thousands of boosted, and in some cases tracked miles… is there any way to compare the old chain and new? +2

Rob Siegel Corey Dalba, I will do that +1

George Murnaghan Rob Siegel What a fine fettle you have me in, Lolita! +1

Andrew Wilson Nixon on the beach! Such a funny image like a fish riding a bicycle. Used to reference Nixon's awkwardness all the time. 😊

> **Doug Jacobs** Andrew Wilson—you forget San Clemente. https://goo.gl/images/xBntnr
>
> **Andrew Wilson** ^ Brilliant

Scott Aaron Oh God what a bummer. Sorry about that.

Bob Sawtelle When you say the crank gear hasn't been removed from the crank, do you know for a fact that it is on there correctly? As you said the motor was seized when you got it, could PO have done something to this before you opened it up? +5

> **Rob Siegel** Bob, you are correct that the whole «well that's how it was and I didn't change it»

saying his skating on very thin ice. +4

Jeffrey Miner Hmmm. My SpecE30 engine builder rebuilds a lot of M20 cores before they blow. He just did one after it blew, and it's got a serious vibration problem and power issues on the dyno. They're both shocked and stumped, as all the crank machine work and new rod and main bearings appear to have assembled fine. But yet, the crank doesn't appear to be spinning true in the block. A pan drop, rod bolt torque check reveal nothing suspicious. He commented to me that the he's exceeded the limits of his experience by attempting a rebuild on a catastrophically failed engine. How much do you know about this engine's mode of failure? Could there be crank clearance issue due to a block deformation? Are the main bearings new? Nothing on the message boards about needing to shim the cam or crank sprockets? +4

Joey Hertzberg Jeffrey Miner very interesting post

Rob Siegel Jeffrey Miner, the engine supposedly «ran when parked, seized when sitting.» But there's evidence that it ingested something at some point in its past and was rebuilt once, so anything's possible. I need to look for the shimming topic on the forums. +1

Federico Sierpien M42 swap it!

Collin Blackburn ...M10 swap it?

Rob Siegel shut the fuck up +16

John Wanner Man this is brutal to watch. I feel for you. +2

Doug Jacobs I feel like this is going to be a stupid question, but is that chain too fat? I'm flailing here because I don't know shit about British—but there's only a limited number of choices here.

Rob Siegel I eyeballed it next to the original chain and didn't see anything amiss. I'll do a caliper measurement. It was purchased from one of the two reputable Lotus parts houses in this country (RD Enterprises), not some eBay source of no provenance. +1

Ben Greisler I am unclear on where it is actually hitting. It looks like there is a plate that bolts to the block and then the front cover goes over the timing stuff. The head bolts on top of all that. Is the chain hitting the head casting or the plate bolted to the block? Makes me wonder if something wasn't machined properly or not matched to the other pieces and now it is out of tolerance. +1

Rob Siegel Ben, on a BMW engine, the front of the block and head are flush, then the timing chain goes on, then there's a lower and an upper timing cover. In contrast, on the Lotus-Twin-Cam engine, the front of the head itself contains the upper timing cover. It's one piece, part of the head. There are then two lower covers, an inner and an outer one. The inner one bolts to the front of the block, then the chain goes on, then the outer cover. The chain is actually barely touching the inner surface of the upper part of the «cover» that's an integral part of the head. The only way I can see that happening is if the crank isn't far enough forward. +1

Ben Greisler Ok, got it. Is it possible the crank thrust bearings are incorrect pushing the crank back? +1

Mike Savage The part that's now rubbing is NOT related to the new timing cover with the cartridge water pump right? So this is all original engine mechanicals (though you replaced gear and chain maybe) very strange… if this is true it would seem parts you replaced are different or something in the installation is off.. +3

Ben Greisler Mike, I wouldn't expect the timing covers (lower) to impact the relationship of the crank and cams, but I could see an issue with the crank sprocket being too deep due to wear or tolerance stacking to cause this.

Rob Siegel Ben Greisler, it has new symmetrical thrust washers that give it the correct end play. I can give it new asymmetrical thrust washers that'll move it forward very slightly. +1

Ben Greisler I'm just brainstorming things that could potentially impact the sprocket alignment.

Martin Meissner I'm really amazed at how troublesome this Ford-based engine is to get back together... I know it's a Lotus head, but man, they couldn't all have been this finicky...

Rob Siegel I'm told that the problems I've been having are not at all unusual. +5

Martin Meissner That's just crazy. Wow. Imagine if M10s were this temperamental... none of us would be in this hobby. Lmao +2

Rob Siegel Oh, I know. I've spent two freaking months on something that on an M10 or M30 would be a ten-minute job. +1

Martin Meissner "From the people that brought you notoriously unreliable vehicle electrics... complete engines!" +6

Joey Hertzberg Martin Meissner yeah, this situation, which to a degree Rob says is fairly common, really highlights the engineering decisions that not only produce an engine that performs well, and lasts, but also, importantly are easy and intuitive to maintain like an m10. BMW is not making engines like that anymore. One might think this easy-fixin' quality is an epiphenomenon that arises simply from the technology of the era, but clearly no, look at the Lotus engine, it's a bear. +1

Joey Hertzberg These qualities are why the M10 is always on those 10 best engine design lists… +2

Rob Blake Rob Siegel hence I don't see too many of these Lotuses on the road?

Scott Linn They only made 9k of these, 3k of Rob's model. +1

Steve Nelson Rob Siegel reason enough to avoid temptations to get an old English car! +1

George Murnaghan The M10 was designed as a complete engine top to bottom, while the Lotus Twin Cam head was added (as I understand it) as a performance modification to the Ford Kent engine in place of the factory engineered original head. Aside from the vagaries of, ahem, British engineering, This opens the path for the mischief and trouble that Rob is working his way along. My observation with post WW2 British cars is that they were conceived and built at a time when labor was cheap and capital expensive, so proper engineering gave way to the need to hand build and intensively service everything (e.g. adjustable ball joints on an E-Type.) Good luck, Rob. I am keeping the faith and enjoying the story-sorry for your suffering! +2

David Shatzer Steve Nelson Another old English car… +2

Russell Baden Musta Without being there, I got nothin Rob… I didn't have any of this when I did mine… but I didn't change the water pump… used OEM…

Jamie Eno Good news is that other than some gaskets and sealant, you're not in trouble. Your methodical approach will get you through this, even if it means taking a step backwards to move forward. Edited for grammar since you are a writer and have certain standards… +6

Luther K Brefo wish I knew more about Loti to help. :(

Luther K Brefo How much has the head been skimmed versus how thick is the new head gasket to counter the change in compression? Are tolerances really that tight in these things?! My thinking is if it's been skimmed/decked, would you need to run a thicker MLS gasket or copper? Gasket to keep the distance the same as before skimming?

> **Rob Siegel** Luther, the block was decked and the head was skimmed, but I don't know by how much. But those are both vertical displacements. The problem appears to be from a horizontal displacement of… something

Kim Dais [gif of hundred dollar bill being lit on fire to light a cigarette] +2

> **David Weisman** Kim Dais 😂
>
> **Maire Anne Diamond** That's just mean. [my darling wife sticking up for me]
>
> **Kim Dais** Maire Anne Diamond the truth hurts! And I can certainly relate!

Scott Winfield Wishing you peace and remedy.

Jeff F Hollen You have a patience of a saint. I can't deal with test fit and take apart wash repeat 😂😂 😂 +2

Rennie Bryant It will machine its own tolerance

> **Rob Siegel** yes, and not in a good way +1

Rob Koenig If that's the only area of concern, I'd just file it down for additional clearance. Though, frankly, it's not actually making contact now. Why would it later?

> **Daren Stone** Rob Koenig Clutch operation puts a fwd load on the crankshaft, so depending on the thrust washer clearances, the crank may actually shift fwd a tiny amount. +2

Michael Cari My opinion is that it's too tight, and not designed to be that way. I would look at the crank. +1

Chris Mahalick I am assuming that you are on this forum (lotuseuropa.org)?

> **Rob Siegel** Yes, but of the roughly 9000 Europas, only about half have the Twin-Cam motor. The Elan forum is actually better for engine issues. I've posed the issue there. +1
>
> **Chris Mahalick** Rob Siegel I also have a Twin-Cam that I am restoring. Mine is a 5-Speed, which is supposedly a good thing. Mine leaks oil, and I am having issues with setting up the carbs. My Porsche 911 and BMW 530i are so much easier to work on.

Paul Garnier Thoughts and prayers? +1

Clyde Gates You have convinced me that I never want to own a Lotus! (except possibly an Elan… but there is the Miata for that) +2

Robert Alan Scalla Next time buy a nice Bugeye Sprite. Simple as pie. +2

Joe Eaton It's often close there. Mine even has a slight groove where the chain has rubbed over the years. Make sure the tensioner is not bent. Have you tightened the tensioner to 1-inch deflection? +1

> **Rob Siegel** Joe, the chain is new, zero stretch, so there's very little deflection even with the tensioner not tightened at all. Yes the tensioner is not bent.

Ned Gray How deep is the bead of sealant? Can you actually see that the links of the chain are making contact with the head? From the one image, it really looks as if the cam gear has to move back... thrust washers?

> **Rob Siegel** Ned, the bead of sealant is just barely oozing out; I was very careful with the sealant. The links are either barely touching or barely missing, I can't really tell which. Either one means it's too close for comfort unless someone tells me "yeah, they're all that way." I will look at the issue of spacers behind the cam gears. (1)
>
> **Steve Woodard** Rob Siegel Won't you need a spacer on the crank as well, to keep the chain square?

Ned Gray Wish I was there to give you a hand. It's hard problem solving from a computer! +1

John Jones A crazy, uneducated, uninformed question: Would it make sense to measure the offset and have the cam gears machined a bit to get the alignment you want?

Carl Witthoft Honest I'm not poking fun, but this situation screams for access to a 3-D laser sintering printer. Produce a new head with 2 mm additional internal clearance.

Eliot Miranda Can you draw a diagram of a plan view showing crankcase, chain, cams etc, showing where it rubs, and which way things would have to move to resolve the issue?

> **Rob Siegel** I love you folks, but hearing about 3D printing solutions is right up there with "LS swap it" solutions.
>
> **Eliot Miranda** Rob Siegel I just want to visualize the rubbing issue properly. I can't figure it out from your pictures. I'm not proposing any solutions at all :-)

Alessandro Botta As it has no dowels, could be the head installed too back?

> **Rob Siegel** As per the details in the post, I tried to install it as far forward as it would go, but I may have not been successful.
>
> **Alessandro Botta** Rob, could have slipped back while torqueing ? It happened to me with a similar engine

Andrew Wilson Personally, I'd file the head to get the clearance. These were made by English cabinetmakers. May not be pretty when done but then only you'll know it. +1

Vic Badrinath It might be a good to remove the new chain and reinstall the old chain and see if you have the same problem. If so, then it's likely the crank that is out of position. You could always, of course, "hack" a solution and slightly grind down the inside of the top cover on the head to give the new chain added clearance.

> **Rob Siegel** I can't remove the timing chain or anything else without breaking the now-sealed engine open, but I can and will measure the new and old chain widths. +1

Paul Wegweiser Maris Mangulis: Let's hope that when we put your engine back together, it takes ONE day and there won't be 39 comments about the trouble we're having with it. BMWs are nice. Your car will be running 24 hours later. it will be OK. I promise. +1

> **Maris Mangulis** Paul Wegweiser never a doubt.

Lindsey Brown Do you have access to the spot that's touching the chain? It shouldn't require much in the way of added clearance by way of machining.

Lindsey Brown Can I come over and play? I live for this kind of shit. +1

> **Rob Siegel** Lindsey, I love you, man. Heading out for the evening. Will correspond tomorrow.

Rick Ramsey Ahhh dude your F...ed ! Seriously it is British made with worn out WWII machinery. Just bite the bullet and buy some new gaskets they should be there by the time you get it apart and grind the clearance you need. You know there is not a magic bullet.

Steve Park Sorry to hear of another difficulty. Many of us have been there, some give up. If an one can pull this off, you can. We're pulling for you. Good luck.

Michael Cari It's been my experience in designing these things and assembling them, that this is not correct. It is not engineered to be that tight, British or not. It's my understanding that this head and block were together and running at one time, without machining itself the clearance it needed (no witness marks.) It should go together without it too. Decking the mating surfaces shouldn't be the issue. The lack of a dowel might be. Ask your guru friend on Monday is my advice. If yours is doing this, others have as well and there is a known solution. Good luck Rob! +3

Russell Shigeta Perhaps this isn't the case, but aren't cams ground with a slight bevel? So that they spin the cam followers? And doesn't that then force the cam either forwards or backwards? (in SBC country, where we are not, I believe the cam thrust button is on the front)

Chris Roberts Damned frustrating. Sorry, Rob +1

Joe Eaton I saw your post on Elan.net. Wait until a guy named Rohan posts a reply. To my mind, he seems to know his shit more than anyone else on the site. +2

Rob Siegel Yes, I have seen his posts and responses. Agreed +1

David Holzman I was a good bicycle mechanic. Never got beyond tuning with cars, and that basically ended when I sold my '77 Corolla and bought a new '93 Saturn. Except for changing the plug wires on the '99 Accord almost two decades later, after being told what the options were after giving a mechanic the scan code. Like Dave Borasky, I get vicarious satisfaction reading your accounts, and I'm a bit envious of your skills in this dept, but I don't envy you when it comes to the Lotus, and I hope you quickly solve these problems. +1

David Shatzer Hopefully you are taking contemporaneous notes for the next time you find another Europa Twin Cam that ran when parked... a great book too...

Michał Maciągiewicz I've seen this before in one of my Cosworths that blew up. This was the cause of it. Cams were not aligned and floated back and forth so the chain alignment was always moving. Someone removed the SLs pump on the Cosworth and didn't get a kit or anything. There was a bolt and an alignment washer or sorts to keep it from floating. Mine was missing that. Took out the tensioner and oil pump and at highway speeds caused to spin bearings. Motor stopped. Just a thought. You may be missing some type of alignment fixture. Whether it be a cap or some bolt and washer setup like was on my Mercedes.

Also a car I picked up not running with known engine failure. It was so stupid and it was simply negligence. Some "mercedes benz tech" did the sls delete. It was a matter of a shorter bolt and it would have been still running.

Gene Kulyk This is way out of my league but I join the "thoughts and prayers" group wishing you a speedy recovery.

James O'Brien Hang in there, man. I realize this is agony on your end, but it's great storytelling from this side of the glass. Carry on. The Lotus must live. +1

    Rob Siegel oh, sure, this is all about YOUR pleasure... :^)

    Matthew Zisk 50 Shades of Lotus . . .

    Rob Siegel well, her name IS Lolita +1

Brad Purvis British cars are a form of masochism. I have two. +2

Dale Robert Olson So, when all is said and done, can you just Dremel out some clearance??

Brad Purvis Life with British cars is a total fucking nightmare, but God help us, we do love it so. +1

TLC Day 2086: No spacers

FEB 24, 2019, 11:51 AM

The Lotus Chronicles, Day 2086:

I can find nothing on the Lotus forums about spacers behind the cam gears. One person said that there's supposed to be a spacer behind the jackshaft gear, which sent me scrambling to verify its presence. It's there.

I have a theory. (For you Python fans, it's not about the brontosaurus. Of course, since THAT theory talks about thickness, it may well be applicable here.)

– The new backplate (inner timing cover) is thicker than the old one by 0.008"

– The new paper gasket between the block and backplate is thicker than the old one by 0.01"

– The new German timing chain (sold to me by one of the reputable Lotus shops in this country) is indeed slightly wider than the old one by (depending where you measure it) about 0.014"

– If you add those together, taking half of the difference in the chain width (only one side rubs), you get about .025, which feels non-trivial to me.

Below is a slightly fuzzy photo of the clearance at the bottom end, between the cam gear and the backplate. Even down there, it's tight, about .046", but if I had that much room at the top, I'd call it done. Note that the .025" I calculate I'd gain is a little more than half that, so it's significant.

I need to think carefully about what to do next. Replacing any of the named components requires tearing the engine back apart. But put another way, if I have reason to tear it apart, I can't see not sourcing a thinner paper gasket and possibly a thinner timing chain.

26 LIKES, 47 COMMENTS

Jeff Hecox How much end float on the crankshaft do you have and is it pushed forward or back at the moment?

> **Rob Siegel** I don't recall the number, but it is in spec. I measured it after installing new thrust washers. I pulled the center cap off this morning and the thrust washers ARE symmetric, so I could gain a little bit by using asymmetric thrust washers. I posted this suggestion on the Elan forum but no one commented on it.
>
> **Rob Siegel** Although the thrust washers are symmetric, the crank does look to my eyeballs (looking at front and back gaps at the other crank and rod caps) to be slightly biased toward the back, so, again, yes, I could gain a little bit by using asymmetric thrust washers.
>
> **Jeff Hecox** Rob Siegel if everything else checks out you could always mill down the inner timing cover plate some. +3
>
> **Rob Siegel** Jeff, I DO need to be careful doing that, as there was a minor clearance issue of the jackshaft sprocket hitting the inside surface of the outer timing cover. I clearanced the point of interference with a Dremel tool, but it's on the backside of a water passage. If I shave the inner cover, it'll bring this surface closer to the jackshaft sprocket and I'll have to clearance it again.

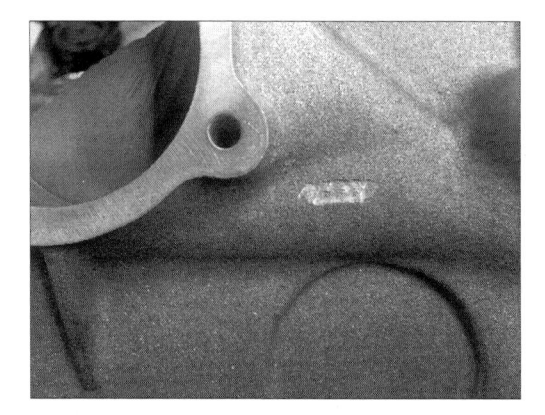

Jeff Hecox Maybe just mill a groove in the plate for chain clearance and a little extra side clearance for the chain to whip around a bit. Is there any way to get a straight edge on the sprockets to see how well the centerline of the gears are lining up? +1

Rob Siegel I tried to do that with the jackshaft and the crank sprockets. I couldn't really find a machine surface to lay a straight edge (they are after all round). I have not yet tried that with the cam sprockets. With the engine is assembled as it is now, there is no way to lay a straight edge there. If I pull everything apart, it might be possible to test-fit the head with the sprockets on the campus but without the front cover obscuring the crank in the jack shaft. I never thought about doing that. +3

Joey Hertzberg Man, I really hate tearing things back apart, it takes a different flavor of motivation. +2

    Rob Siegel I know, right?

    Chris Roberts I got used to it a long time ago :D

Marc Bridge Engines are engineering marvels. It always amazes me how fine the fit is on everything and generally speaking they assemble and work. Then of course there is the other side that coin where the slightest change tips the dominos against you. If there's anyone who can solve this it's you! +3

    Scott Winfield I too often reflect on this. And they last for hundreds of thousands of miles with the right care. God bless the engineer :) +1

George Saylor From what I recall there are no spacers behind the cam sprockets. The timing cover does look a little thicker than normal.

    Rob Siegel George, yes, agreed that there aren't any. What I was looking for was whether people introduce them to solve alignment or clearance issues like this.

    Ed MacVaugh Rob Siegel How much wider is the opening in the chain as compared to the width of the teeth?

    Rob Siegel I haven't measured that. Next time I'm out in the garage.

    Ed MacVaugh Rob Siegel When you said German timing chain, my mind suggested metric measurement. When you said Ford Twin-cam, my mind says inch measurement sprocket. +1

Scott Winfield Measure twenty times, start her up once. +5

Eric Wayte A great visualization of tolerance stacking. Good luck! +1

Skip Wareham How would Edd China solve this dilemma? +1

    Rob Siegel Quit the show 🙂 +7

    Doug Jacobs Rob, or, set it on fire. 🙂

Rob Koenig I'm still trying to understand if it's truly making contact or not. The asymmetric thrust washers could help. But, if you're able to move the crank to its full limit currently and don't have contact I think it'll be fine. I don't mind living on the edge so long as Clarence says there's clearance. +1

    Rob Siegel But I would still need a vector, Victor +3

    Rob Koenig Roger, Roger. +1

    Rob Siegel To answer your question, it's so close that I can't tell if it's touching or not, which means it's too close +1

    Andrew Wilson Need more than a sheet of typing/printer paper for clearance Clarence. Looks like less than that from the photo. +2

Charles Morris Run it until it clearances itself a couple thousandths where it needs, then change the oil? 😊

Thomas R. Wilson 💯

Glenn Stephens My heroes. Their Esprit broke the night before a rally. They went home and returned with... A Jensen Interceptor!

Carl Witthoft Python fans? I prefer R, and maybe Julia.

Ric McGinn [pic of sad face]

John Goddard Yep, you're into custom engine building here. If you do take it back apart, perhaps use an RTV, instead of the paper gasket—might be thinner assembled.

Rob Halsey I have mixed feelings about your struggles. There is a part of me that is completely sympathetic about the frustration and struggles you are having. On the other hand, I take great joy in reading your words and learning about the complexity and precision of this little engine. Thank you for the posts. +2

> **Rob Siegel** Your mixed feelings mirror mine. +2

Scott Aaron (Sigh)

Brad Purvis At least now you have a name for the Lotus...

Scott Fisher On a more practical (hah!) note: could your friend the machinist relieve the part of the head that's interfering with the timing chain? That question prompts another of my long stories, which I'll post separately to keep the derisive comments about "practical" solutions here.

> **Rob Siegel** This has been solved; I'll post it tomorrow. +3
>
> **Scott Fisher** Oh, THAT is good to hear.

Scott Fisher So on MY Lotus-Ford Twin Cam, which was in The World's Crappiest Lotus Cortina (a whole story on its own). I spun a rod bearing and had to have the crank reground and the block line-bored. At this point, I learned something that you may already know about the Lotus-Ford Twin Cam: line-boring the crank involves skimming off a few thou from each of the main bearing caps, installing them onto the block, and THEN grinding them slightly oversize. Which means you have to get main bearings which are, say, 0.020 undersize AND 0.015 oversize. The shop that was doing the work had no luck finding such bearings in Los Angeles. I contacted a friend who had a shop specializing in other British cars. He had gone to high school with Dennis Ortenburger AND knew the Vandervell distributor personally, so I got the Vandervell guy's number. They had no record of the bearings in that size anywhere in N. America.

I had a business trip to San Francisco where I hooked up with a co-worker, Frank Pohl (who is probably reading this, the world being as small as it is) who had an Alfa Romeo Berlina, the little four-door that looks like a slightly smaller Volvo 144 but has a twin-cam all its own. Frank and I played hooky on Friday of the week I was in town. THAT was one of the best car days I've ever had in my life. Frank had to deliver a twin-plug head from an Alfa TZ to Ron Hasselgren. Yeah, the guy from Hasselgren Racing Engines, at the time the leading builder of Formula Atlantic engines (before they went to Toyota.)

When we got to Ron's shop, the man himself was fly-cutting relief notches in the tops of pistons for a Formula Atlantic engine. I happened to ask if he had a set of 0.015 over/0.020 under main bearings for a line-bored Twin Cam. He walked into his storage area, came out, and handed me the set. I then made a bunch of collect calls (this being some years before the invention of cell phones) to my mechanic asking if I should buy the bearings right then and bring them home. Mechanic (who sounded EXACTLY LIKE Jackie Stewart) said no, he thought he could find some.

So I left the only bearings in stock in North America at Hasselgren's shop in Berkeley and went home to Southern California. Where the Wee Scot sound-alike failed to locate them. We bought them from Hasselgren (who still had them) and got the Lo-Cort running.

That August, I packed the Cortina full of camping gear and drove to Monterey for the Historics. I had brought our little Weber "Smoky Joe" mini-charcoal grill, packed in the trunk. I quipped that while there were certainly many cars at Monterey that week with three Webers, I was most certainly the only one with two under the hood and one in the trunk.

On Saturday, the loudspeaker announced a special guest, Sir John Young Stewart. He came over the PA and was interviewed by the track announcer, who asked when Stewart had first driven the track.

"I first came to Lagooooona SEECA when I was racing for Colin Chapman, and I drove a Lotus Cortina here in 1964."*

Brush with greatness, that

* I *think* it was 1964. Rob's friends being who they are, there's quite likely someone reading this who will chime in with "No, Stewart first drove Laguna Seca for Chapman in 1963." Or '65. The story happened in 1986, and I'll claim leaky physical memory for the lapse. +4

> **Rob Siegel** wow... +2

TLC Day 2088: I am a big doofus

FEB 26, 2019, 7:50 PM

The Lotus Chronicles, Day 2088:

Clearance problem SOLVED!

Let me foreshadow the explanation with two quotes:

1) "What I was seeing wasn't what was happening at all"—Jackson Browne (ironically, from the song "Fountain of Sorrow")

2) "I am a big doofus"—Rob Siegel

I ran the problem by The Lotus Engine God, along with my lengthy theory of "tolerance stacking" (that the thicker gasket + the thicker backplate + the thicker timing chain added up to the root cause of the problem), and floated the suggestion so many made about shimming the cam gears outward. This was his response:

"Remove the sprockets, drop the chain in, stuff in a bunch of paper towels to catch the debris, and grind a little clearance with a carbide burr. Vacuum out the chips. NO SHIMS BEHIND THE SPROCKETS!" Well, okay then.

I started to do this, but there wasn't really enough space to get my Dremel tool down there. So I used a flat file.

As soon as the file hit the surface, I immediately realized that I wasn't filing metal. I was filing cork. The back of the timing chain was never hitting metal at all. It was hitting the cork gasket where it had squeezed out from between the head and the timing covers when I torqued the head down.

With the cork filed, I could see that, as far as metal to metal, I had clearance. I ALWAYS had clearance. (Clarence :^)

With hindsight, I know, it seems silly. The fact that there was clearance at the bottom but not at the top had made me suspect an alignment issue. Then, when it looked like things were correctly aligned, I thought I had it nailed with "tolerance stacking," but never asked myself how that would explain clearance at the bottom but not the top. I never once asked myself if I was sure about what I was seeing.

Do I still get points for solving the problem even if the solution is determining that there IS no problem?

I shall take this hard-won knowledge and, uh, file it.

On to pressure-testing!

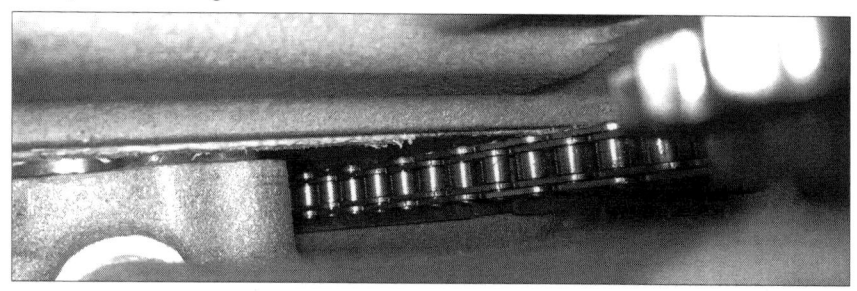

144 LIKES, 71 COMMENTS

Greg Lewis Double points for still being in the game with a bonus for Jackson Browne reference. +3

Bo Gray Surely you would have figured it out. And yes, I'll stop calling you Shirley. +4

Doug Jacobs Live and fuckin learn! +1

Jake Metz Hell yeah! +1

Tanya Athanus Just in time for spring. Rob you are all about forward motion. I have a lot of admiration for your persistence... +2

Luther K Brefo You do realize I haven't been able to sleep well since you first posted about this. Tonight I will sleep like a baby, which is to say I will wake up several times to feed the baby and deal with the random needs of a four year old. +3

> **Rob Siegel** I was about to say "You know nothing about my clearance-related dreams," but perhaps you do +2

Dohn Roush Tolerance stacking sounds like how I got my wife to agree to marry me... +10

> **Luther K Brefo** Dohn, me too... I'm afraid of the day she realizes what she actually signed on the dotted line for. +1

Larry Johnson first we panic +2

John Margherita All this knowledge will come in handy when you rebuild your next Lotus engine. 😊 +8

> **Rob Siegel** HAHAHAHAHAHAHAHA oh god that's funny +3

Rob Blake So... you never really had a problem to begin with.

> **Rob Siegel** I never had THIS problem to begin with. I had several other clearance problems. +1
>
> **Rob Blake** Rob Siegel just a mere pigment of your imagination. 😊 +1
>
> **Jeremy Novak** Well, I bet he was seeing red over the whole thing. +1

John Graham So indirectly I was right in saying since the car is British it had to do with wooden shims, or in this case, cork shims? +2

> **Rob Siegel** I was a big doofus. In return, anyone can take whatever credit they feel they deserve :^) +1
>
> **John Graham** Rob Siegel The word doofus isn't used enough anymore...we need to really bring it back into vogue...I am quite positive with things I do, I could make it a household word.

Scott Linn Awesome

Jim Strickland And it's good to see your smiling face tonight. Great news Rob! +4

> **Rob Siegel** Jim gets it :^) +1

Steve Park I think anyone who's worked on cars much, have had that doofus moment, whether they admit it or not. Congrats on being able to move on. +5

> **Rob Siegel** I revel in my doofus-dom. Especially when I learn from it, it's cost me no money, and hasn't resulted in an emergency room visit. +7
>
> **David Geisinger** as he looks for nearest piece of solid wood to knock upon +1
>
> **Clyde Gates** Rob Siegel I sense a series of articles forthcoming... +1
>
> **Jean Laffite** Rob Siegel If I had a dollar for every time...

Ben Greisler Better to err and be ok than err and not be. Call it a win. +1

> **George Murnaghan** Ben Greisler In other words, in the tug of war between the risk of error and the risk of delay, this was a case where the risk of delay wins hands down. Well done, Rob!

David Geisinger I must say patience is clearly one of your virtues. Great work! +1

> **Rob Siegel** David, you KNOW this is a zero-sum game, right? I was patient here, this time, but I'll pay for it somewhere else :^) +2
>
> **David Geisinger** Rob Siegel take the wins when you can I guess 😊 +2

Christopher Kohler Hey better to approach it with an abundance of caution than just going for it. +2

Clay Weiland It turns out that the clearance was inside you, all along. +3

> **Rob Siegel** [video of secret ingredient soup from Kung Fu Panda] +1

Dave Gerwig Hope the pressure checks: compression and water jacket I suppose represent a tight seal for the long block. You've earned it with your cork working like a windage tray for the Lotus Timing Chain. +2

> **Rob Siegel** yup, conventional cylinder leakdown test, then jury-rigging something up that clamps off the water outlet and allows me to pressurize the water inlet and spray soapy water just like an a/c pressure test. +1

Collin Blackburn Oof. I'm so glad your rabbit hole was just a small crater! +1

George Zinycz Rob as you know I've had a great day, but reading this makes it even greater!!! Very happy for you... +1

Ernie Peters I applaud your efforts in solving the (non)issue. Like you did in this scenario, I, too, often problem solve by starting with the worst case. I read this article a while back, but do not necessarily agree that catastrophising" is a bad thing. [link to article on Business Insider about catastrophizing]

Bret Luter It's cork, dork. ;) +1

Andrew Wilson Glad it was much ado about nothing. Best to err on the side of caution. Glad the file showed you had clearance. +2

Bill Snider What Great News Rob! Cheers! +1

Joey Hertzberg You get credit from me, because this example is analogous to my entire life experience. +1

Rob Koenig Toldja. 😊 +1

David Weisman There always was clearance. Reminds me of the tin man

Brian Stauss From one doofus to another, congrats on getting the problem solved!

David Shatzer Occam's Razor... you did everything you were supposed to. Simplest issue won out. After revisiting the hard stuff. I would have done the same until I got to day xx85. Then I would have packed the cylinders with some high energy substance and... +1

Bill Ecker Sometimes believing is seeing. You were so anxious about a clearance issue you saw one that wasn't (really) there. That outside objective voice cleared the fog. Well done for hearing it! +1

Scott Aaron Oh my God what a frickin relief +1

Jeremy Novak I've seen cork HEADGASKETS on some of those old British things. They don't hold up real well in vintage racing. I don't recall the reasoning behind not upgrading to a more modern material. +1

Lindsey Brown And you had a brain this whole time. [photo scarecrow] +2

Steven Bauer Better to be cautious than careless, especially where your experience is limited. +1

Derek Barnes Eureka!!

    Derek Barnes And probably at this hour "you reeka" +1

David Ibsen Don't become a flat rate lotus mechanic just yet! Congrats! 🍷 🔧 🏁 +1

Thomas Jones Roger the Airplane reference. Glad you're on the right vector now. +1

Anuraj Shah This post made my day!! +1

Franklin Alfonso So now you are seeing problems where there are no problems. This car has you bewitched lol. +1

John Jones Full credit for a victory, particularly since you sorted it out before grinding off a bunch of metal unnecessarily. +1

Gary Beck Ask me to tell you my story about compression ratio sometime. Oh the trauma we go through as engine builders. +1

Lou Millinghausen Follow the chain of events to success... +1

Corey Dalba So the irony is, despite your requests for actual advice, "Send it, you're fine" was, in fact, sound advice. I love it. +1

Seth Connelly this is like weekly episodes of Flash Gordon. Nice job Rob, like Flash you will always survive to move on to "the next" episode. Just a thought... don't buy parts from a parts dealer named "Ming"... ;-) +1

Sam Pellegrino I hope you «Soaked the Cork» before installing that gasket... [video of "Corksoakers" on SNL]

Mike Hopkins I just spent a year trying everything to get a Laycock overdrive to work on a Volvo P1800, when I discovered that it was a sheared 55 cent woodruff key on the pump drive cam. It feels so much more special driving the car now that it would have if the overdrive had worked from the start, or at least I keep telling myself that! +1

TLC Day 2089: Leakdown test

FEB 27, 2019, 7:37 PM

The Lotus Chronicles, Day 2089:

Fighting a winter cold, so just two quickies today.

– I did a quick leakdown test on all four cylinders just to make sure all the valves were seating. No issues.

– I glued on the plate with the rear crank seal. I'd been avoiding this for weeks as it required repositioning the engine slightly on the stand (the arms attached to the bell housing bolt circle were in the way), but by moving one arm at a time, I was able to do it without having to independently support the engine, which is good, because it'll take me several hours of crap-moving to get the lift table out from under the back porch. I'll eventually need to do that to get the flywheel and clutch on, mate it with the transaxle, and get the drivetrain in the car, but that comes later.

Hopefully, tomorrow I can start pressure-testing the coolant passages (meaning making sure that the covers and water pump I just installed actually seal). The hard work is plugging up the ports that plumb things like the heater when the engine isn't in the car. This means finding parts I haven't seen in 5 1/2 years. Wish me luck.

57 LIKES, 37 COMMENTS

Karel Jennings I feel you on the rear main seal. I have to do the same thing on a 600lb cast iron dodge flathead... Safest thing will be to disconnect it from the stand and put it on its side on the ground to bolt the rear main seal doo-dad back on.

    Rob Siegel I got it on in-place, but will need to deal with it again come flywheel and clutch time. 'Course by the time I'm at that point, I'll be nearly ready to install the drivetrain. +1

    Karel Jennings As much a pain in the ass these jobs are, at the end it's worth it. Next time though I'll rebuild an M10. Go with what you know. ;) +1

    Rob Siegel yes, I thought dozens of times during the two freaking months it took me to fit the timing covers and the water pump that I'd have this completed on an M10 in 20 minutes.

    Karel Jennings It's so true! But the weirdness of it, the differences is what I enjoy. Different engineering answers to the same question. +1

    Karel Jennings I mean, sometimes it's a drunken answer. but still. +2

    Shad Essex Karel Jennings Gotta love flat bottom girls and piano dollies. I envy this "stand" thing of which you both speak. :p [photo of an absolutely mammoth V12 on the floor] +1

    Rob Siegel I feel so small and unworthy

    Karel Jennings Is... that a Lincoln flathead v12?

    Karel Jennings Shad Essex I am also unworthy

    Shad Essex Karel Jennings Good eye. It is. +4

    Rob Siegel Shad, man, that's just showin' off :^) +1

> **Shad Essex** Nonsense. Nothing to it, but to do it. "Righty tighty, lefty loosey"...just a few more times. +1
>
> **Karel Jennings** :) I fell in with these hot rodders who taught me a thing or two. :) +2
>
> **Brad Purvis** Shad Essex I believe the term is "Fat Bottomed Girls." 😎
>
> **Shad Essex** Brad Purvis I know, I know...just trying to be punny. ;)

John Whetstone Check the sealed Tupperware container marked "Lotus"? +2

> **John Whetstone** Oh, right.
>
> **Rob Siegel** if life were only like that
>
> **John Whetstone** Rob Siegel not like I do it.
>
> **Bob Sawtelle** its right next to the coffee can of 10mm sockets I'd guess. +1
>
> **John Whetstone** As a cocky yout', id never think of organizing parts. But if I was to tear an engine down today, I'd have to have labeled boxes. And be recorded by a Wyzecam or something similar.
>
> **Rob Siegel** I did some of that, a lot actually, but not 100%.
>
> **Scott Aaron** John Whetstone you've seen those pics of the garage, right?
>
> **John Whetstone** I just assumed it was staged +3
>
> **Steve Park** Every time I do a project without proper organization I regret it, only to continue to do it to some degree the next time. The definition of insanity I guess. +1
>
> **Rob Siegel** Steve, that's me too
>
> **Steve Park** Rob Siegel Great minds and all that! +1
>
> **John Whetstone** So, who here IS organized?!
>
> **Bob Selfridge** Nope, not here. I have so many tools and car parts for the 6 cars I Try and maintain in the garage, I can no longer fit a car in the garage to repair it. Spring Cleaning day is coming soon.
>
> **George Murnaghan** John Whetstone Hey I know where those parts are...or at least which sector of the garage!

Scott Linn Hey, progress! +1

Ernie Peters Luck! +1

Sam Schultz I have the same cold so good luck. +1

Brian Stauss "Several hours of crap moving"... I can relate to that! +1

Robert Alan Scalla When you have to move heavy parts, get assistance. FU the back and it's all over.

Chapter VI: Under Pressure

TLC Day 2090: Stripped threads on neck

FEB 28, 2019, 9:39 PM

The Lotus Chronicles, Day 2090:

Started to pressure-test the cooling system. This requires blocking off all of the coolant passages except one, which you pressurize. Was planning on doing it with a cut-up wheelbarrow tube, stretching the Schrader valve across one of the water necks so I could pump it up with the compressor, but by pure chance I found that I had a fitting for my a/c pressure-testing nitrogen bottle that threads directly into one of the ports on the head. That's great because I can use the a/c manifold gauges. Stunningly lucky.

From then on, it's just like a/c pressure-testing. First you find the big leaks by listening for air escaping, address those (loose hose clamps, thermostat housings that need RTV, etc), then you do it again by watching to see if the gauge drops and, if it does, find the smaller leaks using soapy water.

At 15 psi (pretty low pressure; The Lotus Engine God recommended I go up to 40), I found a leak that, to my surprise, was coming from where the threaded neck screws into the front of the Dave Bean custom timing cover. This is a unique/cool thing with the Bean system. Elans, Cortinas, and Sevens have the water neck on the side of the cover, but Europas have it on the front. With the Bean kit, they can use the same part and just put the neck in one threaded hole and a threaded plug in the other. Both the cover and the neck are aluminum. I ordered the kit for a Europa, so the neck came on the front. No big deal it leaking a little bit there in this first pressure-test, I thought; I'll just unscrew it, put a little sealant on it, and try it again. Hey, you test to unearth these things, right?

The neck didn't unscrew with any alarming resistance, but when it was out, there were shavings on the neck threads. I looked at the threads in the cover, and they appeared irregular. When I cleaned them, I could see that they were clearly damaged. Were they cross-threaded, then re-tapped? Was the wrong tap used and then the neck force-threaded in? Don't know. The threads on the removable neck don't look damaged in the same way, but show some cuts and rounding.

I immediately called Bean, and they said that, if necessary, they'd send me another cover, but that would be days of extra work, plus re-doing the custom machining that went into this cover. For now, we've agreed to try sealant on the threads and a crush-ring under the seating surface of the neck.

I just spent an hour and a half chasing the threads in the cover with a pick. All of the black spots on the paper towel are pieces of metal dug out of the threads. It's much better now than it was. I'll wait until I receive the crush ring before I do anything else.

It's just a coolant neck. It just has to seal against coolant (and this pressure test). It'll probably be fine.

I'm going to eat some ice cream now. With a brownie. That makes most things better.

53 LIKES, 33 COMMENTS

Paul Wegweiser If I ever need a new cover for a Lotus motor, I'm making one myself from a pile of rocks and a head of lettuce. Oughta be less trouble. +19

Tom Egan This isn't one of those straight thread/pipe thread problems, is it? Been burned by that one before. +1

>**Rob Siegel** That's not a bad guess.

>**Paul Wegweiser** Rob Siegel I'm guessing that cutting proper threads would have made this more costly. At least it was cheap, right? Right? +4

>**Rob Siegel** Fuck you. +8

>**Paul Wegweiser** Rob Siegel Sorry... I'd have had a complete meltdown about three weeks ago. +4

>**Rob Siegel** all good +3

Ed MacVaugh Was that opening cast on the Lotus factory front cover, or was it threaded, too?

>**Rob Siegel** Cast... or possibly drilled and a neck welded on. Most Twin-Cam engines were in Elans. Fewer of them were in Europas. Which is why the cartridge-style pump for the Europa is currently only available from Bean. The one with the side opening is available from Burton Engineering.

>**Ed MacVaugh** Rob Siegel Of course, the vast majority of those engines were in Lotus Cortina Fords and race cars, but the Europa was alone with the straight out water fitting. +1

Christopher Kohler Man so much trouble due to one part! Hope it is sorted without a whole lot of effort.

Steven Bauer I was thinking that using a smoke tester would make the leaks much more obvious except for the residue in the cooling system from the smoke.

>**Bob Shisler** You're thinking of British wiring. The cooling systems are different. :D +7

>**JP Hermes** don't let the smoke out of your wires! +2

Bob Shisler Use JB Weld as the sealant. /solved. +2

Steve Woodard What you're seeing is the difference between an automotive casting and an aerospace casting. That said, I don't need to hear the $450 toilet thing. Sometimes you get what you pay for. +1

>**Joey Hertzberg** Steve Woodard had a couple guys come into my cafe. A customer was having trouble with the cable that actuated his toilet on his Learjet. They had the stretched cable they were hoping to source used and threw it on the counter. I said, man, Jeff down at the bike shop could make this out of great quality brake or gear cable for maybe 15 bucks, it would last far longer than this shit. They told me they couldn't do it, it wouldn't be FAA approved cable, and a new one from Lear was $20,000 +4

>**John Harvey** Steve Woodard airplanes run on fuel and paperwork. It has to be paperwork right.

> **Steve Woodard** Making an object that looks like the casting you want is relatively straightforward. If it needs to pass pressure testing, let alone FPI & x-ray, well, that's different. I fly in these things, as you may also. It's not all bullshit.

Roy Richard Any chance this is BPT V NPT? +1

> **Rob Siegel** sure
>
> **Ernie Peters** Rob Siegel If I recall from my Elan, the threads for the temperature sensor @ the thermostat housing are BPT. If your Bean housing was made in the US, I suspect the threads are NPT. But, it's a Lotus. Who knows? +2

Daniel Senie This is so Rob. Not only does he buy a Lotus, he buys one of the few with this engine, just to increase the risk of failure, so that he can ultimately defy the odds, and write another book about it (and maybe a song, when he ultimately decides to come to his senses and torch the thing).

Jake Metz Hey Rob, you're the man. Keep it up. +1

Chris Rouse This sounds like it could be solved pretty easily with some of that Permatex aviation sealant. The thick, brown gooey stuff that goes on like molasses. +1

Brad Purvis It's British. You need a cup of tea with your brownie & ice cream. +2

Duncan Irving Teflon tape a couple times around too.

> **Rob Siegel** I've found that for these applications, Teflon tape isn't anywhere near as effective as some sort of liquid thread sealant. +3
>
> **Duncan Irving** Good to know. I always worry about using too much or the wrong kind. I need more knowledge./

Brian Ach "Was planning on doing it with a cut-up wheelbarrow tube." Sounds about right.

David Stupar Remember this is the same country that gave us Lucas Electronics.

Brian Hart Shoddy workmanship is not relegated to the third world. Calling it out is important.

TLC Day 2092: Found... something

MAR 2, 2019, 3:51 PM

The Lotus Chronicles, Day 2092:

Still waiting on the crush washer I need to help seal up the leaky threads on the coolant neck so I can pressure-test. If it doesn't come in the mail by COB today, I'll seal the threads with Yamabond4 and test without the crush washer.

In the meantime, been installing more and more ancillary parts like the oil pump on the engine, assuming that the pressure-test results won't tell me to tear it apart. On the downside, I was horrified to find that I'd forgotten to install an oil drain tube from the head to the block that really is supposed to go in place before you drop on the head. Fortunately I learned that there are a few workarounds to this issue.

And on the plus side, I can't tell you how satisfying it is to be far enough along in the engine reassembly process that I can open the zip-lock bag that literally was labeled "Something" and figure out that it held the three bolts to the oil pump I was installing.

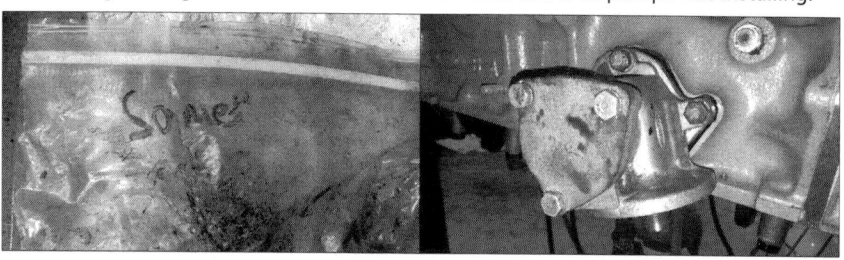

51 LIKES, 32 COMMENTS

Steve Nelson I'd say seal up the pipe with your Yamabond now so you can further test for leaks...then when the crush washer comes in, you can test that single item out... might find other leaks testing now.

> **Rob Siegel** If it doesn't come in the mail today, that's what I'll do.

Sal Mack I have money that says Yamabond will seal that up if the threads even decent. That stuff is amazing.

> **Rob Siegel** That's what I expect, but when you ask the manufacturer for their advice, and they say "sealant and a crush ring," you want to take the advice, at least I do. +2

> **Ed MacVaugh** Rob Siegel Are you aware that you seem to be heading far far away from being a hack mechanic? +3

Frank Mummolo When I do restos, I usually have a number of bags labeled "Something". You're doing great, Rob! +1

Neil M. Fennessey Good to see that you use freezer ziploc bags. The non-freezer bags don't cut the mustard. +4

Robert Alan Scalla I put the dashboard back together on my Eurovan and had 12 screws leftover. Is that bad? +1

> **Robert Boynton** Robert Alan Scalla suspicious

Glenn Stephens Surely local hardware has a copper washer.

> **Rob Siegel** Went to five hardware and auto parts stores. it needs a crush ring, not a washer, that's about 1" inner diameter, about 1.125" outer diameter, and very thin.

> **Glenn Stephens** Rob Siegel And stop calling me Shirley! +2

Steve Park I agree. Why ask for their advice if you aren't going to listen to it? +2

Lindsey Brown If I'm allowed to come over and play tomorrow, we can go to my shop, and have access to a large stockpile of crush washers. +1

> **Rob Siegel** Lindsey, you are welcome to come over and play tomorrow; I will be here all day. But I just laid the coat of Yamabond down on the threads, screwed it together hand-tight (I'm afraid to do anymore with all ready partially-stripped aluminum threads) and will let them sit overnight. I'm not going to want to pull it back apart. But thanks for the offer for the crush ring.

> **Lindsey Brown** Perhaps we'll both just have to live to play another day.

Paul Wegweiser Maris Mangulis and I just filled 7 bags of "something" while tearing his motor apart. *wheeeeee!!!!* [photo wegweiser motor #1] +5

> **Maris Mangulis** How do you make Paul Wegweiser happy? Give him an M10 to work on...many thanks, my friend. [photo wegweiser motor #2] +3

> **Ed MacVaugh** However, you can likely identify each of those parts blindfolded, in the dark, with your hands behind your back. Plus, you will be polishing each one on a wire wheel anyway and fondling them multiple times. +2

> **Brian Ach** Paul Wegweiser the best is when you take it apart and end up with 7 bags, but when putting it back together you find 8 bags, assemble it, and have 2 left over. +2

> **Rob Siegel** Parts matching is an approximate science.

> **Jonathan Selig** Paul Wegweiser is the miracle grow to increase displacement? +1

> **Maris Mangulis** It's an attempt to put some curl back in his hair. I think. +1

> **Paul Wegweiser** Jonathan Selig Not my garage... but now that you mention it... :D

Alan Hunter Johansson Sev reminded me today that my brother Chris used to talk about the point in any project when you flip the ratchet from "OFF" to on "ON" +5

Richard Koch 🐌

Stuart Moulton As long as you don't have a pile of "something" hardware left over when you're done.

> **Rob Siegel** It's a Lotus. If you don't have parts left over, you aren't channeling Colin Chapman! +2

> **Stuart Moulton** Rob Siegel he'd be happy you lost weight from the assembly +3

> **Garrett Briggs** label the bag of leftover parts "adding lightness" +1

Harry Krix Jesus, (I know you have no idea who this guy was...) I've built SCCA National Championship race engines for Austin-Healey Sprites in less time, Perhaps we should go back to the Cubit system... (These definitions ranged between 444 mm and 529.2 mm). just a thought, trying to help. love ya bro. +1

TLC Day 2092: My impatience sets me back

MAR 2, 2019, 10:48 PM

The Lotus Chronicles, Day 2092, part B:

The crush ring didn't arrive in today's mail, so I went ahead and coated the threads on the neck with Yamabond4, the stuff I used on the timing covers. It has a great reputation for metal-to-metal sealing. I laid it on pretty thick, including at the outer sealing face. I let it cure for four hours. The outer lip was certainly firm to the touch. I kept thinking I should wait until morning, but NNNnnnnnooooo mister impatient decided that, if I pressurized it tonight, I could let the pressure sit overnight and imagined seeing the triumph of an unmoved pressure reading on the gauge in the morning.

Big mistake.

Immediately on pressurizing, and not even that high, like 15psi, I could hear air escaping. With horror, I saw the gray Yamabond running out of the threads.

Fuck.

I just took it all apart and spent an hour cleaning partially-cured Yamabond out of the threads with paint thinner, a pick, and an air nozzle. What a pain in the ass.

Next time, patience, and I think I'll try this.

15 LIKES, 16 COMMENTS

Drew Lagravinese You could just wait for the crush washer that will probably arrive tomorrow 😀 +2
Rob Siegel Monday at the earliest.
Rob Blake Rob Siegel yeah... but a better solution.
Glenn Stephens [meme Kung Fu "patience young grasshopper"]
Jim Strickland The 300F upper limit should cover most things automotive? Looks great. And you can undo it +1
Scott Winfield Oy vey. You have my sympathy. +1
Scott Aaron You lay it on thick, all right... +2
Bob Shisler JB Weld +1
  George Zinycz Yeah Rob why isn't this a job for tried & true JB??
  Rob Siegel Seems a bit extreme. and if you pressure-test it, and it leaks, you've got a real problem, as once you've put JB Weld on it, it's not meant to be disassembled. +1
  Bob Shisler Rob Siegel JB weld won't leak (unless you user-error it like you did with the other stuff). In this application would work like a charm. Just mix it up, apply to the threads male and female, screw it together and you have a permanent bond of both parts.

Scott Lagrotteria JB is too permanent.

Dave Borasky when you're on day 2092, waiting another day won't hurt

Franklin Alfonso Lolita won't be rushed. Wait for the crush ring… +2

Bart Collins Patience is a virtue, of which I have none. +1

Bill Mann I just had a trip in the Wayback Machine to my high school garage and a bottle of "Indian Head" gasket sealer. Real gooey gummy stuff. +1

TLC Day 2093: Carbs while waiting

MAR 3, 2019, 7:04 PM

The Lotus Chronicles, Day 2093:

Since I'm in waiting mode for the crush ring, then will wait several more days for the adhesive on the leaky threads on the coolant neck to set (package says partial cure in 24 hours at 72 degrees, full cure in 72 hours, and it's 50 degrees in my garage, so I WILL wait the full three days), I've been doing odds and ends. More and more I'm just assembling this motor, assuming that the pressure test won't unearth anything problematic, which basically means "just don't glue the oil sump on yet."

Was NOT successful at removing a snapped stud from the exhaust manifold despite using so much heat that the end of the stud melted, but tomorrow is another day.

It's incredibly satisfying to have reached the point where I'm opening boxes of parts I haven't seen since June 2013. I remembered the car's being complete, but it's great to see that I DO in fact have the valve cover, alternator, distributor, air cleaner housing, etc.

I had no memory whatsoever what condition the Zenith Stromberg garbs were in. They could've been rusty seized garbage. However, I was VERY pleasantly surprised to find that they look pretty good, throttle shafts all rotate smoothly. I'll pull the tops off and inspect the float bowls, but… this guy's beginning to think thoughts about Lolita that just aren't right.

49 LIKES, 33 COMMENTS

Lindsey Brown Let me know if you need anything for the carbs, or want to see my patented technique for broken stud removal. Just come by the shop to play anytime. +3

Scott Linn One thing about ZS throttle shafts. They have trivially simple bushing replacement, unlike SU carbs. +1

John Goddard Yep, I'd take a close look at those carbs before install...

Jeff Hecox I happen to have a factory Lotus Elan + 2 work shop manual that devotes about 30 pages to the Zenith Stromberg carburetors. I could be persuaded to scan some pages if needed. +1

>   **Jeff Hecox** It sounds like your carburetors are due for the "Red Service." [photos of Strombergs]
>
>   **Rob Siegel** Jeff, thanks, I believe I have this as well.

Andrew Wilson As Cheryl Crow sang "You've gotta run, baby, run baby, run" +2

>   **John Whetstone** Andrew Wilson Away? +2
>
>   **Andrew Wilson** That ship sailed when Ben Thongsai proclaimed "Congratulations, you own a little brown thing" +2
>
>   **John Whetstone** Andrew Wilson You know he'll pull this off!
>
>   **Andrew Wilson** I've been encouraging him for 5 years,
>
>   **Rob Siegel** yes, totally Ben's fault! +2

Bill Snider S'a good thing you're getting over being unwell as thinking THOSE thoughts about Lolita... well, you wouldn't want to Nab a Cough. Grins. +1

>   **Bill Ecker** Bill Snider don't make me throw kuBricks your way. +1

Scott Linn There is a special tool to remove the jets. I think they were originally considered "non-replaceable," at least by the home mechanic. Volvo made a very nice jet removal/replacement tool and you apparently used to be able to find them on eBay but I haven't found one. Luckily a mechanic friend loaned me his. +1

>   **C.R. Krieger** The needle pops into the bottom of the slide. The jets are removable by taking the choke system apart.
>
>   **Scott Linn** I have a Europa and I recently replaced the jets in the ZS carbs They are not easily removed. I used the Volvo tool. I tried without it and was unsuccessful. Probably would have cracked the carb body.
>
>   **C.R. Krieger** My apologies. I figured that when Stromberg copied the SU that they would make it equally serviceable. Apparently not. I remember having a Stromberg on my 2002 and getting the advice from a BMW service advisor to carefully place it into a bucket of acid so I could have the pleasure of watching the SOB dissolve before my eyes. Then get a Weber.

Jeff F Hollen If you confirm a lotus at The Vintage, I'll meet you in VA and caravan my Jensen Harley with you. +1

Scott Aaron I was more into Quasar vs. Zenith. Also liked Quisp cereal. Had a thing for the letter "Q" I guess. +2

>   **Rob Siegel** Quisp and Quake! +2
>
>   **Caleb Miller** Rob Siegel don't forget their brother Special K! +2

C.R. Krieger The float bowls aren't on the top of SU type side drafts. They're on the bottom. +1

>   **Rob Siegel** Thanks, I did not know that.
>
>   **Scott Linn** Uh, these aren't SUs, but Zenith Strombergs. +1
>
>   **C.R. Krieger** They're the same basic design. The top is the housing for a dashpot that has a fluid damping piston inside the sliding throttle. The float bowl is on the bottom. That's why I said "SU type."
>
>   **Scott Linn** I've never seen an SU like that so I had to look it up, the SU HIF carb (Horizontal Integrated Float chamber). Every single SU carb I've ever seen (and I've seen a lot) have had an external float chamber that sits next to the carb. Looks like they weren't used very extensively. You learn something every day...
>
>   **C.R. Krieger** Scott Linn, we both did. Thanks! +1

Scott Linn There are o-rings on the bottom of the fuel chambers. Failure of these has caused fires.

Clyde Gates I preferred Skinner Unions to Zenith Strombergs. Play Steppenwolf while installing. Or Kenny Loggins? +1

Robert Alan Scalla Clean up those carbs on the outside now while you can.

Dan McLaughlin Oh bring those children inside to let them cure... I'm sure Maire Anne won't mind or is that crossing a line?

TLC Day 2094: Grrrrrrr...

MAR 4, 2019, 7:05 PM

The Lotus Chronicles, Day 2094:

Let's just say... Grrrrrrr... and leave it at that.

[meme image of two men in a car with captions "OH MY GOD" and "WHAT A FUCKIN' NIGHTMARE!"]

8 LIKES, 7 COMMENTS

David Kemether I wonder how many M10s you could have rebuilt in 2094 days.
Robert Myles All of them.
Rob Siegel In fairness, I did not spend 2094 days rebuilding this engine. It sat untouched for most of that time. However, it IS completely fair to say that I've spent the last 2 1/2 months fitting the front timing covers and the water pump, something that, on a 2002, I'd have done in 20 minutes, maybe less.
Isabel Zisk The beauty of that clip is that is applies in so very many situations... +1
    Rob Siegel And this is one of them. +2
Rob Koenig [link to "Always Look on the Bright Side of Life" from The Life Of Brian]

Interjection: "Grrrrrrrr...?"

When I read the post above while putting together this book, I didn't remember exactly what it was that I was referring to, and was surprised that, given the fact that I generally let everything else hang out on Facebook, I didn't explain it. But looking through my emails, the cause of the "Grrrrrrr" came roaring back to me.

There had been a quartet of new issues with the Bean water pump—four of them in rapid succession—that made me so frustrated and angry that I was ready to explode. These were:

1) Inexplicably, the hub on the front of the spindle in the custom Bean water pump cartridge was slightly too large to fit through the hole in my original pump pulley. I routed the hole out with a Dremel tool. By itelf, not a huge deal, and I have learned over the years not to automatically jump to "your new pump won't fit my original pulley" because, on a 47-year-old car with unknown provenance, you don't really know that any part on it is in fact original.

2) Ken at Bean had sent me a new threaded neck to replace the one that was mis-threaded on the timing cover. Its threads were different from

the one that was originally installed in the cover, which indicated to me that the parts were mismatched and cross-threaded. I tried it in the cover, but it threaded in with even more resistance than the one that was in there, indicating that the cross-threading damage had been done.

3) But the biggie was that when I mounted the routed-out pulley on the flange of the Bean cartridge pump, it was mis-aligned from the crankshaft pulley by a substantial amount. And to make matters worse...

4) When I spun the water pump, the pulley visibly wobbled. On examination, I found that the shaft and/or flange on the new cartridge water pump, the one that was part of the kit I'd spent nine hundred bucks on, clearly was bent.

I was livid, and wrote Ken the email below.

Why didn't I post on Facebook about this latest round of issues? I believe that, at the time, I felt that it would've been unprofessional of me and unfair to Bean to explode on Facebook in a public fashion, which is what I would've done. But I also felt embarrassed for my own decisions. I'd delayed the project for nearly an extra year by my insistence that the engine rebuild include a cartridge-style water pump (remember: the Bean package was the only one I could order that worked for the Europa, and it had gone out of stock), and I knew that writing about *another* set of problems with it would unleash a hail of comments about my having made a bad choice, to which I'd only be able to say "you're right."

---

email: Dave Bean Engineering

```
From: Rob Siegel
To: Ken Gray, Dave Bean Enterprises
Mon, Mar 4, 2019 4:11 pm
```

Ken, this just gets worse and worse.

I mounted my routed-out pulley on the front flange of the Bean water pump to take a measurement on how far off it is from the crank pulley, and noticed that the pulley appeared warped. However, on closer examination, the problem is that the flange on the face of the Bean water pump is not square on the shaft. See unlisted video here (https://youtu.be/SxGtxX9bezM). Note that in the video, I have the pulley mounted backward so I can spin the pump and you can see the flange without my hand getting on it.

This pump has been in my possession since it arrived Christmas eve. It has not been dropped or mishandled in any way.

What I would like to do is try to take measurement for the minimum and maximum off-center distances for the two pulleys, and then have you send me another water pump that has been 1) checked for trueness and 2) offset by the am ount needed to mount it on my engine. I will send you those measurements in another email. It's not obvious to me that, if the pump's shaft is pushed back far enough to align the two pulleys, the back of the pulley will clear the Allen-head bolts. I also don't have a press. I would rather that Bean be responsible for it fitting than me.

--That being said, if you *can't* pre-configure a pump in this way, and if I have to adjust this spacing myself, let me know.

--Also, let me know if it is possible for me to adjust this without a press by supporting the pump on the mounting flanges and tapping the head of the shaft. I'm game for trying it, but I'd like to try it first on this current pump before you send me another one.

I am an enormously fair and patient person. It is an understatement to say that these quality control issues are trying my patience to the breaking point. At a minimum, I recommend you remove the claim "It actually FITS!!! So many others require various fettling and hacking about with files and reamers, etc" from the product page on your website (http://davebean.com/NewProd.htm), as it clearly isn't accurate.

(Regarding the coolant neck, for the record, I re-mounted the original one, with the crush washer and o-ring, using Permatex Seal+Lock on the threads. I will wait the recommended 72 hours before pressure-testing.)

--Rob Siegel

....
From: Ken Gray, DBE
To: Rob Siegel
Mon, Mar 4, 2019 4:34 pm

Hi Rob. I understand the difficulties and I am sorry. Yes please measure the distance for the flange on pump and I will make one to spec. Please let me know, so that I can get the replacement cartridge on its way.

```
From: Rob Siegel
To: Ken Gray, DBE
Mon, Mar 4, 2019 5:57 pm
```

Okay, I did this pulley offset measurement two different ways.

1) I used a straightedge (a wooden ruler in the photo, but I switched to something metal) on face of the water pump pulley and measured from the straightedge's rear face to the front of the crankshaft pulley, in what appeared to be the furthest and nearest approaches. They were 0.31 and 0.21 inches, respectively. I don't know, though, how centered the grooves are in the pulleys; again, this is a face-to-face measurement.

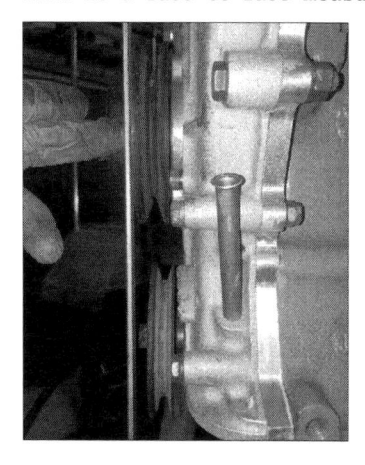

2) I then did it a second, more seat-of-the-pants way as a sanity-check by seeing how far I'd need to pull out the crankshaft pulley to make the grooves line up. I measured from the back side of the crankshaft pulley to a fixed point on the block. This measurement was .233 inches. I then undid the crank pulley nut, pulled the crank pulley forward, and aligned the pulley grooves as best as I could to about the middle of the wobble position of the water pump pulley (in other words, a compromise measurement considering the pulley isn't square on the shaft), and measured again . The second measurement was .478 inches. The difference between the two measurements is .245 inches. It's not right in the middle of the above two measurements, but it's not too far either.

So I'd say that the shaft has to be about a quarter of an inch further into the pump than it currently is.

Let me know if you need to measure it some other way.

--Rob

```
From: Ken Gray, DBE
To: Rob Siegel
Mon, Mar 4, 2019 6:06 pm
```

Hi Rob. Nice work. I can move the flange to help. Just one more measurement. Please give the distance from the back of the flange to the shoulder of the cartridge casting. Want to make sure my distance of .315" matches yours.

Thanks

```
From: Rob Siegel
To: Ken Gray, DBE
Mon, Mar 4, 2019 6:30 pm
```

Ken, as per the photo, I'm not seeing .315 inches, I'm seeing about .215 inches, in which case there's not a quarter inch of room to push it back. Please advise.

```
From: Ken Gray, DBE
To: Rob Siegel
Mon, Mar 4, 2019 6:49 pm
```

Hi Rob. The most I can do is a .155" correction. I think that .060" would be a good clearance for the flange to shoulder. Could move to .040". The v-belt should handle the .100" deflection, but I will make a shim to place on the nose of the crankshaft to move the crank pulley out .100" I will send this with the replacement cartridge. I will include two new Allen bolts with rounded heads for more clearance to the back of the pulley. Let me know if this sounds OK.

```
From: Rob Siegel
To: Ken Gray, DBE
Mon, Mar 4, 2019 6:56 pm
```

Ken, please do what you think will solve the problem on the basis of the measurements I've provided. But why is this so far off? The only thing that appears to be unique about my engine is the fact that it has the RDent backplate gasket, which is more like shirt cardboard than paper (0.015" instead of 0.005").

--Rob

TLC day 2094: Removing snapped exhaust stud

MAR 4, 2019, 8:58 PM

The Lotus Chronicles, Day 2094, Part B:

As per prior post, a number of things that I won't go into went wrong today. I really wanted to end the day with one success. I thought I would revisit trying to get the snapped stud off the exhaust manifold. I've had this Millermatic 141 welder in the garage since August and had not even turned it on. Paul Wegweiser told me to weld a nut on to the end of the snapped stud. I tried three times, and three times the nut twisted right off. I messaged Paul asking what I'm doing wrong. "Call me," he said. He gave me some pointers, including simply using a larger nut, speeding the wire feed up slightly to get more metal into the weld, and unscrewing it while it's still hot. Fucking BOOYA! Thanks Paul!

[search YouTube for "The Lotus Chronicles Unscrewing Stuck Stud"]

106 LIKES, 38 COMMENTS

Bret Luter this is one of my favorite things to do with a mig welder. :) +2
    Daren Stone Bret Luter I'm with ya.
Rob Siegel It was one of my reasons to buy it!
Frank Mummolo When finesse fails, brute force the MF'R!! Good on you , man! +1
Buck Hiltebeitel Heat and torque in large quantities work wonders. (Or destroy shit.) 😄 +2
Clay Weiland Paul Wegweiser is so good, he knows how to weld things apart. +16
Greg Lewis We are in the presence of greatness. Who knows these things? Paul Wegweiser does. +2
Paul Wegweiser Yay team Kook! 😄 +3
Joe Eaton Better call Paul. Sounds like a sitcom. +4
    Chris Ryan Joe Eaton or an add tag line for an injury lawyer! LOL
Brian Stauss You are da man!
John Harvey Throw that in your neighbors pool and start fresh with a nice ceramic coated header.
Rob Siegel If you're buying, I'll tell you my address.
John Harvey Rob Siegel don't tempt me. How much could they be? F'n IE bundle of snakes was like 600.
John Harvey Oh and I'm not buying.
John Harvey I mean, I'd like to but, well, you know, can't.
Paul Wegweiser Hey Rob Siegel! Maybe someone makes a sweet header for this car! [photo of the worst hacked-up exhaust welding job you've ever seen]
Bill Ecker [meme of "tool expectations" ("it can't be tight if it's a liquid"]
Scott Winfield That's great news. "You got to have friends"—Buzzy Lindhart
Paul Wegweiser Well... I guess SOMEBODY has to post it: [link to Dionne Warwick "That's What Friends Are For" +2
    Scott Winfield Nothing wrong with a little corn in the morning :) +2
John Turner Just give it a thorough rogering! Haha
George Zinycz "Harlan Pepper, will you stop welding nuts?" #namethemovie +2
    Scott Aaron George Zinycz Best In Show. +1
    Scott Aaron Are you watching Schitt's Creek? If not, you should. It's hilarious.
Robert Alan Scalla Victory!
Scott Lagrotteria Did you try a stud remover? Much less effort if you have a good one, and working on older cars you need one. 1/2" drive with breaker bar removes most of them.
    Rob Siegel Yes, I have an tried a stud remover. It used to be a good one but it may be losing its grip. (like me :^)
Collin Blackburn What's gonna happen when you finally get this thing running? 😄 Other than the Lucas electronics going.

Collin Blackburn [gif of massive H-bomb explosion] +3

John Whetstone Collin Blackburn seriously, let's think positive! Oh, and that bomb seemed to have worked. +1

Brian Ach Wegweiser and Siegel FTW +3

Paul Wegweiser This is my "I'm helping Rob Siegel" face: +2

Paul Wegweiser ...and for the record: Rob Siegel has helped me wrap my feeble brains around assorted electrical gremlins many times. I'm a visual learner so topics like mathematics and electricity occasionally escape my grasp. (refer to above photo) +2

David Weisman "We're gonna need a bigger nut" +1

Carl Witthoft Nuke it from orbit --- you know...

---

email: Dave Bean Engineering

```
From: Ken Gray, DBE
To: Rob Siegel
Tue, Mar 5, 2019 11:57 am
```

Hi Rob, I will make the replacement cartridge and the crank spacer. Can I have them shipped on Friday? It is hard sometimes for me to see the problems. I want to make the corrections, but harder for me to explain. I have made notes to take these problems out of future kits. I wish Dave was still here to see this or explain, so I am doing my best. Sorry for all the bumps and short falls. [Note: in prior phone calls, Ken had said that the cartridge water pump kit was one of the projects that Dave Bean handled himself, and that the updating of the pump kit was one of the projects that got back-burnered when Dave passed away.]

```
From: Ken Gray, DBE
To: Rob Siegel
Tue, Mar 5, 2019 1:47 pm
```

Hi Rob, I can do a spacer for the crank pulley that is .100" thick. I have a w/p insert that I can adjust the flange distant .150". Please double-check me. I think this will align the two pulley grooves. PS: I will check the flange for runout.

Target now to ship on Wednesday.

```
From: Rob Siegel
To: Ken Gray, DBE
Tue, Mar 5, 2019 3:11 pm
```

Ken:

Looking at the numbers you gave me a bit more carefully...

The spacing on my current water pump from the back of the flange to the front of the body that the shaft goes into is about 0.215. This, and other factors, appears to put the water pump pulley about 1/4" (0.25") too far forward of the crankshaft pulley.

If you send me a 0.1" spacer to put behind the crankshaft pulley, that should reduce the out-of-alignment measurement from about 0.25" to about 0.15".

Assuming that the geometry of another pump is the same as the one I have, if you push the flange down on the shaft from the current 0.215" to 0.15", that will move the pump pulley in by 0.065″. Subtracting that from 0.15" leaves 0.085". So it won't close the entire gap, but it'll close much of it.

Let's try it and get some sense for where the alignment and clearance issues are with the new pieces, and adjust from there as necessary.

--Rob

....
From: Ken Gray, DBE
To: Rob Siegel
Tue, Mar 5, 2019 4:33 pm

Hi Rob. The .085" clearance is Ok with me. I figured .090" so we are within .005 of each other.

....
From: Ken Gray, DBE
To: Rob Siegel
Wed, Mar 6, 2019 7:43 pm

Hi Rob. I was able the finish the .100" spacer for the crank pulley, install the hub to .092" from the cassette casting, measure the hub for runout (.002") and the two Allen bolts (heads reduced for clearance to the pulley). This should enable us to finish the install. I have included a UPS shipping label so that all the questionable parts could be returned to DBE. I appreciate your patience and detailed feedback is invaluable. Please let me know if you have any future questions or requests.

Ken Gray

TLC Day 2095: That way lies madness #3

MAR 5, 2019, 9:30 PM

The Lotus Chronicles, Day 2095:

After I did my one thing per night (chasing the threads in the head for the exhaust manifolds and making sure that the stainless Allen-head cap bolts that just arrived from McMaster-Carr fit), I hopped onto Searchtempest and saw what I already know—that, no matter how I may pull my hair out with this project, I never said "fuck it," and many other people have. (Pro tip: NEVER disassemble a car thinking you're going to "restore" it unless you're sure you have the stomach for it, and I know that I don't. I pulled the drivetrain, disassembled the motor, got into hot water, and stopped there for years until I figured out how to get out of hot water. During that time, I never laid a finger on the rest of the car. In terms of knowing my own limitations, I'm a freaking genius.)

[Craigslist ad for 1969 Europa Type 54, "complete, needs restoration"]

37 LIKES, 66 COMMENTS

> **Paul Wegweiser** I bet if it were turd brown, they'da stuck with it and finished it! :D $_{+7}$)
> **Rob Siegel** yeah red sports cars suck $_{+6}$
>> **Rob Siegel** oh, wait $_{+3}$
>> **Russell Baden Musta** Rob Siegel 😂😂😂😂 [photo of Russell's two red Loti] $_{+3}$
>> **Rob Siegel** [photo of Rob's red 3.0CSi] $_{+4}$
>> **George Zinycz** Rob Siegel oh yeah...
>> **Steven Hussein Bernstein** Rob Siegel Car porn. Please warn us first. :) What color is the interior? $_{+2}$
>> **Rob Siegel** Of the E9? Cream leather $_{+2}$
> **Luther K Brefo** Too late, the M3 is going out of the garage.
> **Ali Jon Yetgen** Anyone need a set of eyes, I'm not terribly far... $_{+1}$
> **Christopher Kohler** I've got a 2002 here that will likely sit for several more years ruining my life but I'm not going to disassemble it for that same reason. My E12 M535i sat mostly disassembled for five years while I got distracted by other projects but it's all back together now. $_{+3}$
>> **Steve Nelson** Christopher Kohler and this is why I never 'started' restoring "Precious" (my '72 Inka) and finally farmed it out to Vintage Bimmers... #vintagebimmers
> **Rick Roudebush** From what I have been reading lately I would rather buy a gunny sack full of assholes and try to make something out of that. ;-) $_{+4}$
>> **Rob Siegel** be fair. that's a "gunny sack full of assholes with British pipe threads." $_{+7}$
>> **Rick Roudebush** I was assuming various sizes and makes. $_{+1}$
>> **Rob Siegel** You know, maybe we all had Colin Chapman wrong. Maybe what he really said was "Simplify, then add a gunny sack full of assholes with British pipe threads of various sizes and makes." $_{+8}$
>> **Rick Roudebush** How many beers deep are you?
>> **Rob Siegel** I've been sober since New Year's, which is maybe why I'm a little, you know, EDGY... $_{+3}$
>> **George Zinycz** African or European assholes Rob Siegel? $_{+5}$
>> **Jeremy Novak** Man, working on my dad's MGB is like that. We have to have a complete set of American Standard, metric, and Whitworth tools. You just never know what standard the next nut/bolt you need to remove will be, and while they're similar, metric and American Standard do not have good equivalent sizes in Whitworth all the time.
>> **Steve Nelson** Rob Siegel and positive ground. $_{+1}$
> **Jake Metz** I distinctly remember reading your wisdom about restoring cars in your first book, right after beginning my own ill-fated restoration. "This Rob guy doesn't know what he's talking about, this is no problem," I thought. Oh what I'd give for a time machine to go back and set myself straight. I'm a lot better at knowing my own capabilities now than I was then. $_{+2}$
> **Paul Wegweiser** February: "Imma LS1S2 this sumbitch with a big ass turbo and shit and lay it down to 450 hp!"
> September: "Project for sale. Comes with expensive dumpster fire and wiring harness" $_{+5}$
>> **Paul Wegweiser** Forgot to add in there somewhere: July: "Gauging interest. Might sell it." $_{+4}$

> **Dohn Roush** Paul Wegweiser ...no lowballers; I know what I got.. +2
>
> **Jake Metz** Paul Wegweiser someone needs to embroider that on a pillow +1

Vince Strazzabosco Thou shall not tempt!

Paul Garnier I'm at the tipping edge with my GMC RV. I gutted about 1/3 before I got scared and yelled stop! I haven't touched it mechanically and it drove here. So I'm going to get it habitable again, before I take anything else apart. Still have many mechanical issues to sort. +1

> **Chris Raia** Yeah for my MGA I'm tacking one small chunk at a time (disassemble/overhaul/reinstall) so it doesn't wind up a carcass in one corner, parts in the other, and me blackout drinking in the middle. +4
>
> **Rob Siegel** For most people, "rolling restoration," if you and the car are capable of it, makes a lot of sense. It allows you to enjoy the car during the project instead of it being years of deferred gratification and a big sucking sound in your left rear pocket. +2
>
> **Joey Hertzberg** Paul Garnier I love those GMCs, Groundbreaking design, Maybe when my kids move on, Brittany and I will downsize to one. Maybe a little small for 4 people.
>
> **Rob Siegel** They ARE very appealing. 26' is a nice size. Unfortunately the fuel economy on any gas RV is atrocious. +2

Scott Winfield This post needs to go to a "Hall of Fame" somewhere. I am sure there is room in that gunny sack for us all. If not I will sleep in my disassembled E23. +1

Chris Lordan By now we're all familiar with the makeover TV shows. These invariably feature a) the impossible task; b) the unrealistic deadline; c) the heroic host, savior of all he surveys; and d) the undisclosed budget. If it's a 5-couse meal prepared in 45 minutes, a redesigned car with custom features, or a renovated house, it all ends happily, and then the crew departs for the next conquest. But it's all fake. I used to think I'd open a restaurant and I'm very happy that I did not, even, try to actually do this. As the old joke goes, "How do you end up with a small fortune?" "Easy," many have cracked wise, "start with a large fortune, and open a restaurant." Funny, but only because it's true.

One of the cable channels used to air a show called "Restaurant Impossible" where the host had, allegedly, two days and $10,000 to renovate the property, train staff, attract customers, and renew the menu. There was an episode aired that was a behind-the-scenes look at one of these makeovers. There must have been 25 or 30 tradesmen and vendors aplenty delivering just about everything that a rebuilt restaurant could ever need. The first two real-life hours must have blown the whole ten grand! And with permits, delivery time, and the twin imperatives of sloth and indecision, I'm sure that these projects actually took weeks, if not months, of expensive, frustrating effort.

It's expensive and difficult to restore a car, do it properly, and not fall into financial ruin. The starting point often predicts the end result even as the owner (actually sponsor) happily dreams of winning Best in Show at Pebble Beach. And for anyone who has the usual adult impediments—work, family, the need to rest mentally and/or physically—the time to restore the car just never seems to be available. The alternative, farming the work out, is usually a worse idea than running a bespoke restoration shop in one bay of the garage.

In my case, my oft-mentioned Porsche 928 still carries flaws that it had before I ever laid eyes on the thing. I'm sure that my estate will be settled before these flaws are addressed. And... tough shit for anyone who thinks that this is less than stellar ownership. The car runs far better than it did when I acquired it. All of the major systems work, though an honest assessment reveals that not everything works as well as it possibly could. It ran better than I could have hoped, and better than I expected, on a coast-to-coast trip in 2017. But it's 34 years old in the middle of the coming July. I'd be a fool to try and make this car new. Part of the charm (and a big part) is that it's old. Just the way I like it.

So, take heart Rob. You did not acquire more than could be consumed (bite off more than you could chew.) And the resulting book will break out of the BMW community to become a modern classic. Take note: You're on a budget of 3 "Fs" for the entire text of this much-anticipated tome. +5

> **Rob Siegel** Chris, as you know, there's a chapter in Memoirs that talks shit about cable shows like Overhaulin'. Whether it's in writing or in songwriting, I am someone who is very driven, both as a writer and as a consumer, by a sense of authenticity. There's more than enough drama in life as well as in this little lotus project by having it spin out in real time as opposed to trying to artificially create drama and compress it into a 30-minute segment. And, while I am thrilled for the people who seek perfection and spotless engine compartments and shiny paint and achieve it to a far greater measure than I ever will, we all need to seek our own path in terms of what we are able to tolerate for forward motion, the rate at which the bank account drains, and what gives us pleasure. +5
>
> **Pete Lazier** Agreed... we bought a '91 318is for cheap, fixed a few rough spots, had Bill Arnold stuff an S52 in it, and drove it from BC to Newfoundland, competed in the Targa Newfoundland, and drove it back... best road trip ever... and we still plate it 3 months a year. Doesn't have to be perfect, just has to be safe and run good. [photo of 318is] +2

Chris Lordan Pete, my 318is, as bone stock as bone stock could be, accomplished the rarely-seen 300,000 miles on the original clutch. So there's all kinds of perfect out there! +2

Pete Lazier Chris Lordan They are tough as nails little cars... we just wailed on ours in Newfoundland. [photo of car]

Pete Lazier Chris Lordan… and got married in it too! [photo of car]

Chris Lordan Pete Lazier All of this rates a "Yay!"

Chris Lordan Pete Lazier I'd dearly love to run my 928 in Targa; anything to suggest re: getting ready? If I can pull it off it may be the retirement flog for the old hag.

Layne Wylie Apparently Overhaulin' really did do the cars in a week. But sponsors shipped in every conceivable part for the car well in advance, and they assembled it like a toy model. You'll notice they never did a Lotus... I assume Summit doesn't have crate engines for those. +1

Rob Siegel They DID to a Lotus. I wrote about it in my book. It was kind of ridiculous. The guy had traded it years ago for a set of wheels for a Camaro. He barely even fit in it when they gave it to him. [link to video of episode] +2

Layne Wylie Oh wow. I missed that one.

Rob Siegel No disrespect for the craftsmanship that went into the build, but if you watch the episode, the prior owner was DONE with the car. This was not a car that had been sitting in his driveway waiting to be restored. HE TRADED IT FOR A SET OF CAMARO WHEELS. When they did the big reveal at the end, his reaction was a little muted. Europas are tight cars to fit into under normal circumstances, and the guy had put on weight over the years, and despite their having installed a removable steering wheel, it was clear that his fitting was an issue. They all laugh it off, but I imagine he was thinking "what the hell am I supposed to do with THIS?" +1

Layne Wylie It's entertainment, but the very premise of the show is off the mark for a real car person. It's like if you went out and bought shoes for your wife... not real likely to be exactly what she always wanted. +4

Duncan Irving Rob Siegel If you have 50 odd people working on a car 24 hours a day on and off you could probably accomplish a lot in a week. Specially with knowledge, tools and parts at hand. Imagine the expense. I don't know I could be wrong. Referencing Overhauling here. And yes the Lotus owner seemed a little Underwhelmed. I guess it was nice of Jay though. Probably he was more excited about it. +1

George Zinycz Chris Lordan this is all exactly what I love so much about reading Rob's articles… it's repair and renovation in real time, at its real pace. He might drop three articles on a stuck bolt—because it really takes three days to unstick it. That doesn't play on TV, but it's what happens in a real garage. +3

Joey Hertzberg I was watching a random Jay Leno video recently and Jay met this guy for lunch, and he was driving the Europa! They talked Lotus and he loves his Overhaulid Europa.

Rob Siegel Okay, so, I'm wrong :^)

Joey Hertzberg He's a pretty low-key guy, so what we saw at the reveal was probably as hyped as he gets.

Wade Brickhouse Rob Siegel that was cool!

Joey Hertzberg What's cool, was that I loved that Europa, but I didn't realize that it was an Overhaulin build, so I was happy to watch the episode +1

Duncan Irving It's all just a process. No one is wrong.

Daren Stone Chris Lordan "... all kinds of perfect", is perfect.

Scott Linn I just saw an almost basket case Europa on eBay for over $5k. Needed body work, new panel(s), painting (of course), had a seized engine, etc. You are miles ahead on yours! +2

Rob Siegel Thanks Scott. I assume you're talking about that green one. FYI I paid six grand for mine in 2013 (in WAAAAY better shape than that green one), plus about $1200 to ship it from Chicago. In total I have about six grand in the motor, including my last order to RD Enterprises which included a new clutch. That will probably just about wrap up the motor per se. My total sunk costs are currently about $13,600, which is more than I'd like, considering that the car will still need everything (radiator, hoses, brakes, suspension, on and on), and that prettier running Twin-Cams seem to go for about that. But it could be worse.

Scott Linn Well, at least you know the status of all of those critical parts vs. relying on someone. I paid ~$16k for mine in 2011, just coming off a home semi-resto. Still had a bunch of detail work to do, resto work that needed re-done, resto work that got damaged because other work (like the windshield install) was done wrong, plus interior, etc., but was fully drivable, which saved me quite a few $ as I could drive it from south of LA to Oregon vs. ship it. And of course I had to find issues like where the DPO hacked into the wiring harness and added a direct 12V/GND short when the interior light was turned on manually, remove the rocks(!) that were installed in the carbs to disable the warm-up devices, etc.

I missed a ~$12k Europa online in a discussion group by about 15 minutes... Fully running, new paint, great interior, etc. And of course the new owner, who got a steal, decided to sell it on after 6 months for $19k because he decided he wanted a more modern Lotus vs. a classic Lotus. I kick myself for missing that one even though it was just bad timing that I logged on 15 minutes after the guy put it up for sale. +1

Scott Linn The prices HAVE been climbing the last 2-3 years, however, so you should be able to at least recoup all of your costs if you decide it's not for you. I had only seen one in the wild before purchasing mine, so I hadn't even sat in one before committing to buying it and driving 1200 miles back home. But I'm glad I did, it's a really fun car to drive, and it's fairly rare to see another driving down the road... +1

Rob Siegel I was surprised that this one went as high as it did, considering the water pump needed to be done, but it DID look like a well-cared-for car.
https://bringatrailer.com/listing/1972-lotus-europa/

Scott Linn I agree, that seems a bit high. 4 speed slightly decreases value. I almost bought a 4 speed, but I'm really glad I held out for a 5 speed. Also that it's a Twin Cam vs. Special so it's not a "big valve" engine as in the TCS you have. Plus the dash layout doesn't look as nice as in a TCS either.

Scott Linn I have to wonder about the 1973 MG Midget that sold for $14.5k too! Crazy money.

Isabel Zisk See now, if car reassembly were an Olympic team sport, I'd watch. Rob might get a 5.3 from the East German judge for some of his kluges, but imagine the finale where they score the purring of the engines!! +1

Tim Lavery Rob, say what you just said again, but slower. I feel like it might be important. This CJ2A is in my garage right now. [photo of tim's jeep] +1

Brian Ach I would buy it and literally hang the body on the side of my garage and sell off every other part +1

TLC Day 2096: More work while waiting

MAR 6, 2019, 10:35 PM

The Lotus Chronicles, Day 2096:

So close I can taste it.

Tomorrow morning will be 72 hours since I put the sealant on the threads to the water neck, so I can again try pressure-testing the coolant passages of the engine. A bunch of parts arrived today including the carb gaskets, some stud hardware, and the (pricey) clutch. Chased all the intake and exhaust holes with a tap and Loctite-d in the studs. Began test-fitting the carbs. I would've gone ahead and mounted the carbs and exhaust manifold except that, once it passes the pressure test, I'll need to rotate the engine upside down on the stand so I can fit the oil sump and its cork gaskets, and it's hard enough rotating the engine on the stand with the head on the block making it top-heavy, and will be harder still with an intake and exhaust on it, so I'm holding off.

Wish me luck tomorrow, because if it holds pressure, this sucker can be buttoned up, mated with the transaxle, and dropped back in the car. And if it doesn't, well, there's always ritual suicide with a Whitworth spanner.

110 LIKES, 60 COMMENTS

Doug Jacobs Man, just hurry up, cut corners, go for it late into the night. It'll be okay. =;^) +5

Paul Wegweiser I'm totally pulling for you. Looking forward to the triumphant video! +9

    Collin Blackburn This will be just behind you driving around a naked Gullwing chassis on my list of epic videos. +1

Andrew Wilson ...or the ritual tossing it in the neighbor's pool +6

Isabel Zisk Will you be live streaming the pressure test? +6

    Bob Sawtelle I agree! YouTube Livestream the bitch! +2

    Rob Siegel That's a big nyet. My anguish, if it occurs, will be private. +5

Neal Klinman Good luck! I would really like to rebuild an engine someday. But NOT one of those. +1

Karel Jennings Rob, what sealant did you use?

    Rob Siegel Yamabond4 along with Dow 1200RTV primer on the metal-to-metal faces and the chocolate-shake-in-the-oil-pan-if-it-fails paper gasket and on the cork gaskets, Permatex Aviation Form-A-Gasket (what I use all over BMW engines) on some of the other paper gaskets, Wellseal around the press-in surfaces of the oil seal and on the copper head gasket (yes) as recommended in Myles Wilkins' book I'm generally following for assembly.

    Karel Jennings Brilliant. I had to figure out what to use on the copper headgasket on the flathead. +1

Harold Simpkins Best of luck, Rob. +1

Scott Aaron Can taste what, though?

    Rob Siegel BMETIIB. [for the uninitiated, that's "bite me, Euro tii boy"] Twice. +2

    Scott Aaron WOW. New heights. Or something.

Collin Blackburn If yellow chicken feathers come out of the exhaust pipe on the first start, you're doomed. +11

David Kemether [page of gifs of people eating gross foods]

Jeff Hecox Your aluminum intake manifold can be cleaned up with some CLR, Calcium Lime Rust remover. Use with a soft white or blue Scotch Brite pad and a tooth brush in a dish pan if you haven't bolted it up yet. just rinse well with water when done. +3

Dave Gerwig Not looking like a boat anchor any longer, it looks sweet! Amazing journey. +3

Tony Pascarella Bottoms up. +1

Scott Linn What do you own that requires Whitworth tools???

    Rob Siegel Nothing. It's just funny.

Stuart Moulton Largest pulley to engine ratio ever +2

    Rob Siegel Stewart, that's because on a Europa, because there's no room for the alternator at the front of the motor, the alternator is on the back and is driven off one of the camshaft's in the head rather than the crankshaft. Hence the larger pulley. +2

Marc Schatell Good luck—you can do it! +1

Tim Palmer Fingers crossed 👍 +1

Chris Roberts Putting the kettle on for a celebratory cuppa Earl Grey. +1

Scott Linn Some people put an aluminum oil splash guard in to keep it from being sent out the oil filler cap. I bent some very thin aluminum and put it in the oil filler pipe until the next time I pull the cover. Now would be a good time to put one in.

    Daren Stone Scott Linn agreed. +1

    Scott Linn Hmm, it appears that the photos are missing. That's too bad. I think Bean used to sell a splash guard at one time.

    Daren Stone Scott Linn we did offer them in the past and have it on our project list to tool up and have them made again. +2

Nicholas Mav The magic eight ball says: "signs point to yes." +2

Tom Egan I see you have kept the secondary butterflies on the intake manifold. Are you keeping the crossover tubes also? My understanding is that the emissions equipment slapped on the car for the fed requirements don't actually do anything to reduce emissions and just rob power and create problems. Though if you are going for total originality…

    Rob Siegel Tom, what I read seemed to say that if you simply delete some of the stuff (like eliminating the crossover pipes using block-off plates) without knowing what you're doing, the two outside cylinders can run very rich. I've read other things that say that the federal-spec Europas with Strombergs, despite the carburetor's bad rap, are really fine in terms of low to mid RPM drivability. I don't know any of this for sure. This is a car that hasn't seen road time since 1979. It is going to need a systematic sort out of everything. It is in that context that I am happy to put the stock intake system on it and address issues as I find them. +2

    Tom Egan Hmmm, I clearly fall into the "without knowing what you're doing" category, but my DPO (Deceased Previous Owner) seems to have done a pretty thorough job on everything else. I do get intermittent rough running sometimes that feel like it might be an unbalanced mixture problem, but it usually goes away before I can diagnose it. I pulled the plugs after one episode hoping to do the "read the tea leaves" thing (too lean/too rich/too hot/too cold) but it looked like I had "one of each". Where did you read about running rich wo/crossover pipes? I'd like to check that out.

    Rob Siegel Tom, I don't remember, but probably on the Lotus Elan forum.

    Jeremy Novak MGB originalists (yes, they apparently exist) will pay big money for functional original emissions equipment these days. Air pumps, catalyst exhausts, single Stromberg carb setup, etc. All the stuff that a normal person took off and tossed in the garbage where it belonged decades ago. It's strange.

    Rob Siegel In the BMW 2002 world, I believe that only exists in California.

    Jeremy Novak Yea, you never hear about people putting the thermal reactor back onto their e12 once it's gone either. That 70s emissions stuff was not very good. +1

Chris Mahalick I bought a Facet Solid State electric fuel pump for mine. Are you going to blank off the mechanical pump and go this route?

    Rob Siegel Probably. +1

Bart Collins What sized Whitworth? When I first started working on my Norton Motorcycle, I would take all of the wrenches in the tool pouch and all my SAE wrenches and dump them in a pile on the floor. Commando was an eclectic mix of Whitworth, British Standard Fine and SAE. Drove me crazy until I learned what size they were.

    Rob Siegel Well, what size is used to commit ritual suicide? Preferably something large and dull.

    Bart Collins Absotively!!! I remember one in the Norton's tool kit that could do it.

    John Morris The Norton tools don't hold a candle to BMW for ritualistic use!

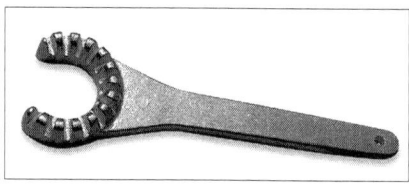

+1

    Bart Collins O M G!!! Ouch.

Jim Strickland This looks great Rob. It would be cool to see (on here or in article) pictures of your nitrogen tank hookup to the engine, and how you plugged all of the cooling ports to do the pressure test. +1

Ernie Peters Looks like you are retaining the secondary throttle mechanism. At the dealer we used to disconnect it, remove the shaft/plates, and plug up all the holes, and this is on the TCs when they were new. If I recall, it helped resolve driveability issues. I did the same on my Elan, ran perfectly without all that stuff. Of course I did not drive mine in cold weather, which is where the secondary throttle system helped.

Rob Siegel Tom Egan asked me this just above in this thread; see answer there. +1

Scott Linn Mine came without the crossover tubes and at some point had received a Euro intake manifold (no secondary throttle). It ran well, but rich. After rebuilding the carbs I now get 31+ mpg on the highway and it runs really well across the rev range. In order to help debug a rich/lean tuning problem on my TD, I bought two sets of Gunson colortune plugs (they were on a great sale) so I could see the ignition color in two cylinders at once. Ended up using it on the Lotus too, which led me down the path of rebuilding the carbs.

Rob Siegel I actually have a Gunson!

Scott Lagrotteria Last summer I almost bought a package of 3 Europas, none running. One was a race car. 1) Was I lucky NOT to buy them, & 2) You have cured my desire for a lotus. Thank you Rob. +3

Rob Siegel I provide this service free of charge. +1

Lindsey Brown Someday, you will be in a position to seduce Tara King. +1

TLC Day 2097: More leaks

MAR 7, 2019, 7:56 AM

The Lotus Chronicles, Day 2097:

The pressure test is graded as an "incomplete," which, as everyone knows, beats the shit out of "fail," but means you have to beg your professor to let you do work over winter vacation.

It isn't leaking anymore at the threaded coolant neck (yay!), but it's leaking in several other places, all of which I should've found when I found the first leak, but I can sometimes think and act ridiculously incrementally and linearly. It's a problem. I'm working on it.

I first found a leak at the head where the temperature sensor threads in. I hastily sealed it up with plumber's pipe dope just to continue the test. I then found it leaking at the OTHER threaded part on the Bean water pump cover, the threaded plug on the side where you would screw in the coolant neck if you had an Elan instead of a Europa. I don't blame Bean for this. They put these threaded parts together with anti-seize, not sealant, for flexibility of use. Fortunately, when I took it out, it didn't have stripped threads like the threaded neck did. It also is leaking out the thermostat housing flange. I had put this together with a gasket skim-coded with RTV just for now, as I don't yet have a new correct thermostat. This was clearly a mistake; it needs to be put together with proper sealant.

But the good news is that I saw no such bubbling around the new cartridge-style water pump, or, more important, along the sides of the engine where the backplate and front cover are sealed up.

I am testing at 20 psi with the nitrogen bottle I use for air conditioning work. The stock radiator caps are rated to seven or 10 psi, so 20 feels about right.

I will seal the offending areas and do it again in another few days.

I can wait. Even though, as a great songwriter once said, it's the hardest part.

53 LIKES, 33 COMMENTS

Steve Park You are really earning your stripes on this job. I am looking forward to the end result though. +1

Andrew Wilson Hmm...Can the garage be made any warmer?

> **Rob Siegel** Andrew, right now, the garage is warmer than the damn house is! +1

> **Andrew Wilson** OK. I equate sealants with epoxy floor cure times, depends on heat, barometric pressure and relative humidity.

Doug Jacobs So, you end up with a leak free Brit motor. Is that the moment the universe sucks back into itself in a reversal of the Big Bang? Be careful what you're playing with dude. +17

> **Bill Ecker** Yes this IS a legitimate concern.

> **Rob Siegel** Doug, I'm just testing that it holds coolant pressure. It still has cork gaskets on the oil seals of the upper timing cover and the oil pan. No worries. The universe most definitely will not end :^) +3

> **Sal Mack** Rob Siegel you know what they say about air-cooled Porsche engines: If it doesn't leak oil check that it has oil!

Jayme Birken This rebuild seem like you decided to rediscover the Bronze Age just for giggles. +3

Wade Brickhouse I'm seeing lots of reasons these might be rare. 😊 +4

John Whetstone Can't help but wonder whether it would have leaked at 10 or 12 psi

Rob Siegel 20 psi seems good testing headroom. The Lotus Engine God says he test to 40, but that's with the head off and using a block-off plate and a rubber gasket. +2

Coleman Maguire I hate when I have to ask my professors for work over spring break especially when I wanna finish my 02 for this season 😊😊😊😊 +1

Bart Collins Keep after that S O B. You'll get it solved.

John Whetstone Been sharing this saga with old school Beezer type and Norton motorcycle owner Mike Cinnamon, I mentioned that it seems that British technology could make just about anyone look like a hack mechanic (note lower case). He texted back: "Got that right. My Norton has British Standard, Whitworth, metric and SAE fasteners. Positive ground and a continuously 'on' alternator that when the battery is fully charged it conducts the excess charge as 'heat' to the frame. No shit. Somebody actually thought that up. Norton owners know all about that huge Zener diode on the frame by the rear foot peg. Sucker gets real hot. Must wear boots. Sneakers would prob melt!" Is it the warm beer? +7

Neil M. Fennessey Two elderly British gentlemen watch a Jaguar on a flat-bed truck go by. One says to the other, "I say Humphrey, isn't that Jag-you-are in the Lotus position!" +4

Rusty Fretsman Lotus: Leak Or Test Until Satisfied..? +3

> **Rob Siegel** Brilliant. +1

> **Rob Siegel** or, to combine that with the usually-quoted expansion of Lotus, "Lots Of Testing, Usually, Unsatisfied" :^) +2

Russell Baden Musta Go forward... move ahead... ! (DEVO) well done Rob! +1

Marc Bridge Ugh!

Dave Gerwig Tiny bubbles: gotta get that song out off my head. Sounds like there's heat in the garage, you New Englanders have it all. Saves draining the water passages so they do not freeze 🙂 +1

    Rob Siegel Because I have been doing things like leaving sealant curing, I've had the heat on in the garage all week, which is unprecedented. I will probably have a stroke when I see this month's heating bill. +4

Brad Purvis Hey! It's the Lawrence Welk Show! Ah one, and ah two…

    Jim Gerock Thanks to our sponsor, Preparation H… +1

Tim Warner Do you dare think about what it will be like to drive this thing?

    Rob Siegel No, I don't. I can't see that happening until the fall. The short-term goals are:
    --Engine fully pressure-tested
    --Engine fully assembled
    --Engine mated to transaxle
    --Drivetrain installed in car
    --Cursory cleaning of radiator and gas tank and replacement of coolant and fuel lines to allow in-car test-start of engine
    --Re-installation of rear suspension to allow car to be taken down off the jack stands where it's been for 5 3/4 years
    Once those things are done, the car will need the full sort-out commensurate with any car that's sat for 40 years. On the one hand, I recently did just that with Louie, then Bertha, then The Lama, but this is an unfamiliar weird-ass low-production British car, and it's going to need everything. I work best when I think in bite-sized chunks. +2

    Tim Warner Good idea! Easier to set short term goals. I better do that with my 3 car projects that are stalled: e30 M50 swap, tundra v8 timing belt full of mouse nest that changed the timing, Meyers Manx front nose crunch, and 3 other minor car projects.

Franklin Alfonso Well, that all sounds easy peasy! +1

    Rob Siegel (Monty Python's "How To Do It" sketch. Can't find an audio. Here's the script. So much of life feels like this, especially when other people give advice.)
    (Cut to a sign saying 'How to Do It'. Music. Sitting casually on the edge of a dais are three presenters in sweaters—Noel, Jackie and Alan)
    Alan: (John Cleese) Hello children.
    Noel: (Graham Chapman) Hello.
    Jackie: (Eric Idle) Hello.
    Alan: Well, last week we showed you how to be a gynecologist. And this week on 'How to Do It' we're going to learn how to play the flute, how to split the atom, how to construct box girder bridges and how to irrigate the Sahara and make vast new areas cultivatable, but first, here's Jackie to tell you how to rid the world of all known diseases.
    Jackie: Hello Alan.
    Alan: Hello Jackie.
    Jackie: Well, first of all become a doctor and discover a marvelous cure for something, and then, when the medical world really starts to take notice of you, you can jolly well tell them what to do and make sure they get everything right so there'll never be diseases any more.
    Alan: Thanks Jackie, that was great.
    Noel: Fantastic.
    Alan: Now, how to play the flute. Well you blow in one end and move your fingers up and down the outside.
    Noel: Great Alan. Well, next week we'll be showing you how black and white people can live together in peace and harmony, and Alan will be over in Moscow showing you how to reconcile the Russians and the Chinese. Til then, cheerio.
    Alan: Bye.
    Jackie: Bye-Bye. +2

    Franklin Alfonso There ya go. I mean, what could possible go wrong? It is British after all…

Frank Mummolo If you decided to build your own engine from scratch and had never done it before, it would go something like you are experiencing. Except this was supposed to be an engine produced by a real company. Watching this saga unfold makes me immeasurably grateful I never bought a Lotus! +1

TLC Day 2098: Yet more work while waiting

MAR 8, 2019, 6:54 PM

The Lotus Chronicles, Day 2098:

No, I'm not done with pressure-testing. Yes, if things fail in a certain way, I'll have to rip the oil pan back off. But as a young snot-nosed Tom Cruise once said, "sometimes you gotta say what the fuck." And look where it got him. Having sex with Rebecca De Mornay on a train. Just sayin'

58 LIKES, 30 COMMENTS

    Doug Jacobs Yes, on a subway, with his car in the lake. So there's that. +3
        Rob Siegel this is what I am saying
        George Zinycz Doug Jacobs "his car" has a name. It's the Nine-two-eight.
        Doug Jacobs George, well, if you're gonna get picky, "his dad's car."
Tom Samuelson Last I checked, there's no subways in Newton Mass, Rob.
    Rob Siegel Rebecca rides the subway of my imagination +6
    Paul Makinen MBTA commuter rail can substitute :-)
    Bart Collins He also said "Porsche, there is NO substitute" while hammering his daddy's 944 (?).
    Tom Tate Just drag the wino off the train 😀
    Ken Sparks Bart Collins That was a 928.
    David Shatzer Paul Makinen But If he rides the MBTA, "he may never return" per the Kingston Trio...
    Russell Baden Musta Bart Collins U Boat commander!
Benjamin Shahrabani Is this your "white whale" Rob Siegel? +1
David Shatzer Rob "Call me Ishmael" Siegel... +1
Greg Bare You could be done with an M10 swap by now 😀 😀 +1
Bill Snider Gosh Rob, hope it all pans out for ya. +1
Rob Koenig Sounds kind of risky... 😀 +1
David Alan Rob, what carbs are you running on this?
    Rob Siegel The original Strombergs
    David Alan Rob they were the carb that got me into working on classic cars. Rebuilt a set when I was in school. Car ran great after a full fuel rebuild, carbs, fuel pump, and new tanks.
Roy Richard You're going to have to pull the pan to paint it anyway. Joining Scientology may be one way to deal with Lotus. +1
Keith Roth What ever happened to Rebecca?
John Harvey But what about the egg? U-Boat commander.
Jay Ford Whoa that went to a whole notha level
Russell Baden Musta As Wayne Campbell referred to her... Rebecca De Hornay! +2
    Rob Siegel "The hand that rocks the cradle will be rocking something else!" +1

Russell Baden Musta Rob Siegel I don't want...anybody else... When I see Kim Bas-in-ger I touch myself.. +1

Andrew Wilson Careful of "Guibo" the killer pimp +3

Thomas Jones I have a deep loathing of cork gaskets. Especially as a professional, in having to warranty oil leaks. Where I can, I like to use paper gaskets instead. M10 oil pan gaskets come to mind. +3

Chris Roberts Hell yeah +1

TLC Day 2099: Lucas ignition

MAR 9, 2019, 11:48 AM

The Lotus Chronicles, Day 2099:

Psst... wanna see something really scary?

59 LIKES, 62 COMMENTS

Kevin McLaughlin Lucas...the reason why the British drink warm beer 🍺 +6

    Eric Heinrich That and they couldn't figure out how to make refrigerators leak oil. +10

    Jeffrey Miner Their electrical is as bad as their cuisine 😀

Anthony R DeSalvo Point taken +9

    Brandon Fetch [meme "I see what you did there"]

Blair Meiser I enjoyed my '67 Lotus Elan SE, when Lucas chose to cooperate! There were times I had to pop the hood and knuckle-rap a relay or two to make headlights or wipers function! So much fun! Good luck!! +2

Paul Wegweiser Don't install it until you shake it upside down and let all the broken parts fall off of it. +12

    Ed MacVaugh The distributor or the car? +6

    John McFadden No, the owner! :-) +9

Frank Mummolo Little known fact: Joseph Lucas invented the short circuit! +8

    Daniel Senie Frank Mummolo and magic smoke +2

Bob Sawtelle Are you going to fit a Pertronix into this scary dizzy?

    Rob Siegel I just replied to someone else on this. I will probably use this unit as-is to try to test start the car. There is a "Pertronix distributor" available which looks like a whole unit, not just an igniter bolted in place of the points, but I regard that along with any number of other general sort-out issues that are still months ahead of me. +7

    Philip D. Sinner Rob Siegel, I'd think you'd feel obliged to try with, first off – aye +1

Trent Weable Is there no vacuum advance?

Eric Heinrich Just the thing to keep it "working as designed"

+22

    Philip D. Sinner That's fuqn phunny, don't care who you voted for!
John Wanner Angel of Death +1
Andre Brown This brings back childhood memories. Living in England in the late 70's, early 80's, with my old man swearing in the garage and not being allowed in the house because he smelled like exhaust. The chest freezer behind the car had black speckles all over it from the Marcos/TR6/Stag blowing oil so bad 😁 +2
Timothy U Ketterson Jr Rob, you let the smoke out! 🤯 +1
Chris Roberts Bonus points 😊 for being the most entertaining car project on the interwebs. +5
Tommy Arnold Those things wore out a lot of walking shoes 😊
Roger Scilley Lucas. Loose Unsoldered Connections And Splices. +14
    Rob Siegel I actually had not heard that one! Combined with Lotus (lots of trouble, usually serious), I'm fucked, right? +7
    Dave Borasky think of it as multiplying negatives +1
    Scott Linn New one to me too, and I've owned one of my British cars for 44 years now.
    Wade Brickhouse Dave Borasky won't they eventually cancel out to a positive? 😊
    Rob Siegel Wade, incredibly, I think that's what he is claiming! +2
    Daniel Senie enough negatives and the Lotus will be sucked into a black hole. +2
    Dave Borasky Rob Siegel, as a reader, all of the misadventures with the Lotus are a perverse source of pleasure +1
Christopher P. Koch I like the graunched adjustment screw...
John Morris I'm a fan of points ignition—simple, easy to maintain/troubleshoot and long-lived. Plus low cost to replace. My Norton can actually run without a battery! +3
    Rob Siegel In general, for a car that you road-trip, I'm not a fan of points. https://www.hagerty.com/media/maintenance-and-tech/ignition-debate-points-vs-pertronix/ +1
    Daniel Senie John Morris points are extra exciting when the condenser fails. +1
    John Morris Never had one fail! Replaced them regularly in preventative maintenance
    Jeffrey Miner Daniel Senie I recall our '74 tii developing and "intermittent" condenser failure. The car would just stall. The mechanic couldn't reproduce it. My father sent me over to take the mechanic for a spin to replicate the problem. 6 minutes in the car was dead. "Well, I guess if you drive it that enthusiastically, it causes the problem," he said. Finally determined that "enthusiastic driving" caused the condenser to overheat and cut out. +2
    Daniel Senie Jeffrey Miner in our case, the condenser failure resulted in 3 foot flames out the tailpipe, passing between Faith's legs. +2
Brandon Fetch Came for the smoke comment. Leaving satisfied. Thanks Eric Heinrich +3
Carl Witthoft PIker. A good EM pinball machine has hundreds of those. +1
Brian Hart How about something positive (yes) about Lucas dizzys—there is help... http://www.distributordoctor.com/ +1
Mark DeSorbo Lucas...the very word is scary.

Alan Hunter Johansson Mark DeSorbo Happy I read through the comments before duplicating this one. I was going to say that the first photo was the scariest. ₊2

David Ibsen Vacuum cleaners? They didn't suck? ₊1

Scott Adair Lucas! Run!!!! 😀

Douglas Wittkowski The horror!

Ben Greisler Dude. Seriously? This is a family show. ₊1

Lenny Napier The dark side…

Andy Veedub Yes! Nos Lucas smoke required for refill!

Chris Raia "Simplify, then add darkness." ₊2

Robert Alan Scalla My 64 Healey is full of original wiring and components that still work perfectly. ₊2

Sam Schultz Lucas. Now I won't be able to sleep tonight…

Russell Dejulio 😱😱😱😱😱 I owned 2 MG's I know the Terror well! 😱😱😱😱 ₊2

Franklin Alfonso Do you have a rebuilt flux capacitor too?

Rennie Bryant Being an MG owner as well as the Bimmers, the Real Lucas ignition parts are pretty good. The reproduction Chinese ones suck.

> **Rob Siegel** This is why I will keep the original ignition parts in it and try to use them to test-start the engine. ₊3

Lindsey Brown Lucas, Ltd. started out making hurricane lamps for British sailing ships. I suspect many an unfortunate British mariner went down to Davy Jones' Locker in the dark. ₊1

Marc Bridge Lucas, where electrons go to die. ₊2

Peter Gleeson Rob, I have spent my life debunking the Lucas myth—every story has a journey, stay with me here—and happily told everybody how many English cars I have owned and never been subject to electrical problems. Oh how I should have kept my rather large mouth shut :-(Six months ago a friend was selling a very nice full history from day one 1984 Jaguar XJ6, I thought what a cool little train station car (as we say back home) more an airport car here. Anyway, in six months I have had to fix, rear window motors, sunroof motor, electric antenna motor, original stereo gave up, new distributor, oh and just for good luck a brake master cylinder… Please correct me and just say "XJ6" if ever I am found guilty of defend Lucas again. ₊4

> **Dohn Roush** Peter Gleeson Sounds like a perfect train station car, as in, drive it to the station and push it in front of a train… ₊2

TLC Day 2099: Carbs and exhaust manifold on

MAR 9, 2019, 7:06 PM

The Lotus Chronicles, Day 2099, part B:

--Got the exhaust manifold bolted up using stainless Allen bolts that can be reached easily with a straight shot (unlike the nuts on studs which are almost impossible to access).

--Dropped the float bowls on the Strombergs. They were amazingly clean, just a small layer of varnish at the very bottom. Test-fitted the carbs. It's looking more and more like an assembled engine.

--Cleaned the valve cover. The bits of flaking paint on it was well within the patina envelope of the car, but when the kerosene in the parts washer hit it, about a third of the paint said "I'm outa here" Took it into the machine shop for a proper tanking and glass beading. I may try to paint it myself, but I may decide to take it in for powder coating MY GOD WHAT AM I TURNING INTO?

--Fitted a replacement cartridge water pump from Bean (don't ask). [note: this is my only reference on Facebook to the issue that nearly drove me over the edge]

--Found a little corrosion on the coolant neck and the part of the head that it bolts to. Patched it with J-B Weld SteelStik and sanded.

--Fitted an oil drain tube between the head and block that was supposed to be put in before the head was installed. D'oh! Read a tip in a Lotus forum that if you boil it, the hard rubber gets soft enough to bend. Was thrilled that it worked.

Pressure-testing (again) in the morning. Let's hope that changes the "incomplete" to a "pass" and I can graduate from Lotus engine school :^)

75 LIKES, 39 COMMENTS

Mark Thompson Amazing... I don't know much about cars but I am amazed at how hard you go at it every day! +1

Mark Thompson Deconstructing Cure songs for a jam session... same details! +2

Frank Mummolo Powder coating is cheap. No brainer, man! +2

William W Lewis III You've got this! Love the bit about your new triumph in the hack mechanic book.

> **Rob Siegel** wait, did I buy a Triumph? +1
>
> **William W Lewis III** Rob Siegel indeed! I understand why you may have repressed the memory though! [photo of chapter from my first book "The British Piece of Crap (This is your brain on adolescence)"] +1

John Webster Nickols It has been both fun and educational to follow your progress. You are doing 5 times a better job than the factory did . +2

Steve Nelson SS Allen's for the exhaust sounds like a great idea...I'm going to use that when I rebuild something 👍 +1

> **Rob Siegel** I didn't invent it; I read about it. And when I saw the lack of access for nuts on studs (which I didn't remember; it was 5 1/2 years ago that I took the thing off), I thought, yeah, I'm totally doing this. +1

Roy Richard Hopefully you used anti-seize on the stainless screws. Stainless loves to gall. +4

Rob Siegel Roy, I did not, but I still can. Thanks. +1

Scott Linn Make sure the rubber washers on the carbs are in place and not hard, else vibrations will mess up the mixture +1

Rick Roudebush We used 3/4" stainless steel bolts and aluminum nuts on paddle connectors on substations. There is no getting those damn things off after a few years. Bring your pipe, breaker bars and best impact sockets 'cause they just have to be broken off. No wonder I'm on my second rotator cuff repair. +1

> **Roy Richard** I would use steel screws rather than stainless. Stronger and not prone to gall.

Dave Gerwig Awesome progress 😀 😀 +1

Chris Lordan #goRobgo +3

Joey Hertzberg I thought about the comment I made on your other post about the water pump. Tearing something apart that one had lovingly put together this afternoon. I unwittingly pumped thick, reddish, gas tank varnish through my perfect running weber 38 and down into my ported intake. Today was the day I removed the intake after all that care with hoses, and fittings, and sealant, and on and on. The weber is clean, chased all the tiny passages and chiseled out the bowl, Chem-dip, sandpaper, 3 cans of carb cleaner, countless glove changes. You know the drill. Tomorrow, the intake gets the wire bottle brush and its own baptism of petrochemical solvents. I had to share. Your engine build has been a fun ride! +2

Clyde Gates All I can say is that you have eliminated any lust I might have to own a Lotus of any form! +2

Alan Alfano I had paint/powder-coat issue with my 1988 325iX. Internet search showed BMW airhead motorcycles using a product called "Rub n Buff" (silver). Comes in little tubes. [photo of powder-coated valve cover] +3

David Holzman It must be nice to know the innards so intimately. btw, is this a low carb engine?

> **Rob Siegel** It's British, so everything it eats is wrapped in newspaper +7
>
> **Steve Nelson** Rob Siegel to absorb the grease 😀 +2

Scott Linn It's amazing how clean the carbs can become with just shooting with some carb cleaner and wiping down. Mine went from grimy to nearly new. People were asking me if I blasted them. +1

> **Rob Siegel** That's literally all that I did to mine. +1

Francis Dance Any galvanic coupling concern on the SS bolts in the aluminum head? +2

> **Rob Siegel** I would suspect that there's that issue with any kind of bolts or studs in an aluminum head
>
> **Ed MacVaugh** Only if an electrolyte is present. Keep it dry. I don't see anyone using stainless on exhaust ports, but I think that is likely an expense issue or expansion rate concern. BMW uses Inconel on down pipes, at extreme expense, but I don't know their justification.

Blair Easthom Next up. Replacing Strombergs with proper Webers. Or Dellortos.

> **Rob Siegel** You can't. The manifolds are a cast part of the head. There are Stromberg heads and Weber heads. I have a Stromberg head. +1
>
> **Blair Easthom** Dedicated Lotus guys know the value of a bandsaw and a MIG welder. Don't try this at home. +1
>
> **Rob Siegel** Blair Easthom, this car has been sitting since 1979. First I have to get it back together. Then I have to get it functional. If I reach the point where the choice of the carburation is the most critical thing on the docket, I will be very happy 😀 +6
>
> **Doug Jacobs** Rob, I've always said, you don't do the liposuction until after you get the pacemaker going. +4

Brad Purvis There is no "pass." There is only purgatory in Lotus Land.

> **Rob Siegel** and I'm there; see new post. +1

Scott Aaron Yeah, "the Powder Coating Guy" is already out there +1

Bill Cardell Whatever you do, please ultrasonic any aluminum that gets beaded, especially if aluminum oxide is used. Don't ask how I know. +1

Andrew Wilson I hope you put anti-seize on the nuts. Stainless nuts on stainless bolts will gall and strip the threads off the stud. +1

TLC Day 2100: Head and plug are leaking

MAR 10, 2019, 10:25 AM

The Lotus Chronicles, Day 2100:

Well, it's not good.

Pressure-tested the system to 20 psi. On the positive side, unlike previous attempts, the gauge stayed steady, I heard no hissing, and saw obvious big bubbles being blown on initial spraying of soap solution. However, when I came back an hour later, it had dropped by a few psi. I can see small bubbles at the flange of the removable water pump. That's not a big deal, as this time I inserted the water pump with silicon lube on the o-rings instead of RTV, so maybe it really DOES need the RTV. I'm also seeing some bubbling through one of the threaded ports (the non-stripped one) on the casing that I just sealed up with proper thread sealant, so I need to redo that.

However, far more troubling is the fact that I'm seeing cocooning of bubbles at several places along the edge of the head gasket. This is most surprising since, obviously, the head and block were machined, it's a layered copper head gasket, and I coated the head gasket with Wellseal, a solvent (non-silicone)-based adhesive. There's debate about whether this is helpful, but that's what it says to do in Myles Wilkins' "Lotus Twin-Cam Engine" book, which I'm generally following for engine assembly, and I haven't read anything that says it's harmful. I suppose I can say that, if I HADN'T coated the head gasket and saw this, I'd be kicking myself right now.

I'm leak-testing this like I leak-test a/c systems. I'd imagine that very few people do it in this way. I need to make sure that I'm not pulling another Jackson Browne like I was when I though the timing chain was hitting the head ("What I was seeing wasn't what was happening at all"), but If this was an a/c system and if someone described this to me, I'd say they were in denial if they thought it wasn't leaking.

Stepping away for a few hours to think. Then I'll probably seal up the water pump and try it again and see where I am. If it's still not holding pressure—and I expect it won't be—I'll seek guidance from The Lotus Engine God before I rip its (or my) head off.

37 LIKES, 69 COMMENTS

    Chris Roberts Just out of curiosity, how many PSI is the system supposed to hold normally? +2
        Rob Siegel Great question. The radiator cap is 7 or 10psi depending which one you have. So 20 psi seems a good testing threshold to me. The Lotus Engine God says he tests to 40, which seems high to me. +2
    Jeff F Hollen Have you considered Evans waterless coolant? It doesn't pressurize. Obviously you want the system to not leak, but the not building pressure would help. +6
    Daniel Neal Was wondering the same thing Chris Roberts

Adrian Radu Operating temp will help with sealing things up. Waterless coolant is a great idea for your troubles and this type of car. +5

John Whetstone If I remember correctly, you're pressurizing the system to 20 PSI. The LEG pressure tests at 40? And the system normally runs app 10?

> **Rob Siegel** yes
>
> **John Whetstone** And he sees ZERO leaks at 40? Crazy.
>
> **Rob Siegel** I'll talk with him on Monday. +2
>
> **John Whetstone** Okay, sleep well.
>
> **Rob Siegel** right...
>
> **John Whetstone** Sounds like these British cars are designed to leak a little! Lest they explode. +3
>
> **Rob Siegel** right, so instead, their OWNERS explode +8
>
> **John Whetstone** Rob Siegel I'm not even sleeping well following this one. +1
>
> **Sean Curry** Maybe it's like the old airplane radial mantra: If it's not leaking it's empty? +1

Layne Wylie I'm not familiar with layered copper gaskets. But on the last engine I assembled with an MLS gasket, coolant poured out like a bucket with a hole, until the first start up, then it completely stopped. +4

> **John Whetstone** Layne Wylie that's a wonderful thought

John Goddard I wonder if the cold, static temperature is a factor here. You have heat there almost immediately after starting. Gotta think expansion would seal +2

Ed MacVaugh I want to be the first to remind you that water is bigger molecules than air. There has to be joints that will leak air but constrain water. I understand that it is expedient to test with air, since puddles of air on the floor don't cause you to slip and fall, but maybe you should be testing with the fluid you are constraining? +9

> **John Whetstone** Ed MacVaugh ooh I like that
>
> **Rob Siegel** Understood
>
> **Paul Forbush** Ed MacVaugh Actually, I believe Rob is testing using Nitrogen, which would have larger molecules than water or air. This is the theory behind nitrogen tire inflation that tire shops sell. +1
>
> **Ed MacVaugh** Paul Forbush https://www.getnitrogen.org/pdf/graham.pdf
>
> **Rob Siegel** Yes I AM pressure-testing with nitrogen, but I'd be lying if I said that it's for any reason of molecule size. It's just the equipment that I have. +1
>
> **Bill Ecker** Dude, PUDDLES OF AIR? I am SO stealing that. +4
>
> **Ed MacVaugh** Rob Siegel That's certainly OK, I am now struggling with the effusion rate of water versus N2. :) It isn't about size, you know :) +2
>
> **Rob Siegel** Ed, well, you SAY that… +1

John Whetstone Also wondering, wouldn't the pressure relief on the radiator cap protect any of these leaks you're looking at now, since it would release at 10 psi?

Jeff Hecox How cold is the room in your test area? How is the test at 10 psi? Thank about the engine temperature when the cooling system is at 20 psi, that engine would be at 180 degrees or more. A stone-cold engine is never going to see 20 psi. With all the heat expansion that cylinder head will probably seal right up. +3

> **Rob Siegel** Understood.

Paul Forbush I have to wonder if you have ever assembled and pressure tested a BMW engine (or any other, for that matter) to this standard. I never have and have never had a problem, aside from a few times I used an incorrect sealant that I knew better than to use. +1

> **Rob Siegel** Understood.
>
> **Paul Forbush** Rob Siegel Maybe you are "over-engineering" this job, just a bit?
>
> **Rob Siegel** see comment below
>
> **Paul Forbush** I generally pressure test a freshly assembled engine with a hand-pumped coolant pressure tester, but this AFTER installation and coolant fill. I can understand your thoroughness, given how much time you have into it, and the PITA factor if it had to come back apart once it was in the car. +1

Rob Siegel Thanks, everyone, for your comments and cautionary minds. I am going to address the two minor leaks in the cartridge-style pump and the threaded plug in its case, as in those I can actually see active bubbling, whereas at the edge of the head gasket I see accumulation of bubbles (cocooning), which is a more subtle thing. +5

> **John Whetstone** Rob Siegel that was intervention—like! Great comments from the car guy

community, I think you have this. +1

Russell Baden Musta Rob Siegel just put it in the car and refill the fluids as necessary! Join us!! +4

Rob Siegel Russell, thanks for that. That made me laugh out loud. Obviously there's a line between "they all do that" leaking and finding an oil pan full of a chocolate milkshake that I could have and should have troubleshot before the motor was in the car. +1

Russell Baden Musta Rob Siegel fully understand!!! Thinking back, I did none of that when I put mine back together... I forgot to glue (hylomar) the cam cover gasket to the cover and the front squished out within a year and just soaked the front of the case! Do you have or need those rubberized star washers for the cam cover nuts?

Rob Siegel Russell, I believe they came with the gasket set, thanks.

Russell Baden Musta Rob Siegel should be 8 of em... RD has 'em if you're missing them. Were you good on shims for valve adjustment? I have a small stash to trade if you need any.

Rob Siegel The preliminary valve adjustment was done when the machine shop had the head. I've read that all valves should be adjusted to 009 to .011, but the guy who did the head adjusted the intake valves to the tighter .005 spec in the manual. I was going to redo them, but a) I have no shims, and b) The clearances will change (and almost certainly widen) after the engine has run in. I wish they weren't as tight, but I think it'll be OK. Thanks for the offer. I'll holler when it's time to take you up on it. +1

Russell Baden Musta Rob Siegel once you do that, just make a chart as to what size under each bucket, mix and match if you can, then let me know what you need and we'll trade if I have them! +1

Rob Siegel I've been reading the forums to try to learn what cars have the same size shims. I understand that Saab 99s do. I have a stash of Alfa shims because they are used to gain some time on old BMW motor is when the rocker arms and shafts get worn past the point of adjustment, but I haven't read that those are the same size as the lotus shims.

Russell Baden Musta Rob Siegel that I've no clue... I have I think about a dozen Lotus shims...

Jeremy Novak Valve clearances sometimes tighten as the seats wear in. I've heard of race engines burning a valve because the clearance was set on the tight side and then tightened more. Loose valve clearances never burned a valve. +1

Alessandro Botta With layered gasket is normal to have leaks before first start, heat the engine with a thermo gun up to 150 C, then test it again. +2

Charles Morris Have you considered rigging up a test stand to run and break-in the engine before you put it in the car? If there's a major (or even minor) repair it'd be a lot more accessible than having to pull it again...

Rob Siegel "The Lotus Engine God" who is advising me has a dyno. If it came to wanting to run it in, I would probably do that rather than trying to rig something myself. +3

Jim Strickland Proof of your success! [photo of "The Lama," the E28 535i I bought not knowing it had a broken rocker arm, fixed, and sold to Jim] +5

Jake Metz You're the man Rob. Keep it up! +1

Rob Koenig I'd bet the farm it'll seal up once warmed up and run in. +3

Luther K Brefo Railroad diesel locomotives and steam are notorious for this. Back in the 90s and early 2000s, one couldn't keep a GE leak free until it warmed up. Alco was the same in the 60s/70s. +1

Luther K Brefo And EMDs never stop leaking.

Rusty Fretsman Leaks Out Tenaciously? Use Sealant! +3

Christopher P. Koch You may have to take the junkyard dog approach and use a gallon of stop leak...

Scott Chamberlain Only slightly tongue in cheek... if you want a Lotus Twin-Cam that doesn't leak, runs reliably, won't break, runs all day at redline, and doesn't cost you a fortune... get the Mazda twin-cam out of any Miata. It's what Colin Chapman was trying to do. Sorry for being cheeky; you may now return to your crisis, already in progress. +3

Jayme Birken https://www.sensorprod.com/fuji-prescale.php

Ken Sparks What is the "Jackson Browne" reference?

Rob Siegel The line "What I was seeing wasn't what was happening at all" is from the song "Fountain of Sorrow." How ironic 🙂 +2

Paul Wegweiser [gif of Amy Poehler face]

Steven Bauer Be glad that you're doing this now instead of when it's back in the car. +1

Steven Bauer Also, it makes me wonder how they manufactured these things in the first place.

Philip D. Sinner So not for nothing… that blue plug sure is racy looking ! (?)

Jon-Paul DeVore Assuming you don't have a bigger issue like a machine work problem etc, call your local GM dealer ask for cooling system pills. We used them for race engines with multi-layer metal head gaskets that seep. Cured it every time. Most had considerable boost . As for head gasket coating, copper coat never failed us yet. I hope it doesn't come to that. Water pump will likely seal itself after some run time. Burn of some silicone etc. Good luck.

TLC Day 2100: The plug strips

MAR 10, 2019, 7:19 PM

The Lotus Chronicles, Day 2011, Part B:

Went to do what I said I would do earlier today, which was to address the two small leaks coming from the custom cartridge-style water pump and front timing covers. The larger of these was the one at the threaded plug in the side of the cover. This already had leaked and I already had sealed it up with Permatex Seal+Lock, so I assumed that the fact that it leaked meant that the sealant somehow had not taken in the threads. It did come out a little hard, but it's a big Allen key so I was able to put a lot of torque on it.

Bad idea.

As you can see from the photo, it took some of the threads with it. I should've use heat on it first.

Fortunately, the damage looks much worse on the plug than it is on the female threads in the cover (where it looks like damage is limited to the first two threads), so I'm not panicking. It's certainly not as bad as the thread problem I had last week on the coolant neck. Hopefully I will be able to get another plug, may be a tap to chase the threads if necessary, and try sealing it up again.

But hokum smokum, can't a hack mechanic catch a break?

(Yes, this is right on the edge of my creating problems by trying to address small leaks that show up in pressure-testing that may never actually leak coolant, but I'm not second-guessing chasing this one, except that, as I said, I should've used heat on it first.)

As the great songwriter Robert Earl Keen said, the road goes on forever and the party never ends...

38 LIKES, 61 COMMENTS

Michael Cari This build is just crazy. I have built German, American and Japanese engines, and never have I encountered so much trouble with fitment and sealing. You have a stronger constitution than i, my friend. Good luck, I hope it's loads of fun when you finally get to open it up. You deserve it! +2

    Ali Jon Yetgen Michael Cari I feel exactly the same... I would have gone insane long ago if things worked out like this.

> **Rob Siegel** who says I haven't? +4
>
> **Jamie Eno** If you're going through hell, keep going. +2
>
> **Scott Linn** You're still going, so the diagnosis is insanity for sure. +1

Frank Mummolo Rob, Job could take lessons from you!!

Ed MacVaugh You aren't suggesting that the sealant took out aluminum threads, are you? +1

> **Rob Siegel** I am. It's Permatex Seal+Lock. The threads were fine when I applied it. The mystery is how it did that and still leaked. +1
>
> **Ed MacVaugh** I would accept the hypothesis that the threads looked regular and neatly formed when you applied it, but I am skeptical that that sealant bonded so well that the threads were torn off the adjacent material due to the bond strength. I would suggest a flat topped plug and sealing washer (crush ring) for reassembly. +5
>
> **Rob Siegel** That was part of the solution on the threaded neck. +2
>
> **Ben Greisler** That would assume a milled flat area for the crush washer to seal against. (and being British I doubt anything is truly milled flat) +1
>
> **George Zinycz** Ed MacVaugh it does seem impossible but then posit a more plausible explanation...
>
> **Ed MacVaugh** George Zinycz The threads were crap in porous aluminum and they formed for the tap, or machine tool, but were ready to peel out of there long before he installed either the plug or the sealant. Since they were porous, they leaked, despite the sealant. Note that he visually observed crap threads in the other threaded joint in the same part. +4
>
> **George Zinycz** Ed MacVaugh ok but how did the threads come out if not because of the sealant gluing them to the plug?
>
> **Ed MacVaugh** George Zinycz That is exactly how they came out, but they would have come out regardless (in my hypothesis).
>
> **George Zinycz** Ed MacVaugh oh ok I understand now. Thank you
>
> **Ed MacVaugh** George Zinycz If the bond was indeed strong enough to tear machined threads from the substrate material, my thought is that it would have pulled them all. Actually my position is that no sealant is a strong enough bond to tear out properly machined threads.
>
> **Rob Siegel** Ben Greisler, it has that
>
> **Rob Siegel** Ed, George, I have no other explanation. The plug came threaded in when I bought the kit. I unthreaded it because it was leaking. It unthreaded fine. I then cleaned the anti-seize out of the threads and applied the Permatex Seal+Lock. It threaded in fine. I didn't feel like I was over-torquing it when I reinstalled it. If anything I was concerned that I didn't put enough torque on it. I used the Permatex primer compound to speed and add strength to the adhesion and then waited the full recommended 72 hours for it to cure in a garage that I've been keeping at about 60° so all of these sealants can cure. It was surprisingly hard to take out, but so are things that you install with regular Loctite red. Then I saw the threads on the blue piece in the photograph. +3
>
> **Ed MacVaugh** Rob Siegel The image suggests that the plug is slightly tapered. Is that the case?
>
> **George Zinycz** Rob Siegel did the threads come off the plug more easily?
>
> **Rob Siegel** I don't know. Certainly the undamaged threads are higher than the damaged ones. If I were to guess, I don't think that it is tapered, but I am not certain
>
> **Tom Egan** Rob Siegel Hmm... Something is fishy here. Are you sure these are NPT tapered pipe threads? Not straight threads that should be plugged with something like this: [link to straight-threaded plug]
>
> **Rob Siegel** Tom, no, I'm not certain, but enough people have raised the issue that I've ordered one of these straight-threaded plugs with sealing washer: https://www.mcmaster.com/4936k449

Jackie Jouret This is a quintessentially British tale. +4

Kevin McGrath milled goodness

Ben Greisler If you never need to get it out again, seal it up with some JB Weld. +1

> **Wade Brickhouse** At this point I might put the whole thing together with JB Weld! Or use it as garden sculpture. 😀 +1
>
> **Rob Siegel** It may come to that +1
>
> **Brian Lowder** Ben Greisler JB Weld! JB Weld!

Dave Gerwig Might consider a thread gauge. Could that be a US Threaded plug up against Metric tapped hole?

Rob Siegel Dave, the first thread mishap with the threaded coolant neck I think probably was something like that, but this one threaded perfectly fine when I put it together.

Dave Gerwig Rob Siegel "Pipe threads" are cut to swell, for lack of a better description. With steel going into aluminum, if they go too deep they really bind or gall as a toolsmith would say. Hope it seals up next try!

Rob Siegel Dave, I'm not sure if the blue plug has pipe threads or not. If I were to guess, I would say not, but I'm not certain.

Layne Wylie The blue plug is NPT. Jegs sells them +1

Scott Aaron 😂😂😂😉🫡

Steve Nelson I've been entertaining the idea of an old British roadster (MD-TC)…I think you've broken me of that idea! +1

Dave Gerwig Steve Nelson TCs and TDs are easy to work on by comparison to the Lo-tush +1

Scott Linn I own a 52 MGTD, 68 MG Midget, and (horrors) a 73 Lotus Europa TCS. The TD is simple and easy to work on. The Midget slightly more difficult but still dirt simple. Then there's the Europa… Order of magnitude more difficult. The later T series cars (TD/TF) have much better steering and suspensions. But the TC looks nicer IMO. More of a classic square-rigged design. +2

Joey Hertzberg Steve Nelson, lots of people swapping Japanese engines in those old girls, and enjoying them instead of "enjoying British car ownership"… +2

Joey Hertzberg I had a restored 1969 Datsun roadster with a warmed-up engine and a 5 speed, SU carbs, some Triumph engine parts could be used in the Datsun engine. I had a local British car guru tune it up for me. He only agreed to do it because it was so clean and close enough to being British. Man that car would scream and I really loved those carbs when after he dialed them in.

Steven Bauer And I'm rethinking my interest in buying an E-type.

Rick Roudebush My wife and I were going to look at a really nice 1970s E type as this Rob adventure was coming to a boil…bailed, no thanks. +1

Rob Siegel Steven Bauer, Rick Roudebush, please don't pass on an E-type on my account! +2

Tom McCarthy Steve Nelson my dad has a 53 MG TD that is quite reliable, not a daily driver but relatively trouble-free.

Rick Roudebush Memories of trying to find simple bolts to replace missing or lost ones and the level of frustration that caused me in a small town back in the 70s when I went through my MG and GT6 days was more to blame for the bail on the E type Rob, but the need for three tool boxes, adjustable wrenches, vice grips and small pipe wrenches and the very high price of said E type (with automatic) was the final deterrent. +1

Buck Hiltebeitel Also British standard pipe threads are the same pitch as National (US) standard pipe threads, but the angle of the flanks is different, so they'll go together but never seal. +1

Layne Wylie NPT is tapered and is supposed to seal on the threads (not really, without some sealant). British pipe is straight and seals on an O-ring.

Layne Wylie Since the water neck fits both locations, and it seals with a crush ring, it's possible they have tapped both holes with British pipe threads, and then plugged the unused one with an NPT plug. +3

Zenon Holtz Layne Wylie BINGO! That seems like the best explanation given only the threads nearest the opening stripped—straight hole, tapered plug means only the first few threads of the hole could engage. They deformed/detached on assembly which is why they leaked and then came off. No way compatible threads would have leaked at 20PSI with that much high-strength sealant. +3

Buck Hiltebeitel Layne Wylie there are 2 British pipe threads, one straight and one tapered. The straight threads require a seal ring or crush washer.

Rob Siegel BSPP (parallel) and BSPT (tapered) +1

Rob Siegel I actually just ordered this straight-threaded BSPP plug from McMaster-Carr. It comes with a sealing washer. It's steel, so at a minimum I can use it to chase the threads in the cover at a fraction of the cost of a tap. I think I'll try it with some removable adhesive like Yamabond4 and see if it seals things up. If not, the nuclear option is J-B Weld. https://www.mcmaster.com/4936k449 +4

Zenon Holtz Rated for 3000 PSI! It just might be enough…

Harry Krix Ouch…

Brad Purvis SPFCCSMF&Ts.

Robert Alan Scalla Patience and persistence is what I find gets car repairs complete. And a bunch of other stuff too.

TLC day 2103: Munged-up plug threads

MAR 13, 2019, 7:50 PM

The Lotus Chronicles, Day 2103:

I think that those of you who said the fact the threaded hole on the side port of the outer timing cover was only munged-up on the first two threads was evidence that the hole is straight-threaded and the plug is an NPT thread were right on the money. My goal was to replace it with a straight-threaded plug. I first measured the threaded opening and found it to be almost exactly one inch. Then I looked on McMaster-Carr and ordered a 1" British Standard Pipe Parallel (BSPP) straight-threaded plug with sealing washer. Because clearly I needed a 1" plug, right?

Wrong, Lotus breath.

The plug arrived last night. As soon as I saw it, I knew it was too big. Did I measure wrong? No. Did McMaster screw up? No. Then why is it wrong? Because pipe threads appear to be one of these things, like certain a/c fittings, where the size that's used in the name of the fittings don't correspond to the physical dimension you'd actually measure. The 1" opening takes not a 1" fitting but a 3/4" fitting. I ordered a replacement which will probably be here Friday. If it seems to thread in (which I don't know for a fact that it will), I'll use it to chase the damaged threads, then try sealing it up with Yamabond4.

I know that some of you rely on me for suspense, resolution, and entertainment, but you're simply going to have to wait a few days.

Talk amongst yourselves.

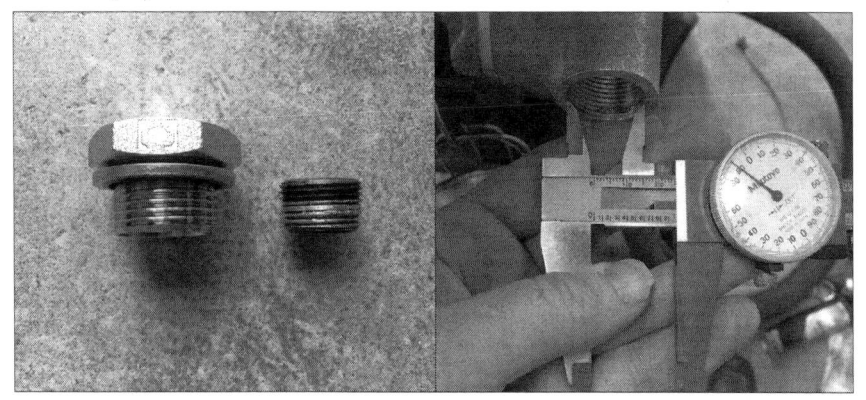

45 LIKES, 43 COMMENTS

Scott Winfield Your rebuild is inspiring many of us. "The struggle itself towards the heights is enough to fill a man's heart. One must imagine Sisyphus happy." -Albert Camus +8
    Rob Siegel Mad points for Camus! +1
    John Biesecker Al Camus built some damn fine engines in his day!
    Ken Sowul I wonder if Godot had a Lotus Twin-Cam?
Josh Wyte Your rebuild is cementing my belief that I never want to own an English car! +8
    Tim Warner Josh Wyte agreed
Paul Garnier I knew this 3 days ago. Didn't tell you because words... +1
Ernie Peters You are providing feedback to DB, right? I'm sure they would be interested to know of your travails so that they can pass the info to other buyers of the conversion as well as the provider of the part. And for the record, based on my Elan rebuilding experience (and my MG Midget for that matter), the challenges you are facing with the Lotus are unusual, even for British cars. I welcome your

sharing how you handle the challenges. +2

> **Rob Siegel** Believe me, they hear it all :^) +3

Bob Ball When in doubt tap it out +2

Bob Ball Like there is nothing standard about if you have Wentworth and or Brit standard wrenches the sizes just don't correlate to anything you can measure sensibly. +1

Steve Nelson British logic 😀

Andrew Wilson Subtly cursing under my breath, as I'm sure you have done this whole time. +2

Scott Linn Sorry. I drove mine today in support. Contemplating my LED headlamp replacement. My way of adding lightness, and not needing to add relays... +2

Derek Barnes Whenever I need plumbing parts for my house, I take a deep breath, meditate, say a prayer, read my tea leaves, take out the dried chicken feet THEN measure. It works half of the time 😀 +1

Richard Koch Pipe size dimensions are nominal to the ID of each size and wall thickness determines the "schedule" as in schedule 40 or schedule 80. Tubing on the other hand is labeled as to the OD of the tube. Hope that clears up your confusion.

> **Roy Richard** Richard Koch. Well Richard you have this wrong. Regardless of the schedule the OD of the pipe is the same ie a 3/4 pipe of schedule 40 or 80 has an od of roughly 1 inch. For tubing it depends on the type, copper water tubing is designated by ID where as "engineering" tubing is measured by OD. +1

> **Jeffrey Miner** So is this all a scheme to keep people from doing this themselves? Without proper apprenticeship to obtain "the secret decoder ring" of pipe size dimensions, you're forced to hire a professional. I suspect this traces to the 1500s in England 😀 The more modern version or equivalent would be having to pass grueling actuarial tests to enter a profession now dominated by computer that perform the calculations anyway. 😀 +1

> **Roy Richard** Jeffery, I hear you but I doubt there was any effort to dis DYI. These nominal sizes were adopted long before DYI was a thing. I have worked my whole life in machine design and believe me it confuses even engineers and technicians. In the end I think we can blame it on the British 😀

Richard Koch You are incorrect. Having used the description I provided for 40 years of designing machines, I think I should know this.

> **Brad Purvis** Richard Koch—Did you design British machines?

Marc Touesnard So because you used the wrong plug and it wouldn't go in correctly, could it be called a Threadnought? +1

Dohn Roush Rob strings us along with this thread

Anuraj Shah My gosh, Rob. You have incredible patience. Let me know if you ever need a hand. I am in your area all the time. I am very good at listening while someone vents., as he slowly unravels... +3

Anuraj Shah My gosh, Rob. You have incredible patience. Let me know if you ever need a hand. I am in your

> **Jeffrey Miner** Anuraj Shah please put a GoPro on your forehead first, thank you. 😀 +1

John Harvey Bang it home with an air impact!

Bob Sawtelle In before the Thread Lock! [photo of "thread lock or not" meme] +1

Jim Denker I can't take it! Sell! Sell!

Marc Dobin Reading this makes me feel better about a sink repair that took two weeks as I floundered my way through the Kohler parts catalog. Thanks, Rob. +1

Brian Ach Christ.

Walter Eschelbach Glad you are not doing our dinner party this week. +1

Rick Roudebush I hate plumbing. Three trip minimum to get all the right shit and then they are out of the final dofloppy when I get it figured out then have to get an adapter to adapt to the street ell that could have been an elbow in the first place. I woulda shot this sumbitch by now. +1

> **Walter Eschelbach** Rick Roudebush Don't be our guest chef

Tim Warner I was able to get a 3/4" pipe thread tap at a local old timey hardware! Had to chase the threads on a water main shutoff valve in the bottom of a trench!

TLC Day 2107: Sealing... munged-up plug threads

MAR 16, 2019, 7:48 PM

The Lotus Chronicles, Day 2107:

Confirmed beyond a shadow of a doubt with Dave Bean Engineering that, as many of you suspected, the threaded plug that wouldn't seal in a pressure test is an NPT tapered plug in a straight-threaded hole. I don't think there's anything inherently wrong with that in this application; I just want things to pass the pressure test, and a straight-threaded plug that threads all the way in, with sealant on the threads, seems like a better bet to me. The straight BSPP plug and its sealing washer arrived from McMaster-Carr. I used it to chase the damaged threads in the cover, cleaned everything out, put a healthy amount of Yamabond4 on the threads, and snugged down the new plug. I'll pressure-test tomorrow. If it doesn't seal, I'll entertain irreversible solutions (e.g., JB Weld).

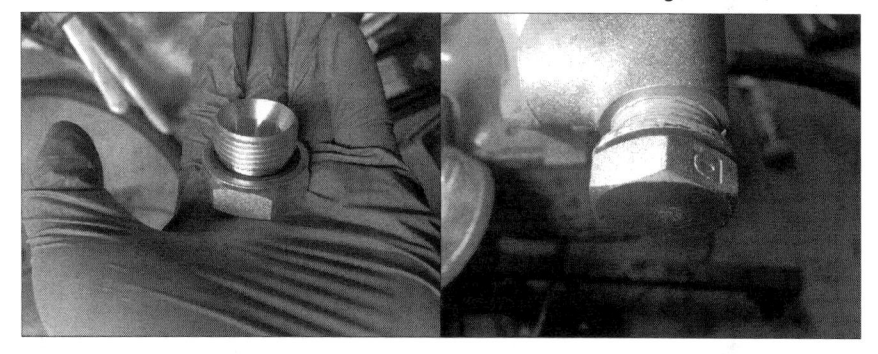

61 LIKES, 29 COMMENTS

Steve Nelson Hope you don't have to resort to irreversible solutions 😊 +2

John Whetstone You'll at least satisfy the huge JB Weld crowd here! +2

Buck Hiltebeitel Makes me wonder what other screwy "features" are in that water pump design. Mixing thread forms is a really bad idea. +1

John Turner I like how you're holding the extremely clean brand new piece with gloved hand, Doctor. +2

Chris Lordan That part looks "smooth and unremarkable." I think we heard a bit about another smooth and unremarkable part (though not for a Lotus) a while back. +2

> **John Whetstone** Chris Lordan Finest kind!
>
> **Chris Lordan** John Whetstone Surely—and forever—I am!

Daniel Neal As a former heavy equipment mechanic this irks me just a little bit, solely because BSPP is designed to be used without sealant +1

Paul Muskopf Straight threaded plugs seal at the face with a sealing washer, not the threads. That's why your NPT plug didn't work. +4

Rob Siegel I don't really care what's designed to do what and what isn't. I simply want it to pass a 20 psi nitrogen pressure test and then not leak any coolant. Neither of those seem like difficult goals, and yet this is now my third attempt to pass the pressure test. +6

> **William W Lewis III** Rob Siegel man! Just wondering what circle of hell are you presently in with the lotus do you reckon? +3
>
> **Daniel Neal** Rob Siegel I understand your frustration with it, but as someone who worked a lot with BSPP (and I mean a lot, Kubota uses it for 95% of their lines), that sealing washer would have no problem holding 20 psi, the systems I worked on that used that ran at 2300 psi operating pressure and 3500 psi peak pressure. +4
>
> **Buck Hiltebeitel** Daniel Neal and I have seen them hold at 5000 psi on underground mining machinery with a water-based fluid. +1

Paul Wegweiser Your mamma is an irreversible solution! 😊 +6

> **Paul Wegweiser** I kid! I kid! I'm sure your mother is a lovely person! 😊 +2

Franklin Alfonso Saint Bim to you… ₊₂

Brad Purvis Just remember, anything built by man can be overcome by man. ₊₅

Paul Garnier Nice

David Shatzer Brad Purvis Or permanently altered with a suitable application of high explosives… 😆 ₊₂

Franklin Alfonso I don't understand why Rob is working so hard to assure the Lotus does not supply him with its standard level of British leaking and smoke. After all, it's tradition! ₊₃

Russell Baden Musta Franklin Alfonso correct! Having a thick supply of cardboard under my Europa is tradition! ₊₁

Scott Aaron So. You're getting screwed by a giant screw. Got it. ₊₂

Dave Gerwig Looks like a permanent solution.

Rob Siegel We shall see. ₊₁

Alessandro Botta Finger crossed ₊₁

Jim Gerock Looks like you need (another) air cooled car 😆. I know one for sale…

Rob Siegel I wish

TLC Day 2108: Close, but still leaking

MAR 17, 2019, 7:59 PM

The Lotus Chronicles, Day 2108:

I am asymptotically approaching having an engine with a bona fide demonstrably sealed cooling system. It's very, very close, maybe even good enough.

The new BSPP threaded plug with Yamabond4 on it appears to finally be pressure-tight. I'm not seeing any bubbling of the soap solution on it when it's pressurized like I was before. Whew.

However, the system still isn't as tight as an a/c system would need to be. If I pressure-test it to 20 PSI, the gauge doesn't move while I'm watching it, and there's no obvious visible bubbling, but the gauge DOES drop slowly by about 2 PSI in 30 minutes, and there ARE several groups of tiny bubbles that look like insect egg sacks.

One such group is on the connection to the hose on the thermostat housing (photo below). Who cares, right? I can pull this off, clean it, re-seat it, use a different hose clamp, and get it to seal, and even if I don't, if it weeps a drop of coolant, no big deal, right?

The problem is that if I admit that THAT cluster of bubbles is a real leak, I have to also admit that the similar clusters of bubbles at the head gasket seam on the back of the head (photo below) may also be a real leak. Yes, I've read the opinions that it may seal up once the engine is heat-cycled a few times.

As others have noted, I may be testing to a standard that's too high. I'm going to seek wisdom from The Lotus Engine God before declare it good enough or do anything more invasive.

As Bill Murray said in Meatballs, "I'm very close… stick your finger in my ear…"

59 LIKES, 57 COMMENTS

Marc Touesnard Rob, I do believe the opinion that you may be testing to a standard that is higher than required. I'm sure someone else has suggested this, but now that you have the major leaks tightened up using nitrogen, perhaps you could fill the cooling system with fluid and check for leaks under pressure? +1

Brandon Fetch T-bar clamps all the way for me anymore. Tired of the worm style ones. Hope you can get it to a point where you're comfortable putting it back in its hidey-hole. +1

Rob Blake Good enough ain't close enough except in horseshoes and hand grenades my friend. +4

> **Brandon Fetch** And according to my dad, a weapons loader in the USAF during Vietnam, 122mm rockets. 😀😁 +4

Joseph Hower Hang in there!

Lindsey Brown All Cadillac Northstar engines required special sealant (known as grams of hash) added to the cooling system whenever any service was done, or they would leak everywhere. The hash works. In small doses, it won't clog anything, but any pinhole-sized leaks will be sealed up in no time. Something to consider before resorting to a total tear-down. Swing by the shop, and I'll give you a couple to keep on hand. +9

> **Rob Siegel** Lindsey, I think you're talking about these. I kept some of these tabs in the glovebox of my last Suburban when it began to leak coolant at the intake manifold gaskets. [Link to GM cooling system tablets]

Rob Koenig Send it!

Brad Day Nice use of asymptote. +3

> **Rob Siegel** Yo. Math/physics major. I LIVE for using asymptote. +6
>
> **Ed MacVaugh** Rob Siegel If I recall correctly from 1975, asymptotically means approaching but never reaching. No? +1
>
> **Rob Siegel** you are correct sir!
>
> **Paul Forbush** Ed MacVaugh Gold star to you sir! I had to look it up. My 1975 math memory is not as sharp any more.
>
> **Ed MacVaugh** Paul Forbush Good for trivial pursuit, but not much else, I am afraid. +1
>
> **Neil M. Fennessey** Mares eat oats and asymptotes but little lambs eat ivy. 😎 +4
>
> **Bill Ecker** Ah, the Xeno's paradox of repair. +1

Tom McCarthy My guess is things will seal better after engine has warmed up. Maybe move it all inside your living room to get it up to room temperature ;-) +1

David Shatzer Agree^^, and must ask what is the design operating pressure? 15-16 psi?

> **Rob Siegel** 7 to 10 +1
>
> **David Shatzer** Rob Siegel Oh. Over-testing a bit, but I do understand 110% +1

Steve Kirkup Asymptotic approaches to problems never quite get there, but always get closer. At time equals infinity, you will achieve perfection. Meanwhile we will all be dead. +8

> **Bob Coffey** The whole Asymptotic Postulate is discredited by the Indelible Chair-Back Marks on the Wall Corollary. +2

Chris Langsten It's English, so even with your Jedi electrical skills, will it ever run long enough to get to full cooling system pressure before you "accidentally" set it alight in a "freak geophysics PowerPoint lecture rehearsal using a laser pointer"? +4

Corey Dalba Head gasket definitely should go through a proper heat cycle before spending an outrageous amount of effort to remedy that "leak." But I'm completely sharing your mindset with it at this point. 20 psi is more than the radiator cap so idk. Trust your gut. +2

Steven Bauer At what pressure does the gauge stop?

> **Rob Siegel** Have not checked.
>
> **Steven Bauer** Rob Siegel If it stops at 15, ship it. +1

Roy Richard "It could be worse. It could be raining." +6

Tom Egan Perfect is the enemy of good enough. Like Patriots footballs, 2 psi in 30 minutes could be an adiabatic phenom. +2

> **Charles Morris** I always thought Gisele had something to do with deflating his footballs...
>
> **Tom Egan** Charles Morris you're thinking of Florida massage parlors. +1

Dave Bentz ...it always gets back to a/c... +2

George Zinycz Those bubbles don't look like they'd add up to volume needed to drop the pressure -2 psi. Maybe your gauge is leaking. 😀😜 +1

> **Rob Siegel** Interesting.
>
> **George Zinycz** Rob Siegel nope—it's a trap! Don't overthink things like I do!!! But always account for test setup integrity and calibration.

Luther K Brefo Prophylactically. ;) +3

> **Scott Aaron** Luther K Brefo YES

Doug Jacobs And as he says in "Stripes", "$20 shine on $3 boots." +2

Paul Wegweiser You can't say "demonstrably" without saying "demon" or "monster"... Just sayin'. +3

> **Rob Siegel** Coincidence? I think NOT! +3
>
> **Clint Carroll** AND you can't pronounce "asymptotically" without saying, well, you know… +3

Scott Aaron Should it stay or should it go? 🎱 +1

> **Rob Siegel** we already know that it's, uh, trouble +1
>
> **Scott Aaron** it's double...
>
> **Rob Siegel** the more I think about it, the more I am particularly troubled by the applicability of the line "should I cool it or should I blow?" +2

David Holzman Then all you'll have to do is polish and anodize the muffler bearings, right Rob? +2

Brian Hart Heat cycle & re-torque... fingers crossed. +1

Stuart Moulton This one is kicking your butt for sure +1

Jeremy Novak Next thing you know Rob will want it to not leak oil! +1

Andy Veedub Use Norma "constant tension" hose clamps on all hoses. Rob Siegel

Mark Manasas asymptotically or asymptotically?! +1

Mike Miller Your hose clamp etiquette has improved vastly over the decades, yet this is one is too big. I have more than I'll ever use. Rob, ask and ye shall receive. +2

> **Rob Siegel** Mike, this was just to fit a block-off hose for pressure-testing. I promise that, for final installation, I'll use one that won't offend your length sensitivities :^) +1
>
> **Mike Miller** Also, hose clamp orientation along with precisely laid hoses. Gotta keep 'em separated. [photo of Mile Miller's OCD hose clamp installation] +2

Brad Purvis Personally, I would not count on the head gasket "sealing" with heat cycles, but then I'm anal that way.

Franklin Alfonso Meh, like tuning a guitar. Close enough for rock and roll... 😀 +1

Clyde Gates Which finger? +1

TLC Day 2109: Mounting the distributor

MAR 18, 2019, 8:51 PM

The Lotus Chronicles, Day 2109:

One thing a night. Even if it takes 30 seconds. Which of course this didn't. There's no visible mark for #1 plug wire on the distributor rim. And, counter-intuitively, when you read the manual, it says that the spade connector to the condenser is supposed to be located up against the block. They then refer to that spade connector as being at 12:00, and then, relative to it, the four plug wires are at 10:00, 2:00, 4:00, and 8:00. And the distributor rotates counter-clockwise. Because Lotus.

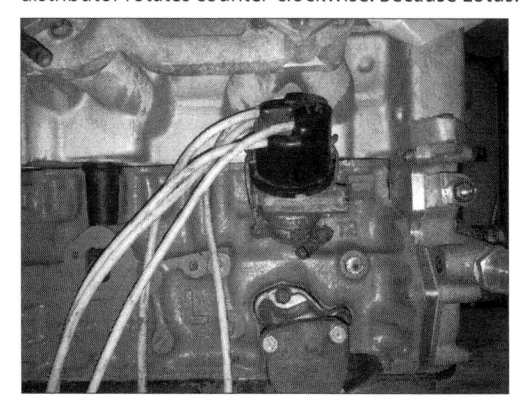

41 LIKES, 28 COMMENTS

George Zinycz The CCW rotation seems more "Because British" to me. +3
> **George Zinycz** But I'm probably splitting heirs. 😂😂😂😂 +4
> **Rob Siegel** It's driven off the jackshaft, which dates back to its Ford lineage, so it may be "because Ford." +6
> **Russell Baden Musta** Rob Siegel actually, it's a Tri-Cam! +1
> **Karel Jennings** Rob Siegel Because not a 2002. +4
> **Rob Siegel** Karel Jennings, TOTALLY not a 2002! +1
> **Clyde Gates** You are Far More Dedicated to this project that I would be!
> **Rob Siegel** Clyde, it's dedicate or abdicate, and the cost of abdication is very high. +2
> **Bill Snider** Rob Siegel Abdication costs, Yeah, and then there's the money. ;-) +1

Robert Myles Obviously your problem is one of basic comprehension. It doesn't rotate "counterclockwise," it rotates "anti-clockwise." I'm going off for some warm beer now. +12
> **Richard Stanford** Oi oi oi! Not warm. Cellar temperature. Which means that whatever you think it is, it's wrong. +2

Dave Gerwig The ultimate challenge: assembling someone's basket case! +4
> **Scott Linn** So true. Which is why I bought something that was "mostly" there. If I had seen all of Rob's troubles and weirdness of this engine first, I might not have bought mine! Sometimes ignorance is bliss. I've been lucky, 8 years of Europa driving minus a year down due to windshield and dash work. +2
> **Franklin Alfonso** Beyond lucky I would say... +1

Daniel Neal That makes my head spin/

Roy Richard "Abby who?" +3
> **Brad Purvis** Roy Richard Abby Normal.

Steven Bauer Of course then there's the M10 distributors that rotated one direction when they had points and the opposite direction when they went electronic starting in the 1980 model year. +5
> **Rob Siegel** I did not know that!

Frank Mummolo Rob, not sure about you, but I'm beginning to feel like you've been dropped into the middle of an avant-garde play, where down is up, Left is right, etc. My hair hurts just reading this!! +4

Mike Miller The Brits also built nuclear-powered ballistic missile submarines, although I believe we supplied the reactors. Anyway, they can't afford to drive them anymore. +2
> **Steven Bauer** Mike Miller I think we supplied the Trident missiles on the latest generation as well +1
> **Mike Miller** Well that's good to know. We can rest assured that they are not positive ground! +1

Richard Stanford They also use the space for a tea kettle in every tank. Because priorities. +2

Chris Mahalick No real mystery, LOL. Just remember that this engine is installed "backwards" from the initial intended orientation, so everything is opposite.

Steve Park This saga seems never ending. I admire your fortitude. I would have thrown in the towel a long time ago. +1

TLC Day 2110: A strange dream

MAR 19, 2019, 9:15 AM

The Lotus Chronicles, Day 2110:

I woke up at 3 AM after having had a very vivid strange fantastic voyage-style dream in which a tiny me was flying around under a pair of large mechanical assemblies with levers and screws, being shown where all of the fasteners go. I realized after I woke up that I was seeing the Stromberg carburetors on the Lotus motor, which I am planning on finishing installing today.

As REM said in Driver Eight, I've been on this trip too long…

86 LIKES, 33 COMMENTS

Andrew Wilson May you reach your destination… +3

   Rob Siegel …though it's still a way's away +1

Brian Hemmis Are you sure it really was a dream?

Sal Savino And, as Jerry Garcia said, lately it occurs to me what a long, strange trip it's been. +4

David Weisman But finish and get out before you rebigulate +1

John Turner Great song +1

Lee Perrault Wow, have never known angels to be flying around with wrenches and screws, but it seems they are there 😊 thank heavens!

Faith Senie They say you're not really fluent in a language until you start dreaming in that language. Guess you're starting to be fluent in Lotus… +7

Jay Ford It's not the destination, it's the journey.

Joey Hertzberg ONE OF MY FAVORITE SONGS!!! Damn those guys were amazing back then, bunch of art students, in the south no less. Thanks for the reminder. +2

Scott Aaron Dude, you have issues. +2

Franklin Alfonso As they say in Maine. You can't get thare from heer.

Roy Richard "Damn your eyes." "Too late!" +2

Jeff Dorgay I just have dreams about aliens! +1

Steve Kirkup On my Model AA I had suffered for weeks on trying to replace a bearing seal. Each time I tried to replace it, it mysterious collapsed on itself beyond salvage. In a dream I was presented with the solution, that the inner ring of the previous seal from 1931 was still on the spindle and cloaked by the visual geometry. I had almost exhausted the nation's inventory of Model AA axel seals when the dream brought me in check. I got up in the middle of the night, went to the barn that I rent, and confirmed it was still there. [photo of Model A bearing seal] +2

Rob Siegel Damn, man, that's good.

Roy Richard Steve Kirkup. Been there done that. When I have a design I go to sleep visualizing the design and often woke up with the solution. It really is about internalizing the design. Wish I was as good about that as Tesla who claimed who claimed to run a machine in his mind and tear it apart and inspect for wear!

Ed MacVaugh I agree, some minds continue noodling on problems while the body sleeps and next thing you know, a solution is at hand. +2

   Rob Siegel Or sometimes you just feel like shit the next morning because your fucking brain wouldn't shut the fuck off. That would be me. +4

   Karel Jennings Me too.

   Karel Jennings Even worse when said brain doesn't solve the problem.

TLC Day 2110: Is the head gasket leak normal?

MAR 19, 2019, 9:44 PM

The Lotus Chronicles, Day 2110 (part B):

Squelched the small leak from the hose at the thermostat neck and re-tested. Unfortunately, there's no doubt that there are a series of leaks not just from the edge of the head gasket at the back of the head as I posted yesterday, but coming from all around the head. At first, I thought it might be an artifact of squeezing the trigger of the soap solution too hard and it splatting and bubbling on impact, but I laid the soap solution on with a finger, and it's denial to think it's the bubbles aren't real. If I look carefully, I can see the bubbles coming from between the layers of the copper head gasket. So the pressurized nitrogen is likely traveling up through the block, through the coolant passages from the block to the head, and through the layers in the head gasket.

I tremendously appreciate all of your input. In my post on the Lotus Elan forum, some folks chimed in on using "stop leak" tablets as folks did here. The idea that there may in fact be nothing wrong—that it's much easier for it to leak pressurized nitrogen than coolant—is obviously appealing. I may see, as some have suggested, if it's possible for me to fill it and test it in that fashion. I'd need to be convinced that the head is actually full of coolant. I'm not sure how to do that without pulling out the thermostat. If it IS full of coolant, the presence or absence of coolant coming under pressure from the head gasket would be a pretty definitive test.

The "even though you can see it leaking with your own eyes, nothing really is wrong" scenario is to be weighed against the fact that, if I want to pull the head and use a different more modern head gasket (e.g., not layered copper), it'll never be easier than right now.

Once again, more thought, more care, more time.

42 LIKES, 60 COMMENTS

Karel Jennings I do appreciate the amount of care you are putting into this. My 1949 dodge flathead engine rebuild is going well. Nowhere near the need for that level of precision and care though. 0lb coolant system. +6

Tom Melton Fill with coolant and invert. +7

> **Rob Siegel** I can be so fucking dense sometimes. The engine is even on a stand. Flipping it over would be trivial. Thanks Tom! +2
> **Paul Muskopf** ["brilliant" meme] +1
> **Neal Klinman** Eureka!!!

Isabel Zisk For the ignorant: cost of pulling the head and using a different more modern head gasket?

Rob Siegel probably about a hundred bucks and the better part of a day

Charles Morris Doesn't sound like fun, but how much more trouble is it to change the head gasket once the engine's back in the car? +1

Rob Siegel It's probably not that bad. I pulled the head off with the motor still in the car 5 1/2 years ago because I couldn't get enough height to drop the motor and slide it out with the head still on it. +1

Daniel Senie Rob Siegel ummm, wait. Are you going to be able to get the engine back IN this beast with the head on?

Rob Siegel Daniel, yes. I can either drop it in with the engine hoist, or I can use a winch suspended from the ceiling of the garage to lift the back of the Lotus higher. I am likely to try the second method first. The goal is to get the engine off the stand, put it on a small lift table that I have, attach the transaxle, roll them both under the back of the car, raise the table, and install them. This is the reverse of what I did 5 1/2 years ago except that, then, I removed the head first because otherwise there wasn't enough clearance to roll things out from underneath.

JP Hermes New HG, and Hondabond that thing shut (scrape the stuff that spreads out so nobody can tell) +2

Collin Blackburn Hondabond ALL the things

JP Hermes Hondabond looks close enough to aluminum. Take off excess, it looks perfect

John Whetstone I'm guessing that never has a group of Lotus enthusiasts changed a club member's failed head gasket in the parking lot of a national event? Safe to say not in less than a week and half anyway. Paul Wegweiser +8

Rob Siegel Who knows? All groups have their experts and their heroes. +1

Paul Wegweiser I'm no longer taunting Rob Siegel. I just don't have the heart. But once this Lotus thing is finished... WhoooooBOY! Lookout! +4

Paul Wegweiser Rob Siegel The real expert / hero [in the "changing the 2002tii head gasket in the parking lot episode] was Ben T for inferring I was a wimp if I didn't GO for it. That jerk. I love him. +3

Rob Siegel Hey! Who had the important job of ripping up paper towels and twisting them into little dowels that you could shove down the head bolt holes and clean out the oil? THIS guy! +3

Jake Metz Rob Siegel ^ not all heroes wear capes +1

John Whetstone Tyvec? +1

John Whetstone I guess Tyvec Man already exists. [stock photo of man in Tyvec suit, looking like Tyvec Man :^)]

Bob Sawtelle I am so invested in this damn British torture device... I need this to run... whatever you choose, I stand behind you!! +4

Bill Ecker Never succumb to paralysis by analysis! +4

Rob Siegel No paralysis here. Lots of forward motion. If the engine is still sitting in my garage in a month because I haven't made a decision, THEN you can slap me. +3

Russell Baden Musta Rob Siegel and I will too!! I'll hunt you down... you know it! +1

John Harvey Heat cycle will fix that. +3

Paul Wegweiser And there you have it... John Harvey has the answer to all your Lotus problems. I like the way he thinks. Heat Cycle... or maybe Heat Lotus.

+8

Rob Siegel And the meme works for "Hot Wheels" as well! +4

Paul Wegweiser Rob Siegel I think I found the source of your problem. Inside your engine, there's one of these. Cute little fella ain't he? [gif of Dumbo bathing. Paul's kind of a strange guy.] +3

> **Dohn Roush** Paul Wegweiser Paul, you didn't have to get Huffy about it... +1

Jonathan Poole Do you think it is a physical leak, or potentially molecular permeation? A nitrogen molecule would be relatively small compared to ethylene glycol, right? I like the idea of getting everything full of coolant and then pressurizing it to see what happens. +8

> **Rob Siegel** Me too. +2

> **David Shatzer** Jonathan Poole I hate to violently agree, but I believe we are applying 21st century technology and Teutonic engineering to a mid-20th century machine... I am somewhat confident a heat/cool cycle and re-torquing of head bolts will work wonders. Find a big commercial oven and heat the completed engine, minus carbs, to, say 165 F, allow to cool. Retorque and re-test. Please forgive if I didn't read the previous posts(s) that said same. +1

> **Marshall Garrison** There did seem to be a number of comments in earlier posts suggesting heat helping or even being necessary for the metal head gasket, if I recall. Good luck Rob, hope u get it sooner than later!

Chris Mahalick Jeez Louise, Rob Siegel, this is like climbing Everest! LOL +1

Roy Richard Rob Siegel Stop obsessing with the nitrogen leak. Get the engine running and see if there is a leak. +1

Tim Warner Yes, a floor bench test, start that bastard up! Gotta hook Tranny up anyway. Used to do it all the time with air cooled VWs

Richard Memmel How much psi will this coolant system create naturally?

> **Rob Siegel** 7 to 10

> **Richard Memmel** Rob Siegel why are testing so high then?

> **Rob Siegel** See prior posts about recommendations from The Lotus Engine God.

Skip Wareham The nice thing is you've got the time to think this through without a customer waiting to get his Lotus out of the shop +1

Bill Ennis Will it seal under engine operating temps? Had the same problems on a Chevy 283/301 bracket motor. When I reached 210 there were no more leaks. Hasn't leaked since. +3

Clyde Gates so the Lotus IED still bites you 😀😰😰😰

Steve Park This makes my various car projects look so inconsequential. I'm getting soft and lazy in my old age I guess. +2

Ed MacVaugh Has the Lotus God been consulted and weighed in? Isn't he the original genesis of the pressure test?

> **Rob Siegel** Yes. He recommends testing to 40psi. If it leaks at 20, I don't see the point. He's also offered me a good rate on his dyno to run the thing before dropping it in, but the logistics are involved. +1

> **Ed MacVaugh** Rob Siegel You know, being the lawyer that I am, there is a long ways between "40 psi he recommends testing to" and the "40 psi he tests all of his engines to meet". :) However, watching you in action, and having obtained nitrogen gear for AC work, I plan to test my first M44 BMW engine to 20 psi just to see what it does. +1

Brian Hart With all the effort going into replying to FB posts, could you not take a moment off and think about firing that engine up on the stand? Fuel, 12V and some coolant. Let it idle and build up some heat and see what happens.

> **Rob Siegel** It's a fair question. I routinely start long-dead engines using a gas can, and then, once they're running, clean the fuel system from stem to stern. Hot-wiring the starter and ignition are trivial. Jury-rigging a cooling system is a bit more involved. It becomes a trade-off of work upfront versus risk avoidance. If what I'm most concerned about is the head gasket leaking, it doesn't really make sense, as the head isn't really much harder to pull with the engine in the car. If, however, I'm concerned about coolant leakage into the pan, then it DOES make sense. +1

> **Jake Metz** Rob Siegel can't believe this is found at Walmart... [link to Mobile Engine Testing Station]

Marshall Garrison For some reason or another, I am reminded of this... [link to video "Let's light this candle"]

Ned Gray Stupid question but I assume you planed the head and decked the block?

> **Rob Siegel** that is correct

TLC Day 2111: The trade-off of doing nothing

MAR 21, 2019, 7:55 PM

The Lotus Chronicles, Day 2111:

Two days ago:

--Isabel Zisk asked: "What's the cost of pulling the head and using a different more modern head gasket?"

--Charles Morris asked: "How much more trouble is it to change the head gasket once the engine's back in the car?"

These are crucial questions.

Nearly six years ago, I had, in fact, yanked the head off with the engine in the car because I wanted to drop the motor out from underneath, and the way I had the car jacked up, the motor wouldn't clear the body with the head on. So it's not really that big of a deal to pull the head with the engine in the car.

So the question then becomes: What am I trying to find in pressure-testing? If I'm ONLY concerned that the head gasket might be leaking, I should just put the motor in and see what happens, and if the head gasket leaks, pull the head and change the gasket. But if I'm ALSO concerned about coolant leakage into the oil pan that might require disassembly of the front covers to fix, I should test to the max now.

And the answer is: I have no indication of leakage into the block, so maybe 85% the former, 15% the latter, but 15% is significant, because the cost of being wrong is tearing apart the motor.

So, as per the other comments about filling with coolant and THEN pressurizing, that's what I'm doing. Tom Melton suggested that, if I'm concerned about making sure that I've gotten antifreeze up into the head to fill the coolant passages to the head gasket, I could flip the engine upside down. So that's what I've done. I've pressurized it to 20 psi and will check it in the morning. Then I'll flip it right-side up, let it sit for the day, and check for coolant in the sump. If it passes these tests, the sucker is goin' in.

If it fails, you'll find me facing off with the motor, holding a Hattori Hanzō sword.

59 LIKES, 33 COMMENTS

Daniel Neal I think you will find that all the hair pulling has been for null. It is better to be 95% sure though. We can never be 100% sure; anyone that says you can is trying to sell you something, Usually a Lotus of some kind... +1

JP Hermes Rob, if this doesn't work I can get you a deal on a Tesla rear end; $3400 for motor and axles, $1300 for drive computer for 395hp +6

    George Zinycz JP Hermes SOLD! +2

    JP Hermes For those who liked my response I'm not joking. I'm honestly thinking about this for my NK because a M10 doesn't seem like enough +3

    Bob Shisler JP Hermes Would love to see that!

    John Harvey JP Hermes and the batteries go where?

    JP Hermes where the engine normally sits

    Christopher Kohler So it still won't roll then. Got it. :D

Samantha Lewis I remember every bolt, but staring at it from underneath, right side up, often in the rain, many miles from nowhere, ...memories. God I want to see this thing in the car and running. +3

Stan Rose If the Hattori Hanzō sword doesn't work, try the Five Point Palm Exploding Heart Technique. +3

    Robert Myles This might be more satisfying… [gif of massive H-Bomb explosion] +1

    Paul Fini Nuke it from orbit... it's the only way to be sure. +1

    George Zinycz Paul Fini #namethemovie

George Zinycz It's not a big engine is it? It sure looks wee from underneath.

    Rob Siegel yes, but what it lacks in size it makes up for in being A COLOSSAL PAIN IN THE ASS +8

    George Zinycz Rob Siegel yeah, I picked up on that.. +1

Steve Nelson Wow, that's a deep oil pan 😀

    Robert Boynton Steve Nelson Has to be large enough to hold all the coolant as well as the oil. +4

Steven Bauer A couple other things came to mind: Did you try a compression test to be certain the cylinders are sealed? Did you check that the head was flat before you installed it? Are the head bolt threads binding somehow and giving you a false torque reading?

    Rob Siegel Yes, yes, and I have no indication of that. +2

Tom Schuch I took your troubles down to Madame Ruth. You know that gypsy with the gold-capped tooth... I asked her to look into the future for the outcome of this project, and she shared this pic. I said, "No, no, no, there must be some mistake... that is the past." She said, "No, the Past is the Future, and the Future is the Past, especially when it comes to British cars." I had no idea she knew so much about cars! ;) But good luck, anyway! I am enjoying the write-up.

+9

Doug Jacobs For a guy who lives a pretty sweet life, I'm starting to feel for you. Make no mistake—you invited this shit into your own life, still. Stiff upper and all that... +1

Marshall Garrison I presume swords and what-not are part of what will be the Advanced Lotus Position Sutra...

Paul Muskopf I didn't know the Lotus engine was Australian... +4

Chip Chandler You expected an English ANYTHING to retain fluids? You should name this car Rocinante! +1

TLC Day 2112: Upside-down pressure test

MAR 22, 2019, 10:02 AM

The Lotus Chronicles, Day 2112:

Let the engine sit overnight, pressurized to 20 psi, with antifreeze in it, upside down to be certain that the head and gasket are full of coolant. This morning, I don't see a drop of antifreeze around the head gasket or anywhere else. Pressure is down to 17 psi. I have to let go of the idea that this should be like an a/c system that holds pressure indefinitely. Dry is dry.

I've flipped the engine back over (well, my son Ethan flipped it back over :^) and will let it sit until this evening, then I'll undo the oil plug and check for any antifreeze in the pan. Will report back tonight.

112 LIKES, 73 COMMENTS

Jamie Eno Seems like a great outcome. Gasses are smaller molecules vs liquids, this result supports that. Good exercise to complete while the engine is out. Now you know! +1

David Weisman If corporations and government put a fraction of give-a-shit as you are with this Lotus engine we wouldn't need a wall and plane co's wouldn't sell optional tilt warning.

    Bill Ecker David Weisman we don't need a wall

Joseph Hower Good luck Rob!

Franklin Alfonso Seems to be above British standards for sure lol. +2

Chris Mahalick I think you are going to be fine. Just stop worrying and throw that lump back into the car! LOL.

    Rob Siegel I've gone this far. There's zero sense in not doing the final right-side-up pressure test with coolant. +1

Skip Wareham eLotus? +1

Frank Mummolo Archimedes is smiling! +2

    Rob Siegel Archimedes reportedly said "Give me a lever long enough, and I'll shove that fucking Lotus engine the fuck out of your garage for you!" +12

    Frank Mummolo Yes. I recall reading that somewhere! 🙂 +1

    Rob Siegel it's one of his lesser-known quotes :^)

    Frank Mummolo Aw, who'd he ever fight, anyway? +1

    Dave Borasky Does Archimedes kiss his mother with that mouth? (2)

Kent Carlos Everett Yahoooo!

John Wanner Heyyy!

David Kiehna You're really milking this for all it's worth +3

Rob Siegel I'm branching out my Hack Mechanic brand from humor into suspense and horror. +8

Russell Baden Musta Rob Siegel NOW PUT IT IN THE F@&KING CAR!!!! +3

Craig Lovold I was feeling smug about how much easier/better rebuilding my S52 was. No need to pressure test or waste time with other things like that. I started it yesterday for the first time and saw coolant leak out of the head gasket between 3&4. I should learn from the master. Do the testing first when it's on a stand. It should be fun and you will know it's right when you are finished. I can't wait to see it back on the road. Enjoy! +2

Rob Siegel I am so sorry!

Bill Ecker Craig Lovold Ah, hubris. +1

Ernie Peters Hoping it cooperates this time. I want you to experience the thrill of driving a Lotus. +1

Robert Boynton Don't forget to fill it with oil when you're done 😎 +1

Rob Siegel but… but… but then it'll leak! +5

Robert Boynton With my Model A Ford, you only worry when it stops weeping fluids. +4

Tom Egan Show some respect for the finest cooling system ever designed on a Hethel pub bar napkin. +7

Rob Siegel I'll admit that I had a eureka moment when I understood that the design stems from taking a pushrod engine with the coolant passage in the front of the block and a conventional bolt-on water pump, then slapping a twin-cam head on it and having to design timing covers that enclose the timing chain as well as allow a water pump to push coolant into the existing water passage. But it doesn't make the whole thing any less of a kludge.

+3

Tom Egan I knew it was a kluge when I realized our "twin-cam" engine had a vestigial 3rd cam.

Rob Siegel yes! the "jackshaft!" +1

Rob Siegel "is that a high-lift jackshaft, or are you just happy to see me?" +3

George Zinycz Is it called a jackshaft because it doesn't do jackshift? +3

Rob Siegel "Dat's the fact, jack! +1

Steven Bauer Tom Egan But the M20 has a Jack excuse me, intermediate shaft…

Tom Egan Steven Bauer Sure, but does it have cam lobes that don't do anything? +2

Rob Siegel they're, uh, "oil massagers." +3

Tom Egan Let's not bring Robert Kraft into this… +3

Steven Bauer I can pull the cover for where the distributor used to be and see if it has a vestigial drive gear.

Steven Bauer Tom Egan According to Rob, Kraft would be the oil, not the lobes…

Rob Siegel I am NOT going to weigh in on splitting that particular analogy. +1

Rally Mono Based on this build I surprised any lotus car ran after a few years. +2

Scott Linn Every time I drive mine I feel lucky. 8 years and counting… +2

Rob Siegel These 70s cars are, reportedly, one step removed from being kit cars. The engines are NOT bullet-proof and daily-driver durable like 2002 motors. Even considering that people bought them as second or third cars, it was reportedly not uncommon for them to need major work, such as a water pump replaced, at 20,000 miles, and to be rolled into the garage and left there for decades. Mine reportedly was a "ran when parked" car that the original owner could no longer drive due to back issues. It was rolled into a garage in 1979 and the engine reportedly seized from sitting. But when I bought it, I thought that a 24k Europa would be a rare thing. That turned out not to be true at all. +4

Bill Ecker "Ran when parked, the sequel" +2

David Kerr Rob Siegel well, 24k may be rare in the sense that few logged that many miles +6

Neil Simon For a while, Lotus's importer was based in Millerton, NY (or, perhaps, it was next door in Amenia). I remember an unpaved lot filled with all these wondrous little Europas! +1

Rob Siegel that must've been quite a sight +1

Neil Simon It was and filled me with Europa lust. However, you're curing me! +2

Rob Siegel Apparently I provide this service, free of charge :^) +3

Neil Simon Millerton, it was! http://www.fredstevensonlotus.com/Lotus%20East.htm +3

Rob Siegel very cool. +1

Ernie Peters When I worked for the Lotus dealer in St Louis, I flew to NY to pick up a new white Eclat for a customer who wanted his car delivered "broken in." I recall Lotus East being a not very large operation. I know for a fact they did not stock a lot of parts for the 907-engine Esprit, Eclat, or Elite (or Jensen-Healey). I actually went to a local gasket maker/supplier to get 907 cam cover gaskets made since Lotus East had none to supply. Ones I had made were of better material and cost less than our cost from Lotus.

The Eclat was a blast to drive, even at break-in limits. I had a CB radio with me, and the truckers voluntarily did recon for smokies all the way home, although I never exceeded the speed limit. They also had several compliments for the car along the way. It was an awesome road car, very stable, quiet, comfortable. After a detailing, fluid changes, and valve adjustment check and carb adjustments. we delivered the car. Customer was ecstatic.

Sorry for the digression, the reference to Lotus East in Millerton brought back a fond memory. +4

Brian Ach winning

Collin Blackburn If it doesn't run, I'm shipping you a Toyota 5SFE. 😊 +2

Rob Siegel If it doesn't run, I'm shipping a dead '74 Lotus Europa Twin Cam Special to the highest bidder. +2

Collin Blackburn What crazy bastard would want to own one of... oh. +6

Karel Jennings Still waiting for the "screw this and put an M10 in it"

Rob Siegel that ship has sailed, caught fire, and sank +1

Ed MacVaugh I considered something similar, but BMW never really made a suitable engine. Although that engine looks mighty on the stand, and it was Ethan that inverted it, it is still a much smaller engine than BMW made for an automobile. +2

Rob Siegel Although I never expected the rebuild to be this persnickety, the problem with any engine swap, even the Ford Zetec that has the same bolt pattern as the Lotus' Ford 1600 block, is that you pay for it in adaptation of every system—mounts, intake, exhaust, cooling, electrical, etc—and there doesn't seem to be a kit, just websites of six guys who have done it, each rolling their own. +2

Chris Langsten For the love of all things automotive, please put this engine in the Flotus (Short for F*&ckin Lotus) and promptly bury it under book boxes for another 1053 days. I want my FB feed to go back to BMWs, I'll even entertain the errant feather finding expedition! ;) Glad you had success with the fluid test. Fortitude, is nothing without thought. Keep going, you're closer than you ever have been before! +2

Rob Siegel Gee, and I always thought that Flotus stood for "First Lady Of The United States." Who knew that both the executive branch of government and the press had such painful experiences with Brit bits? +3

Marc Bridge Wishing you luck!

Kevin McGrath [photo of Betty Crocker Pineapple Upside Down Cake] +1

Derek Soltes So you'll be good in those loopty-loops +1

Rob Siegel Yes. It'll be the only the car that only runs AFTER you've flipped it by heading into a corner too hot.

TLC Day 2112: Sealed, signed, and delivered

MAR 22, 2019, 7:08 PM

The Lotus Chronicles, Day 2112, part B:

It's good! Chez Siegel is now filled with car guys named Nigel, women dressed like Emma Peel and Tara King, kegs of warm beer, and odd British lubricating products.

Unfortunately, I've decided to take a different direction. I'll be using a three-cylinder 55 hp motor from a Geo Metro. For, you know, lightness and fuel economy. The ghost of Colin Chapman came to me in a fever dream.

Seriously folks, thanks. Some of your suggestions were crucial in helping me think things through and come to closure. But if one more person said "LS1 it" or "M42 it," I swear I was going to fucking clock them.

Next:
--Get the valve cover back from where it's being powder-coated (yeah...) and button up top of engine
--Get the engine off the stand and onto the lift table
--Fit the clutch
--Decide whether to do any seal replacement on the transaxle
--Mate transaxle to engine
--Raise the back of the car to slide the drivetrain under
--Lift drivetrain into place and secure the mounts
--Clean-out radiator and connect cooling system
--Fit starter and wire up ignition
--Test-start with a gas can
--If it starts, clean-out and connect fuel system and start for real
--Run and check for leaks
--Reconnect "stressed member" rear suspension to transaxle, and decide how much of suspension to renew while doing so
--Reconnect clutch and shifter
--Let car down off the jack stands where it's been marooned for almost six years

And that's all before doing the basic sort-out that ANY car that's been sitting for 40 years will need. So it'll be a while before it's moving under its own power. Probably not before the fall.

PLUS, if I'm going to drive Bertha to MidAmerica 02Fest, which I'm seriously considering, she needs some attention.

Although future posts may lack the horror and suspense of these past few months, I'll certainly work my hardest to continue to supply you with the finest in Lotus-related entertainment.

251 LIKES, 77 COMMENTS

Kevin McLaughlin [gif of Brazilian soccer "GOOOOOAAAAALLLL"]
Kevin McLaughlin Sorry Rob. Had to do it 😊
    Franklin Alfonso Don't be silly, Everybody knows that's a touchdown!
Frank Mummolo Lazarus has nothin' on this! +1
Ernie Peters Woohoo! Thought about the fuel pump: my "glass-bowl" original looked great and pumped fuel, but also leaked internally into the engine. I caught it before I drove too far, ordered a replacement that had a priming lever, very helpful with the Strombergs, and never looked back. +1
    Rob Siegel Forgot to add fuel pump to the list. Will probably just use an electric one. +1
    Rob Siegel And yes I love the look of the glass one. It looks like the head of the robot in Forbidden Planet, whose poster I have in the garage. I may leave it in place for looks and just run the gas line around it. +5
    Scott Linn Mine leaked but hasn't since I tightened it up. I love the look too, Forbidden Planet for sure. +1
    Tim Warner Rob Siegel I have a glass one on my 1950 Ferguson TO-20 tractor. A new one! Don't think I'd use one on a car... +1
Robert Boynton Don't forget that oil...!!! 💦 +2
Barry Gross Congrats maestro! +1
Rob Koenig Screw the Mueller report. THIS is the most epic news of the day. Congrats! +9
Russell Baden Musta All that above should take about 20 minutes... +3
Bill Ennis Grit and determination pays off!
Ken Sowul HUZZAH! HUZZAH!!
Coleman Maguire Hell yea
Tom Melton Perfect. Look forward to more posts about it.
Kieran Gobey Good job mate!
Emery DeWitt Seriously? Emma Peel and Tara King, and we get a photo of YOU? +6
Tom Tate Who took that photo? And are they free yet? 😊
    Rob Siegel My lovely Maire Anne +1
    Dave Borasky she just dropped a large insect in his pants 🐛 +6
    Dave Gerwig Lotus only hatch every 40 years, distant cousins of their kid species known as 17-year Locusts. Great work and dialogue: hope to see it sitting on all 4 rims soon! Tires are probably trash by now +4

Rob Siegel Dave, yes, among the mundane issues are the shit tires and the cracked windshield
    Dave Gerwig Rob Siegel 165x13 ZX Michelins
David Weisman 😂😂😂
John McFadden "kegs of warm beer, and odd British lubricating products." Isn't that redundant? 😂 +3
Clyde Gates For a minute or two, I thought you might have totally (as opposed to marginally) lost it. But then, I recalled your history of EOD and that mentally You Never Had It To Lose! +2
Marshall Garrison So, in the sutra, this position is... the "Un-Leaky Lotus"? +2
Bill Snider So a bit of premature Exultation lol
JT Burkard And somewhere in this list you forgot a perfectly good aged bourbon. 🥃 +2
    Dave Gerwig JT Burkard and that will be coming from Kentucky +2
Roy Richard I suspect Lucas still has some surprises waiting for you.
    Rob Siegel no doubt
Delia Wolfe It's astounding...Time is fleeting...

+6

Daniel Neal This is wonderful news!
Mark Thompson Love this picture... and thank you for showing us the size of your penis! Sorry... I meant "Touchdown." +2
Scott Sturdy [gif of Emma Peel]
Bruce Kwartler 3 cylinder turbo has a nice ring to it! :P +2
Wade Brickhouse Whew, I'm safe from getting clocked. I said that thang needs a Hemi! :-) +2
    Rob Siegel You get watched, not clocked :^)
Tim Warner Glad this phase has passed
    Rob Siegel Ya think?
Alessandro Botta Great!!! 👏👏👏👏👏😂👏👏😂 may I suggest some penne all'arrabbiata with a real Italian receipt?
Bill Foley Mrs. Peel... we're needed.
Alan Hunter Johansson Mmmmm... Emma Peel... I'm sorry, did you say something after that? +4
Caleb Miller He. Could. Go. All. The. Way! +2
Franklin Alfonso How does Emma Peel dress now? She is 80 years old after all...
    Alan Hunter Johansson Franklin Alfonso Still 100% class act.
    Franklin Alfonso yup
    Franklin Alfonso Well when she was young we all wanted to see Emma peel...
    Rob Siegel oh behave!
    Alan Hunter Johansson Rob Siegel We ARE behaving! Badly, perhaps, but still...
    Franklin Alfonso Like Mae West always used to say, when I'm good I'm great, When I'm bad, I'm better... +2
Greg Lewis Advantage Siegel! Glorious ness abounds. The winter of our discontent has ended. Sorry that's all I got. Oh yes one more: YOU ROCK! +2
Licia Sky You are an artist!
    Rob Siegel I am a head case is what I am. Just a very persistent head case. +2
Lee Highsmith Congrats Rob! I know this is a big milestone in returning to the misery of British car

ownership ;)

David Layton 3 Cylinders: Looks like 1 Brexited.

Robert Boynton Keeps you off the streets (in more ways than one)

Franklin Alfonso Well, this has all been so easy...When are you gonna buy another Spitfire?

Brian Ach AMAZING. For the record I never said LS1. I said 4A-GE. +2

Mike Hopkins Congrats on the milestone!

Rob Halsey I have thoroughly enjoyed reading your posts about this project and look forward to the next challenges you face with this little beast. BMWs seem easier to work on, but a Lotus tells a good dramatic story. It has suspense, sorrow, irony, and many other wonderful story elements. Maybe I should have my Language Arts class restore a British sports car.

David Ibsen Fear & loathing in lotus Vegas. You poor thing...

Pamela Cormier Greene You are one great and crazy dude! 😎

    Rob Siegel I have my moments.

Brad Purvis I'll start to worry when you begin eating bubble & squeak and drinking Horlicks before bedtime.

Steve Park I should have my girlfriend read your posts. I'd appear sane in comparison.

Rob Siegel As I tell people, I provide this service free of charge.

Kai Marx Very excited to see this car rolling soon!

Russell Baden Musta Love this Rob!! Keep going forward my Friend!!

Interjection: "Smooth and unremarkable"

There were occasional references in posts over the past year to something "smooth and unremarkable." The context of this is that, in a Facebook post of mine from 2018, I said "Happiness is having the 65-year-old South African urologist say that your prostate feels 'smooth and unremarkable.'" My PSA, however, was high enough that the urologist recommended a prostate MRI. That showed irregularities, which resulted in a biopsy, which showed, to our surprise, that I had prostate cancer. They used all the "good words"—contained, low grade, low volume—and treated it with brachytherapy, which is insertion of over a hundred tiny radioactive seeds into the prostate. Unfortunately, following the procedure, I couldn't urinate the required volume, so they sent me home with a catheter. This all happened in early March.

So, for much of the Lotus engine pressure-testing adventure, I had a pee bag strapped to my ankle and a catheter the size of a spark plug wire going into the middle of my dick.

This is one of the reasons I was content to be so incremental about the pressure-testing. In contrast with the rapid strides I'd made in prior weeks, for a while, hobbling into the garage for ten minutes, putting sealant on a threaded plug, waiting a day or two or three for it to set, starting up a pressure test, and returning to check on it several hours later was about the right speed.

I'm fine, cured, done. But when you read about being kind to people because you never know what battles they're fighting, that's the truth.

WELCOME TO

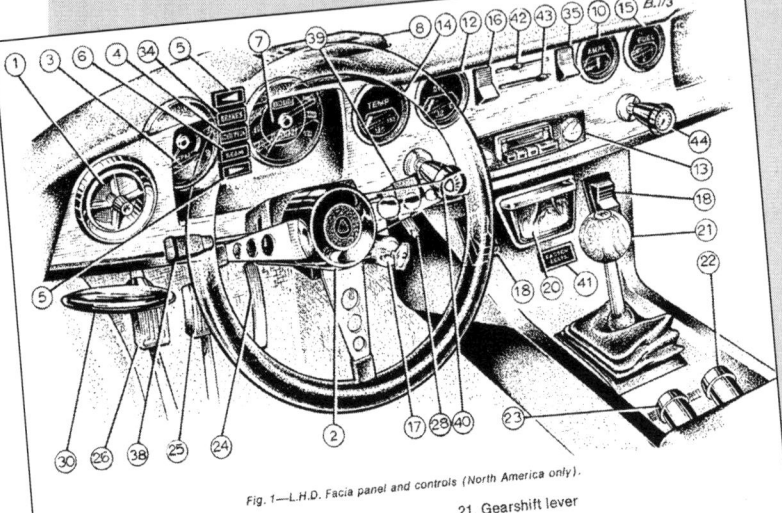

Fig. 1—L.H.D. Facia panel and controls (North America only).

1. Face level ventilator
2. Steering wheel
3. Tachometer
4. Ignition warning lamp
5. Direction indicator warning lamps
6. Headlamps main beam warning lamp
7. Speedometer
8. Windscreen demister vent
10. Ammeter
12. Oil pressure gauge
13. Radio (when fitted)
14. Water temperature gauge
15. Fuel gauge
16. Heater fan switch
17. Ignition/starter switch and lock
18. Window operating switches
20. Ashtray
21. Gearshift lever
22. Choke control
23. Heater temperature control
24. Accelerator pedal
25. Brake pedal
26. Clutch pedal
28. Speedometer trip control
30. Handbrake
34. Brakes warning lamp
35. Hazard warning switch
38. Horn, direction indicators and headlamps switch
39. Windscreen wiper/washer control
40. Lighting switch
41. Seat belts "fasten" warning lamp
42. Heater air direction control, driver
43. Heater air direction control, passenger
44. Panel lamps switch

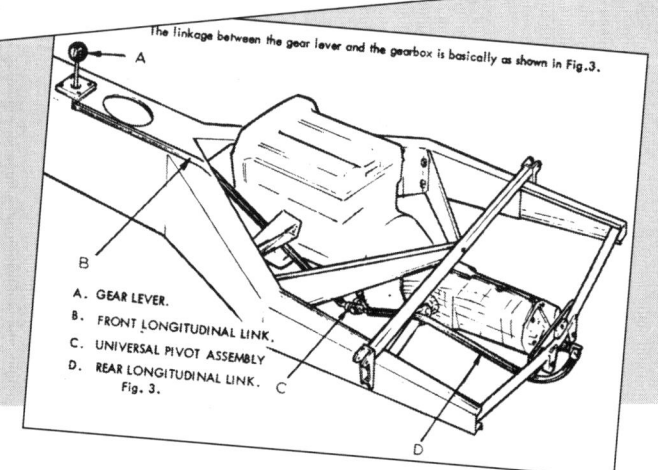

The linkage between the gear lever and the gearbox is basically as shown in Fig. 3.

A. GEAR LEVER.
B. FRONT LONGITUDINAL LINK.
C. UNIVERSAL PIVOT ASSEMBLY
D. REAR LONGITUDINAL LINK.
Fig. 3.

LOTUSLAND!

The logo's yellow circle signifies the sun.

The triangle is British Racing Green.

ACBC letters are for the founder's full name, Anthony Colin Bruce Chapman.

Unoffical Lotus acronyms include "Lots of Trouble, Usually Serious" and "LOT UnSold."

Lotus is based in the small village of Hethel in Norfolk, England, about 10 miles south of Norwich.

1970s badge

— COLIN CHAPMAN QUOTES —

"Simplify, then add lightness."

"Adding power makes you faster on the straights. Subtracting weight makes you faster everywhere."

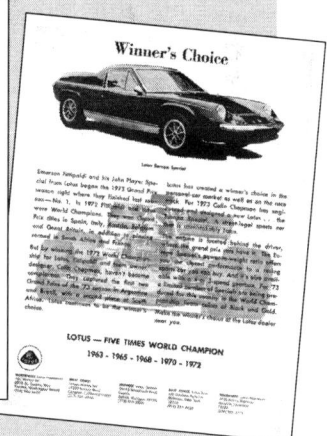

Series 1 Europas had the fiberglass body bonded to the frame, fixed side windows, and fixed seats; the pedals were adjusted to fit the driver.

S2 fiberglass is so thin that, with the headliner removed, you can see the outline of a shop light through the roof and the body can be lifted off the chassis by two people.

Rumor has it that Europa ashtrays came from Cessna aircraft.

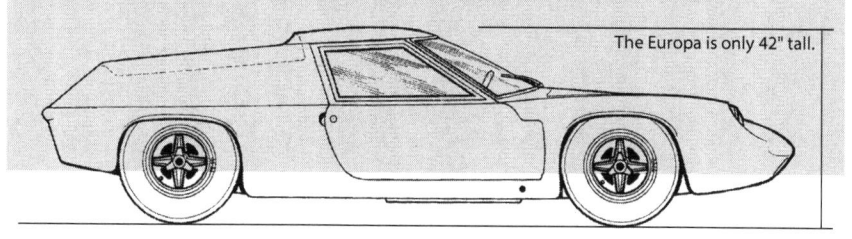

The Europa is only 42" tall.

Chapter VII: Putting Lolita Back Together

TLC Day 2113: Introducing the drivetrain components

MAR 23, 2019, 5:21 PM

The Lotus Chronicles, Day 2113:

Lotus engine, meet transaxle. Transaxle, meet Lotus engine. I thought I would introduce them before I, uh, mate them. It's the polite thing to do.

90 LIKES, 43 COMMENTS

    Kevin McLaughlin Don't forget to put on some Barry white or Marvin Gaye. 😊+4
        David Weisman Kevin McLaughlin 😂 👍 👍 best comment here!
        Robert Myles It's English. I'm thinking some Brothers Gibb might be appropriate.
        Bob Coffey 'Ceptin they woz Ozzies 😉

Clint Carroll Did (Lucas-induced) sparks fly? +1
Rob Capiccioni Don't forget the lube! +2
Andrew Wilson May they make beautiful music together. +2
Paul Garnier Your garage makes me itchy +4
    Rob Siegel My garage makes ME happy +8
Robert Boynton How much coaxing did it take to get that shy transaxle out to the dance floor? +1
Tim Warner How come Kugel is floating on your lift?
    Rob Siegel Because it watched Repo Man too many times. +4
    Raymond Wright Rob Siegel Can you really watch Repo Man to many times?
    Rob Siegel WE can't, but IT did +1
    Bob Sawtelle Now I feel like getting sushi and doing crimes. +1
    Chris Roberts Because floor space for more stuff.
Tom Egan A little scrubdown of your tranny's (!) nether regions is also a polite thing to do before mating. Though I suppose some do prefer it dirty.
    Rob Siegel Just finished doing that. Damn it, dude, you're jumping my punchlines!
Lindsey Brown A proper gentleman always ensures that the pilot shaft is sufficiently lubed. +4
Blair Meiser Be sure both provide explicit consent before any "connection" is made! +2
Roy Richard Still remember: a NO IS NO.
Chris Lordan To quote Austin Powers, "Oh behave!..." To which I add, "... all of you." +1
Thomas Jones They are married ya know. You're the mad scientist who separated them, so many years ago. The engine has been in a coma for years, got a new heart, had its head examined, been experimented on and soon will be awakened from said coma a new person; hopefully. The tranny has patiently been waiting in the room the entire time. The doctor who performed most of the procedures, while not a licensed professional, has taken every precaution, even beyond that normally practiced and has taken advice from some of the best... Let's all cross our fingers and pray that the Gods of Lotus and Lucas shine down on this creation, union, and bless it with life. +8
    Roy Richard I've heard rumors that the tranny has been fooling around with a 2002 engine. +1
David Kemether A nice dinner should be in order first. +2
Collin Blackburn Give 'em a good pint before you let them bang gears together. +2
Samantha Lewis Pretty soon, before you know it, you'll be hooking up the most ridiculous shift transfer contraption that the world has ever known, and I'm waiting for that turnover of this little Twin-Cam. +3
David Shatzer Whipstocking is now a new skill... +1
Jim Gerock Please tell me you at least tried to degrease that tranny before bolting it up to the engine... +1
    Rob Siegel Dammit, Jim, you're ruining my next post! +4
    Thomas Jones Well the tranny should have a bath before they wake the patient up from its coma. It has been sleeping in the corner for years. +1
John Whetstone You're telling us you're trying to have MORE? +2
Bob Sawtelle Every time Rob posts a garage pic, I do a where's Waldo on the pic looking for a 10mm socket... no luck this time but did see a wrench and a roll of shop towels! +4
Cameron Parkhurst Cigarette? +2
Brad Purvis Unless you're Robert Kraft +1
Dohn Roush Hell fellows, well met.. +1
Clyde Gates Formal Introductions, I hope... +1
Christopher P. Koch Clever of you to paint your platform truck bright orange to avoid taking a flying fall over it... +1
    Roy Richard Christopher P. Koch That won't stop Rob.
Franklin Alfonso Is not that damn thing running yet? I really hope all this trouble is worth it and the car drives and handles make it worth all the effort and grief it's put you through +1

TLC Day 2114: Cleaning the transaxle

MAR 24, 2019, 7:51 AM

The Lotus Chronicles, Day 2114:

I emptied six cans of brake cleaner at the transaxle while scrubbing it for nearly an hour with a wire brush. That's my limit. It's actually now pretty clean. It's not Jim Gerock clean, but let's just all admit that that was never going to be its fate.

59 LIKES, 89 COMMENTS

Doug Jacobs It looks like a six-year-old did that casting. +3

Russell Baden Musta I'll be driving my Europa up to you to clean mine!! +1

Kyle Duquette I've found diesel actually works amazing at cleaning grime off cast aluminum. Have you ever tried it? I've only done it on an oil pan and a couple other random covers but it worked unbelievably well. +1

    Rob Siegel I have a small parts cleaner that I bought at tractor supply. Ironically, kerosene in it. It works pretty well. But I can't fit the transaxle in it. I just had to shoot it in my driveway and hose it off. +1

    Wade Brickhouse Rob Siegel the EPA may like a word with you. 😊 +1

Matthew Zisk Still looks like something Rey might have picked up at the beginning of The Force Awakens. +5

    Leslie Goldstein Are you building a robot? +4

    Rob Siegel No, a '74 Lotus Europa Twin Cam Special. The opposite of a robot. So unreliable that it's guaranteed to never rise up and take over ANYTHING. +8

    Leslie Goldstein If it is so unreliable, does it need to work? Hindsight is always 20/20. You are very determined and patient.

    Doug Jacobs Rob, I think you could make the argument that it's successfully taken over a small part of West Newton. +5

    David Weisman Rob Siegel 😂😂😂

Craig Fitzgerald So this is perfect timing for this question: If you hosed it off, would you be opposed to hitting it with engine cleaner, scraping it and then blasting it with a pressure washer? I'd cover the parts I didn't want to get wet and avoid blasting the seals and stuff, but I'm going to replace those and change the gear oil anyway. [photo of the transaxle from Craig's Corvair]

    Rob Siegel Craig, I used to have a pressure washer, but it died. Things that are really greasy and small I clean in my parts washer. Things that are really greasy and medium-size, like a differential, I will do exactly what you say--Carry them down to the car wash, hit them with brake cleaner, scrub them, and pressure wash them off right there. The transaxle was simply too big and heavy for me to deal with moving, so I just did the best I could in my driveway. +3

Craig Fitzgerald I bought a small one last year. I think I'll go that route today if it warms up a bit.

Craig Fitzgerald For the record, these posts have really gotten my ass in gear on the Corvair. I figure if you can get a Lotus engine to stop leaking coolant, with some help I can get an air-cooled six running again. +7

Rob Siegel Craig, there's running badly, there's barely running, there's dead, and then there's 40-year-Lotus-dead. I am glad you were not facing the latter. Still, I understand that every project is a Gordian knot of time, money, space, priorities, sanity, on and on. One of my big lessons is that forward motion, in ANY volume or form, is a thing to be treasured. +11

Tim Warner Craig Fitzgerald me too! +1

Dohn Roush Rob Siegel Forward motion is ideal, as long as you're sure you're not standing on the edge of a cliff. But if we had that sort of judgement, far fewer old cars would get saved... +3

David Weisman Rob Siegel damn that's good.

Larry Johnson Have you tried Easy Off?

Rob Siegel I have not, but I am aware that some people swear by it. +1

Larry Johnson I find it cost effective and works well on stubborn stuff +1

Karel Jennings Rob Siegel it works. [before and after photos of engine block]

Karel Jennings Easy off and a wire brush and an hour +2

Karel Jennings It was like near 32f in there... if it's warm in your garage or you can leave stuff in the sun. It works. I usually get the most nasty extra strength one I can find. I try for the gel one. Leave sit for 2-6 hours depending on grease thickness. Then have at it. +2

Peter Potthoff And hopefully the EPA doesn't test the runoff off of your drive... Hope you use hazmat suits... +2

Rob Siegel Peter, this is why I prefer to do this at a car wash +3

Robert Myles Just never use EZ Off on aluminium (You all are pronouncing it as the British do, aren't you?):
"Can EASY-OFF® Heavy Duty Oven Cleaner be used on any aluminum? We do not recommend using this product on aluminum, as it may pit and discolor it. However, it does not penetrate into the metal or remain on the surface after the recommended rinsing instructions have been followed. Utensils and appliances cleaned with EASY-OFF® Heavy Duty Oven Cleaner are completely safe for use in cooking or serving food after they have been thoroughly washed and rinsed in a vinegar and water solution." +2

Neil Henry Rob, It's getting hard to find a true self-service car wash in my Rt 2 burbs (wand gun w/pressure)

Chris Raia NEVER ON ALUMINUM particularly if it's a structural part of the suspension :P The lye will eat the hell out of it +3

Brandon Fetch I've used EO on painted AL wheels. Gets the brake dust right off tuit suite!

Brad Purvis No Q-tips? No toothbrush? 💀

Rob Siegel no. it's not bloody personal hygiene.

Brad Purvis Rob Siegel 😄

Thomas Jones Brake clean doesn't do shit anymore... That purple clean works wonders, on grease, but wear gloves! +1

Rob Siegel I find that three buck a can Autozone non-chlorinated brake cleaner works pretty well +3

Trent Weable Rob Siegel have you tried BMW brake parts cleaner? Up until recently was $3 a can and now it's like $5 but you can literally spray the grime right off. +1

Thomas Jones The transmission and engine are about to renew their vows and you send the tranny to the event looking like this?!? +2

Rob Siegel sucks to be them, right? +2

Thomas Jones While the engine is wearing the equivalent of a tux?

Rob Siegel "He was a dandy of an engine from a good British family. She was a trashy little French transaxle from the wrong side of the tracks." Hey—It's "Pretty in Pink," but with drivetrain parts! +14

Hans Batra Haha. I remember that scene when Andie's dad is talking to Duckie:
Andie's dad: "It either will or it won't. It's all in the heart." Duckie: "Yeah, sure. Cardiovascular." +1

Franklin Alfonso Meh, it's gonna be under the car so with luck you won't see it again too often. +1

Eliot Miranda isn't that engine a sophisticated British head on a robust American body?? +1

Chris Mahalick That needs to be bead blasted, LOL

Rob Siegel Unless someone wants to show up in my driveway and do it for free, that's simply not going to happen. +6

Chris Mahalick Rob Siegel Don't worry about it. At this point, you really need to keep up your awesome momentum and get the engine in the car. Come on man, I am within a week of a test ride in mine, and I still think you could beat me. Put some Dead Kennedy's on the stereo and get it done, LOL

Rob Siegel (Chris, mum's the word, but I did pay money to get the valve cover bead-blasted and powder-coated. Shhh… don't tell anyone…) +3

Chris Mahalick Rob Siegel Wait until you see how hard it is to shift these cars. And I thought the Porsche 914 was too hard…

John Morris Ah, the patina of a vintage Lotus Twin-Cam valve cover… wait… what?!? +1

Frank Mummolo Rob, many in the Beemer crowd swear by WD-40 and a stiff nylon brush on aluminum castings. I can't verify as I tore my R90S completely apart and vapor honed the castings. This is the bike's trans cover before and after.

+6

Scott Hopkinson Yes. Airhead BMW in the group following Rob. R100RS & R50/2

Frank Mummolo Awesome, Scott! R90S, R90/6, R100S, R100RS! Nice to connect. I will friend you! Let's chat! [photo of T-shirt saying "I never knew BMW made cars until I passed one."] +1

Corey Dalba On-demand propane water heater hooked up to a pressure washer, soaked in purple power = insta-clean. +2

Dave Borasky When will you post the "after" pic from the cleaning? 😊 +4

Rob Siegel go away. +2

Mark Greene Look in to Vapor Blasting +1

John Morris Vapor blasting is the best! Amazing what it does for vintage motorcycle engines.

Rob Siegel Mark, we're all different in how we tackle this sort of stuff. Me, seriously, there is zero chance I'm going to vapor-blast something that sits underneath the car.

Frank Mummolo Rob Siegel plus, you'd have to disassemble the trans. There is no way to protect the seals, bearings and internals from that media. It gets into EVERYTHING, believe me! But it sure does produce a nice result. I vapor hone everything that I have apart now. +1

Mark Greene I saw example of vapor blasting while shooting an episode of my Cars Yeah TV show at Canepa and was amazed. You can vapor blast almost anything, except your children. They'll just have to remain messy. +1

Frank Mummolo Mark Greene they say you can literally take the ink off a business card with it, though I've never tried that.

Mark Greene Frank Mummolo—Yes! It works on rubber too. Incredible for restoring cars. +1

Joey Hertzberg I wonder if you can build a diy vapor blaster or whatever that water/blast medium machine is

Lee Highsmith Let us not talk falsely now—the hour's getting late! +4

Rick Roudebush There are many here among us. Who feel that life is but a joke ;-) usually at Rob's expense. (4)

Lee Highsmith Rick Roudebush I knew if I lofted it up there somebody would bring it home... +1

Rick Roudebush I'm doing Jimmy's version in my head, how about you?

Rob Siegel But face it, NONE of you can say "but you and I have been through that!" +4

Mark Valentino Cover it with paint and nobody will notice.

Scott Linn Getting so close you can taste it... +1

Marshall Garrison Good enough! Did you change trans fluid already? Wouldn't remember from prior posts if you'd mentioned that already...

Rob Siegel Not yet, but I will.

Marshall Garrison Hopefully there will be no sealing or leaking issues with the drain plug!

Marshall Garrison or the rest of it for that matter...!

Jim Gerock Allimgoingtowrite: power wash 👊😂 +1

Paul Amonds why didn't you use a pressure washer

Rob Siegel I don't own one, and it's too big and heavy for me to take to the car wash without help.

Joey Hertzberg What manufacturer made that transaxle Rob?

Rob Siegel Renault +1

Kevin McGrath That table is awesome! +1

Jim Gerock Looks similar to the one I bought from HF. Very handy for storing engine blocks, transmissions and even supporting subframes. +1

Rob Siegel I bought it when I still had the 911SC. Having the mid-rise lift and the table made engine drops almost unfairly easy. +1

Jim Gerock Mine's yellow.

TLC Day 2114: The engine comes off the stand

MAR 24, 2019, 4:52 PM

The Lotus Chronicles, Day 2114, Part B:

Well that was terrifying. I thought I could just roll the lift table under the engine stand, use it to support the engine, and undo the bolts holding the block to the stand, but I found that the table won't clear the leg of the stand because of the location of the hydraulic cylinder. I thought about getting the engine hoist, but it's buried under the back porch; it'd be hours of work getting it out. I wondered... how did I do this when I pulled the motor apart? I looked at the photos and realized that I never had the engine on the stand; I tore it apart right there on the lift table.

I looked at the lift table and realized that if I slid the engine stand under it from the side not the front, the stand's leg would clear the table's hydraulic cylinder. So I did that, raised up the table, and test-balanced the engine on it, supporting the shallow part of the pan with blocks of wood, and convincing myself that it was stable and wouldn't topple over and crack the rare and expensive Lotus twin-cam head.

I began to undo the bolts holding the block to the stand when I realized... holy shit IT'S SINKING. Turns out the $120-on-sale lift table I bought at Tractor Supply eight years ago now has a slow leak in the hydraulics (no surprise, really). I pumped it back up and watched it. I gauged that I had about 45 seconds to undo and disconnect all four bolts before ugly bending toppling catastrophe set in, at least in my mind.

So what do you do?

Work fast.

Once it was down and I was convinced it was stable, Ethan and I spun it 90 degrees to the orientation it'll need to be in to mate the transaxle and get the pair in the car. Guess I'll have 45 seconds to do that too. Boy, that'll be fun... (flywheel and clutch next!)

58 LIKES, 23 COMMENTS

Stan Rose And at the same time taking pictures! +2
Ernie Peters I bet someone here will blame it on that Lucas hydraulic cylinder. 😀 +2
Neil M. Fennessey Best to keep Ethan in your Last Will and Testament.
Paul Garnier [meme "Some people gamble in the casino, others in the shop"] +2
Alexander Wajsfelner And I wouldn't trust the engine hoist. I replaced the hydraulics on that back for the POS engine swap and that was at least 6 years ago... sitting under the porch. +2
Bob Sawtelle A Q-Tip and either a brake wrench or some bent beyond recognition wrench... that's my takeaways from these pictures. It's never easy with this Lotus... Namaste.
Daniel Senie Yikes!
Russell Baden Musta Now, there's another project to leak check!!! You're great at that!!! +2
Drew Lagravinese I don't suppose the Hack Mechanic could fix a small hydraulic leak in the lift table and do the job safely? (I usta was a safety professional for an insurance company, so I had to ask.)
    Rob Siegel There are no hydraulic lines. Everything is internal to the cylinder. So it's probably an internal o-ring. +2
    Robert Myles Brake fluid has always worked remarkably well for rejuvenating failing jack seals. +2
    Drew Lagravinese Robert Myles or that power steering that stops leaks 😀 +2
    Robert Alan Scalla Try STP. +2
Frank Mummolo Remember that show "Beat the Clock"?? You would ace it if it were on today! 😀
Franklin Alfonso Remember, safety first! That Lolita can be dangerous!
Jef Scoville Cut a 2x4 to right length so you can put it in diagonally but slightly short. Jack it up, insert the 2x4 and lower the table so the 2x4 is taking the weight. Do your work, jack it up again and remove the 2x4. +6
Bill Ecker "Hurried work is worried work" as my old man used to say. +1
Steve Nelson Have Ethan slowly pump the lift while you work... should get you more than 45 sec. +3

Joey Hertzberg Steve Nelson exactly what I thought

Joey Hertzberg Good time for Ethan to extort Dad for something +1

Brad Purvis Or... and this is just a thought. You could take some time and dig out the hoist. But hey, you're the Hack Mechanic. What could go wrong?

Scott Aaron If Louie is your Claim to Fame, this Lotus is your Claim to Pain.

TLC Day 2115: Pilot bearing installation

MAR 25, 2019, 8:07 PM

The Lotus Chronicles, Day 2115:

Got the flywheel installed, but not without having to solve a mystery.

As part of the clutch replacement, I'd ordered a pilot bearing (or, as it's called on the Lotus, a "spigot bearing"). Like any other car, it lives in the end of the crankshaft, and receives the tip of the transaxle's input shaft (oh baby). I unwrapped it, and, to my surprise, found that the part wasn't a roller bearing like it is on any other car I've ever worked on, but is instead simply a brass sleeve. I googled it and saw images of roller bearings, so I called the vendor to check that I was sent the right part. "Yes," he said, "on your car, it's just a brass sleeve." Okay then.

So I tried to tap it into the end of the crankshaft, and it wouldn't fit. It was close, though, so I put it in the freezer for 90 minutes to let it contract, then tried again. No dice.

I scratched my head a bit, then realized that I must have the old spigot bearing somewhere for comparison. The machine shop probably removed it from the crankshaft. I looked around, and to my surprise, found it with the freshly-machined flywheel, along with the dowel pins that hold the clutch, all in a little pill bottle and taped to the flywheel.

So... the spigot bearing goes in the flywheel?

I have an original paper copy of the factory Europa manual. I looked in it, and it clearly describes pressing the bearing into and pulling it out of the end of the crankshaft. But when I searched on Europa-specific forums, I found references to the spigot bearing being pressed into the flywheel's center bore.

I tried it in the flywheel, and there it clearly fit. Apparently on an Elan, it's a roller bearing and it goes in the crank, but on a Europa, it's a brass sleeve and it goes in the flywheel, and the Europa manual is simply dead wrong about it.

"Like any car, it lives in the end of the crankshaft." I forgot that the Europa is not like any car. Silly me.

So the flywheel's installed. If I wasn't missing one of the three little dowel pins, I could get the clutch on. At noon I placed the physically smallest auto parts order I've ever placed—one flywheel dowel pin the size of a Tic Tac. So clutch goes on maybe Wednesday, at which point the engine and transaxle can be mated together.

This means that I have little choice but to spend the next two days cleaning the nearly six-year accumulation of crap out from behind the Lotus, as it's looking like I'm going to need that access pretty soon.

70 LIKES, 35 COMMENTS

Paul Wegweiser Regarding that rear oil seal flange, that'll never leak. EVER. EVER. EVER. You can tell, because they used FOUR bolts on that giant cover... er... four bolts on HALF of the giant cover. And yet: probably 19 bolts to hold the transaxle to the engine, if it's a typical British car! Seriously? Is this a goddamned joke with them? +7

> **Jeff Hecox** Paul Wegweiser don't forget the British tendencies to use fine thread bolts to hold that transaxle on +2
>
> **Bob Shisler** I see six bolts.
>
> **Paul Wegweiser** Threaded HOLES? You know... so you have to get the timing of installing them "just right" FUT DAT!
>
> **Paul Wegweiser** Bob Shisler Well... sorta... underneath... to hold that piece of horse hide or marmalade or whatever that crank seal is in place. +2
>
> **Tim Lavery** *snort* marmalade... +1
>
> **Shaun Doherty** Marmalade reminds me of Paddington Bear. Paddington is British. Maybe he worked for Lotus? +1

John Wanner Progress!

Chris Mahalick Your progress is good. A little each day. You are SO close!!! +2

Daniel Neal Is it Brass or is it Bronze Oilite?

> **Rob Siegel** I don't know.
>
> **Daniel Neal** I am just curious because I have never seen a brass pilot bushing before, I have seen the Bronze Oilite before and I know there is plenty of debate of whether a Needle Bearing style or Bronze Oilite style are superior when both are available for the application. +2

Bob Coffey "Spigot... pressed into the flywheel's center bore." 🙂

> **Robert Myles** Bob, "centre" +4
>
> **Bob Coffey** Robert Myles quite right +4

Brandon Fetch Miatas have their pilot bearing in the flywheel. Welcome to the new world. 🙂

Rick Drost Oh, Baby!

Harry Bonkosky [meme "If there is no struggle, there is no progress"] +3

Jeff F Hollen I was surprised to find that on e9xs the pilot lives in the end of the trans input shaft and is a hugeeeee pain to change without the proper "special tool"... thanks BMW. +2

Peter Wright So how much longer is it going to take for you to realize that everything you knew is wrong, and everything you've learned since then will also be wrong, sooner or later? Thanks for the journey. +3

George Zinycz Rob Siegel you are in Bizarro World!

Steve Kirkup When you are done, will this be part of the fleet or a fleeting parting? +1

Rob Siegel who the hell knows +2

Tim Warner Steve Kirkup one step at a time! Don't get ahead!!!

Steve Kirkup I accidentally spoiled everything... damn it

Bob Sawtelle Whew, another Lotus Europa special feature. This is a book for sure. It will of course be in the adult section with the other S&M books but a book for sure. +11

Bart Collins Just like my 1960s GM—bronze bushing +1

John McFadden From vocabulary.com: "In the U.S., most of us call an indoor valve (in the kitchen or bathroom) a faucet, and the outdoor one a spigot. In other English-speaking places, a spigot is a plug inserted in a cask, or one end of a pipe. It's not uncommon for this word to be pronounced "spicket," with some dictionaries including that as an acceptable way to say it." Here in Philadelphia, it is not officially summer until you drink wooder from a hose connected to the spicket. +8

George Zinycz John McFadden the spicket goes in the bung then. But just the tip. 😁😁😁 +1

Bob Shisler Do yourself a favor and don't sit at red lights with your foot anywhere near the clutch pedal. +2

Steven Bauer Every clutch I've done on Chevys and Japanese cars had a bronze pilot bushing in the end of the crank. Along with the joyful exercise of filling the cavity in end of the crank with grease and hammering clutch tool into the bearing to force it out. That one looks awfully thin walled compared to the ones I remember. [photo of thicker bushing] +1

Scott Linn Another hint... There is often resistance in the ignition system so a "click" and no starter action is common. The solenoid is the high current switch, but the lower solenoid current is also often too much. Fitting a relay for the solenoid is very common.

Tim Warner I'm on day two of a little work every day on my tundra timing belt removal, to clean out the mouse nest and dead mouse that changed the timing so it wouldn't start back in August! Your "bite at a time" inspired me! Thanks again. +1

Russell Baden Musta That looks right Rob... if it fits the flywheel, as I remember, mine is just like that... I didn't soak it... changed a throw out bearing in about 2007... no problems... I took the trans off in the car... engine in... put it back in by myself... tricky, but I did. Careful balancing with a floor jack... GO! +2

Alan Hunter Johansson Clearing that path looks like a two-day project. +1

Brad Purvis "Dead wrong?" Now we know Brits are never wrong (just ask my wife), but in this case they may have been a little weak on being right. (If you know what movie that last line was taken from, then you can have a scone* with your tea at elevenses.) *If you pronounce "scone" correctly, you can have some jam in your scone as well. +1

Franklin Alfonso Screw the scones, Rob prefers cold toast with Marmite spread. It tastes more like used axle grease and that is more appropriate with working on the Lotus at this point in the assembly... 😁

Charles Morris He cuts down trees. He eats his lunch. He goes to the lavatory. On Wednesdays he goes shopping, and has buttered scones for tea.

Rick Roudebush I feel like you should wear this every time you go suffer on that thing. It seems like torture. [link to Amazon for Strict Bondage Hood with Breathable Gag Ball] +2

Jeff Hecox So British. The Triumph TR6 pilot bushing is in the flywheel also. In the Triumph it's a loose fit and not a tight press fit.

Rob Siegel Well, as many men get older, they prefer that :^)

TLC Day 2116: Working on THE CAR

MAR 26, 2019, 7:00 PM

The Lotus Chronicles, Day 2116:

Quite a day here in Lotus Land—in its own way, even a milestone. Just not in the way I thought it was going to be.

My goal was to install the clutch release bearing. This looks like a small donut and slides over a little collar. But it didn't fit. Turns out the $60 bearing I bought on Amazon to save $25 is six thousandths too small on the inner diameter. I had to return it and order the $85 version from RD Enterprises, a Lotus house where I buy many of my parts. Hey, I try to save where I can, but this time it bit me.

So with the release bearing taken off the table for today, I actually worked on the car—not the motor, the CAR—for the first time since June 2013. I've been absolutely adamant that I'm not going to do anything on the car, not spend a red cent on parts for it, until the drivetrain is in, but I'm well aware that I need to deal with the cooling system. After all, I can run the carbs off a gas can, but the cooling system has to be connected. On a car that sat for 40 years with coolant in it, particularly a weird-ass car with the motor in the back (well, the middle) and long metal coolant pipes that run the length of the car, if you think you're not going to need to clean corrosion off every metal neck and replace every rubber bit, you're delusional. Plus, it's a shit ton easier to reach everything while the drivetrain's still out. So I began disconnecting hoses, and immediately found trouble.

I pulled off the reservoir tank, shook it, and discovered that it was filled with big crusty rust flakes. I threw some nuts and bolts in it and shook it around to knock off the worst of it, and will do more later.

The tank gravity-feeds a small metal pipe with a right-angle bend, so of course that was filled to the top with rust. When I tried blowing it with compressed air, nothing came out. I had to spend half an hour with a coat hanger until it ran clear.

I got all of the hoses off except the two that connect the front ends of the metal pipes to the radiator. Those connections are under the nose of the car, and with the ass end of the car in the air as it's been for nearly six years, I can't get under the nose. I may need to see if I can live with the two original front hoses until I get the back of the car down.

Chasing the rust all the way to the radiator, I pulled it and the electric fan out of the nose (yes; mid-engine car with no pulley-driven fan in front of the radiator, so it needs an electric fan; in its own odd way, ahead of its time). The radiator looks surprisingly good, and the fan actually works, though it looks big enough and primitive enough to have propelled Churchill across the Channel.

And oh! Oh! And the temperature sensor on the radiator isn't threaded in—it pushes into a rubber grommet. I can just yank it out with my fingers and the radiator drops its antifreeze. Are you fucking KIDDING me? All the work I did pressure-testing the cooling system, and… THIS? I had assumed that I'd just flush everything out, replace all the hoses, and put it back together, but now I may opt for a $170 Chinese-made aluminum radiator. At least it has a fucking threaded sensor port.

My "I can run the carbs off a gas can" thing notwithstanding, I can see how much easier it'll be to pull the twin gas tanks when I can just stand up in the engine compartment, so when I'm done with the cooling system, I may have a go at that.

So, the drivetrain won't be going in the car in the next week or two. But hey, I actually worked on a Europa and not a Twin-Cam engine :^)

73 LIKES, 55 COMMENTS

> **Doug Jacobs** You are a broken man, aren't you? +6
>> **Rob Siegel** This is just standard sort-out stuff. I've encountered the same need-to-ream-it-out-with-a-coat-hanger thing in BMW fuel systems. +4
>> **Peter Wright** Would that be a metric coat-hanger, decimal inch, or the universal version (as found on pg 723 of the McMaster Carr catalog)? +4
>> **Rob Siegel** It's British, so it's actually a really long stale French fry that LOOKS like a coat hanger.

+7

Steven Bauer Metal parts in a plastic bag filled with enough Evaporust and 1/2 an hour or so in the ultrasonic cleaner does a number on rust. +4

Russell Baden Musta Okay... that temp switch that pushes in the grommet sure surprised me too!! Mine popped out near home after being fine for 39 years!!! Bought a new one from Ray, used some kind of sealer he recommended and I think it's good for another 39 years!!! It works... didn't need all that pressure for the leak tests I'm pretty sure!! Move forward!!!! +2

Alessandro Botta Rob, I would favour myself restoring the whole brake system before putting the British masterpiece ehr engine in the car +1

> **Rob Siegel** I'll go through the braking system and give it what's required. If that's just pads, it's just pads. Odds are it'll be nearly everything. But I approach things from a maximum-reuse philosophy to try and contain cost. +3
>
> **Scott Linn** Same here. Sometimes the starter spins without engaging the flywheel. It has been that way for the last 8 years I've owned it. Finally pulled the starter this weekend and took it apart. Removed a bunch of crap & oil from the inside of the starter, a bunch of grime from the solenoid, oiled the suspect places, cleaned the contacts/contact nuts/bolts/washers, etc. After hours of work, it acts the same... +3
>
> **Robert Myles** Scott Linn, at least after hours of work it isn't more broken. Small victories are still victories. +3
>
> **Scott Linn** Well, it was more broken overnight...!
>
> **Roy Richard** Rob, 6 thou is an adjustable reamer job. Use to have lots of em in Waltham 😀 Don't be too hard on the fan; it's probably Lucas 😀 +1

Daniel Neal That push in sensor is cool. The car should retain it for originality sake. +1

Rob Siegel like the 40-year-old coolant? +4

Daniel Neal Rob Siegel not quite that much originality +4

Fred Aikins You are a braver man than I, Gunga Din. I think I'll stick with my air-cooled Porsche, thank you. +2

Scott Linn That Otter switch sure is strange. I found an extra in the box of random parts that came with the car. I was like, what the heck is THIS? So I read through the parts manuals and there it was... You've got to be kidding me. It just pushes in??? +2

> **Rob Siegel** I'm totally in shock. +1
>
> **Rob Siegel** $170 is cheap enough for a Chinese-made radiator with a real threaded switch port. But then the triple-core ones are more, and are you really going to install that 50-year-old fan? No; you'd buy a new modern small fan. So before you know it, you've topped $300. +2
>
> **Scott Linn** I've been tempted to do some updates, then I ask myself why, because everything is working (or seems to). And a bit risky sometimes to rock the boat (it IS fiberglass after all). +1
>
> **Robert Myles** A decent radiator shop might be able to both clean the existing unit and weld a bung on the rad to run a proper switch. The next issue then becomes determining what "a proper switch" is and how it interfaces (interferes?) with the Lucas electrical system. +3
>
> **Rob Siegel** I actually hadn't thought of the welded bung, thanks. The "proper switch" path is well-trodden because people use Chinese-made aluminum radiators. +2
>
> **Robert Myles** Sometimes not having to reinvent the wheel is A Good Thing. +1
>
> **Rick Roudebush** It's not hard to solder a bung on your old rad if it is good otherwise. Don't powder coat it afterwards though, the solder might melt if you do. +2
>
> **Brandon Fetch** All this talk of bungs makes me giggle like someone we all know...
>
> **Scott Hopkinson** FYI. two places I found recently that did great work, both one person shops: Lowell Auto Radiator Service, 134 Congress St, Lowell, MA 01852 (978) 453-1710 Jay's Custom Exhaust, 113 Congress St, Lowell, MA 01852 (978) 441-9248 +3
>
> **Rob Siegel** Thanks. I haven't had good luck rodding-out or re-coring radiators for many years; on BMWs, I just replace them. But it is good to have a shop for radiators and gas tanks. I was using a place on 135 just east of Framingham, but the tank they fixed began leaking again in short order, so I stopped using them. +1
>
> **Scott Hopkinson** The radiator shop reworked my 1951 Willys radiator with custom arrangement (no off the shelf would work). +2
>
> **Robert Myles** I've has the hyper-expensive aluminum rad on the Miata racecar cleaned a few times over the years as basic maintenance with no issues. +2
>
> **Jeff Lavery** Rob, I've had great luck with Roger's Radiators in Medford. Found them through the Jalopy Journal page and they've been outstanding. They handled my 320/6's dual tanks and the

big 70L from my Cosworth. +1

> **Rob Siegel** Jeff, thanks. I've spoken with them over the years but for some reason never wound up using them.

Scott Linn You definitely want to get those coolant tubes sorted with the engine out as you're doing. I wouldn't want to try and fish them through with everything in situ. I've heard too many horror stories of getting pinhole leaks inside the tunnel… I've also heard of people who have run hoses the entire way. That seems non-optimal, but I can understand. +1

> **Rob Siegel** As much as I can, yes. As per the photo, there was a lot of corrosion in the inlet to the tube feeding the water pump. It appeared to be clumps of corrosion around the inside. I took it off with a file. I'll inspect the nose ends when I have access under the front of the car. I'll flush everything. But I won't be able to tell if there are any pinholes until I run the car. The system had a certain amount of coolant in it, but I pulled the motor nearly six years ago, so not every pipe was still full. +1

Brad Purvis MGBs have the same shitty temp sensor. I had an '80 with rubber baby buggy bumpers some time ago, and I must have replaced that damned grommet about every six months because it leaked. So you have THAT to look forward to, which is nice. +1

> **Scott Linn** Mine hasn't leaked a drop in my Europa in 8 years… Maybe it's the Volvo grommet.

Charles Morris Why not solder a threaded bung in that hole and use a sensor with matching threads? Or TIG weld if it's aluminum. Originality doesn't do you much good when standing next to a steaming overheated car, a long way from home... :) +2

Paul Garnier you should have a warning on those pictures. They are NOT for minors!

Robert Alan Scalla You are the correct guy for this job. +2

Charles Gould That first picture looks like a ratchet and a rolling form and papers to roll a huge joint. +1

> **Rob Siegel** which, believe me, is what I need. +1

Roy Richard Rob Siegel you may want to consider replacing those long tubes with PEX tubing, the oxygen barrier type. It's flexible so easy to run and good for 180F at 100 psi.

> **Roy Richard** Or soft copper.
>
> **Scott Linn** It's a low pressure system. Is PEX rated to >212F at lower pressure?
>
> **Roy Richard** Scott Linn Its max temperate rating is 200F. May be too low which is why I added copper 😀
>
> **Scott Linn** The "idea" of PEX is appealing due to the bends and difficulty of getting access once the engine is in place. People have put in shortened tubes or even welded up solutions in place. Something like a high temp PEX could make things a lot easier. I'm keeping a list of these types of ideas for possible future use in my own Europa.
>
> **Roy Richard** Scott Linn soft copper is easy to bend and neat if you use a bending spring.

Roy Richard Rob Siegel That temp sensor pushed into the rubber grommet is dual purpose. It is also a pressure relief valve. 😀 +1

Peter Gleeson I apologize on behalf of my homeland Rob, we build our houses to last for hundreds of years, but our cars are meant to just reach the end of the race. +2

> **Scott Linn** A beam in my wife's grandparents house near Sutton Valence had a beam dated to ~1250… That's when it was apparently cut down. +2
>
> **Peter Gleeson** My wife, when I meet her working at an American school in England, the school had a 10th century church on the grounds :-) +1
>
> **Scott Fisher** My son said, as we passed St. Paul's: "It's humbling to see a building that's older than your entire country." I replied: "Wait till tomorrow when we see Stonehenge." +1

Scott Fisher Get a new reservoir tank. Lotus was such a small manufacturer, they almost certainly sourced the reservoir tank from some cheap and easily obtained source, like a 1963 Triumph Herald.

Franklin Alfonso Truly, you are a man with patience. Way before now I would have dropped off the Lotus at one of those Army artillery ranges and let them practice on that miserable thing…I guess it's a good thing it was not in a rolling state. +1

Mark Gascoigne Nuf said ! [photo of Europa fiberglass body on a motorboat] +1

Neal Agran When you go to replace the hoses, be aware that the supplied replacements may be somewhat longer than the originals and require trimming. Check the clearance around tight spots closely and trim to accommodate. Sounds obvious, but this caught me out on the water pump inlet hose for my Esprit last year. I'm a Porsche guy, but I've found Lotus ownership to be a pleasant experience and challenge. It's definitely next level difficult given the limited production of the cars, but it's kind of a nice new step in my DIY car guy evolution.

This was the initial dump from the cooling reservoir. 1/2" ratchet shown for scale.

16 LIKES, 9 COMMENTS

George Zinycz Dude let's roll that up and smoke it! Oh wait... +4
   Claudia Keller When I was scrolling by I thought that's what this was!! Ha ha ha! Had to stop and see what Rob was up to... +4
Scott Linn Gotta use the right terms here. That's a Swirl Pot. +1
   Rob Siegel Well, as I understand it, yes and no. Coolant doesn't flow through it, so it's not technically a swirl pot. But you're correct that it is referred to that way. The parts book says "swirl pot (header tank)."
   George Zinycz Scott Linn Pot again! I was right... +4
   Dohn Roush Scott Linn More like a Neti pot... +2
Lee Highsmith I think I see a spider +1
Robert Boynton Looks like "scale" on the paper towel too.
Wade Brickhouse OH NO! you've dumped the anti-leak material you twit!!! Rap your knuckles with a spanner! +1

This shows the left side of the engine compartment. The gas tank is immediately to the right of the coolant reservoir. One of the two metal pipes that run the length of the car is shown sticking up like the Loch Ness Monster.

(4 LIKES, 18 COMMENTS)

Ricky Marrero Dual tanks, right?
Rob Siegel yup

Scott Linn At one time Lotus used plastic tees to connect the two tanks together. A few fires changed that... +2

Wade Brickhouse Rob Siegel you get to run out of gas twice. cool. +1

Robert Myles What is that brake booster looking like assembly in the upper left corner of the picture? (Oh, Flying Spaghetti Monster, please tell me Lotus didn't put the brakes in the back of the car.)

    Rob Siegel DUAL brake boosters, baby!

    Robert Myles [gif Jon Stewart mind blown]

Scott Linn Do you have any idea if your brake boosters work? That bottom one looks a bit "bad". Many people have just bypassed them without even touching the M/C.

    Rob Siegel not a clue what the status is of any of the braking system. yes, I've read that people delete/bypass the servos.

    Samantha Lewis Dual vacuum assists in a 1300 lb car. You need that. +1

    Scott Linn Well, 1600lb car. But the point stands.

    Rob Siegel I thought the TCS was 1500

    Scott Linn I've seen 1631, 1665, and 1570...And 712kg from 73 R&T. Just checked my reference book from period magazine testing and it says 1665 for the 5-speed TCS. R&T says 712kg or around 1570lb. So I just use 1600...

Scott Linn I commented on another photo first, but those large tubes at the bottom connect the shifter to the transmission.

    Samantha Lewis Indeed, for that precision British feel, by bargepole, whilst shifting one's Renault transaxle, from the very back, of course. Shall we just say that it doesn't shift with quite the precision that would be desired, but in men in brown lab coats in sheds tradition, it does work. +1

Jim Gerock Holy cow. Are there any good Engineered parts to the Lotus Europa? Wonder if the shifter is any better/worse than a 914 Porsche...

Chris Mahalick Just as bad. LOL

A GROMMET IN THE RADIATOR?? ARE YOU FUCKING KIDDING ME?

58 LIKES, 65 COMMENTS

Ricky Marrero And you're worried about leaks from the engine... +6

Rob Siegel I know, right? +1

Corey Dalba [gif Robert Redford Jeremiah Johnson nod] +3

George Zinycz Please start and keep a list of all the wtf design issues you learn about on this project.

Steven Bauer They must not expect too much pressure in that system.

    Rob Siegel 7 psi +1

Corey Dalba This is wild.

    Rob Siegel "Wild" is not the term I reached for. +1

    David Shatzer Rob Siegel like I asked awhile back "what is the radiator pressure cap/system rating?" Yep, that's about a 7-10 pound push. Probably a safety feature for the bad antifreeze of the day for frost plug in radiator? +1

    Corey Dalba David Shatzer idk why, but my first thought when I saw it. As a drain plug. +2

Craig Lovold That is my favorite thing ever. +2

Steve Park Now we know why the British Empire isn't around, amongst other things. +2

Rusty Fretsman Say it again, but in a Cockney accent. #bollocks +1

Lindsey Brown It's a British thing. You wouldn't understand. +1

Josh Wyte English "engineering"

Dave Klink That's added lightness... +1

Chris Langsten "I have pressured tested the engine at 20psi..." LOL! ;) +3

    Rob Siegel I know, right? +3

    Grice Mulligan Chris Langsten I just tried to explain this whole debacle to my German wife, who was far more interested in eating a fresh French baguette than hearing about a shitty British design failure. Even she saw the humor and irony in the situation. +2

Andrei Fenner Volvo's used to have exactly that for a six fan switch. We used to have to tie wrap the switches in so they wouldn't blow out of the radiator. +6

Andrei Fenner In fact from the picture it looks like the Volvo seal might be the same size! +1

    Rob Siegel I've just been reading up on this. The Volvo seal has a longer lip than the Lotus seal. It's a Lotus trick to use the Volvo seal to hold the "otter switch" tighter. +3

    Andrei Fenner Nice! It came in 240's and 740's. I am sure I could come up with a part number if needed when I back at the shop tomorrow.

    Rob Siegel 1378869 +1

    Andrei Fenner That is the one! Haven't had to use one in a while and forgot the number

    Collin Blackburn Lol. I had a friend with a 740 that blew the damn thing out of the radiator on a road trip. Still boggled to this day about that design. I like my S70 so much more. 😊

Thomas Jones Any old school radiator shop worth their salt oughta be able to weld/solder/braze a bung in that. +1

Paul Wegweiser ...*bung*... te heeh heehe eheheheeee... +5

Thomas Jones [gif of Beavis and Butt-head] +5

John Turner BUNG

Bill Snider Problem solved! [grommet thumbs up gif] +8

    Jennifer Wyman-Clemons I was just humming their tune today! Lol. +1

    Matthew Zisk My thought, exactly +1

Jake Metz Because why not? +1

Paul Wegweiser Think of it this way... You're not repairing a car. Instead, you are creating a work of art... like a painting… or a sculpture. A work of art... one that will burst into flames and leave you stranded… in the rain... on the worst day of the summer... after you spent $10,500 on materials to create it. +8

Grice Mulligan Paul Wegweiser I'm thinking Rob's Europa could be licensed as a playa vehicle at this year's Burning Man. Umm, I mean some future year's Burning Man. There's no way in British hell this car's gonna be roadworthy in time for Burning Man.

Paul Muskopf [gif Grommet eye roll and face palm] +1

Brian Stauss So my FB news feed tonight. The first five entries are the last five days of Lotus Chronicles. I can't get enough. Keep them coming Rob! +1

Scott Hopkinson "The Hack Lotus Mechanic: 3000 days to a running Europa"

Christopher Kohler Hey the sensor was still in the grommet, right? +1

    Rob Siegel yup.

John Turner "Built to Spill"

Wade Brickhouse Just pressure test it and move on. It's the secret coolant division of Lucas. 😊 +2

    Rob Siegel Wade, so, FOUR settings on a Lucas product? (dim, flicker, off, and trickle?) +1

    Matthew Zisk Sounds like a prostate issue [Matt, this was truer than you possibly could've known :^)] +2

    Wade Brickhouse Rob Siegel Don't attach it to anything important

Glenn Stephens Threads are so new age. +1

Daren Stone Agreed, the grommet is one of the more facepalm-worthy design details of the Europa

T/C (and late Elans). The "fix" is to run a loop of safety wire through the radiator core and across the otter switch. Or find a radiator from a pre-T/C as they have a screw-in switch. (Second facepalm). +2

Daren Stone While you're in there, ensure the end tank baffle is still in place by removing the air bleed valve and sticking a piece of wire down through the hole. The wire should stop midway between the hose connectors when it hits the baffle. +2

> **Rob Siegel** I didn't know that, thanks. Daren, I probably will just re-use this radiator for now, but do you have any opinions on the Chinese-made aluminum ones? I see the two-row ones for as low as $170, two-row ones with electric fans on them for as low as $210, and three-row ones for as low as $200 without fans.
>
> **Daren Stone** Rob Siegel At the present time I am opinionless wrt the imported aluminum units, but at that price I may try one out if I needed a replacement.

Mark Gascoigne Safety blow off valve!

Bill Howard You know, Rob, you could just get a nice end-of-life-sale Chevy Cruz. Sorry, Cruze. Save you a lot of screwing around on your back. +1

> **Rob Siegel** I doubt I could get a book out of THAT. +1
>
> **Bill Howard** Rob Siegel Maybe a wellness book. You know, reduce stress points. Deal with only calming things. +3

Layne Wylie I would just solder up the hole and switch to a fan controller that either has the fin probe, or fits a small NPT thread. I once used a dime to cover the hole left by a broken drain petcock. Coins solder easily. Looks like you might need an old English penny for this one.

Delia Wolfe Grommet in the radiator = lime in de coconut +1

> **Rob Siegel** you're such a silly woman :^) +3
>
> **Delia Wolfe** I saw what you did there...

Russell Baden Musta It works... ask Ray at RD which sealer to use... I know I bought a small tube. I cleaned that hole with Scotch Bright pad, stuck it in... thought, hmmm, really...? Well, it's still in and it's been overheated a couple times in travel and on Talladega!!! +1

Chris Pahud [gif of Grommet throwing bread out the window of a breadvan, which, for a kind soul like Chris, seems especially cruel :^)] +2

Brian Ach Better than fucking aluminum bolts used to hold a water pump on +1

The infamous "otter switch," the press-in coolant temperature sensor that controls the radiator fan.

The mound of rust peeking out from the top of the metal pipe that goes into the bottom of the coolant reservoir.

Corrosion at the neck of one of the fat pipes.

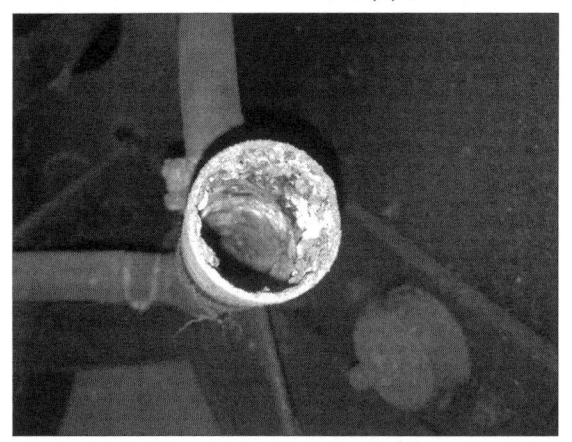

TLC Day 2117: Fuel pump

MAR 27, 2019, 7:13 PM

The Lotus Chronicles, Day 2117:

Didn't do much in Lotus Land today, as I spent most of the day writing. (I actually DO have a job, if what I do qualifies as a job. Wait... what do I do?)

Dealt with the fuel pump. As much as I love the fact that the ancient glass-domed pump looks remarkably like the head of Robby the Robot on the giant painting of the movie poster of Forbidden Plant that lives in my garage, it's broken, and an electric fuel pump that'll easily fill the float bowls of the Strombergs makes more sense for testing anyway. Will probably order a $30 EMPI (popular in the Volkswagen world). Block-off plate will be here tomorrow.

43 LIKES, 39 COMMENTS

Dave Klink Coincidence? I think not. +1

Scott Linn What broke on it?

> **Rob Siegel** The dome is cracked in one place, and there are seals that look bad. Yes I realize that I can rebuild it, and that does have a certain appeal, but the ability to just fill up the fruit bowls on the demand when I'm trying to get it running that is something that I want. +5
>
> **Charles Morris** Fruit bowls? Must be an English thing... :) +3
>
> **Paul Wegweiser** Rob Siegel Aren't you afraid that this car will continue its habit of rejecting any and all "user friendly" upgrades? +9
>
> **Rob Siegel** Crap, I had not thought of that! +2
>
> **Paul Wegweiser** Rob Siegel I'm in the "rebuild the old fuel pump" camp. FREE ROBBY! FREE ROBBY! FREE ROBBY! 8)
>
> **Scott Aaron** RAH-BEE RAH-BEE +3
>
> **Scott Aaron** Rob Siegel whatever your plan is, will totally not work. You have to call some random guy in South Africa that is mostly a dairy farmer but also makes a fuel pump designed for the Europa. He refuses to make one for any other model. You know it's true. 👹 +9
>
> **Roy Richard** Rob Siegel a fuel pump replacement is like a heart transplant. You have to get the right match. Does Lucas make an electric pump? I'm sure the Lotus wouldn't reject that. +2
>
> **Carl Witthoft** Rob Siegel I hear you can get quality replacement seals from Morton-Thiokol. +2
>
> **Jeremy Novak** I know Lucas makes electric fuel pumps for the MGB (OE item shockingly). They're pricey for what they are, but they're actually pretty good pumps. +2

Chris Mahalick Go to your local NAPA Store. They have a low power (3-6PSI) Facet Solid State pump for not a lot of money. You really want solid-state. Mine works like a champ. +3

> **Rob Siegel** The EMPI pump is also solid-state but about half the price of the Facet. I try to save money wherever I can. +2
>
> **Chris Mahalick** Rob Siegel I cannot argue saving money!

Tim Lavery That looks identical to the pump on my '47 CJ2A +2

> **Tim Lavery** [photo of identical looking pump]
>
> **Rob Siegel** Tim, I just spend an hour on line looking at this. The pump DOES look identical, but the lever-arm that's driven by the in-block camshaft is a different shape and comes out at a different angle..
>
> **Tim Lavery** I just noticed the new base gasket for my pump says "Made in England." Also, the lever arm is replaced pretty easily as we discovered when rebuilding, so possible it's a similar casting with unique lever arms for each application.
>
> **Ed MacVaugh** From 1920 onward, everyone used AC Fuel pumps, or licensed variations of them, Stewart Warner, Lafayette, and Hitachi all were licensees. As noted, the body is very similar with a different arm to suit the camshaft lobe location.
>
> **Ed MacVaugh** AC Delco is still with us today.
>
> **Ed MacVaugh** http://www.classicpreservation.com/fuelpumpkits.html
>
> **Rob Siegel** Yup, this rebuild kit is sold by RD Enterprises as well. At some point I may rebuild this

pump, as I love its look and feel, but for the install and start-up I'll hopefully be going through in the next few weeks, which will probably include feeding the Strombergs with a gas can, an electric pump is the way to go. +1

Ed MacVaugh Rob Siegel I saw you say that elsewhere, but I have seen real problems with the Facet brand electric fuel pumps and carburetors. They appear to easily overpower the needle valve for the float. I also see that Empi is labeling Facet pumps with their name and a disclaimer about needing a fuel regulator. If you have to buy both a pump and a regulator . . .

Ed MacVaugh [link to EMPI 1.5-4 PSI fuel pump]

Rob Siegel Ed, that is the one I just bought.

Jeremy Novak When we switched to a solid state pump on my dad's MGB we had to add a low pressure FPR to the system to prevent the pump from over-filling the bowls. The Lucas pump didn't need that because it was much weaker.

Jim Strickland Will make a cool lamp +1

Rob Siegel For a very small nightstand +1

Paul Wegweiser Rob Siegel Perfect for your Barbie Doll playset! (PS: why isn't Ken wearing pants, Rob? Are shenanigans afoot?) +3

Scott Linn Darn it, now you've got me lusting after a bad mech fuel pump for that cool lamp... +1

Jim Strickland Scott Linn car parts can make some fun lamps. Susan Brown Strickland found this one for me, which was done by a local artist [photo of spring lamp] +3

Christopher P. Koch Looks somewhat like our aviation units although ours are split into the mechanical pump bolted to the engine with a gascolator ahead of it. Yours is similar but "all in one."

Alan Hunter Johansson There's a Mitsubishi pump stock on CVCC Hondas that I use partly because of the fact that its sound changes noticeably when it's sucking air (making diagnoses easier), but mostly because in the decades since we started using them, we've never had one fail. In fact, when Chris decided to buy a new one as a spare from the Honda dealer sometime around 2010, the parts guy asked why he wanted it. It turned out he'd never sold one. +2

Andrew Wilson In Rush's "Freewill" there is the lyric "There are those that think they were dealt a losing hand, the cards were stacked against them, they weren't born in Lotus Land." +5

Rob Siegel Yes. It's a little-known fact that Geddy Lee doesn't say "I will choose free will." He says "I will choose a 45-year-old 1600 pound 42-inch-high boutique vehicle that can drive beneath parking lot barriers and probably tractor-trailers." +3

Dave Borasky Wrong. He chooses his uncle's red Barchetta.

Andrew Wilson Different song.

Thomas Jones The resemblance is campy. +1

TLC Day 2118: Clutch installation

MAR 28, 2019, 7:02 PM

The Lotus Chronicles, Day 2118:

The smallest automotive package I've ever ordered—one clutch retaining pin to replace the one that had been lost in the nearly six years since I bought the car (so far the only thing I've lost, which is miraculous)—arrived.

So, on went the clutch. Three pins, six bolts, 20 ft-lbs, and everything fit flawlessly. It was almost anti-climatic. I thought I must've done something wrong.

Friday: The great Lotus Twin-Cam Engine / Renault 5-speed transaxle mating. Sounds like a date, right? Too bad I couldn't have gotten it done on hump day.

78 LIKES, 45 COMMENTS

Scott Linn Shiny parts!

Marc Touesnard At least you didn't lose the car body in the garage. That would have been a difficult one to explain. +3

Carl Witthoft What you got against climate? #autocorrectstinks +1

    Rob Siegel Whoops! Thanks!

    John McFadden He's a climax change denier +4

Steven Bauer Waiting for the other retaining pin to drop?? +2

John McFadden Rob, do you know what model (or models) Renault the transmission/transaxle was sourced from?

    Ed MacVaugh I'm not Rob, but think I know the answer. None directly, but similar cases used in Renault cars with specific gears to suit Lotus' needs:
    Early Europa Twin Cam: Type 336 (identified by side mounted gear shift mechanism) was in Renault 16
    Europa Twin Cam: Type 352 4 speed (identified by rear mounted gear shift mechanism.) was in 1972 Renault R17TL
    Europa Twin Cam Special Type 365 5-speed (identified by extension housing at rear for 5th speed. Rear mounted gear shift mechanism) was in Renault Alpine A310

    Rob Siegel Part of the appeal of buying a rare Vixen RV (which I didn't do), in addition to the BMW M21 turbodiesel engine, was the fact that it has a Renault transaxle, which would've meant that I would've owned about the two most different Renault transaxle-equipped vehicles you could imagine. Hey, we all have to do SOMETHING for fun. +9

    Ed MacVaugh Rob Siegel Then throw in the DMC 12 and you could time travel. +2

    Rob Siegel Those French and their silly transaxles show up when you least expect it. +4

    John McFadden Ed MacVaugh thanks!

Robert Myles NOBODY expects a French transaxle... or the Spanish Inquisition. +4

John McFadden Robert Myles please commence with listing our chief weapons! +2

Robert Myles John, fear, surprise, these comfy pillows and a Lotus assembly manual! +2

Rob Siegel Amongst our weaponry are such diverse elements as... +2

> **Robert Myles** Cardinal Biggles, read him the directions! +2
>
> **John McFadden** Don't forget THE (steering) RACK!!! +1
>
> **David Kemether** They go together like escargot and Yorkshire pudding. +2
>
> **Vince Leo** You could have saved it for next hump day! +1
>
> **Brian Stauss** Is that a new flywheel? If not, you did an outstanding job cleaning it up! +1
>
> > **Rob Siegel** I brought it to the machine shop to be re-faced. I think they tanked it and glass-beaded it as well. +2
>
> **Steven Bauer** Of course, you know that now you've purchased the replacement, the original pin will make its presence known. +1
>
> > **Rob Siegel** And that's fine. Sometimes it takes a sacrifice to summon the gods. +3
>
> > **Sal Mack** Rob Siegel let's cut the gothic game of thrones or other paganism crappola that is not directly related to worshipping at the idle of lightness please 😊 Order a new radiator fan. Put one in an amazon wish list and maybe one of us will buy it and send it to you to help the car feel more contemporary. +2
>
> **Marc Schatell** I can't wait to hear about that next step!
>
> > **Rob Siegel** sicko +1
>
> **Marc Schatell** BTW, I'm pretty sure people would pay to watch this stuff +2
>
> **Rob Siegel** I'm trying to think if there's some Lotus version of the "I'd like the porn disabled" "We only have regular porn, you sick fuck" joke. +2
>
> **Marc Schatell** Rob Siegel surely there's something there...
>
> **Scott Aaron** Marc Schatell I can't believe this show is still on. But people watched the Gong Show, so there's no accounting for taste. +1
>
> **Steven Bauer** Marc Schatell One word: schadenfreude. +2
>
> **Jonathan Poole** This is moving along so nicely that if I show up here Monday and you say the thing is running I *might* actually believe you... +2
>
> > **Rob Siegel** I appreciate that, but that's not going to happen. *I* wouldn't believe *ME*!
>
> **David Holzman** mechanical things have soul, unlike electronics
>
> **George Saylor** You'll find the old pin when you move the car. Not that it's anywhere near the car but shit just works out that way.
>
> **Steve Nelson** 20 ft-lbs to hold on the flywheel 😊. Doesn't sound like much! I think beetles had only 1 nut, but was torqued to 217!
>
> > **Rob Siegel** The clutch, not the flywheel. The flywheel had much more torque on the bolts. +1
>
> **George Zinycz** I'm getting excited about this!! Can I be there for the initial startup?? [link to First Start Up Engine Rod Knock Prank] +2
>
> **Mike Miller** Better go kiss the Blarney Stone!
>
> **Steve Park** It looks too good to cover up! +1

TLC Day 2119: Grease me up

MAR 29, 2019, 10:50 AM

The Lotus Chronicles, Day 2119:

It's mating season. Lunch lady Doris, grease me up. I'm going in.

102 LIKES, 63 COMMENTS

Sean Curry [gif "I'll be in my bunk"] +3
Bob Gronberg I really want that Forbidden Planet poster... willing to trade '02 stuff for it +1
Tom Melton Don't forget to dim the lights to set the mood. +5
    Alan Hunter Johansson Tom Melton Lucas sees to that. +12
    Rob Siegel HA! All those years everyone's gotten it wrong! It's not unreliable, it's MOOD-ENHANCING! +5
    Alan Hunter Johansson Rob Siegel Never the right mood that was enhanced. +2
    Franklin Alfonso Buying a Lotus guarantees you get screwed.
Paul Wegweiser I bet you have the WEIRDEST hard-on, right now. +13
    Rob Siegel Cue my best Jeremy Irons voice: "You have no idea." +9
    Philip D. Sinner Rob Siegel priceless ! +1
    Alan Hunter Johansson Rob Siegel Or your best Ron Jeremy +1
    Rob Siegel Alan, see two comments down ^)
David Weisman Paul Wegweiser 😂😂😂
Roy Richard Light a few candles a little mood music and don't forget the cigarettes for after. +2
    Robert Boynton Roy Richard ...and two stemmed crystal glasses of Girling Crimson. +2
Ernie Peters Cheers!

+14

    Alan Hunter Johansson Ernie Peters Oh. Dear. God. +2
    Brian Hemmis [cocktail glass]
Steve Park It's good to know there are people sicker than I am. +2
    Rob Siegel who? +1
Jeremy Novak [gif scrubs "Why do you hate me when I show you nothing but love?"] +1
    Jeremy Novak ^ ^ later today +1

Greg Bare [gif of thwanging doorstop] +3
Samantha Lewis Enough innuendo, crack on, there's a Lotus to reassemble and get running. +1
Rob Siegel I'm all over it. But I DO have adolescent entertainment responsibilities. +8
Rick Roudebush [gif Wayne's World SCHWING!] +1
Thomas Jones You're playing the part of a mechanical sex therapist today... "Ooo Tranny, you're so dirty, come and stick that shaft right here." +3
Steve Park I only mean that in a good way. Sanity is overrated.
Christopher P. Koch All this for 113 horsepower? +5
> **Rob Siegel** shut up. +4
> **Russell Baden Musta** Rob Siegel yeah, shut up Man! 😂😂 +1
> **Christopher P. Koch** I hope they are at least Percherons or Clydesdales...
> **Scott Linn** Too bad British hp doesn't translate like their gallons do... +1
> **Andrew Wilson** 113 Shetland pony gallops. +3

Paul Garnier suddenly everyone's 14
> **Rob Siegel** when HAVEN'T we always been 14? +4
> **Charles Morris** "You can only be young once, but immature forever" :) +7

Bo Gray Literal "Auto-Erotica"—I think there is a 12-step group for that.
Lindsey Brown [suggestive still from Forbidden Planet] +1
> **Rob Siegel** nice!
> **Gordon Arnold** Lindsey Brown just watched that a few weeks ago.

Jim Dawson I'm hearing banjos! +2
Franklin Alfonso Paddle faster!
Vijay Malik ...maybe some Barry White +1
> **James Pollaci** Vijay Malik was thinking 'Let's get it on' by Marvin Gaye +3
> **Vijay Malik** Strauss waltzes work too...
> **David Weisman** Vijay Malik 🎩
> **Bob Sawtelle** Some mood music by a local Cambridge band from back in the day when the Europa was new: [link to "Let's Live Together" by The Road Apples]
> **Franklin Alfonso** [link to The Cars "You might think I'm crazy"] +3
> **Rob Siegel** Perfect. +1

Walter Eschelbach Scares me to death
Don Bower I hope somebody is archiving these. Should be a book in here somewhere. +2
> **David Weisman** Don Bower def

Dirk Rasmussen Slip on some edible coveralls and torque until the sun comes up +4
David Weisman Ain' never gonna do it without my fez on... +1
David Weisman Hey there's Robby the Robot poster!
Andrew Wilson Punchline is Wood I? You look 'armless +1
David Kemether

+4

Dohn Roush Redefining the term "autoeroticism"... +1

TLC Day 2119: Mated!

MAR 29, 2019, 9:04 PM

The Lotus Chronicles, Day 2119, Part B:

As Carl Reiner said in Ocean's Eleven, "That is the sexiest thing I have ever seen."

I found that the back of the engine was a little too low for the transaxle to mate to the clutch. Ethan helped me lever it up with a 2x4 and shove a few blocks of wood under it. Then it was the usual slide and shove until the holes in the bell housing and block were close enough that I could insert the bolts and draw the two together. Done. Mated. Woo-WHOOOO!

Until I saw the spacer plate that's supposed to be between them sitting two feet away on the table and had to do it again. Grrrr.

It actually wasn't too bad. The spacer is U-shaped, not a closed circle. I was able to separate the pieces the bare minimum amount and slip it between them.

Then, extra innings, I tested the starter to make sure it worked, and mounted it. Installed the alternator too, though the bearings sound a little loud. It's at the top of the engine, easy enough to remove if it's bad.

So, I have a mated drivetrain. I could put it in tomorrow, but I won't. While it's still out of the car, it'll never be easier to reach the two big rubber bushings for the radius (rear trailing) arms. You can see the right one in the center of last photo. It's really soft, like almost gooey. $34 each, and I can probably get both of them changed in a day. Aren't you proud of me?

So, drivetrain in, with some luck, maybe sometime next week. It'll probably be another few weeks before I'm ready to test-start it, as I still have things to do with the cooling and fuel systems.

But, with any luck, April. Let's hope it doesn't live up to its reputation as being the cruelest month. T.S. Eliot was an optimist, you know.

98 LIKES, 30 COMMENTS

Scott Winfield A thing of beauty at this point. Congratulations to you. +1

Russell Baden Musta Rob! Get new bushings for the trailing arms pivot you see there... also, note where the washers are, as that's how you adjust the tow for the rear! I think you want the head of the bolts inside... so you can remove the nut and slide bolt in to move washers for tow adjustment... +2

Rob Siegel They're on order from Ray.

Russell Baden Musta Rob Siegel And Ray has that tool for adjusting the thrust collars where your stub shafts stick out of the trans! You're doing great sir! Ignore these negative people!!! You will soon realize good handling! +1

Scott Linn Dang, I'm getting excited and enjoying seeing the progress. Awesome. +1

Jake Metz You're the man Rob. Love it. +1

Steve Park Great to see so much progress. It must be a good feeling for you. Congrats. Looking forward to the startup video. +1

Blair Meiser "lever it up" "slide and shove" "mated" "had to do it again" "It actually wasn't too bad" "mounted it" "big rubber" "soft, like almost gooey." Should this be R-rated??? +3

    Brad Purvis Blair Meiser R-rated like Bevis & Butthead.

Kevin McLaughlin [link to Austin Power "Yeah, baby, yeah!"]

Rob Blake 👍

Chip Chandler Congratulations! Always a great day. Got big step parts for my own project today. One thing every day, no matter how small, right? +2

    Rob Siegel Right! +2

Rob Halsey Just love that you are taking us along for the adventure. +1

Pamela Howe I know nothing about any cars and I'm enjoying reading along. +2

Elizabeth Marion I love the reality that even after all these years of 'experience', you still do stupid sh#$ like the rest of us mere mortals and then you fess up which makes the rest of us mere mortals feel a tad bit less inadequate! +2

    Maire Anne Diamond That's what makes even non-car people relate to it. +1

Franklin Alfonso Way to go Rob! Who knew it was going to be so easy eh? +1

    Franklin Alfonso Lolita is almost ready...

    Rob Siegel well, I don't know about THAT...

    Franklin Alfonso She's hot to trot!!! It's been a looooooong time she has not been running around.

    Franklin Alfonso If after all this effort, she does not perform beyond expectations and she's not fast and handles well it will be disappointing.

Ian Sights I almost hate to ask but have you changed the lube in the tranny? Or at least checked the level?

    Rob Siegel No, but I will before it's installed. Thanks for the reminder. +1

    Russell Baden Musta Rob Siegel can do once it's in the car too. +1

Chris Langsten I sense the next book title: "How I rebuilt the 4th most Archaic Engine ever; and lost my sanity with the electrical system." With the subtitle "One man's journey through Lucas Electronics, and beyond" as the Amazon caption. +2

Dave Borasky I thought you cleaned the transaxle.

    Rob Siegel It got dirty.

Chris Lordan "Vincerò, vincerò, vincerò!" = "I will win, I will win, I will win!" And so will you. +1

Brian Ach FTW

TLC Day 2121: Why THIS car?

MAR 31, 2019, 8:31 AM

The Lotus Chronicles, Day 2121:

Someone asked me a few days ago why I'm going to all this trouble for this particular car. There are three answers, the first two of which are from my first book.

Answer 1: "Men are total sensation junkies. We spend a good portion of our life chasing sensation, be it with women, skydiving, rock climbing, drinking, drug use, or fast cars. I'm not saying that women aren't enamored of sensation; I'm just saying that men are. As men get older, the realistic limits on our bodies and the desire to remain faithful to our spouses may keep us in line, but the craving for sensation often continues. When I was in my teens, I worked for a forty-something guy who had an impossibly low unbelievably angular red Lotus Europa. I overheard two forty-something friends of his talking, and one of them said, "A car like that, YOU CAN GET SEX OUT OF." I thought, yeah girls like guys with cool cars, I get it. But I didn't get it. Not for years. They weren't talking about using the car to pick up girls. THEY WERE TALKING ABOUT THE SENSATION."

Answer #2: "As a guy who tends to overanalyze everything (you can tell, can't you?), it is such a simple pleasure for me to go into the garage and fix something, or to displace job-related stress with the thrill and anticipation that comes from looking at another car. And the metrics for success are so straightforward. I can't make a client award me a geophysical survey contract, or book me as a performer in some coveted venue, but I can replace a water pump. [Note: The ease with which I could replace the water pump in the Lotus, obviously, was sorely tested :^)] In a world where I can't fix health care, I can diagnose and replace a bad alternator and know the problem is fixed, completely. I can take something that wasn't working yesterday and make it work today. And while one may not have complete control over anything in this life, Jeez Louise I have an awful lot of it in my garage. I can make decisions without seeking approval from anyone. And most of the time they're such gloriously inconsequential decisions–do I buy this part or try to fix the one that's in the car?–with so little real consequence even if I'm wrong. By occupying my hands, working on a car occupies my mind. Cars give my brain the vacation it needs from itself."

Answer 3: As is the case with anyone, there's more going on in my life right now than you know. Focusing on the Lotus, getting shit done, seeing it inch toward closure, I can't tell you how much clarity and pleasure this has brought me.

Gotta go. Removing the old radius arm bushings this morning.

(The photo was taken June 23rd, 2013, nearly six years ago, when I was removing the drivetrain. As I ready the engine compartment to receive the drivetrain, I am, um, assuming this position again.)

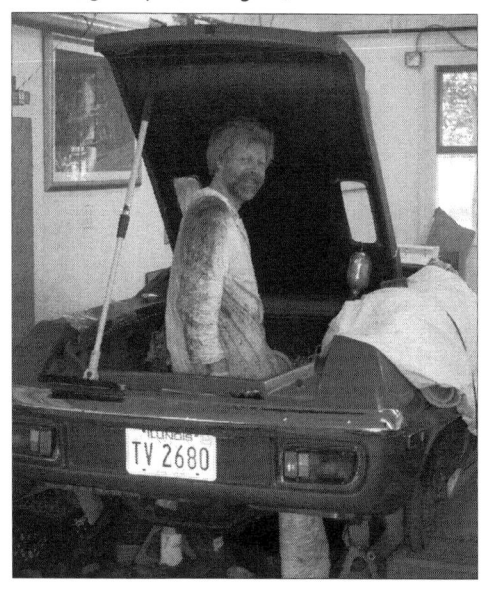

230 LIKES, 74 COMMENTS

Scott Winfield Beautifully articulated.
Carl E Swanson Best pic ever!
Tony Fornetti 6 years ago! I can't wait to see a picture of you cruising down the street in this car.

Andrew Wilson You look so young, full of piss & vinegar. +4

Phil Salmon "Cars give my brain the vacation it needs from itself." That pretty much nails it... +7

Bob Sawtelle "Assume the position"... heh heh "Thank you sir may I have another"

Clay Weiland Please tell me that this picture isn't of you "getting sex out of" the car. +17

> **Rob Koenig** Clay Weiland He's obviously making it his bitch... +2

Douglas Wittkowski I have the highest respect for Rob, unlike most DIYer's but this thing would have been ditch-fill many years ago. +1

Ernie Peters I should never need a therapist: '56 Chevy, '63 Avanti, '70 Lotus Elan, '95 BMW M3, '74 MG Midget, all completely rebuilt/restored by me. My brains. My hands. My garage. My satisfaction. +13

Marilyn Rea Beyer Love the Illinois plates! Driving it back to the Land of Lincoln? 😊 +1

> **Rob Siegel** Has to get out if my own garage first :^) +1
>
> **Marilyn Rea Beyer** Rob, when you do, you've got a place to stay in Chi-Town. +2
>
> **Rob Siegel** Which IS the Lotus' home town! +2

Carl Nybro Looks like an OHR (Over Head Rob) engine in that fancy furin' car! 😊 +1

Tom Tate What? It's brown? Of course. 😊 +1

> **Franklin Alfonso** British Racing Brown... 😊
>
> **Paul Hill** Tom Tate skid mark brown
>
> **John McFadden** The brochure called it "Mink"—sounded fancier than brown +1
>
> **Tom Tate** Porsche called it Togo Brown. We called it sh*t brown. 😊

George Zinycz The Lotus Position! +2

George Zinycz Up to your ass in the project +2

George Zinycz And I 🖤 #2 +2

Steve Park I agree Rob, but you said it far more eloquently than I could. It's hard for the non-dirty hands group to understand. +1

Franklin Alfonso Maybe while you are restoring the Lotus you can put that bumper back below the tail lights where it belongs... lol +1

Russell Baden Musta Answer #1 is EXACTLY my feelings with my Europa and Evora... exactly. Nothing gives you sensations like a Lotus! (With respect to your BMW friends). It doesn't matter to me being seen, or the car dirty, I drive them just for the sake of driving them. It is just a simple pleasure that releases my mind. I love to just get lost and see where I end up. When I clean them up and they are sparkling, they are ready for another day! +3

Skip Wareham What are your long-term plan for the Lotus now that have all this sweat equity in the car?

> **Rob Siegel** Who knows? I'm fixated on:
> 1) getting the drivetrain in
> 2) getting the rear suspension reassembled so the car can be let down off the stands and can be rolled for the first time in six years
> 3) getting it started
> Getting it driving, much less what happens after that, is months away. +4

Sean Curry Obviously you must re-take this picture for a side-by-side. +2

> **Rob Siegel** to show how much grayer my hair has gotten? +5
>
> **Dave Borasky** rub some grease in it +2
>
> **John Schanzenbach** or to show how clean the garage once was. +4
>
> **George Zinycz** That was probably grease in it then! +1

Charles Morris Don't feel bad about six years... I bought a bare block for my '72 Dart in 1997 and just completed the car this winter. Waiting for actual spring so I can start the shakedown cruises. 21 years and counting ;) +4

Jaime Kopchinski #2 really hits home for me... and reminds me when I enjoyed your books. There is so much satisfaction in fixing something that's broken, especially when it's a challenge or something you were warned about or intimidated by in the past. It's therapy for modern life. +4

Robert Boynton Didn't realize the Lotus had been hoof powered like Fred Flintstone's car. +3

> **Steve Nelson** Robert Boynton That was my first thought when I saw the pic... yabba dabba doo 😊 +1
>
> **Rob Siegel** Unlike a 914 (the other low-budget mid-engine car), there's an astonishing amount of room around the engine. It's easy to straddle, Slim-Pickins-In-Dr.-Strangelove-style. Again, these photos were nearly six years ago when I ripped everything apart. +3

Lindsey Brown Rob Siegel It appears to have aged you like four years in the White House. +2

Robert Boynton Rob Siegel Almost as if they intentionally left plenty of room to conveniently remove the engine. +1

Rob Siegel Robert Boynton, I doubt that was a design goal. Light weight, performance, holding down production cost… that's about it. +1

Steven Bauer Rob Siegel Lotus, unlike Porsche did not succumb to need to provide a "trunk" in the back

Rob Siegel There IS a trunk in the back.

Kevin Pennell Da man in his element! +1

Doug Jacobs Eh, don't mind me, doctor's orders… +1

Daren Stone Never heard it said better. +1

Vince Strazzabosco Well written and a great picture! +1

David Weisman Caption: Jonah and the Lotus. +3

Chris Mahalick You are doing so well!! Your work is most impressive! +1

Mike Miller Here we go with the hood strut again. Rob, I take no responsibility for the Lotus unless you want to sell him to me, but I will happily buy hood struts for your E9 if you promise to install them. +2

    Rob Siegel Hey, the broomstick is a factory lightweight part! +4

Mel Green Your next book title Perhaps? "Zen and my Lotus" +1

    Sean Curry "Let us Lotus!"

Richard Baptiste Inspirational determination (stubborn as she who must be obeyed says) either way I can't wait to read your first road test! +1

Lisa Breslow Thompson Let your freak flag fly! 🖤🚗 +2

Bart Collins I like that picture. +2

Tim Warner Paint looks pretty nice! This has had me working on my two stalled projects! Thanks. +1

David Kemether "Congratulations Mrs. Europa, it's a boy" +3

Scott Estes Hrmmm. 1 Rob Siegel power = maybe 1/30th HP +1

Brian Ach Assume the position. But only when wearing full-body Tyvek.

    Rob Siegel A.k.a. the automotive body condom +1

Robert Alan Scalla All of that for a brown Lotus?

    Rob Siegel Hey!

    Robert Alan Scalla sorry. +1

Brad Purvis Total. Plutonic. Reversal. +1

Mark Manasas Maybe you need to add another project to the garage [Craigslist ad for 1970 Europa S2 project]

    Rob Siegel Thanks, I'm good!

    Robert Boynton Rob Siegel With Model A Fords, 4-drs with a lot of wooden body structures, I've heard it said that you need to rebuild three. First one to learn, 2nd to hone your skills, and 3rd as proof of your craft.

    Rob Siegel Robert Boynton, so this is my "starter" Lotus?! +2

    Robert Boynton Rob Siegel Think you get AP credit on the hydraulic and suspension issues, but seems like at least one more engine rebuild would be necessary (assuming another survivor will even exist by the time this one is done). I don't think taking this one apart a second time would count. +1

TLC Day 2121: Rear bushings out

MAR 31, 2019, 7:56 PM

The Lotus Chronicles, Day 2121, Part B:

Got the radius arm (trailing arm) bushings out. Unfortunately, the flexible brake lines sit at the ends of the radius arms at the end of a box channel in the fiberglass body, and it'll never be easier to replace them than with the bushings out and the ability to drop the entire arm down (which is exactly what I thought about the radius bushings with the drivetrain out). I unscrewed both ends of both metal brake lines and blew compressed air through them to verify that the rubber lines aren't swelled shut, but I'm thinking I may regret it if I don't change them now anyway. Hello, slippery slope. I'll sleep on it.

Repaired the studs that I'd broken off the bottom of the original radiator. They snapped off during removal, and I need them to secure the radiator to the bottom of the hood compartment. I drilled them out and re-tapped the holes. Time-consuming but oddly satisfying. And free.

Not a bad day.

62 LIKES, 25 COMMENTS

Chris Roberts Free wins every time 😊 +2

John Whetstone Quite a spectacular day, by the current standards. +2

Tammer Farid The Queen of While You're In There is a cruel ruler, but a wise one. +8

Collin Blackburn What miserable British bastard designed that crap, though?

Daniel Neal You would be a fool to not change them, you know, Murphy's Law and all. +3

    Rob Siegel Yeah, I know.

    Roy Richard Daniel Neal. Even more important O'Lear's law. "Murphy was an optimist."

Ricky Marrero Did you already cover the subject of door hinges?

    Scott Linn Those can be done at any time though. +2

JP Hermes While I'm in there…

Chris Mahalick My bushings looked OK, so I did all the soft lines using a clawfoot socket. You are FLYING on this! And I mean that sincerely. By the time you are done, that car will be like new. I look forward to future posts. +5

    Rob Siegel I have read that people use the same trick that's used in the BMW world, which is to cut the soft line near its hex fitting and then put a deep socket over it to get it off. Of course the problem is that if you can't remove it, you have cut off your method of retreat 😊 +2

    George Zinycz Rob Siegel 😊

Chris Raia By policy, I'd change anything that is original and rubber on that car. #NeverReuseRubbers +6

Paul Wegweiser Make sure you replace every goddamned fastener with a METRIC ONE… just to totally mindf*ck the next owner. You know you want to. +4

    JP Hermes And some Whitworth. You can't forget those. +5

Jonathan Poole Always fix the brakes. After all that nice engine work it would be a shame to stuff it because a brake line burst. Safety first. +2

Roy Richard [very odd gif of zoom in and out of sour expression of a dog's face]

Roy Richard I have no idea how this gif was posted 😊 +1

George Zinycz Hope you slept on it well. Now wake up and replace those brake lines! +1

Rob Siegel On it. +2

Franklin Alfonso Rob Siegel Yeah, anything that can go wrong will... I vote change them.

Scott Aaron Oh yeah, fiberglass!

Steve Park These are strange little cars. Great to see steady progress. Gotta do the brake lines though.

Rob Siegel removed this morning +3

TLC Day 2122: Flexible brake lines out

APR 1, 2019, 7:37 PM

The Lotus Chronicles, Day 2122:

As I said yesterday, the flexible brake hoses are located at the ends of the radius (trailing) arms, inside a channel with the frame on one side and the fiberglass body on the other. Since I'm changing the radius bushings (removal of which makes the end of the radius arm drop down), it makes total sense to change the flexible hoses at the same time. If you cut the hose, the arm drops down all the way, giving you unimpeded access to the lower connection (first pic). But the one at the top is tougher. On the right side of the car, you can get at it from a gap at the top (second pic), but on the left side, that access is blocked by the battery tray, so you need to reach up from underneath into a channel where you have to put the 9/16" wrench on the fitting, rotate it less than an 8th of a turn, then take the wrench off and flip it around, while holding the back of it still so it doesn't twist off (3rd pic). Of course the rustiest of the four connections was the one that was the hardest to reach. When doing these on a vintage BMW, because you REALLY don't want to strip or snap one of the metal lines, I'm in the habit of heating them up first and touching a wax stick to the threads, but I'm more than a little hesitant to be using an open flame around the fiberglass. Fortunately it came off, maybe only hitting 4 on the pain-in-the-butt-o-meter. (Or "buttometer." Hey. It's British.)

New radius bushings arrived today. Flexible brake hoses should be here Wednesday. A few hours to button it back up, and then—and I can't believe I'm saying this—I can't think of any good reason why the next step isn't raising the back end of the car so I can slide the drivetrain in under it.

56 LIKES, 30 COMMENTS

Rob Theriaque Soooo... I just did the calipers on my 2002. One of the brake lines was giving me a hard time, so I gave it the old fingers-crossed yank. It came free. Are you telling me I would have been SOL if I'd broken it? Note to self... Check these things before applying excess torque next time...

Rob Siegel Even having a roll of copper-nickel tubing and a flaring tool, I really dislike replacing metal brake lines. There are two reasons to replace them: A) they've rusted out and are leaking, or B) you've broken one while doing related brake work. B has been more common than A. You're not SOL, but it's a lot of work, enough that I take maximum precautions to not have it happen (e.g., heat and wax). +2

Daniel Neal I have never seen flare nuts like that.

William Davis You love that car like a fun S&M show... ;) +1

Rob Siegel Actually, I'm just trying to get it whole and moving. It totally remains to be seen whether I love it or not. +2

Clyde Gates I foresee another book: The Agony and the Ecstasy of the Lotus...

Franklin Alfonso Oh there is definitely a book from resuscitating Lolita. +1

Chris Mahalick Jeez... that car is fighting you on every front!!!

Roy Richard Great April Fools post. +2

Scott Linn Looking at the photos of the one in the channel, I'm glad mine were done before I got the car! +1

David Shatzer I always thought I wanted a Europa. Until I tried to sit in one and close the door. Jimmy Clark sizes, not Teutonic Saxon sized. Living the Dream vicariously with you! Here's to the soon to be most sorted out driver Twin Cam on this planet.

Rob Siegel Yeah, it certainly won't be that, but thanks :^) +1

Chris Raia If it were an MG, one of those nuts would be 9/16, one 15mm, and one 3/8 Whitworth. +2

Rob Siegel My selection of SAE wrenches and sockets is shit, but fortunately 14mm is a tight 9/16". +1

Tim Warner Rob Siegel during the early 70's in college, I'd work the metric to inch wrench conversions on a primitive Wang calculator! Fun to see how sloppy wrenches are! +1

Eliot Miranda The British for that would be Arseometer... +5

Rob Siegel much better, thanks

Martin Bullen Or bumometer +2

Christopher Kohler So hate to ask this but do you know what the insides of the fuel tanks are like?

Rob Siegel Yup. Looked at them last week. No rust, just a thin layer of varnish. This arrived in the mail today to help me clean it out. I think letting it soak in gas, a scrubbing, a rinse, done. [link to Brushtech Extra Long Super Flexible Drain Brush]

Christopher Kohler Great news. Guess that's the difference with real gas vs the crap we have today. I was greeted with a similar situation but 2002 tank is a lot easier to remove than whatever program you've you there so I had a radiator shop take care of it!

Rob Siegel Christopher Kohler, They're twin tanks so they're manageable in size, but the rubber filler necks that clamp to the tops of the tanks and to the snap-lock filler lids have hardened from age. I'd need to cut them off and replace them to remove the tanks. Not the end of the world, but I'm gratified that I can just clean them in place.

John Thomas Ratchet wrench wouldn't get in there?

Charles Morris Have you considered the fuel tank sealer kit from POR-15? That stuff never comes off or breaks down, and you won't have to ever worry about the tank rusting either. I've used it in tractors (diesel and gas), cars (ditto) for over 20 years. There are other brands that have a deserved reputation for dissolving or peeling off and clogging everything, but this "US Standard Fuel Tank Sealer" is tough. +1

Rob Siegel Yes, looked into it, but a) they recommend it as the third step of a three-step solution, and b) it's not necessary here, as the tank is not rusted.

Charles Morris I suppose if it's not rusted by now, it will outlast both of us. On the other hand, there's ethanol in the gas now, and it does mix with water... this is another "it'll never be easier" prophylaxis. :) +1

Rob Siegel See comment above about getting the tank out. I'm not restoring this car; I just want to get it running. The bushings and brake lines ARE things that are easier to reach with the drivetrain out. The tanks are pretty accessible either way.

Charles Morris OK, saw the comment. Anything rubber in the fuel system I'd want to replace... especially if it's hardened from age. But if you can get to them later without disassembling half the car, that's an easier decision to postpone their replacement. Assuming they don't leak now, that is ;)

Rob Siegel Well, I'll find that out, won't it? I agree with you about the rubber, except these are just dead straight dead vertical filler necks.

Daren Stone Go Rob, Go! +1

Wrenches on the difficult-to-access brake line joint.

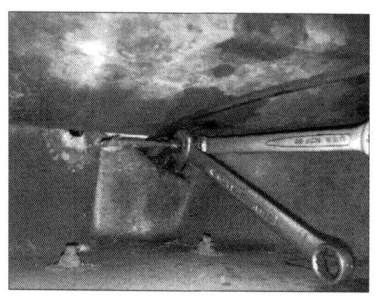

19 LIKES, 40 COMMENTS

Marshall Garrison ratcheting flare wrench too short? +3

Rob Siegel I don't have one

Marshall Garrison I was thinking of a past discussion about them, but then, come to think of it, that would've been metric for BMW too!

Rob Siegel 9/16" is just an RCH larger than 14mm. you can see that one of the wrenches on it IS 14mm. +2

George Zinycz Marshall Garrison there's such a thing?

Marshall Garrison George Zinycz Yeah, that's what I said too when I heard of them—there are a _variety_! This Facom set doesn't swivel, The Stanley Proto looks similar, and looks like Wright Tool ratcheting flare nut wrenches—haven't been able to google up a metric set from Wright Tool, but you'd think they would have it (edit, nope, no metric). [links to sealey, gearwrench, astrotools, jbtoolsales, falcom, tannerbolt] +1

George Zinycz Marshall Garrison a bargain even at twice the price! Now if someone made a power one that runs on compressed air, we would be at the zenith of human civilization! +2

Russell Baden Musta Well done Rob! +2

Drew Lagravinese RCH—a term I have not heard since my Navy days. Well said. 😄 +6

Jeff Alexander Drew Lagravinese ditto

Drew Lagravinese Jeff Alexander holy crap Batman, we were both born on 8/8. I'm 1953.

Roy Richard Drew Lagravinese red pr royal?

Jeff Alexander Roy Richard red of course

Drew Lagravinese Jeff Alexander agreed!

Lindsey Brown When it comes to renewing a stopping hose, the British expend the spanner operator's time like there's no tomorrow.

Craig Fitzgerald Oh, the British!

David Weisman Heh heh, he said RCH. Heh heh

Rob Siegel As Mona Lisa Vito said in My Cousin Vinny, "it's an industry term." +4

David Weisman Rob Siegel 👍

Greg Aplin David, I worked with engineers at Cape Canaveral that used the term! +3

David Weisman Greg Aplin I've heard other awesomely vulgar versions

Jim Strickland I want a tool that doesn't exist (I think). Starts as open end wrench, then closes and turns somehow (motor or air?) Would be fantastic in these close situations.

Rob Siegel Jim, I have something like that in an 11mm size, sort of. It does close around a flare nut like the one above, but it doesn't auto-turn. I'll take a photo of it.

Jim Strickland Rob, it would be great to see. Perfect tool for my lazy self as I work on 2002 brakes.

Sean F. Gaines Ya. I've seen open end wrenches, but decided I couldn't justify buying them just because they would be great when needed. But looks like this might be your chance?

Rob Siegel Jim, this is the wrench. The brand is "Premier."

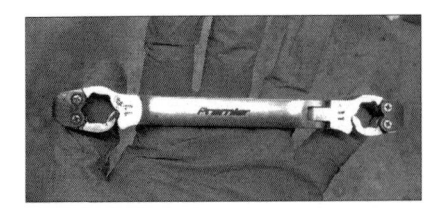

Jim Strickland Rob, thanks! That's going on my shopping list.

Marc Schatell Jim Strickland I have a set of those, and they're awesome.

Francis Macaluso Rob Siegel that looks like a VERY useful brake line tool!

Rob Siegel someone here on Facebook bought it for me. I'm ashamed that I can't remember who it was 😊

Jim Strickland Rob, some great person! Appropriately, the eBay sellers of the tool seem to be in UK 😊

Marshall Garrison Jim Strickland I think they're made by Sealey (apparently a Brit company) & Premier is the name of that Sealey tool line. They have other varieties. I put links in my reply to George, above.

Marshall Garrison Rob Siegel Yep, that's what I was asking about initially. I was remembering the discussion on that whenever it was!

Jim Strickland Marshall Garrison, thanks!

Rob Siegel Right, that's the exact one I have. To be clear, it doesn't have a "ratchet" in it like the gear wrenches do. I think what they mean by "ratchet" is that you can rotate the wrench and the hinged on alarm will let go and grab again when it's back in position.

Bill Ecker ^^Snap-On's infamous "speed-wrench" experiment from the 90's, only improved.

Alessandro Botta Rob Siegel this is the ultimate wrench [link to holy grail of ratcheting flare nut wrench]

Carl Witthoft Pretty sure you can get a power wrench as Jim S. described. Might not be cheap. Alternative: if you don't need that stub of line, cut it off and use a cable-driven driver straight down?

TLC Day 2123: Cleaning the gas tank

APR 2, 2019, 7:39 PM

The Lotus Chronicles, Day 2123:

The flexible brake hoses arrive tomorrow. While I'm waiting, today was tank cleaning day.

--I got as much surface rust as I could out of the metal coolant reservoir tank by repeatedly filling it with small fasteners, shaking it, and dumping it until there was no more scale and all that came out was fine particulate rust. I could've done a more complete, more expensive, and more time-consuming treatment, but in the end I simply blew it out, took a bottle of Rust-Oleum Rust Reformer (some of that stuff that treats and encapsulates rust, turning what it can touch black) I had left over from other projects, poured a bunch in, rotated the reservoir tank to coat the inside, and let the excess run out the bottom. Done.

--The twin gas tanks are, in theory, not that difficult to remove to clean, but in practice, there's a rubber hose that connects the top of each tank to its fill neck. None of the hose is actually exposed to liquid gas unless you overfill the tank. The hoses on mine are likely original and thus hard as rocks. To remove the tanks, I'd need to cut the hoses off and replace them. I looked in each tank with a flashlight and saw no rust in one and just some small flakes in the other, and just a light coating of varnish at the bottom of both, so I decided I didn't need to remove them. I soaked the bottom of both tanks in fresh gas, and ordered a brush with a flexible handle (actually a short drain snake) from Amazon, but when it arrived, it turned out to be way too flexible to bite into the bottom. So I improvised. I took the aluminum rod I use for banging out rockers and shafts, and zip-tied a piece of Scotch Brite to the bottom. It worked perfectly, allowing me to scour the softened varnish off the bottoms of both tanks. Done and done.

What? I half-assed it? Rust in the reservoir might come back? The gas tanks might start to rust due to ethanol in modern gas attracting water and they should've been stripped and painted? Yes. Sure. But I think it's fine. This ain't a restoration. I've got shit to do.

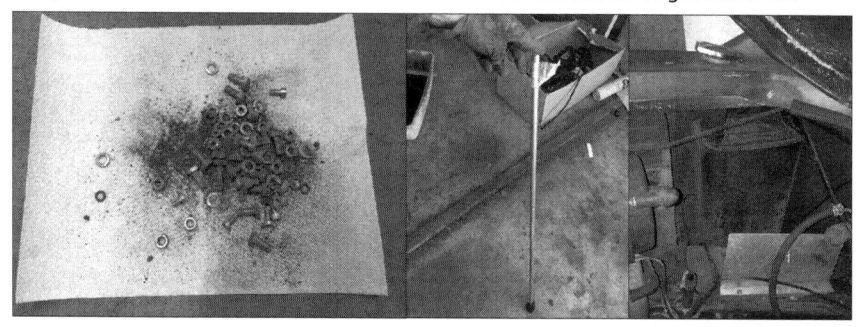

69 LIKES, 47 COMMENTS

Craig Fitzgerald Nobody wants to see you go whole-assed, Rob. +3
    Doug Jacobs Oh jesus, that's the truth. A wholy-assed Rob Siegel—we don't need any of that shit.
    Garrett Briggs Yeah. Fuck the whole ass. Uh, you know what I mean. +2
Kevin McGrath Oh. And I thought that was your breakfast. Add milk… some 30W. And a magnet spoon.
    Rob Siegel Grape Nuts. For real mechanics. +3
    Kevin McGrath Wow! Grape Nuts. Straight out of the Wayback Machine. [photo of Ewell Gibbons]
John Graham Is this the entire car? +1
Daniel Neal Solution to the ethanol problem is simple: Just buy non-ethanol fuel.
    Rob Siegel essentially not available in Massachusetts unless you want to buy race gas
    Daniel Neal Rob Siegel and here I thought NY was bad
    George Zinycz Daniel Neal um I feel weird carsplaining this to a bunch of gear heads, but the 10% ethanol is a *national* mandate, not a state-by-state option (and so you know, individual counties and municipalities mandate different formulations of gas, the refining and distribution complexity that goes with that fact is responsible for a non-trivial (to coin a popular phrase) portion of the cost fuel. Race gas doesn't have ethanol in it, but it costs like $8/gallon and comes in rather inconvenient bucket-sized cans. +1
    Daniel Neal George Zinycz there are 6 or 8 gas stations near my house where you can get Ethanol Free gas, most of them the Ethanol Free is 93 Octane, a couple it is 91 Octane. And it's not just them claiming to sell Ethanol Free, I know people who have tested the fuel (they have a sample kit that tests the level because they have cars tuned on E85) and proven that the amount of Ethanol in the fuel is <1%. Of course all but 1 of the stations selling the Ethanol Free fuel are on an Indian Reservation.
    George Zinycz Daniel Neal well that's different then lol! That's another country. I'd be so happy to have that near me.
    George Zinycz Daniel Neal then that one station off the reservation is lying 😂😂😂 +1
Skip McCauley I love reading these updates, Rob! Keep it up, you're a hero with coveralls! +2
Steve Park If it works, who cares. It's not your first rodeo. If you know anyone with a Miata, the brush you bought will clean the top drains. +1
Rob Koenig How's that rust reformer going to hold up to hot coolant? Think you should top coat it?
    Rob Siegel I suspect it will be fine, but I'll read up on it. +2
    Christopher Kohler Sounds like more shit to do now! +1
    Rob Siegel Christopher Kohler, well it's more polite than saying "fuck the fuck off!" +5
    Rob Siegel "Do not apply to surfaces, when heated, exceed 200°F (93°C)." Right on the line. It may

be okay since the reservoir tank doesn't see circulating coolant. Bloody troublemaker, you are. +5

Rob Koenig Just trying to save you from insanity like the material breaking down and forming a thermal barrier throughout the entire cooling system... 😀

David Holzman Just as long as you anodized the muffler bearings like I told you

Rob Siegel I anodize muffler bearings for no man. +2

David Holzman You're supposed to do it for yourself! +1

George Zinycz David Holzman speaking of which when are we going to discuss blinker fluid options?!? Or is it too soon still?

Bob Sawtelle Rob seriously with the amount of work you have done so far the last words I would use are "half assed." Now, half-baked lunacy for even attempting this is a different story. +1

Maury Walsh Rob Siegel, I'll see your grams of rust from the gas tank and raise you 7 ounces.

+6

Rob Siegel DUDE!

Maury Walsh It was epic.

Rob Siegel Which car is that from?

Jamie Eno How does this tank not leak??

Garrett Briggs My guess is the rust blocks the fill tube.

Franklin Alfonso You can fill potholes with that...

Maury Walsh 1941 Bantam Roadster. Amazingly, everything else was in pretty good shape. I had the tank cleaned and sealed after the excavation. [photo of Bantam] +4

Rob Siegel THAT is a very cool car!

Rob Siegel After I bought my white 2002tii about eight years ago, it was exhibiting signs of a clogged fuel filter. I chased the rust back to the tank. When I removed the pick-up tube and could look into the tank, and then stick a gloved hand in, it looked like someone was making a fucking pot roast inside. +6

Daniel Neal Rob Siegel that makes me think of a dump truck we had at my old job. I had gotten it running in tip top condition and roughly 6-8 weeks after putting it in service it came back to the yard one day with a complaint of a loss of what little power it had. I pulled the filter and the fuel looked great. It was then I noticed a small plastic bowl of sorts that was ahead of the filter. I unscrewed it from the filter head and found the strainer, which was packed completely solid with rust. Cleaned it out and the truck ran beautifully for another 8-10 weeks and it would plug up again, so then we drained the tank and found it was all rust on the inside. Then the owner's son rolled the truck and we didn't have to worry about it anymore. +2

Brian Ach Looks familiar Rob

Rob Siegel Brian, I don't know why you'd say that. [photo of the stuff I pulled out of Brian's gas tank]

Rob Siegel (I actually used that pic in a Hagerty piece that'll run next week on why you don't just dead-start a long-dormant car without changing the oil and cleaning out the fuel system, cooling system, and air cleaners)

Chris Mahalick Carry on and keep your forward momentum. You could always go back later and do some weekend projects once you are up and running.

Rob Siegel you GET it! +1

Alexander Wajsfelner $2.95 in line filter should take care of that. +1

TLC Day 2124: Flexible brake lines in

APR 3, 2019, 8:18 PM

The Lotus Chronicles, Day 2124:

Flexible rear brake lines arrived, so in they went. Did need to resort to using the 9/16" crowfoot wrench to get everything snugged down. Done.

With those installed, in went the radius arm bushings. Found some weirdness on the Lotus forums about how there's a service bulletin that the bushings and bolts should be flipped around relative to their original orientation to allow sufficient adjustment of the tow-in via washers. Looked in the manual and did in fact find reference to it. Hated to change it from how it originally was, but I asked the question on one of the forums and the answer was "yes, flip them." As far as finding a shop to do that kind of alignment measurement and either insert washers or tell me how many I should insert (Lindsey Brown?), who knows. As with so many other things, I look forward to when this is the most pressing issue on the car.

So with the "while you're in there" items of radius bushings and flexible brake lines done, the punch list of things needed before the drivetrain is dropped in is getting mighty short indeed. Stay tuned. The show's about to get good.

77 LIKES, 27 COMMENTS

Charles Morris Just wondering how many ft-lb you set the torque wrench to, with that compound angle :) +1

    Rob Siegel Just did it by feel. Hopefully I didn't crack the flares.

    Charles Morris I figured as much ;) just tighten until it strips, then back off half a turn! +7

    Ed MacVaugh There is a chart in the torque wrench box, if the center is an inch away from the center of the wrench, if affects the torque a certain amount, the greater away from the centerline, the more the effect.

    George Zinycz Rob Siegel I will cry for you if they did +1

John Harvey Don't forget, Vintage Prep. +1

    Rob Siegel Um, I don't think they'd let it in :^)

    John Harvey Your other cars I meant.

    Rob Siegel Oh I know, I'm just fucking with you. +1

    John Harvey I mean, feel free. Do you feel lucky, punk? Well, do ya?

    Jonathan Poole I think Rob could show up in just about anything short of an e36 and Scott would let him in. +6

David Kemether

+3

Rob Siegel aw, it's got a little mustache...
David Kemether

+2

Dave Borasky [gif Animal House "Oh boy this is great"]
Scott Aaron Golly. I thought that cooling system pressure testing stuff was great. This is really going to be 🔥🔥🔥 +1
    Scott Aaron *not saying your car is going to burn!** +1
    Scott Aaron **not saying it won't, either +2
    Rob Siegel you're awfully risk averse +1
    Scott Aaron Rob Siegel I haven't been giving you and the fantastic plastic enough guff lately. Been too busy. +1
Lindsey Brown I'd be honored to put it on my alignment rack once it's running, assuming I don't die of old age first. +8
Brian Stauss Crows foot wrench. I have a complete set of those that I have never used in 30+ years, wrenching on all sorts of cars and lawn equipment. But, like a good number of my tools, I have them "just in case." Glad to know that if I ever own a Europa, I can plan on using the crows foot wrench. Thanks for sharing! +3
Mike Savage Another great demonstration of how/to use the crows foot wrench. I think you may have used it on a master cylinder replacement on Louie...
    Rob Siegel Yes, that's the only other time.
Ned Gray Ditto on the Crows Foot wrenches! I think I've had mine for 40 yrs! +1
Paul Skelton That master cylinder is in a fun spot eh? lol Bloody pain in the ass!
Dave Gerwig Is that the Louisville Crow Foot?
    Rob Siegel It IS! First time I've used it since! +1
    Dave Gerwig Rob Siegel glad to see it still in service! I know where to find them when the need arises in my garage. 😊 +1
    Rob Siegel I love that you know this!! [Note: Dave spent a day with me during the Louie adventure when my clutch master failed. This included driving me to Harbor Freight to buy this very set of crow foot wrenches.]
    George Zinycz Rob Siegel you mentioned buying the set in the Louie saga right? I still do not own any.
    Dave Gerwig George Zinycz it's all in the details 👍 +1
    George Zinycz Dave Gerwig I often say that there's no such thing as a tool I don't need—just a tool I don't need *yet* +2
    George Zinycz Dave Gerwig it's been suggested by friends that I'm only a car guy because I was a tool guy first.
    Dave Gerwig George Zinycz tricks of the trade
    George Zinycz Dave Gerwig one of my favorite things about reading Rob's articles is the minutia or tool use.
Tim Warner I've never ever needed those crows foot sonsobitches. They live in my toolbox...
    Rob Siegel this is only the second time I've needed them, but need them I did! +2
    Tim Warner Rob Siegel I can see there will be a time... maybe before I'm 70 or so... +1

Kevin Sheehy Rob Siegel it's times like this when you feel stupid proud of yourself to have that one inexpensive tool, never used before that makes all the difference! You didn't even have to leave the house. You are king at the moment amongst us mortals but don't despair, the car gods will give you grief another day. +2

Richard Memmel Tim Warner I've worked in the automobile industry for over 30 years my tool boxes are full of tools I needed to make my life easier once or twice. lol +4

Tim Warner Kevin Sheehy. I rebuilt my first engine over 52 years ago... I have many tools that are like that! Some I have moved out of the box into cabinets. +1

Al Sinclair I have the flare nut version of them. Did not us them much but was really thankful when I needed them. +1

Gary Beck Rob, I've been working since November on my coupe. At first just normal work days but I am retired so I can put in as many hours as I can. I didn't write it up that much because it takes too much time. You are killing me with all that you have posted. FB is going to make you a Platinum member. WTF that means I don't want to know. So kudos to great dedication to a car none of us understand. Looks cool but English cars??? So hope it runs like a scalded dog and you go under the deer the dog is chasing instead of hitting it. See you at The Vintage in a BMW? +4

Rusty Fretsman "Bit of a dodgy bodge, cor!" #bollocks +1

Harry Krix Very impressive, RS! My Dad did a lot of aviation air frame and power plant (A&P) work; he used crow foot(s), crows feet? for a great number of things. When we went SCCA racing in the 70s and some of our other friends/competitors came around and saw him at work, they was always came away with a sense of amazement. I believe we in the car world could get a lot of things done more quickly if we knew how to properly use them. However, I my case that ship has sailed... but the complete sets of metric and SAE look awfully impressive in the top of my Snap-on toolbox. +1

Mark Bart Cool tool!

Rusty Fretsman What a phenomenal PITA! :|

TLC Day 2125: Drivetrain tease #1

APR 4, 2019, 8:54 PM

The Lotus Chronicles, Day 2125:

I'll tell you tomorrow. Maybe Saturday. But it's epic. And ongoing. I thought I won, but then the Lotus reached up like the Balrog in the mines of Moria when it grabbed Gandalf with the fire whip and dragged him down into shadow.

But here's a tease :-)

110 LIKES, 128 COMMENTS

- **Brandon Fetch** Run, you fools! +5
    - **Doug Jacobs** Brandon Fetch 'fly you fools'. +5
    - **Brandon Fetch** Fly... run... GTFO ya idgits!
- **Ed MacVaugh** Did you get a satisfactory answer regarding clutch lever movement?
    - **Rob Siegel** I decided to risk it, but yes, I did get answers on both forums. The response was "probably fine, don't worry about it, when it's in the car you'll adjust the cable to take up that slack." +2
- **Bill Snider** Just be glad it wasn' t a BalRob lol +6
- **Tom Tate** You're trusting your life to a Home Depot barrel? What have you been drinking? 😉 +2
    - **Rob Siegel** Tom, I'm not trusting anything to it. It's just supporting the hub assemblies before the shocks and lower control arms are put back on. +2
    - **Charles Morris** That's what I thought at first! Until I looked closer and saw the floor jack plus jack stands holding everything up ;)
    - **John Wanner** Rob Siegel that's a relief you had me worried for a second.
    - **Tom Tate** Ok you're in the clear 👍 As you were 😺 🙊 +2
    - **Steven Hussein** Bernstein It's "The Hack Mechanic", not "The Whack Mechanic"! ;) #itrusthim
    - **Rob Siegel** Steven Hussein Bernstein, no, some days, it really is the Whack Mechanic
    - **George Murnaghan** Steven Hussein Bernstein Best of all, not the Flat Mechanic!
- **Dave Gerwig** Must be mate-ing season! +4
    - **Doug Jacobs** You IDIOT!!! You're using _penne_!!!! Everyone with a brain uses the little

cartwheels. That penne is going to guck the motor up! +6

Jim Gerock Doug Jacobs but it's Barilla!!

Mark Thompson I'm always into this as a history guy. Can't wait for Lotus Chronicles... 2047... Search for Spock fuel liner... +3

> **Steven Bauer** Mark Thompson what if the aliens intended to steal Spock's brain and got a Lotus engine by mistake?
>
> **Mark Thompson** Hilarity ensues in the farce of the year!

Matthew Rogers ...Day 2126... Rob begins to understand the British sense of humor... +5

> **Scott Lagrotteria** Matthew Rogers does anyone truly understand British humor (or humour)?
>
> **Matthew Rogers** Scott Lagrotteria the British do. Just look at their cars.
>
> **Matthew Rogers** (As a long time MG owner, I am authorized to make this statement.)

Chris Roberts Tough crowd, Rob 😀 🍺 Maybe next time use a simile at the fourth-grade reading level.

Neal Klinman Does the epic story involve the Penne pasta on the shelf?!? +1

Bob Sawtelle Dammit already three Penne references, there goes my "Penne for your thoughts joke"! oh well… +9

Jeffrey Miner I love the 2002 snickering in the foreground :) +2

Tim Warner Kugel is a "another level" lol +1

Roy Richard Rob Siegel Do you really want us to believe you're Gandalf?

Drew Lagravinese I know, it's the wrong engine.

Brian Stauss All that engine and transmission fits in that itty bitty space?

John Jones Brian Stauss It barilla fits +6

Rob Siegel Honestly, what the hell is it with you people? I show what's clearly a Lotus drivetrain about to go back into a lotus AFTER SIX FUCKING YEARS and you fixate on Home Depot buckets and a box of Penne? That's it. You no longer deserve this level of entertainment. +56 **DING DING DING! MOST-LIKED POST!**

> **Neal Klinman** I thought the penne was important. You know, because it's an Italian car and everything. :-) +5
>
> **John Jones** Rob Siegel It's like having someone you really like married to someone you really don't who treats your friend like crap all the time. After a certain amount of time, you stop feeling bad for them and realize they are in their shitty relationship voluntarily, and clearly, the friend has a weird kink for all the abuse and humiliation, so we just try to pile on to make you enjoy it all the more in your twisted little way. +16
>
> **David Kemether** Mmmmmmmm, penne pasta. +3
>
> **John Wanner** Rob Siegel we never claimed to deserve it, but it was free so we said "what the hell, I'll watch." +10
>
> **Rob Siegel** that explains a lot
>
> **Clint Carroll** Rob I just noticed the picture frame in the background looks like it's crooked… +2
>
> **Dave Borasky** [gif "Don't be upset… eat some spaghetti"] +3
>
> **Rob Koenig** I can't believe you're still using a 2x4 as a hood prop.
>
> **Doug Jacobs** Rob, oh, I'm sorry, you title a picture "Here's a tease." And you expected...what? +1
>
> **Daren Stone** Tell 'em, Rob Siegel!
>
> **Brad Purvis** Rob Siegel Welcome to social media me amigo. +1
>
> **Lee Highsmith** I have to admit, the buckets surprised me and I wondered if the penne was a hack...

Stuart Moulton Do you use the Costco pasta to clean up oil spills? +4

> **Rob Siegel** yes, soak it in oil 'till it's al dente +4

Kevin Fernandez The Penne Chronicles, Day 1 +9

> **Franklin Alfonso** There is no time to tidy up the garage! Can't you see the man is about to reach the climax of his relationship with Lolita after over 5 years of frustrations and delays? There is a serious case of blue balls here, pasta enough! +1
>
> **Leslie Goldstein** How do we know it is Lolita rather then Lothario? +1
>
> **Rob Siegel** Great, Leslie, but Franklin is right—it IS Lolita :^)
>
> **Leslie Goldstein** Are you Humbert Humbert played by James Mason or Jeremy Irons? +1
>
> **Rob Siegel** James Mason. At least he's quasi-respectable :^) +1
>
> **Leslie Goldstein** Jeremy Irons can be creepy at times.

Rob Siegel [link to video of Jeremy Irons "You have no idea"]

David Holzman Leslie Goldstein Uh, James Mason was head of the Centers for Disease Control under Reagan—but well respected on both sides of the aisle.

David Holzman And Rob, that goes for you, too—about James Mason.

Franklin Alfonso [gif of James Mason "God was wrong!"]

Franklin Alfonso Cars and boats are always female...

Jonathan Poole It sounds like PASTA was actually invented by Toyota. Which leads me to believe there will be a bait and switch for some nice twin-cam Toyota motor at the last minute. [link to video "Toyota presented PASTA Car-Hacking Tool at BlackHat London"] +2

John Graham Actually, what crossed my mind—how much does that car weigh? I am guessing at least 100 lbs but I might be high with that number! :-)

Rob Siegel I believe it's a 1600 lb car. +2

John Graham With that weight, if you only get 2 cylinders running the car will fly!!! +2

Rob Siegel This is one of the reasons I was comfortable with a stock rebuild of the engine instead of spending gobs of money to change the pistons and cams to European Sprint specifications. +1

John Graham Our 02's are chubby little things in comparison.

Franklin Alfonso My brother in law had a Mini-Cooper way back in the 70s and man that light weight and low profile little wheels and power made it fly! +2

John Graham You are in Indycar territory weight wise. In fact, Indycar is here in Birmingham this weekend at Barber Motorsports Park.

Rob Siegel That's part of the appeal. Yes it's a total pain in the ass (and I haven't even driven it), but there aren't a lot of affordable mid-engine cars that weigh 1600 lbs and have a 42-inch-high roof line. +1

Franklin Alfonso Not a car made for a big guy like 200 lbs plus 6' though lol.

John Graham Made me think—200 lbs is 13% of the total car weight! :-) Even small guys like me at 5'4» at 150 lbs is about 10% of the weight!

Tom McCarthy Similar picture if you were putting an engine back in a VW beetle. +1

Rob Siegel Trust me, that's much easier +2

Tom McCarthy Air cooled pressure test was super easy 😀 +2

Steven Bauer The engine fell off the hydraulic table?

Steve Park Even Corvairs are cake compared to this. I had the "pleasure" of helping remove an engine from a Lotus Esprit. What a PIA despite having a number of people to help. I didn't go back for the reinstall. I can't imagine doing this alone. +1

JT Burkard I, for one, enjoy these chronicles. If this wasn't free I would be willing to pay up to a single shilling for this level of entertainment. +1

Luther K Brefo I for one thought the penne set off the right Fung shui, or however you spell it. Wait, I get it now, Rob is pastafarian!!!! +2

JT Burkard R'Amen +3

Carl Nybro [gif of deer eating popcorn "Keep goin'"] +2

Paul Muskopf [gif of Ted Knight in Caddyshack "WELL?"] +5

David Kiehna Did you ask permission before you lifted her skirt? +2

Tony Pascarella I hope that the Home Depot bucket is a fairly recent version, because on those the words "Let's do this" are written on them.

Ezra Haines Don't look up "space docking" it's disgusting and offensive +2

George Zinycz On-topic guessing: The drivetrain won't go in mated?

Robert Olmedo I just watch to see if the hero wins. +1

Garrett Briggs Methinks Rob trapped the Balrog under the Home Depot bucket and is using the Lotus to make sure it doesn't escape. Sounds like Wizard trick to me. +2

Scott Aaron Time to get your pipe and your tweed jacket with elbow patches out of the Bav. +1

Steven Bauer I was considering the amount of shadenfreude inherent in this situation, and I remembered the story about trapping a monkey with an orange in a jar... +2

Marilyn Rea Beyer I am sorry to tell you this, Rob. I don't think that thing will fit inside there. 😀 +1

Roy Richard Rob Siegel I think I see the problem. You're trying to put the engine into the trunk. +3

Steven Bauer Another bit of Lotus quirkiness: the alternator is mounted on the bell housing? Is there a shaft running along the engine to drive that pulley?

> **Rob Siegel** On an Elan, the alternator is in the normal place in the front of the motor. On a Europa, the front of the motor is mashed two inches from the firewall, so the alternator goes in the back. That big pulley is driven off the intake cam. +1
>
> **Steven Bauer** Rob Siegel so, if you tighten the belt too much, you side load the cam bearing and kill your head?? It's definitely a more expedient solution than re-orienting to a Sideways mounted engine

Paul Wegweiser Can you remind us all why you voluntarily purchased a BROWN sports car? "Brown… the color of irrational, unbridled passion" said no one ever. +8

> **Steve Kirkup** I know, I know! Paul Wegweiser as a piece of permanent furniture, one would want it to match the room. +1
>
> **Rob Siegel** because fuck you. +7
>
> **Roy Richard** First thing he should have done is paint it BRG with yellow racing stripes 😜
>
> **Roy Richard** And the 🐎 you rode in on?
>
> **Rob Siegel** ahem… [photo of my dear departed brown 911SC] +3
>
> **Steve Kirkup** Rob Siegel I like brown, and it does match the sliding door, for the record. +2
>
> **Clint Carroll** Paul I'm tellin' ya' Rob's going to have the last laugh when this is all over… that little car is going to be fast; in fact, I heard he's going to name it "The Brown Streak…" +3
>
> **Rob Siegel** honestly, I'm going to start charging for this level of entertainment… +4
>
> **Steve Kirkup** Rob Siegel what's our cut for contributing? Can't have banter by yourself…
>
> **Jim Gerock** Rob Siegel guess this is the "back away from the gorilla" targa. +1

Scott Aaron I just had a premonition. I know what he is going post next 😜

Kevin McGrath Hack Mechanic set-up all the way! Love it!! Weird lifting, 20th century lighting, cramped environment. Skunk Works this is not. It's just like our garages. #torpedoheater

> **Rob Siegel** No torpedo heater. There is a NG-plumbed Modine Hot Dawg suspended from the ceiling 😜 +1
>
> **Kevin McGrath** NE garage requirement, of course!
>
> **Rob Siegel** I actually flipped it on today! I'm getting soft in my old age 😜 +1
>
> **Kevin McGrath** I'm only 54 and starting to see the value of a cardigan. It snowed for a bit this morning here in Valley Forge, PA (~270 miles south of your area) +1
>
> **Robert Boynton** Rob Siegel Separated combustion?
>
> **Kevin McGrath** I should've said 'composite lifting' not weird

Jim Gerock Please don't say he's going to swap the M10 into the Europa and vice versa 😜😜 +1

Brian Hart How about some inspirational pictures from the Netherlands… [photo of black Europa TCS and yellow Europa S2] +2

Scott Hopkinson Lotus Chronicles… even better than Louie Chronicles… +2

Keith Roth Did not see a bucket or penne. Will not post a picture of my shit hole/garage.

Jeffery W Dennis Waiting for the enviable start up … needing my daily dose of LC love.

Michael McSweeny trunk stick… 🦨😜

Bronco Barry Norman Are you telling me the Europa is possessed?

TLC Day 2126: Drivetrain tease #2

APR 5, 2019, 7:49 PM

The Lotus Chronicles, Day 2126:

I'll tell you tomorrow morning.

40 LIKES, 14 COMMENTS

> **Jim Gerock** Oh no. You left the penne on the stove too long. Or perhaps you took a wrong turn at Albuquerque +8
> **John McFadden** I think he left the cake out in the rain +2
> **Luther K Brefo** [gif angry popcorn-eating fans]
> **Douglas Wittkowski** Shall we pray he took the nuclear option for a quick sale? 🖤
> **Scott Aaron** I'll block my schedule 😬
> **Greg Lewis** Taking bets now
> **Doug Jacobs** Now you're just being hurtful.
> **Matthew Rogers** Cryptic Rob is Cryptic.
> **Anthony Neu** Home Depot drum circle +1
> **John Morris** Whatta maroon!
> **Ken Sowul** Waskly Wabbitt
> **Franklin Alfonso** Did you drop an anvil on it? +2

TLC Day 2127: The Divetrain is installed!

APR 6, 2019, 8:55 AM

The Lotus Chronicles, Day 2127:

It's in! Oh my god!

Forgive me for the length of what follows, but this is really big. Huge.

(Wait... are we still doing phrasing?)

Note: In the photos below, you'll see the top of the engine covered in a trash bag. That's because I sent the valve cover off to be power-coated about two weeks ago. I still haven't gotten it back. I was planning on waiting until the head was all closed up to install the drivetrain, but, well, I was at installation's threshold and I got impatient.

As I was rolling the drivetrain up to the back of the car, I realized I had one more thing on the punch list—to verify that the new clutch functioned. I was a little alarmed to find that 2/3 of the motion of the release lever was play (the release bearing sliding on its shaft), and with the lever fully extended and moving the clutch fingers, the transaxle still seemed bound to the engine, but after a phone call to the vendor who sold me all the clutch parts and posting the question to two Lotus forums, I decided that the odds were that it's fine. I hope I'm right.

So then began The Great Lotus Butt Raising. I wanted to put the drivetrain in the car the way I got it out, which was by sliding it under the back of the car on a lift table and then raising it into place. I really DIDN'T want to use my engine hoist and drop it in from the top, as it's much easier to make small changes in height or position by shoving the table than by shoving the hoist with the drivetrain dangling from it. Plus, the idea of swinging all that weight around in a fiberglass engine compartment scared the shit out of me. The problem was that the back of the car was currently about 26" in the air, and the height of the top of the carburetors of the engine on the lift table was about 36", so I needed to get another ten inches (phrasing!). The maximum height of my high-lift floor jack is 29", so the jack alone wouldn't get it high enough. When I removed the drivetrain six years ago, I solved this problem by first pulling the head off, eliminating a foot of needed height, but I didn't have that luxury in reverse.

I'd been thinking about this for months, and decided to lag-bolt a short piece of 2x4 into the ceiling joists, sink a big hook into that, hang my PullzAll electric winch from it, attach it to the mounting rail at the back of the engine compartment, and use it to lift the back of the car. I figured this would be fine (it's not dead-lifting the whole car, just tipping it), but I didn't really know if it would work.

After moving the Lotus for the first time in nearly six years so that the center of the back of the car was under a ceiling joist (itself a harrowing operation involving levering the floor jack with a pipe while moving the jack stands and those pails holding up the rear hubs a couple of inches at a time), I got everything into position, and began hoisting the Lotus' butt into the air while progressively moving the floor jacks further back to compensate for the increasing tilt.

Everything worked perfectly, except that I was hyper-aware that the hook or the lag bolts might give way at any moment. Because of this, I positioned my floor jack as a backup the only place I could—beneath the car at the branch point of the "Y" frame with a block of wood under it. But even with that, because I didn't want to be under the car or in the engine compartment at any time, I had the jack hanging out from the side. I'd lift the winch an inch or two, then raise the jack to keep pace.

When I got to about 32", alarm bells began to go off in my head. As the back of the car got higher, the angle of the cable to the vertical was increasing and trigonometry was doing its thing and lengthening $r*\sin(\theta)$. The hoist cable began rubbing against the fiberglass in the back of the engine compartment. The angle between the jack's lift pad, the block of wood, and the frame of the car looked wonky. I stopped and thought, crap, I'm going to have to undo all this and spend hours getting the engine hoist out from under the back porch.

Then I noticed that entry to the engine compartment from under the wheel well had more clearance than going in from the back. I found that if I swung the unhooked hub assembly all the way out and forward, I had just enough room.

So, under and into the engine compartment the drivetrain went on the lift table. That part was astonishingly easy.

Next, I lowered the car around the drivetrain, skooching the winch and the jack down by a few inches at a time. Again, easy.

But as I got down to mounting height, it rapidly became more difficult. Although there's ample room to straddle the transaxle and stand up, the side-to-side clearance of the front of the motor, where the "Y" part of the frame narrows, is very tight. There's a notch in the right side of the front timing cover, between the timing chain tensioner adjuster and the big threaded plug I had so much trouble sealing up, where the frame has to go. At one point during all the shoving and lowering, I found the weight of the engine sitting on the plug. Yeesh! Soon after, I found it sitting on the tensioner adjuster. Double yeesh!

In addition, part of this involved the challenge of using the Magical Sinking Lift Table that I encountered when I took the engine off the stand, but had of course completely forgotten about. I kept having to pump it up every 30 seconds or so.

The drivetrain is held left, right, and rear. At the rear are a pair of simple rubber mounts that through-bolt the back of the transaxle to the rear wrap-around carrier. I got these lined up and secured in a minute.

The two engine mounts, however, were hell. Unlike on a vintage BMW where the mounts have two bolts, one to the frame and the other to a short bracket on the engine, this motor wasn't originally designed to be in the car, so the mounting scheme is more torturous, with each mount having four bolts, and long stout triangular reinforced brackets, each with six bolts. I got the right one on with great difficulty, but could not for the life of me get the left one on, even taking a Dremel tool and widening the holes.

Finally I realized, after looking at photos I took during disassembly, that where the bracket abuts and bolts to the engine mount, I had the bracket on the wrong side. But with everything settled in, I couldn't get the right one out. This was the point where, Thursday night, exhausted and frustrated, I called it an evening.

Yesterday I disassembled everything and did it again, and it went together with only the usual amount of pain commensurate with the design.

So, after 4 years and 10 months, the drivetrain is back in the Lotus, and the car is no longer at the extreme rake in which it's been sitting. After I tightened everything down, I noticed that the threaded plug is hitting the underside of the frame. I'm not panicking. I'm hoping the engine settles in on its rubber mounts.

Big picture, it's gone from:

--"A 24k Europa Twin Cam Special for six grand! This is so cool! I'll just rip out the seized motor and throw a used one in there" to

--"What do you mean there are no used Lotus Twin-Cam motors?" to

--"It costs WHAT to buy a rebuilt Lotus Twin-Cam motor?" to

--"It costs WHAT to pay someone to rebuild a Lotus Twin-Cam motor?" to

--"Damn it I'll just rebuild it myself" to

--"What do you mean it's been sleeved and there are no pistons this size" to

--"Cool! Custom pistons are about the same cost as new old stock" to

--"I'm changing from engineering to writing and my income is being cut by 2/3 stop spending money on the Lotus" to

--"I'm unemployed stop spending money on ANYTHING" to

--"I'm a self-employed writer, creditors aren't knocking at my door, but how the hell am I going to dig myself out from under this Lotus?" to

--"Just like any other project, one thing per night" to

--I am seriously going to go back in time and bitch-slap the engineer who designed this motor" to

--"It's in! Oh my god!"

I remember reading the "Grover and the Everything In The Whole Wide World Museum" book to my kids. Grover goes through the museum, and at the end, there's a door that opens up, and shows him everything in the whole wide world that's not in the museum. That's how I feel. The drivetrain's in, and I've now got every other automotive task in the whole wide world to do to this car.

But I'm out of the goddamn museum.

As Drama said in Entourage, "VICTORY!"

[search YouTube for "The Lotus Chronicles Lifting the Lotus' Butt"]

143 LIKES, 76 COMMENTS

Chris Leonard Excellent recollection of events, as usual!! +2
Doug Jacobs Congrats my friend!!! +1
Rob Koenig You know, it's going to look really weird to you when it's sitting on its wheels again. Congrats on the progress. The end is near! +1
Chris Lordan Rob, you're a saint because of the BREVITY of what's above. I'd be having the equivalent

of Mardi Gras if I'd accomplished this! Yay!!! +1

Russell Baden Musta Outstanding! You need shims for the axese Rob? You good? Looks like we'll have a new guest at LOG this year!!! And I can take photos of YOU "in the wild"!!! +1

    Rob Siegel Haven't looked into the axle-shimming issue yet, +1

    Russell Baden Musta Rob Siegel roll pins and shims are readily available... and I'll find my spares for you if they are out of stock. +1

    Rob Siegel Russell Baden Musta, probably my next order to Ray :^) +1

Daniel Senie Yay! Now you can put wheels on it and move it around the shop, and put it on the low-rise lift as needed.

    Rob Siegel I'm not there yet, but soon +1

Chris Mahalick Awesome job! You are making such great progress. +2

Wade Brickhouse Ah, now time for a warm beer and fish and chips 😊 +2

    Rob Siegel Last night, Maire Anne and I went out to an impromptu celebratory meal. She said "I'm thinking we should go somewhere British-themed." I said "What, so we can order food that's quirky, frustrating, and leaks oil?" +11

    Franklin Alfonso Look at the bright side, after this, everything else on other cars will seem easy...

    Rob Siegel let's hope so

    Brad Purvis Rob Siegel I would recommend Bangers & Mash, because you will do a lot of both putting this back together. +5

    Franklin Alfonso Rob Siegel What? You buying a GT6 next? Lol

    Franklin Alfonso Brad Purvis Or maybe some Marmite on dry toast because it's British, oily, and gross... +2

    Rob Siegel Brad Purvis, a fine choice!

    Vince Strazzabosco Poor choice of car, excellent choice of woman.

David Weisman Clap Clap Clap. Excellent! +1

    Rob Siegel I've just been slow-clapped on Facebook.

Neal Klinman Congrats. Feels good to be ALIVE. What a journey. Enjoy the outside (of the "museum") air! +1

Dave Klink Awesome work! Thanks for sharing it with us.

David Weisman How do you keep your hands from getting flayed? Do you wear gloves or soak your hands in Palmolive after? +1

    Rob Siegel My close friends call me Madge. +4

    David Weisman [gif of "spirit fingers" from Bring It On]

    Rob Siegel David Weisman, spirit fingers!

Jeff Lutes Reminds me of the rig I used to pull the m10/trans out of my '74 2002 in '85. It's still there in case I need it! [photo of pole clamped and lag-bolted to ceiling]

    Rob Siegel man after my own heart

Jeff Lutes And congratulations! You're nothing if not persistent! +1

Dave Borasky Wow—what an adventurous display of testicular fortitude. +2

Bob Sawtelle In honor of this milestone I give you Michael Fassbender as Lt. Archie Moore doing his best British accent.

Carl Witthoft ...That's what she said...

Rob Siegel (that's actually how it initially read, but my darling wife gently suggested that the adolescent humor had recently been getting out of hand) +2

    Carl Witthoft Rob Siegel "recently" ??? +2

Carl Witthoft Also, that collection of lifters and supports is uber-hack-mechanism

Frank Mummolo "...going where no man has gone before!" Well done, Sir! Cheerio, what! But you KNOW there's more, much more. Rob, if you don't write a book about this, I will hunt you down and write it for you! Put me down for #1 pre-published purchaser! +1

Chris Pahud Bravo, Rob! +1

Matthew Zisk Wow. Nicely done. +1

Daren Stone Shock and awe ~ I've been waiting for this post for as long as I've been following you. Well done, sir! You're now a lifetime member of the Lotus Sufferer's Club. Your hair shirt is on the way. +1

Maire Anne Diamond And I only had to go into the garage once to make sure he was still alive. +11

Ernie Peters Bravo!

+1

Kent Carlos Everett Living on the edge man. It's exciting. Thanks for the post! +1

Steven Bauer One thing at a time may take longer, but it works!!

Jake Metz Way to go! +1

Don McMahon You now have so much invested in this that you are going to sell off all your BMWs and become a Lotus mechanic . . .

    Rob Siegel Dude, clearly I already AM a Lotus mechanic!

Jim Strickland Awesome Rob! When you crank her up, it will be a great sound! +2

Lindsey Brown We need a Europa play date. Call me. +1

    Rob Siegel Lindsey, I have endless buttoning up to do. I still don't have the valve cover. Hopefully next week. Maybe next weekend I'll be ready to try and start it. Stay tuned. Thanks! +1

    Lindsey Brown I'll bring the Elephants, confetti, marching band, and strippers. +3

    Scott Linn Who needs strippers with that Europa body... +1

David Kemether [link to video of Black Lectroid from Buckaroo Bonzai saying "So what? Big deal!"] +2

Francis Dance You really are the Hack Mechanic!! Kudos on this great milestone. I look forward to your future posts! +1

Andrew Wilson F'ing A Bubba! +1

Chip Chandler Congratulations! After a full teardown of the new not cheap truck, I started metalwork this week, and am hopefully blasting the frame tomorrow. Feels awesome making the turn, don't it? +1

Randy Yentsch Wow, I'm through my St Bernardus Triple and both entertained and enthralled. I'll have you know that while you were busy mucking about with the Lotus, I cleaned spilled milk from the trunk of our Infiniti. And hand washed a car for the first time this season. Impressed?

    Rob Siegel Totally. I don't wash anything. So you got me. +1

John Jones Very nice work! An aside on "phrasing," which, apparently we're still doing. Last night I completed a mandated (for everyone, not just me) harassment training on my laptop while binge watching Archer on the tube. +1

    Rob Siegel LANA!!

    John Jones Rob Siegel You shut up +1

Andrew McGowan Awesome!!

Bob Shisler [gif of Jon Stewart saluting]

TE Cole In other news, Aging Man Crushed by Aging Lotus Falling from Home Depot buckets.

    Rob Siegel That's "AGING Home Depot buckets." (and, again, they're only supporting the hub assemblies; the car is on jack stands.)

    Peter Potthoff Rob Siegel but they're aging jack stands... +5

Greg Lewis Eureka! Oops too close to Europa, but awesome progress. You will drive this car in your own lifetime. 😊

Randy Yentsch That is a roomy engine compartment. My first car was a 1971 Pontiac Firebird with a 350 cubic inch engine; one could option a 455 cubic inch engine into that car. The engine compartment wasn't quite as roomy as yours.

The PullzAll's cable attached to the wrap-around transaxle carrier at the very back of the engine compartment.

The piece of 2x4 lag-bolted into the ceiling joists.

Steve Kirkup Leading contender for 2019 Darwin Award! +3

    Rob Siegel Don't you think I was aware of that every time I hit the "up" button on the winch? As I said, I was very careful, also supporting the car with a floor jack so the winch and the lag bolts weren't carrying all of the weight. And I wasn't dead-lifting the car, just raising the back (of a 1600-lb fiberglass-bodied car without the drivetrain in it).

    Steve Kirkup Rob Siegel... simply my attempt at humor. +1

    Rob Siegel With all my rationalization about it not being a dead lift, I thought "this thing is going to pull out, smack you in the skull, send you to the emergency room, and you are going to have absolutely no defense regarding its stupidity!"

    Steve Kirkup ...but, think of the small plastic trophy 🏆 you would have earned. +3

    Franklin Alfonso Maybe a safety helmet might be a good idea...

Jeff Lutes I'm with you, Rob Siegel [photo of bar clamped and lag-bolted to the ceiling]

    Steve Kirkup Jeff Lutes you've taken over first place!

    Jeff Lutes Steve Kirkup these are 6" lag bolts thru the 2x4 and into the ceiling joist. I was not afraid. 💪💪

    Rick Roudebush [link to video of Aerosmith "Livin' on the Edge"]

Thomas Jones You setting up to bring those kinky activities out of the bedroom? +1

Frank Mummolo Looks safe!

Barney Toler Well I'm glad it all worked out, but the whole thing was supported by the wood the

threads on that eyebolt are catching...
>   **Rob Siegel** I am quite aware of that!

Barney Toler Rob Siegel screw the bolt into the side of the 2x4 if needed again... +1

Gabriel Lay Good god man. At least use an eye and not a hook. +3
>   **Rob Siegel** Both points taken, thanks.
>
>   **Bill Snider** Yeah, the Eyes have it. +1

Ric McGinn Next time Robby give me a call they don't call me Ricky the rigger for nuthin'
>   **Randy Yentsch** Ric McGinn excellent point. I'm sure you could have gotten at least one safety observer (camera holder) that would work for beer.

Luther K Brefo Gnarly dude. I commend your massive pair of brass nuts. +1
>   **Rob Siegel** it was right on the hairy line edge between ballsy and foolish +3

Alan Hunter Johansson Simplify...and add frightness. +3

John Whetstone We should start a Go Fund Me page for one of these! Phone Tripod, UBeesize Portable and Adjustable Camera Stand Holder with Wireless Remote and Universal Clip, Compatible with iPhone, Android Phone, Sports Camera GoPro 2018 New Version

John Whetstone Better still... Action Mount Universal Head Mount for Your Smartphone, Operable with Any Smartphone. Strong Hold. Don't Waste Money on a Sport Cameras. Also Compatible with Rugged Cases, or Sport Cameras

Slipping the transaxle in through the extra height clearance in the right wheel well. Note how the wrap-around axle carrier with the chain hanging from it is several inches lower.

>   **Dave Borasky** I like this pic. Best pic of the penne to date. +6
>
>   **Rob Siegel** Note to myself: transfer E39-related spare parts from Penne box to Amazon box so people WILL SHUT THE FUCK UP ABOUT THE PENNE BOX +6
>
>   **Carl Witthoft** Rob Siegel You want them to be... penne-tent? +4
>
>   **Bill Snider** Penne for your real thoughts. Lol +2
>
>   **Steve Teager** That penne box really ties the room together, does it not... +1
>
>   **Brian Ach** I've Penned this post +2
>
>   **Dave Borasky** I mean, at least put the e39 parts in an appropriate container [photo of Maggi Spaetzle Authentic German Dumplings] +1

The drivetrain under the car and rotated into position.

With the drivetrain under the car on the lift table, initially I lowered the car around it, then made fine adjustments by lifting and rolling the table.

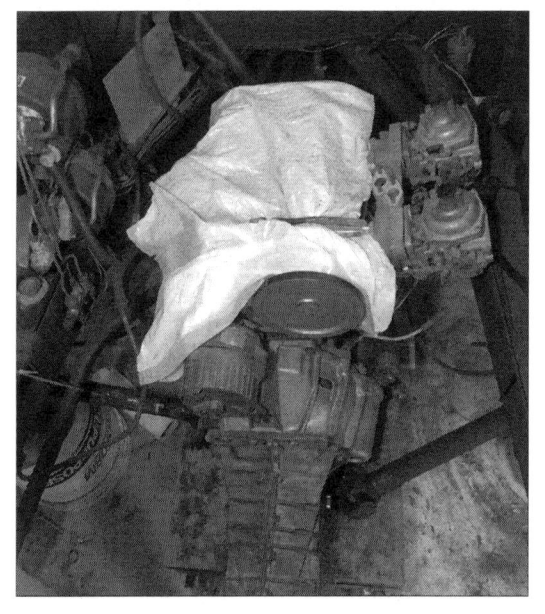

Paul Wegweiser So… you figured out your ideal "mounting height"… *snicker* +2
Rob Siegel Yes. Installation angle is critical 😊
Andrew Wilson This is monumental. +1
Dohn Roush Or perhaps just mental… +1

The rear mounts were trivial.

David Kiehna That transaxle is filthy. +1

David Kiehna After hours and hours with a putty knife, various flat head screw drivers, picks, wire brushes, tooth brushes, and gallons of degreaser and mineral spirits later... [photo of David's incredibly clean drivetrain]

Rob Siegel David Kiehna, I actually had gotten it pretty clean, at least by my standards, but then I spent a lot of time walking and sitting on it 😔

Franklin Alfonso Meh, Who cares if it's a little dirty it's not like you are gonna put it for sale and show on Barrett/Jackson.

Russell Baden Musta David Kiehna it is sparkling clean compared to mine!!!! 😂😂😂

Christopher P. Koch Franklin Alfonso are you kidding? As soon as he slaps some plates on it Maire Anne wants it sold!

Rob Siegel Christopher P. Koch, actually, just last night, she asked me if I would let her drive it. I found that a rather surprising question, as she has always been able to drive any of my cars. She voted the red 3.0CSi "Best car to go to a high school reunion in." +8

Frank Mummolo Rob, Maine Anne wasn't asking permission. She wanted to know if she might kill herself in it! +1

Alan Hunter Johansson I think it is well established that Rob married well.

This is the right front side of the engine. The frame needs to go BETWEEN the troublesome plug I spent all that time on (bottom) and the timing chain tension adjuster (top).

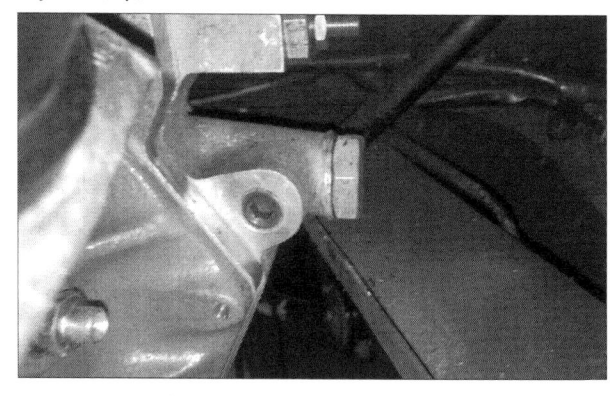

Luther K Brefo What the shit?! +1
Andrew Wilson (sigh) Damned cabinet makers! +2
Tom McCarthy Sawzall, and then duct tape it back together.
Frank Mummolo This is beginning to read like "The bear went over the mountain." Jesus!

The engine positioned with the upper lip of the frame between the threaded plug and the timing chain tensioner. The threaded plug is actually hitting the underside of the frame. I'm hoping the whole thing settles down a millimeter or two on the engine mounts.

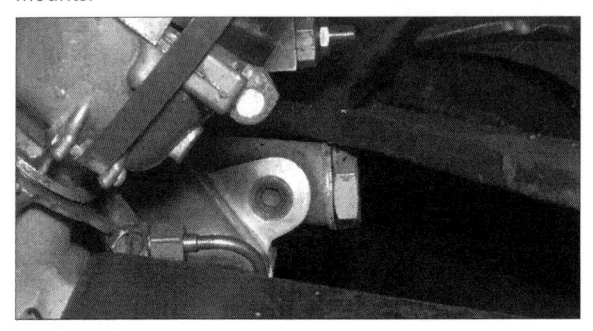

(18 LIKES, 35 COMMENTS)

Dave Gerwig Is there a shim for that? You'll have a Buzz Bomb even with some settling 😀

> **Rob Siegel** Some of this problem is of my own making because I used a threaded plug because it had a sealing washer where is the original configuration of this after-market cover had an Allen plug that didn't stick out as far.
>
> **Dave Gerwig** Rob Siegel the old Inny vs Outty dilemma, should have recalled the sealing challenges. Darn that's a tight fit.
>
> **Franklin Alfonso** A few Boston potholes will sort that out one way or the other... 😀 +3

Christopher Kohler I don't see how the Allen plug would have cleared either.

Rob Siegel Christopher Kohler, it's recessed, so it would've been better but not perfect.

Christopher Kohler Yeah. Maybe it's the angle, but it looks to me like the housing itself is in the way. Could be an optical illusion.

Rob Siegel Christopher Kohler, no, you are correct +1

Rob Koenig Is there enough slack in all of the mounts to clock the engine slightly to that side to gain clearance?

> **Rob Siegel** Not right now. I'd need to remove them and notch out the holes. +1

George Zinycz Not to be negative, but I don't think a couple of mm's will buy you the space you need. When the weight of the car is on the rear axle, will the engine sit lower?

Dave Gerwig George Zinycz in the Lotus world: you never know +1

Rob Siegel We shall see.

Dave Gerwig Rob Siegel I feeling this one: as they said in Monty Python to the dolt, ummmm bolt, off with your head! 😀 If that bolt is just a drain passage why doesn't Bean tap it to go downside? Jeez no help with your situation. Now for the Collin Chapman solution, build in some more lightness: cut the neck down on the Bean casting, mil the bolt head down to 1/2 thickness, throw away the washer and replace gasket with paper or skim coat with sealant. You'll save a gram or two.

Rob Siegel On an Elan, that passage doesn't have the plug; instead it has the threaded coolant neck. The neck on a Europa is relocated to the front of the cover because of this clearance issue. It's that way stock from the factory. On the Bean covers, they just make it so one cover can be used for both. When I put in the cap-head plug, I was aware that there might be an issue because the plug stood out proud of the cover rather than being recessed into it like the original Allen key plug, but the original one didn't seal, it still didn't seal after I put sealant on the threads. Using the BSPP cap-headed plug with an integral sealing washer was part of what finally got it to seal. I'm hoping I don't need to do anything, but if, over time, I do, I'm hopeful that I can just grind a little off the plug where contact is being made. +4

George Zinycz Rob Siegel the frame will grind it for you lol. But it sounds like you're right. +1

Thomas Jones Looks to me like another design flaw in that timing case. +1

> **Rob Siegel** It's a motor that was never originally designed to be in this car. It's just one of dozens of quirks. +1

Frank Mummolo Hope... an essential character trait that develops into a fine art among British car owners. +1

Sean F. Gaines No need to worry. It will just be your way of telling when you've gone over a big pothole. That special clunk that reverberates through the entire chassis. +1

Der Alte Classics Well done Rob. Impressed with your persistence. I think the profits of future endeavors should go to a 2 post lift. Trust me even after installing it you'll be kicking yourself the next 5 years that you didn't do it sooner, I did 😊. Btw just reorganized, it usually looks like your garage 😊 [photo of garage much cleaner than mine]

> **Rob Siegel** Thanks. Appreciate the compliment. But I don't have the room (the ceiling height) for a post lift. The midrise is an acceptable compromise for the real estate that I have in my garage.
>
> **Rob Siegel** And besides, what makes you think that any of my future endeavors have profits? 😊 +2
>
> **Der Alte Classics** Fair point. On both accounts. Just hope it will. Good luck getting the bast#rd back on the road. I'll be following it with interest and a smile +1
>
> **John Wanner** Der Alte Classics tooooouring. +1
>
> **Paul Hill** Der Alte Classics Looking very tidy in that man cave. +1
>
> **Daniel Senie** Rob Siegel You can use all the money earned as a folk singer. 😊😊😊😊 +1

William W Lewis III Is it too soon to talk about a ls3 swap yet? 😊

Luther K Brefo Two questions, Rob, One, how high is your ceiling? Two, what is the purpose of the bung? I know it's too late now but I had a thought…

> **Rob Siegel** My garage ceiling is a little less than nine feet. By the "bung," you mean the threaded plug in the center of this photo? It's because the front timing cover is designed to be used for both an Elan and a Europa. In a Europa, the water inlet port is on the front and a threaded plug goes in the side. In an Elan, they're reversed.
>
> **Luther K Brefo** May I trouble you to check that height once more with lasers? It felt taller than that when I was there. :) As for the bunghole, now I remember. I wonder if the manufacturer will listen to your feedback and make two separate castings, I can't imagine you're the only one that has seen this interference?
>
> **Robert Myles** Heheheh… he said "bung hole." [photo of cornholio]
>
> **Brandon Fetch** Came for the B&B reference, leaving satisfied. +1

Robert Myles Brandon, we're here to please.

Scott Adair You need a bigger hammer Rob. 😊

After nearly six years, the Lotus has the drivetrain back in it and is no longer at the extreme angle at which it's been sitting.

Lee Highsmith Rob, can you give me the part number on those buckets? +3

Nikki Weed Lee Highsmith I was thinking the same thing

TE Cole It comes as a kit with the milk crates. You have heard of Griots Garage… well he didn't get them there. You have to go to Guido's Garage to get this kit. +2

TLC Day 2128: Lotta buttoning up

APR 7, 2019, 8:32 AM

The Lotus Chronicles, Day (well, morning) 2128:

Got a lot of buttoning up of the engine compartment done yesterday.

--Connected the shift linkage, which runs all the way to the back of the car (mid-engine, transaxle behind engine, shift rod at the tail). When it was done, I climbed inside the car for the first time in six years, grabbed that amazing round wood shift knob with the weird 5-speed pattern on it, and rowed through the gears. Heaven.

--But when I put the clutch pedal in, I was reminded that all three pedals in the bucket are seized. I found the replacement pedal bucket in a box on the shelf, the ONLY non-engine part I bought before I shut off spending 5 1/2 years ago.

--Hung the rear coil-over shocks and springs. Before bolting them up, I still need to deal with shimming the rear axles. I wrap that and replacing the bushings in the lower control arms with the larger task of buttoning up the rear suspension, which will allow me to lower the back of the car.

--Got most of the gas lines hooked up. With two tanks, two carbs, two tee fittings, and the usual pump and filter connections, I though ten new band clamps would be enough, but apparently I don't know how to count. More are on the way.

--To do the pedal bucket and complete the cooling system flush and hose replacement, I needed to put the nose in the air and get under the car, so I did that very carefully this morning. Remember, this is a fiberglass-bodied car without traditional subframes; you need to be very careful where you jack.

--With the nose in the air, I inspected the front end for the first time. Tie rod and ball joint boots, to no one's surprise, are rotted and retreated, and the left end of the front sway bar is snapped off. The idea that I'm going to be whipping around the neighborhood anytime soon is fantasy, says the guy who was looking at ordering tires last night :^(

--The hoses are hidden by an access panel. All the bolts were nasty and rusty, but I got 'em off without cutting Thanks, SiliKroil (also, remember, fiberglass-bodied car means not whipping out the torch as the first sign of stuckness).

But lots of progress.

Coffee break :^)

109 LIKES, 23 COMMENTS

Carl Witthoft Reminds me of Slocum's "Sailing Alone Around the World" where he asks: How much of the boat can I repair / replace and still claim it's the original? +5

John Whetstone You don't need heat to use the SiliKroil effectively? Or do you use just a touch of the torch?

Jim Gerock In for a penny, out for a pounding. +6

Tom Melton Did you go vroom vroom vroom when sitting in the driver's seat? +6

> **Rob Siegel** I'm saving that for when I have a working pedal bucket :^) +3
>
> **Pete Lazier** I would have done that for sure... 👍 +1
>
> **Jeremy Novak** I confess that I used to do that as a child while sitting on a milk crate in the tub of my dad's restoration project 54 MG TD. +2
>
> **Robert Myles** I confess that I did that earlier this morning sitting in the Miata race car... +2

Russell Baden Musta Rob, that access panel is the ground plane for the lights and cooling fan... I've fabricated my own from stainless... I'm pretty sure the bolt pattern will be unique to your car into the frame... I can't remember now what I used but pretty sure I either put helicoils into the existing nuts in the bottom of the frame, or I drilled them and put in rivnuts... if you need one, I'll see what I have for material and bend you one up.. you can drill the holes on application to match... sway bar down rod should be easy to find. +1

>**Rob Siegel** Thanks Russell. I saw that connection in the hood compartment and wondered how screwing to fiberglass could do anything until I realized it went through to that metal under plate.

>**Russell Baden Musta** Rob Siegel I can tell you when you hit a groundhog or other similar sized creature it will most likely bend the sway bar and likely bend that plate! I've straightened my sway bar 2 or 3 times... if you have one of those inexpensive arbor presses bolted to your workbench, that will do the trick. I promise my next layover in Bedford, I'll come over and have a look. [photo of arbor press]

>**Russell Baden Musta** Rob Siegel, you want this? It's a spare. I've straightened this one after I made a new one .. you may have to adjust the holes for frame attach... but the fiberglass part just drill new in body... I use bolts but big flat washers for the electric grounds… [photo of spare access panel]

Franklin Alfonso Look at the bright side, it's not gonna take another 5 years before you get it on the road... hopefully 😀 You are making great progress. +3

Chris Mahalick Wow! Yours really is quite the project! I also had a snapped drop link, it was pretty easy to replace. +1

Frank Mummolo Them's some nasty lookin' bushings and seals!

Kent Carlos Everett What a massive job! Well-earned progress. I bet it feels good. :-)

Andrew Wilson Crank this on the Europa radio! [link to Budgie Breadvan] +1

Scott Linn I had to laugh regarding the hose clamps in the fuel system. I had to run out and buy a bunch myself when I last worked on it. I was concerned about the large amounts of rubber hose running around so I ended up bending some line and using metal for the fuel lines with short sections of
rubber hose. +1

Scott Linn The PO messed up my sway bar bushings (he couldn't get them on so he split them). I drove 900 miles without one and then replaced it with a piece of garden hose for the rest of the trip home.

>**Rob Siegel** One of my 2002s is running on sway bar bushings made from an old cooling hose. I bought real replacement bushings but I've had no reason to put them in; the hose rubber is working just fine. +3

Is that a thing of beauty or what?

27 LIKES, 23 COMMENTS

Scott Aaron Wonder how many people that have worked on these things were compelled to OD after a while? +3

Sean Curry This is really cool. Are you going to refinish it?

>**Rob Siegel** I don't refinish anything :^) +5

>>**Ken Sowul** Before you refinish you first must finish. +7

>>>**Rick Roudebush** Ken, you took my answer. ;-)

Mike Miller Will detail for free.

> **Luther K Brefo** Looks like a knob that needs polishing... 👀 +4
>> **Frank Mummolo** Clean it. Spray some Pledge on it. Wipe it off and move on! +2
>> **Rob Siegel** I usually just use spit and my t-shirt :^) +2
>> **Frank Mummolo** Rob Siegel, me too! So this would be an upgrade! 😀
>> **Russell Baden Musta** Rob Siegel that's right!
>> **David Shatzer** Frank Mummolo there are some fine knob polishes out there... I was reprimanded from another car page for posting photos of a can of the polish... and questioning if anyone would polish a knob... +1
>> **Rob Siegel** "Knob polishes." Heh. +1
>> **Bill Ecker** aaaaaand we're there
>
> **Kevin McLaughlin** #patina
>
> **Chris Langsten** "O/D" = "Oh Damn, you really think you'll get it running long enough to need this gear?"???????? Sick joke by engineers? +1
>> **Rob Siegel** you're a cruel man
>
> **William Thibodeau** With the current generation, it is an anti-theft device
>> **Rob Siegel** true, and this one more than most
>
> **Jean Laffite** [photo of incorrect 6-speed tattoo] +3
>
> **David Shatzer** O/D... could that be overdrawn...? In patience? Bank Account? Great job in perseverance over the Lotus blossoms... +2
>
> **Brad Purvis** [link to Steve Martin and Carl Spackler "What the hell IS that?"]
>
> **Paul Hill** A string back leather glove and a few gear changes will sort that out +1

The Europa's dash doesn't totally light my fire like, say, an E-Type, but it's growing on me.

28 LIKES, 18 COMMENTS

> **Scott Aaron** I see a pants-related gofundme in someone's future +3
>> **Rob Siegel** #PantsWithPatina +6
>> **Garrett Briggs** My generation calls that Skid Marks. But if they are old enough I guess Patina would work.
>> **Richard Stanford** Garrett Briggs Skid marks traditionally go under the pants. Or so I learned back in the day. +1
>> **Garrett Briggs** Yeah, as a kid that's how stopped with knees as a kid. Skids marks under the pants was usually signs of an accident... of either variety.
>
> **John Whetstone** Parts money has to come from somewhere, I always say! +2
>
> **Franklin Alfonso** E-type is also slightly better-looking body-wise IMO than a mini British Racing Brown hearse lol. +1
>
> **Rick Roudebush** It looks like something is growing on the steering wheel too. Try a 50/50 blend of white vinegar and some foaming hand soap. Stir the mixture vigorously with a soft dish brush. Dip the brush in the suds and brush the suds on the mold, wipe dry and treat with Leatherique.
>> **Dave Gerwig** Rick Roudebush You've got the words twisted: the owners growing to love this miniature brown sports car! Glad you fit Rob Siegel, even with a human size Shoe Horn, most folks cannot squeeze into a Europa +4

Russell Baden Musta Dave Gerwig me at 140 lbs when I bought it.

+3

Brian Ach did you just compare a Europa to an E-type? +2
    Russell Baden Musta Brian Ach taillights are right from an e type! +2
Andrew Wilson Black shoe polish will make the steering wheel look a bit better. +1
    Scott Aaron Andrew Wilson Lysol is what I was thinking 😀 +4
Russell Baden Musta It works! Sort of...
Paul Hill I hope you will be wearing suitable attire whilst piloting your Lotus Rob... +1
    TE Cole [photo of Level A hazmat gear]

Bought 5 1/2 years ago, and I found them in the garage!

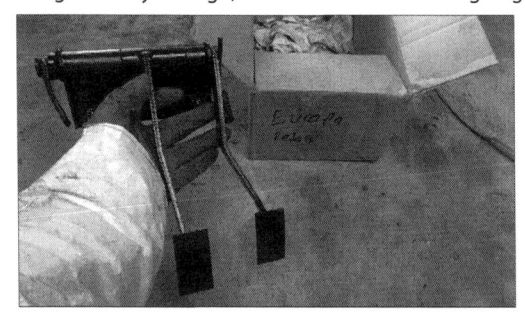

    Dave Borasky Looks like it's being held by a team member of an Apollo mission +4

A bag of 10 hose clamps goes quickly when one connection uses up five.

Broken sway bar link

(10 LIKES, 17 COMMENTS)

Scott Aaron Golly. For a lot of reasons.
John Whetstone Welding machine getting rolled out?
    Rob Siegel It may!
    John Whetstone Rob Siegel Okay then! Cue up Paul Wegweiser and I'll turn on the corn popper! +1
    George Zinycz John Whetstone that might be a little too thick for a Millermatic +1
    John Whetstone George Zinycz That's how much I know!
    Rob Siegel John Whetstone, that's what Mrs Millermatic said... +8
    John Whetstone Rob Siegel 😂
    Luther K Brefo Grind it to a point and build it up +2
    Rob Siegel that's what never mind +4
    Paul Wegweiser How shitty must that metal be, if a 600-pound car can snap it in half. This is reason #2903 why I will never spend my hard-earned money on one of these things. +6
Jayme Birken That looks loud. +1
Frank Mummolo I am getting weary. And I'm just watching!
Sean F. Gaines Oh, that does not make for a happy day. Good luck.
Mark Manasas JB Weld +1
Paul Garnier OMG Rob! What the hell did you buy?! I mean, I've made some impassioned purchases before but, damn! I think I've seen like 4 major busted things in the last 72hrs that would have been deal breakers or major discounts. That said I fear the Elise I'm waiting for depreciation to deliver into my reach will be in a similar condition. +1
Carl Witthoft Or just leave it broken and make sure never to turn? +1
Dohn Roush Even this did not sway him...

Rotted upper wishbone ball joint rubber. They're all like that.

22 LIKES, 4 COMMENTS

Gordon Arnold Now, there's a surprising find…
Anthony Neu Wow
John Harvey Sure it wasn't a leather boot? +2
George Murnaghan The rubber has perished. But can be replaced in situ.

The under-nose access panel that hides the steering rack and master cylinder.

(11 LIKES, 16 COMMENTS)

Jeff F Hollen Looks like it's not built for "easy access." Sundresses are on sale at Target. +2
Brad Purvis I hope you've checked the Kniplinger pins for proper torque.
> **Rob Siegel** No, but I am expecting to have to rebuild the trunnions, and those are real things.
> **Karel Jennings** So is that like bunions but for cars? +2

Jeff Baker Man, is there a single part on this car that *doesn't* need to be repaired, restored, or replaced?? You should get a Restoration Grammy for resurrecting this pile! +1
> **Rob Siegel** It's been sitting since 1979 (fortunately indoors), so, like any long-dormant car, I went into it with the expectation that it would need many things. The stuff I'm doing now is just basic sort-out stuff. It doesn't bother me at all. How hard the engine was to rebuild, however, was a total surprise to me. +2
> **Jeff Baker** Rob Siegel I understand that, however, with all that you're doing, I'm not sure if the "Hack" part of your aka still exists. 😊 +2
>> **Wade Brickhouse** Jeff Baker there is plenty of hackishness present. :-) +2
>> **Russell Baden Musta** Jeff Baker no longer a hack! +1

Wade Brickhouse Smuggler's Box? +1
Steven Bauer Perhaps there's a government official you can pay off to get better access? +1
Ezra Haines That car looks like it was designed and built by a bunch of guys trapped on an island. +3
> **TE Cole** You misspelled retards and salvage yard!

Rob Siegel Ezra, yeah, the fiberglass body is actually built from an outhouse. Just like Tom Hanks' sail in Castaway. Where do you THINK he got the idea?
Ezra Haines And room in the cabin for you and Wilson
> **Rob Siegel** well, maybe Wilson

TLC Day 2128: Flushing the coolant pipes

APR 7, 2019, 7:04 PM

The Lotus Chronicles, Day 2128 (evening edition):

My goal for today was to flush the long metal cooling pipes that run the length of the car and install the four major cooling hoses—the two that connect the engine to the long metal pipes, and the two that connect the other ends to the radiator. Easy, right?

Oh my god what a colossal pain in the tuchus.

As I said, this morning I removed the plate that allows access to the ends of the two radiator hoses... if you call that "access." Both hose clamps were seized. The lower one was easy to cut off, but the other was up high above the steering components, right next to a brake line, and the Dremel tool came uncomfortably close to the line. But, done. With the hoses out, I flushed the long metal pipes with a hose and fresh water, and the rustiest-looking snot imaginable came out. Very satisfying. (I have a piece that'll probably run on Hagerty tomorrow about why you don't start long-dead cars without doing things like this.)

Then I installed the two new radiator hoses, but not before having my Big Doofus of the Day moment. I searched the garage and basement for an hour looking for the Fedex Priority Mail box that the hoses came in and could not find it. I finally found that I'd opened the box, put the hoses in the big box with all the other new parts, and didn't see them because they were black and at the bottom. D'oh!

Access-wise, the two hoses in the engine compartment were similar. The one that connects to the thermostat is up high and easy to get to, but the other connects to that troublesome water neck on the timing cover. The hose is a short elbow that goes somewhere you can only reach from under the car, and even then it's a bear getting it in place. In retrospect I should've installed one end before I put the motor in, but I didn't fully understand how bad the access to the front of the engine is.

And that brings us to this uncomfortable topic. With the engine in the car, I can now see—and it's astonishing that I'm saying this—that I misjudged this whole "I'd be an idiot not to install the removable water pump kit in order to be able to change the water pump in the car." There's so little access that, after all that work, I'm not sure it's possible to pull the pump cartridge. It'd be a bear even to change the belt. When I bought the car, I didn't really look at any of this before I yanked the drivetrain. I recently read a post on one of the Lotus forums expressing this concern, and I waved it off. I thought that, if there ever was an access problem, I'd just go full Hack Mechanic on its ass and simply cut a hole through the fiberglass firewall. However, the problem is that the pump isn't behind the fiberglass firewall. It's lower than that.

Having said that, a) I'm not sure I'm right, b) even if I am, the removable pump cartridge STILL lets you change the pump without having to pull the head, drop the oil pan, and remove the front timing cover, so even if you need to lower the drivetrain and skooch it back, it's still a major win. Note that a nicely cared-for Europa TCS that needed a water pump was sold on Christmas Eve on BaT for a good sum (https://bringatrailer.com/listing/1972-lotus-europa/), so it's not an academic issue.

I then reinstalled the original radiator. A few weeks ago I noted that the radiator, incredibly, has a rubber grommet in the side, into which is pushed an "otter switch." Many people commented that I should take it to a radiator shop to have a threaded bung welded in. Instead, I found posts on Lotus forums that said that a Volvo grommet is a little longer, with more rubber to seal against. I bought one, and sure enough, the otter switch was an extremely tight fit against it. I expect that, when and if the car reaches the point where the most important thing is cool running in hot summer temperatures, I may go for an aluminum radiator with a threaded temperature port, but for right now, this is fine.

Lastly, Mike Miller would be proud of me. The Lotus uses these quirky wire hose clamps that look like they're made out of recycled coat hangers. Hose clamp originality is the sort of thing I never cared about (and up underneath the car where it's hard to reach, I employed the nut-driver clamps I use on BMWs), but where they're in plain sight in the engine compartment, I want to banish any nut driver or hardware-store-looking clamps and be faithful to the original look and feel and use the wire clamps. I thought that I had enough, but the old ones look like shit next to the new ones. Twelve bucks will set it right. I'll place the order on Monday.

But, a few hose clamps notwithstanding, I appear to have a fairly clean, plumbed cooling system. Fuel is right behind, also just waiting on hose clamps.

I'm exhausted.

64 LIKES, 68 COMMENTS

> **David Kiehna** This looks like a fun swap... [photo of insane intercooled turbo engine in Europa] +3
>> **Rob Siegel** I am just a lowly hack mechanic.
>> **Scott Linn** Yeah, but you lose the rear trunk. I use both the rear and front trunks all the time.
>> **Rob Siegel** the rear trunk does offer a surprising amount of space
>> **Scott Linn** And, one is heated and the other cooled... +3
>> **Samantha Lewis** Actually the front trunk is also part of the cabin air system, if you put anything in the frunk, you block the flow of air to the interior vents, luggage, or fresh air, it's a choice, either or. +1
> **John Whetstone** Will you also wire the sensor in place as you mentioned in an earlier post?
>> **Rob Siegel** probably
>> **John Whetstone** (We'd all feel better...)
> **Jim Gerock** Your upcoming book will be as thick as "War and Peace." +3
>> **Rob Siegel** but with more profanity +2
>> **Jim Gerock** OK then, twice as thick...
> **Scott Linn** Hey Rob, guess which hose has to come off during a cooling system flush/fluid change... +1
>> **Rob Siegel** STFU +2
> **David Holzman** If I ever decide to get a project car, it will not be a Lotus. All this leads me to think that women are more practical than men. I know two women who have restored cars. Both restored (old) Beetles. +2
>> **Scott Linn** There is definitely a level of difficulty involved plus parts availability. I've been driving and working on my 68 MG Midget for 43 years now. It's way more straightforward to work on than the Europa. Same thing for the 52 MG TD, although if you find one that needs the wood body replaced, that is a ton of difficult fitting work. +1
>> **Rob Siegel** there is nothing about what I'm doing that makes any sense whatsoever +13
>> **Greg Aplin** Rob, sure it does. It's a part of who you are, although this is quite a project. +1
>> **David Holzman** Rob Siegel Well actually it makes sense in that you're preserving an important piece of automobilia. We're all grateful. +1
>> **David Holzman** Scott Linn Tom Magliozzi (of Click and Clack) had one of those. I don't know what happened to it, but I do remember helping him get the thing started once. +1
>> **Scott Linn** David Holzman I would never have bought a TD on my own, not my kind of car. But I inherited it and did some work to fix it up and now it's actually kind of fun to drive and you don't see many on the road. And lots of people stop and say "I used to have one of those." So the TD is currently staying in the fleet. Been 14 years now. The Europa (a 73 TCS) is definitely a challenge and requires more work than either other British car. But man is it fun to drive. I've done some of the same jobs Rob has, but haven't needed to pull the engine/transmission yet. He has done a lot more in that area and I'm much appreciating the documentation and learnings.
>> **Rob Siegel** As I like to say, if I can't be an inspiration, at least I can be a horrible warning. +12

David Holzman Scott Linn I'd never tell anyone to get rid of a car they like! Drive fun! +1

Russell Baden Musta Rob Siegel simply owning it is the deal... it will be an ongoing maintenance, but once sorted, tire size and alignment, you love the way it drives.

Ernie Peters When I restored an Elan a few years back I became the Elan. I studied how and why it was designed and put together the way it was. I became an electron and walked through the harness to understand how things were supposed to work. My fingertips grew eyes so that I could "see" areas I could only feel. Parts were readily available as was information. To a lesser degree my 74 Midget was the same way. Having a feeling of intimacy allowed me to bypass things like a crude seat belt ignition interlock rather than curse it. Similarly, I learned everything I could about my E36/7, E36 M3, and our 535xi wagon. No car is perfect. Some break more than others. Some are easier to fix than others. Become the car, it makes living with it a whole lot easier. [link to Caddyshack "Be the ball"] +4

Lee Highsmith Rob Siegel I see it more as a "Rob does it so I don't have to even think about it..." kind of thing... +2

Chris Langsten Nice progress Rob! Keep going, BMW season is here, lest your focus be stolen away from the Lotush. 😉 +1

Rob Siegel yes, and I am torn +2

Chris Langsten Rob Siegel no battle, or war, was ever won without sacrifice. You are close. Nobody will judge you otherwise. #firstneighborhooddriveisnear +3

Paul Wegweiser Huh. I removed, rebuilt, reinstalled, and tuned a 2002 engine in less than 5 days. I'm sure that Lotus will be 425.6 times cooler though. +6

Rob Siegel bite me. +4

Craig Fitzgerald Now that there's nothing worth watching on Netflix, this is my favorite limited run series. And I do mean "limited run." +1

Craig Fitzgerald I was just in the garage, sitting in the Corvair's vacant engine bay, pissed off that I'd snapped a bolt holding the u-joint retainer in the yoke, and I thought "At least I'm not working on that fucking Lotus." +12

Rob Siegel Yes. I'm your Portrait of Dorian Gray. With a Lotus. Or something. +3

Craig Fitzgerald Rob Siegel Seriously, I've told you this before, but this whole thing has been an inspiration to get out there and put that car together. You need your own show. I love watching the better episodes of Roadkill for the same reason I've enjoyed reading this. +4

Scott Linn The owner of my Europa prior to the person I bought it from was a retired engineer. He would apparently wake up excited to be retired then think, "damn, I need to work on that Europa again." He finally sold it to keep his sanity. +3

Paul Wegweiser Rob Siegel, you do understand that your very public struggles with Lotus repair and parts is completely fucking the current market value of every other Europa in existence, right? I predict that in 6 months, you'll be able to buy a Europa for about $4. +13

Rob Siegel Yes. Right. That's why I'm doing it. I'm shorting Europa futures. +5

Paul Wegweiser Rob Siegel "Lotus: The $900 750iL of the British car world." +4

Robert Boynton Rob Siegel Imagine the comments you'll receive when you buy Lotus #2 ...and for only $1 +1

John Turner Are you basically saying that 12 is your limit on Schnitzengruben? +2

Rob Siegel it's more like a guideline

John Turner let's face it, everything below the waist is kaput +1

Robert Alan Scalla I have been thinking about a Jensen Healey lately (with a Lotus engine), but maybe now I'll pass.

Bob Sawtelle I was always infatuated with Lotuses and would stop and longingly look at a vintage Elan not three miles from my house every weekend (Emma Peel was my first crush, Tara King couldn't hold a candle to her). And once inquired on an Esprit that was for sale at a "bargain price." Thank god I resisted these urges as your trials and tribulations have scared me away from these lightweight money suckers for good. Of course I still can't wait till this is finished and will probably hound you for a ride in it anyway. Sorry for long rant. +2

David Shatzer Bob Sawtelle Remember how the Elan frames were bait for rust-mites? So delicately designed for light weight and efficiency... +2

Bob Shisler Good lord don't need the toilet shots, Rob. We trust you flushed it, no need to show us. =p

Franklin Alfonso Next car... Jag-e-warr E type! Most beautiful car evah! +1

Chris Mahalick The hose replacement was the easiest thing I have done on mine, LOL [photo of hose near steering rack looking much cleaner than mine]

Steve Park Enjoying your progress. Maybe I'll get the motivation to do something with my TVR in the next 5 years. +1

    Rob Siegel TVR? You'd better!

    Scott Aaron you could probably sell it to Rob in 5 years

    Steve Park I see another book! +1

Ed MacVaugh What is the attraction to changing out a highly thermally efficient brass and copper radiator for aluminum?

    Rob Siegel New, moderate cost, light weight +1

    Ed MacVaugh Rob Siegel I remember when 65 grams weight savings cost $65, but that was a bicycle. Most of us would be better off dropping the kilo off *our* weight and saving the hips, ankles and fuel economy :)

Looking upward from underneath, the cartridge-style water pump is behind the pulley at the top. The firewall of the car is above that. The challenging-to-install elbow hose is at the bottom of the photo. Yes, a drop of antifreeze left over from the pressure test.

I'm so glad this isn't coursing through the cooling system's veins.

25 LIKES, 23 COMMENTS

    Andrew Wilson Looks like a cesspool.

        Rob Siegel lotusbarf +2

    Matthew Zisk Shouldn't that be spelled, "CURSING"?? +1

    Craig Fitzgerald Nutella

        Rob Siegel ew.

    Albert Smith Progresso Lentil Soup?

    Scott W. Brown Where do you get rid of this? I'm looking at a bucket of the same from an E28. Never

found anywhere that accepts coolant.
> **Rob Siegel** Here in my hometown of Newton, the recycling depot accepts oil and antifreeze six days of the week. +1
>
> **Scott W. Brown** Rob Siegel Thanks. Guess I'll press the Town highway dept seeing as we, the people, just built them a new garage :-) +1
>
> **George Zinycz** Rob Siegel mixed together though??

Kevin McLaughlin Looks like balsamic vinegar
> **Rob Siegel** for a really rancid salad +2

Wil Birch Ick.

Jeremy Novak Just imagine what sludge like that might do to, say, a nice new water pump. Yuck. +2
> **Rob Siegel** exactly why I endured what I did to flush it +2

Robert Alan Scalla Your pee pot?

Binki Witherby De Collibus Best looking BBQ sauce. Bottle it and sell it at a farmer's market. +2

Tom Egan Looks like your camel has diabetes! +2

Franklin Alfonso Poutine sauce for the fries...don't forget the cheese curds! +1

Bo Gray Lotus Puttanesca sauce... +1

Stan Chamallas you made chocolate pudding!! +1

Jay Walters Thermacure through "stored" E9 took 20 gallons of distilled water with intermittent runs to clear up. You've got a long way to go, right? +1

Wade Brickhouse I thought it was your coffee cup!

TLC Day 2129: Pedal assembly replaced

APR 8, 2019, 7:24 PM

The Lotus Chronicles, Day 2129:

Replaced the seized pedal assembly with the good one I bought 5 1/2 years ago. I now finally have a working clutch and brake pedal. Clutch is cable not hydraulic, works fine. I'll change the front rubber brake lines, bleed, and see what I've got.

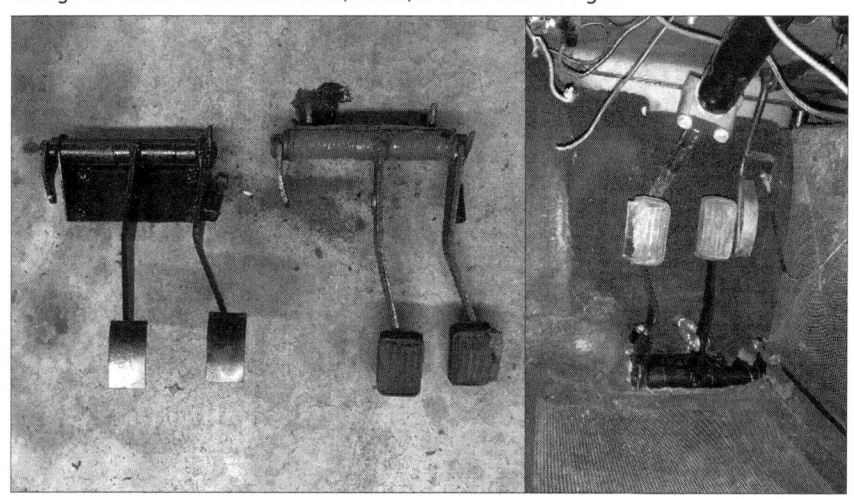

64 LIKES, 20 COMMENTS

Christopher P. Koch I wear at least 5 mil gloves so I don't bleed as much when I do these jobs. +2

Frank Mummolo Into the valley of death rode the six hundred... +2

George Zinycz Is only providing two pedals part of that whole "simplicity and lightness" deal? +1

Rob Siegel Probably! (The accelerator pedal isn't part of the assembly; it's separate.) +1

George Zinycz Rob Siegel Curiouser and curiouser…

Rob Koenig The ultra-rare combo clutch/throttle pedal. Throttle and clutch action are proportionally inverse. +1

Bob Sawtelle A nice positive day!!! Ever forward. +2

Pete Lazier Joking aside, I admire your determination… this is no work for the faint of heart… modern repair seems to be plug and play and pay… +2

Rob Siegel Thanks. It was either git 'er done or bail out. I chose the former. +4

Dave Gerwig Pedal to the Metal is a figure of speech in your glassic. Hit the glass to pass. 😎😎😎 +1

Lindsey Brown Did the pedal cluster come with the requisite pair of ballet slippers?

Rob Siegel It's British. I thought I'd just use Hylomar to glue my feet directly to the pedals. +1

Franklin Alfonso I guess driving this with size 14 winter boots is out… +1

John McFadden Franklin Alfonso no heel and toeing for you

Dave Gerwig Are new pedal pads on the shopping list?

Rob Siegel Right now I'm just concerned about functionality. There will be any number of appearance issues, exterior and interior, later. +2

Daren Stone Rob Siegel although pedal pads are available, I've found adhesive-backed "stair tape" to do a better job keeping your feet from sliding off what little pedal there is.

Russell Baden Musta Daren Stone I use the OEM ones… they work fine… not to differ, your idea is good as well.

Jim Gerock That looks like a bloody door hinge contraption.

TE Cole And that is why Italian shoes are thin and pointy.

Scott Linn Many folks bend the pedals to the left to give more room. I did. Be sure to wear narrow shoes when driving for the first time, it can be quite a surprise. I have size 9.5 feet and drive with slip-on Pilotis.

Paul Forbush Ballet slippers?

Steve Park So much for 12 wides. Reminds me of my Tiger. Hitting the gas and brake at the same time isn't recommended.

Eric Pommerer [photo of giant Converse All Star sneaker]

TLC Day 2130: Fuel system plumbed

APR 9, 2019, 7:10 PM

The Lotus Chronicles, Day 2130:

Got the fuel system fully plumbed. Those who were pulling for the Robby the Robot-looking fuel pump with the great retro glass dome to not be usurped by some modern upstart electric pump got your wish. I took it apart and the diaphragm looked fine. A replacement for the shriveled-up cork gasket that seals the glass dome was available on RockAuto for $1.57. The glass dome appears to be cracked, but it's not currently leaking (at least not in a water test), but if it does leak, they are available on eBay.

Really, the only thing preventing me from filling the engine with oil, dumping a gallon of gas in the tanks, connecting a battery, and test-spinning the engine to see if the fuel pump fills up the float bowls is that the guy who's power-coating the valve cover still hasn't done it. It's been nearly three weeks. I told him it either needs to be done by COB tomorrow or I want it back unpainted. We'll see what happens.

As far as test-starting the car, it's very close, but I don't want to do it with the car still on jack stands. So the rear axles need to be shimmed (shims arriving tomorrow) and then be installed, and the rear shocks and lower control arms reattached. This weekend is—dare I jinx it?—not impossible.

90 LIKES, 26 COMMENTS

Lee Highsmith I guess it's fate you have that poster in your garage...
    Rob Siegel ya think?
John Thomas Anne Francis. Yummy!
Chris Mahalick I have started mine a lot on the jack stands. It is fine. +1
Scott Fisher You should definitely test the glass bowl with a simple alcohol, with traces of fusel oil. Forty gallons should be sufficient. ;-) +1
Scott Linn If the crack is outside->in, I would be tempted to try and pressure fill with something (like from a windshield repair kit, or nail polish, or ?). But I'd probably do what you are doing to try it out first...
    Steven Bauer Scott Linn seal-all is fuel resistant and only $2 or so a tube
Bill Ecker Fun moment coming. Me and my son did a transplant into an e46 last weekend and got to experience the thrill of the first firing from the motor. +4
Andre Brown Fingers crossed 🤞
Daniel Senie So valve covers can't have patina?
    Rob Siegel Daniel, it had flaking paint, and when I hit it with kerosene in my parts washer, about a third of the paint just fell off and it looked just horrible. It's my one non-essential expense. +3
    Bill Anderson One non-essential expense besides the car itself? 😂😂😂😂 Thanks for the play by play, it's an enjoyable daily read!
    John Goddard Oh, just paint it
Frank Mummolo Not sure what the glass would cost, but after all this work I'd be hesitant to risk a gas shower. But that's just me! 😂 +4
Jamie Eno How much for a new one vs a potential leak that causes significant damage down the road. Something about prophylactic replacement comes to mind here... +1
Bill Ecker Or you could encapsulate it in an ACTUAL prophylactic sheath! +2
Paul Garnier Dat thumbnail tho!! Time to put the man back in manicure! +1
    Rob Siegel I'm workin' here!
Andrew Wilson I'd replace that.
Rob Siegel Just ordered.
Bob Shisler JB Weld comes in a clear version.
David Kiehna Flex seal
Jean Laffite I can't even remember the days before the internet... yellow pages, ads in the back of enthusiast magazines, calling all around town for the chance of finding a part for a car or anything else. Now, "I need a glass globe for a fuel pump on a car no one else owns that I found parked next to the Titanic." "Sure, you'll have it tomorrow." Sweet! +2
    Rob Siegel I know, right? +1
Scott Linn Just remember that the fuel pump sits above the starter... I worry about that myself. +4

Rob Siegel Scott, starter comment is very well considered. Replacement glass dome ordered. For thirteen bucks, I don't want to be wrong.
Steve Park Good move. This is getting exciting. Great progress of late. +1
Tim Warner And get a matching glass mixing bowl for your hat to wear for the test start! Can't be too careful! +4
Maury Walsh I've had several of these Robby-the-robot glass fuel bowl filters on vintage cars, but they've all had the bowl with the opening up. When un-fastening the bowl, fuel stays in it. This one seems upside-down to me. A Lotus bonus to have the fuel spill all over when cleaning the filter? +1
    Rob Siegel I'll go with that. Sure. (Actually we can blame that on the Ford block.)
    Rob Siegel Maury, you did just make me double-check that I have it in the correct orientation. It appears that I do.
    Maury Walsh Yeah, I didn't want to accuse you of goofing it up, but I did hope you'd check.
    Scott Linn My first thought when I saw that after buying my Europa... WTH? How do you clean it? And it's sitting right over the starter! While it does allow an easy visual of fuel flow, a transparent gas filter now fills that niche.
Thomas Siegel Do I hear music from Mission Impossible? Excitement builds!! +1
Mike Heyer Glad you are able to get back on that Lotus project. We've heard about it for years and now it sounds like you may be motoring about soon. Congrats. +1
Franklin Alfonso Glad you ordered the glass dome. Why take a chance on it cracking and spilling gas and have the whole thing on fire after all this work.
Roy Richard Oh come on Rob Siegel a big fire would be a great end to all of this. Think Notre Dame.

TLC Day 2131: I'm sorry, Lucas!

APR 10, 2019, 7:01 PM

The Lotus Chronicles, Day 2131:

While waiting for the axle shims to arrive, I had some fun and installed a battery. I, who daily-drove a Triumph GT6+ between 1976 and 1978 and experienced every electrical problem known to man, am astonished to hear myself say this, but I take back (almost) every bad thing I ever said about Lucas. Mario Langsten described this same thing happening when he bought his long-barn-stored GT6: Nearly every electrical system simply came up running. It's amazing what a difference it can make when a car is stored dry.

[search YouTube for "The Lotus Chronicles Lucas ain't so bad"]

95 LIKES, 65 COMMENTS

Neil Simon But... but... isn't it Lucas??
    Sam Schultz Neil Simon it's waiting until he's on the road and preferably cold and raining. +3
    Rob Siegel you're a cruel man. +2
    Marc Touesnard I remember when I was growing up in the 80's, my parents owned an Austin

America. Many times that car left us stranded if someone so much as sneezed next to it and it got a little damp. +1

Kai Marx I'm so excited to see this come together!!! +1

George Thielen Great to see the hazard flashers working… you might need those. +22

> **Rob Siegel** and the hits just keep on coming! +4
>
> **Dohn Roush** Rob Siegel In an English car you could be the Duke of Hazards… +1
>
> **Rob Siegel** *groans* (although a British flag on the roof of the Lotus WOULD be right at home :^) +2
>
> **Dohn Roush** And after all this you may have a desire to fly it over a ravine…

Steven Bauer In other words, Lucas systems are reliable if you don't use them? +4

> **Scott Lagrotteria** That is asking too much of the Prince of Darkness. +1

Neil M. Fennessey Out in Springfield, carpenter ants attacked the wiring in a MINI! Be sure to get the Lotus inoculated if that's what it takes! +1

Ed MacVaugh Speaking of dry storage, have you checked on the rest of your fleet lately?

> **Rob Siegel** It's on my list.
>
> **Christopher Kohler** More shit to do!

John Whetstone Who deserved a break like that more than you? +3

George Zinycz It's not a break, it's a trap. After its long slumber, it's just lulling you into a premature sense of security and confidence before it dishes out the retribution that's been stewing in its heart these last four decades… +3

Charles Morris "The Lucas switch: Dim, Flicker, Off"… +1

Robert Myles Sure, they flash… but you're ASSUMING he's testing the flashers. +5

> **David Kerr** Lucas was smart to over-engineer the component that would be used the most. +1

Dave Gerwig Try again tomorrow 🙂 +3

> **Rob Siegel** more cruelty!
>
> **Dave Gerwig** I'm as amazed as anyone that the grounding has held together after so many years +1

George Murnaghan That is a really good sign! My experience is that easily 2/3 of the problem with Lucas is in the connectors used throughout the electrical system. Bullet connectors are really susceptible to corrosion given their limited contact area. The other 1/3 is crappy switchgear (which often can be repaired or "fettled with.") Amazing what a new wiring harness/sub harness and connectors will do. That and a dry garage. Fortunately though you are off to a great start! +2

> **Justin Gerry** This. I've fixed my share of Lucas. It was always the connection or some crap previous owner repair. +3

Mario Langsten Lucas is the parent company of Bosch. Just sayin' 🙂 +1

> **Ed MacVaugh** Bosch is owned by a charity. I don't think Lucas is a charity :)
>
> **Tim Warner** Mario Langsten that's correct. Bosch became a non-profit back under Hitler. Bosch family gets royalties. As I understand.

Frank Mummolo Mock the Prince of Darkness at your own peril… +2

Steve Park I always get concerned when a project goes better than expected. Invariably the other shoe drops and kick me in the nutz. You've paid your dues and then some on this one. +1

Scott Linn I wouldn't run those dry on the windshield. The windshield is somewhat unobtanium at times. +1

> **Rob Siegel** As you may be able to see, the windshield is already cracked. Replacing it is on my list once I reach the point where the car is otherwise inspectable. That will be a very happy problem. +2
>
> **Scott Linn** Didn't notice that on my small screen. Took me 6 months to find a windshield, then they became available 6 months later, then became unavailable again. Also, chromed plastic trim is mostly unavailable. My car fortunately came with a new set. +1

Andrew Wilson This is awesome! +1

Russell Baden Musta I am supremely jealous! Your wipers work after all these years on a dry windshield! Awesome!! +1

Harry Krix Hush yo' mouph…the Dark Knight is lurking. Never speak of stable English electrics… said the owner of many MGBs, Austin-Healeys and Triumphs… why do you think I switched (see what I did there "switched") to 2002s… +1

> **Harry Krix** You always inspire, motivate and make me scratch my head and wonder… IS THIS

GUYS NUTS OR WHAT?

Rob Siegel what? :^) +1

Roy Richard Rob Siegel you need to do your take on the Beatles Let it Be... Lotus be Lotus be... In my hour of darkness, mother Lucas comes to me, singing words of wisdom, Lotus be... +7

Paul Wegweiser A SMART man would unplug that battery before going to bed tonight. (I did this on all the cars I was working on, when doing full-tilt restorations... until I pretty much test every circuit for a short...it kept me from laying awake at night wondering if the shop was burning down. +8

> **George Murnaghan** Paul Wegweiser I like the disconnects that go on the negative post-just twist a knob and sleep tight! +3
>
> **Rob Siegel** I unplug it every time I left the garage +6
>
> **Jeff Hecox** Paul Wegweiser my greatest fear as I toil away this week on a 47 year old Italian wiring harness 😩🇮🇹😩 +1
>
> **Dave Gerwig** Fiberglass is great kindling +2
>
> **Paul Wegweiser** Dave Gerwig it gets all... "syrup-y" :D +1
>
> **David Shatzer** Rob Siegel Are you certain? (Sorry, OCD moment...)

Bruce Kwartler Those lights are just supposed to be "steady on" right? Cause: Lucas!!! +1

Brian Ach "Daily-drove a Triumph GT6+" All you need to know about Rob Siegel. +5

> **Paul Wegweiser** ...and still managed to snag Maire Anne Diamond. I will never understand that. +1
>
> **Paul Wegweiser** ...and yet Wendy Burtner rides in THIS heap! How do we DO it, Rob Siegel? Do we have magical powers or something? All I know is that we should use our powers for good and the betterment of the world. :D +4
>
> **Maire Anne Diamond** Paul Wegweiser you have to understand that I lived 10 miles off campus and rode the BUS. I dated him because of the Triumph, not in spite of it. It didn't hurt that he was cute.

Jeffrey Miner My earliest memories of British electrical system encounters were from the backseat of my mother's Sunbeam wagon—this was about 1970 or 1971 and I was about three—and she'd pull it into the garage, turn and remove the ignition key, but the car would ignore her and defiantly continue running until she blew the horn and it would shut down. I don't know what's more intriguing about the episode, that the car would shut down with the horn, or that British electrical systems are so idiosyncratic that it doesn't even elude a three-year-old's perception. +3

Brandon Vitello It has a mind of its own!

Chris Roberts My favorite car thing in the whole world is when they first wake up after years of hibernation.. very exciting and encouraging :-) <3 Will the Lotush get a name? +1

Larry Johnson this only proves the beer is warm. +1

Nicholas Bedworth The Lotus has windshield wipers? You're going to drive it when wet? +1

Nicholas Bedworth One late night at Bradley International Airport, I had just arrived from Chicago (this must be 1971) in January, where the E-Type had been parked in sub-zero conditions for a week. Realizing that getting the car to start was an extremely remote possibility, we nonetheless gave it a whirl. Firstly, the ignition switch didn't jam or break; then the red light over the choke glowed; panel lights illuminated, and with the briefest twist of the key, the engine roared to life, all systems were go. My navigator Bradley Harrison and I looked at each other in amazement, but decided we should get headed back to Amherst without a moment's delay. +3

Jose Rosario My only British car was an Austin Marina. Nothing electrical worked when needed.

Franklin Alfonso Maybe you have just not found the black smoke switch yet?

Franklin Alfonso D-day coming soon when it will start, or crush your soul with disappointment and oil leaks, black smock and covfefe...

Russell Dejulio Love the Europa but at 6' 2" 200 lbs I can't fit 😊

Jeffrey Baker No problems with my '72 TR6, once I replaced the harnesses!

TLC Day 2131: Axles and exhaust

APR 11, 2019, 7:39 PM

The Lotus Chronicles, Day 2132:

Got the rear half-axles shimmed and installed. After reading on forums, I finally understood that all this is is making sure that the roll pins are under slight tension when you drive them through their holes.

The goal was to then re-install the rest of the rear suspension, including the "stressed member" lower control arms that bolt directly to the transaxle, so I could let the car down off the jack stands. I began test-fitting the arms when an alarm went off in my head. It said "You'd better be sure that the rear suspension doesn't have to be out to get the exhaust installed." In a perfect world, it doesn't—the headpipe bolts up to the exhaust manifold on the left, and the muffler mates to it and hangs from the right side of the transaxle—but I'm re-using the old exhaust, the two pieces of which are mated together with rust and age, making it a single ungainly piece. It's not in great shape (I may need to hose-clamp some chimney flashing around the back of the muffler like I did on the Bavaria), but it'll do for now. But with the two pieces effectively rust-welded together, there's no way to get it installed with the right half-axle attached. So, having just banged in both roll pins, I had to remove the right one, drop the half-axle, install the exhaust, and then bang the roll pin back in again. Trivial really. I would've been pissed if I'd buttoned up the whole rear suspension.

Not quite as far as I'd hoped to get today, but I have limits.

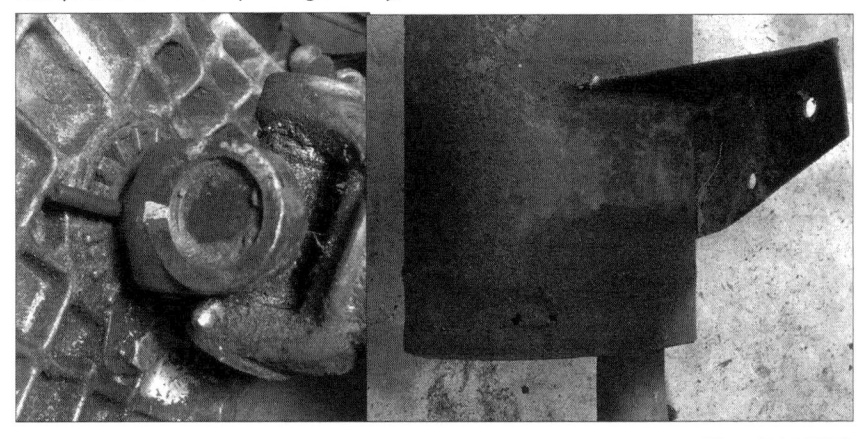

72 LIKES, 11 COMMENTS

Scott Linn Nice! Progress is always good, even if a little slow. +1
George Denoncourt Can't wait to see it on the road! +1
Dave Gerwig In the swing of things now. +2
Russell Baden Musta Good job!! You are kicking ass!! +1
Brian Stauss Anxiously awaiting day 2133 update!
Franklin Alfonso Once you are past the purchase of a Lotus, you are already way past what not to do lol.
Karel Jennings Well... After seeing that valve cover, the Lotus isn't the only one with a "stressed member."

TLC Day 2132: Stressed member reinstalled

APR 11, 2019, 7:39 PM

The Lotus Chronicles, Day 2132 (continued):

Went back into the garage for extra innings. Rear "stressed member" suspension is in. Will lower the car and torque down the nuts tomorrow.

22 LIKES, 10 COMMENTS

Russell Baden Musta Rob Siegel whooooo hooooooo!!!
Tom Egan I'm looking at a transaxle swap (upgrade to 5 speed) in the next few weeks, so I'm counting on you Rob, to make all the mistakes so I can learn what not to do. +1
Scott Aaron OK now I get it.
    Rob Siegel On any normal car, the lower control arm would be attached to a rear subframe. Here, there IS no rear subframe. It's using the transaxle instead. It's a race car trick rarely seen in road cars. +4
Scott Aaron Rob Siegel exactly what I was noticing.
Frank Mummolo That's pretty cool! +1
Brian Ach very F1 +1
    Rob Siegel Race car. Vroom vroom.
Chris Mahalick [photo of his, uh, stressed member]
    Scott Linn The shifter tube makes its appearance.
Steve Nelson That muffler might need patching +1
    Ed MacVaugh 20,00 original miles, over 46 years. +1
    Jim Brannen JB Weld +1
    Rob Siegel chimney flashing and hose clamps +4
    Jim Brannen Rob Siegel you are the "hack mechanic"!!! +1
    Rob Siegel Did it five years ago on the Bavaria. It's still holding :^ [photos] +4
    Steve Nelson Rob Siegel I'll have to remember that trick...any kind of sealant?
    Rob Siegel nope +1
    George Zinycz Rob Siegel that actually looks unhacklike +2
    Rob Siegel George, I'll try and mess it up next time I'm under the Bavaria :^) +1
    Skip Wareham Rob Siegel and I thought you were joking.
Marcia Felth Artistically beautiful +1
Scott Linn Went to a hardware store one Saturday morning to buy some plumber's tape to replace a broken exhaust mount point on my MG Midget. I get behind another guy in line, who is buying the same thing. I say out loud "I know what you're going to do with that." He said no, I didn't. When I said he was going to hang a muffler with it his jaw dropped. +5
    Neal Klinman I'm laughing.

Paul Wegweiser That's so the oil can get out. +9

Stuart Moulton High speed NASCAR duct tape will fix that right up

---

TLC Day 2134: Good with Bean

APR 13, 2019, 8:46 AM

The Lotus Chronicles, Day 2134:

The car is now sitting with coolant and gas in it, but an attempted start is delayed because I still don't have the valve cover back from the guy who is powder-coating it.

So I'm going to tell you a story about the Dave Bean cartridge-style water pump kit. It's a good story.

I wrote quite a bit here on FB about the trials and tribulations of installing the kit. I highlighted the things that were clearly quality control issues on their end, the things that were instead part and parcel of dealing with this strange Lotus-Ford Twin-Cam engine, and the large gray area in between. I had folks here on FB saying that I should demand my money back, whereas The Lotus Engine God whose advice I was taking shrugged at most of the problems, literally saying "well, at least none of them were showstoppers."

Once the water pump and timing covers were installed and passed pressure-testing, I thought about it all carefully and sat down and wrote a long email to Ken Gray at Dave Bean Engineering (DBE).

First, I detailed every issue I had with their kit (which was from a new batch after a lengthy out-of-stock period) and conveyed my impression that I clearly must've been the first person to install it.

Second, I said what I intended to do about it, which was... nothing. I explained that, despite the issues with the kit, Bean's support was outstanding—they were hyper-responsive to every issue and they never left me hanging. Yes, I wrote about much of it on FB and vented a certain amount of frustration because that's what I DO (The Lotus Chronicles would be boring without man-versus-machine-related content), but no one is paying me to write an article that's a blow-by-blow review of the kit, so I wasn't going to, for example, put together a synopsis of the issues and post it on the Lotus forums, with a "don't buy this" slant to it. I said that, if someone on a Lotus forum heard that I'd installed the kit and sought me out and asked me a question, I'd certainly answer it, but would probably do it in the form of a private message. I certainly wouldn't do anything vindictive.

Third, I said what I wanted in return for this restraint on my part, which was... nothing. I explained that in my 33 years of writing, I've never asked any vendor for a quid pro quo and wasn't going to start now. If someone treats me well, I write about it. If someone treats badly, I write about it. If someone's expensive product is great, I write about it. If someone's expensive product is poor, I certainly write about it. If someone's inexpensive product is "you get what you pay for" crap and I was an idiot to use it, I probably write about both it and what an idiot I was to buy it.

Lastly, I explained that, having said all that, I was now going to ask them for something. That something was a spare water pump cartridge, with the pulley offset properly configured for my engine (one of the modifications that needed to be made). I explained that, if, X number of years up the road, the car needs another water pump, and if the

Dave Bean Engineering stops making the pumps or goes out of business, I'll be shit out of luck. I explained that I DID incur real costs and a lot of delays due to the issues with the kit, and because of that, I thought this was a reasonable request to make. I explained that they could say "yes," or they could say "that's too much to ask, but we'll sell you one for a good price," or they could say "no, we're already losing our shirt on these things," and it would have no effect on my other decisions.

Ken Gray at DBE responded the next day, saying (this is a direct quote): "You have been though it all with this water pump. Really, you are NOT the first to fit one of these latest pumps. [But] most are shops that never let us know the effort. You have been the most valuable source of insight. I think the least we can do is provide you with the extra cartridge (set up to the new alignment specs). You have been very patient and kind with us. I will send one the first of next week."

Yesterday a new spare properly-aligned Dave Bean cartridge water pump arrived.

And that is how I like to go through this life.

139 LIKES, 58 COMMENTS

Russell Baden Musta Fantastic Story!! Thank you Rob!!! +2

Jim Strickland Awesome feedback and response! +1

John Morris Nice.

Tom McCarthy Nice, and great response from them too. +2

Layne Wylie Wow. I guess I assumed all this work was so you could use a readily available pump from another application.

    Rob Siegel No, it's all part of the kit.

Jonathan Brush Great story and outcome. You have a really good attitude and I'm sure there's a lot of good in your karma bank. +2

Daniel Neal I love seeing companies like this, that are willing to stand behind their product and accept and use feedback. It's such a rare occurrence these days. +2

Barry Gross Awesome. How much are these things at full price?

Rob Siegel The entire kit including the inner and outer time and covers was 900 bucks. I actually don't know what the individual cartridge-style water pumps cost.

Bill Anderson Wow that's great on DBE. +1

Doug Hitchcock Freakin' awesome story.

Sean Curry And this is how a company should do business! Great job on your part and theirs. +1

Keith Roth Fantastic!

David Weisman You're way too nice. I'd have hired Rabinowitz Rabinowitz and Rabinowitz. +1

    Rob Siegel One of those guys is a dick. +1

David Weisman Rob Siegel 👍

Dohn Roush But they're circumspect about it... +1

Dave Borasky I have to share this one with my kids, because I think a lot can be learned from it about how one appropriately plays well with others. Modern communication (social media, texting) has badly degraded this communication skillset, so it's nice to provide a real-life example that demonstrates what happens when people honestly communicate in good faith, especially in the context of resolving a problem or concern. Thanks Rob. +10

Thomas Jones About twenty years ago I bought a used pair of those red and zinc plated brand name polyurethane isolated Kmac adjustable strut mounts from an eBay seller. When I received them they appeared to be in decent condition as described. A couple of years later I finally got around to installing them. I assembled the mounts onto both struts and began to install them in the car, first by hanging them from the three top strut mount nuts. I went over to my toolbox for something to start connecting the ball joint to the control arm. Halfway there I nearly jumped outta my skin as one of the struts fell to the floor. Upon inspection, I found that the polyurethane bushing had delaminated from the steel housing. I did what I had to do in the situation and reinstalled the standard strut mounts that I had removed from the car and went home from the shop I was working for at the time. Later that night I sat down at my computer, looked up their website, found their contact email and proceeded to write up my findings and opinions of how dangerous it would be if the separation happened on the road or track when a car gets light over a bump, curbing or hill and asked if there was anything they could do to help in my situation. The next day they replied that they would like my address so they could send me a new pair of mounts for free, all the way from Australia, with an explanation that they had improved their process over the years and my used ones must've been an early set. It's good to be honest and calm in explaining the situation to those parts makers who support our hobby, as they're usually really good people. +9

Thomas Jones You do know that there are companies that specialize in rebuilding water pumps for vintage cars, right? We had an M1 water pump rebuilt by a shop in Oregon once. +1

>   **Justin Gerry** Thomas Jones Flying Dutchman water pumps. I've had them rebuild tii water pumps too. Larger bearing is a good thing. +1
>
>   **Jeff Hecox** Thomas Jones I've used a Flying Dutchman Water Pump in my Triumph TR6 with great results.

Scott Winfield That's a beautiful thing. +1

Greg Lewis You are the real deal, Mr. Siegel, my hat is off to you. Well done. +1

Franklin Alfonso Some companies do appreciate good honest feedback on their products. +1

Neal Klinman Bravo. +1

Phil Salmon Almost makes me wish I had a Lotus so I could do business with these folks. Emphasis on _almost_... +2

George Zinycz Put that one on and keep the first one as the spare...

Paul Garnier More of this please

Alan Hunter Johansson And this is one more reason I love you, Rob. In a very manly, appropriate way. +2

>   **Franklin Alfonso** The man's saintliness is only surpassed by his wife and mother... +2

Nicholas Bedworth Being reasonable and balanced is usually a good position... Being totally adorable and innocent helps as well. Good job, Robby. +1

Kai Marx Hey Rob, did you ever considering using an electric water pump rather then going through these fitment and installing issues?

>   **Rob Siegel** It's more complicated than that. If the engine is still in the car, and the water pump is seized but not leaking, then, yes, people use a Craig Davies electric water pump. If however the pump is leaking, then the leak has to be fixed, and the only way to do that is to pull the engine, pull the head and oil pan, and remove the front cover to rebuild the integrated water pump. In my case, I was rebuilding the engine for other reasons (it was seized), so I had to address the water pump, which was also seized. Had I known how many issues there would be in using this cartridge-style water pump, I might have saved the money and simply rebuilt the one in the original timing cover. +1
>
>   **Kai Marx** Ahhh makes sense! The configuration of these engines seem like they create so much unnecessary work!!
>
>   **Rob Siegel** Kai Marx, as I wrote in another post, it's an odd consequence of this being a Lotus twin-cam head on a Ford block for a pushrod engine that originally had an internal camshaft. The weird timing cover is because it needs to enclose a timing chain that was never originally there, and also pump water through the original water passage that's on the front of the block. +1

Franklin Alfonso It was all designed by the Ministry of befuddled British engineers and bafflement bureau of automotive quirkiness… 😊 +2

Kai Marx Gives me another reason to try to make a Mazda rotary fit on a Europa! I've seen a few online, but I don't know how they modify the attachment points to everything.

Rob Siegel this is the problem with any sort of an engine swap. I thought about swapping in a Zetec, but when I looked online, I found the six or so people who had it done it, but no source for things like engine mounts, not to mention everything else needed for the intake, exhaust, wiring, cooling.

Kai Marx A 13b Europa would be like a road going rx500!! [photo of RX500]

Scott Linn Regarding the difficulty and strange stuff on these cars… Back in the day, they weren't building these to last 46 years. They were expecting owners to buy them, use them up, and toss them. I read on the Europa list that a number of years ago a Europa owner met a Lotus engineer who was employed there when they were making these cars. The owner asked the engineer what the expected/planned lifetime for a Europa was. 3 years was the answer. That explains a lot. The water pump issue, bonding the S1 bodies to the chassis, super thin fiberglass (as Rob showed by seeing light through the body). +4

Kai Marx Scott Linn were these considered expensive cars back in their days?

Scott Linn Yes.

Bill Snider Concurring with comments above, my mind wandered to the idea—has anyone organized a wry way to stuff a bunch of yellow chicken feathers in the tailpipe for the 1st start? ;-) w/apologies to Mr. Wegweiser. +3

Dave Gerwig I would be very Bean satisfied 😊 +1

Ernie Peters Ditto on Ken and DBE. Immediately after I bought my Elan, I called and spoke to Ken about parts availability. He explained most parts I would need were in stock or otherwise available, and sent me (without cost) a catalogue. Then, after I put together a parts list, he provided a cost estimate. Needless to say, what I had to buy from a Lotus supplier I got from DBE, mainly because of Ken's willingness to talk and help. Later on I spread the wealth and went to RD for parts DBE did not have in stock, or that I needed quickly (shipping to my house in Illinois was faster from PA than from CA). +2

Jean Laffite I just hope that when you turn the key for the first time there are 30 cameras and mics recording every angle and sound of the first start of this beast. What the hell is wrong with the powder coating guy? He's jacking up this story (or creating a new one). +3

    David Shatzer Jean Laffite His wife won't let him use the oven in the kitchen while she's home… +1

Joey Hertzberg That's how good businesses do business. Good on Bean. +2

Joey Hertzberg Jean Laffite no kidding! Dude, just shoot the powder and bake the damn thing for chrissakes +1

Mike Heyer And that's one of the reasons we like you so much Rob. +2

Gene Kulyk Good story to hear. So much different than my recent ordeal with Dell "customer service."

Martin Meissner I very much appreciate the whole exchange, both your level headed "look, I think you should know that your kit has issues" correspondence and DBE's appreciative and fair response. I always try to approach situations such as this in a similar manner and feel that more often than not it results in a positive outcome. That said, customer service as a whole has been eroding in society and many companies could learn something from this whole exchange. +1

Matthew Zisk Restores one's faith in something… +1

Jonathan Brush I think there's a song somewhere in this story. Maybe a really long "talking blues" one a la Arlo.

Marshall Garrison The Lotus Europa Massacree with full orchestration & 5-part harmony! We can probably all picture the part where Rob is talking to the shrink and after the shrink asks him about the Lotus, he starts jumpin' up and down yellin' "I wanna kill… Kill… KILL! KILL!!" 😊 +1

TLC Day 2134: Engine spun with starter

APR 13, 2019, 5:02 PM

The Lotus Chronicles, Day 2134 (afternoon edition):

I spun the engine using the starter motor for the first time, which was exciting, but other than that, it's been a frustrating day.

--I still don't have the valve cover from the guy who is powder-coating it. This is the third time he said he would have it by a certain end of day. As per this morning's post, I'm a very patient reasonable guy, but it's really starting to piss me off.

--There's still plenty to do, so I nibbled at the edges. With the car still up on jack stands, I changed the front flexible brake lines. Unfortunately, when I tried to bleed the brakes with my Motive power bleeder, the cap I have for the Motive doesn't fit the reservoir on the Lotus. I learned that Motive's standard British adapter doesn't fit either. You apparently have to make your own. Unfortunately, spare caps don't seem to be available, at least not by part number. I broke out my ancient Gunsen's EZ Bleed, which does have a cap that appears to be the right size and threads, but it leaked air and brake fluid everywhere. I toughed it out enough to push fluid out all four corners, but when I was done, the pedal was still soft, so I need to go through it all again. I may order a $25 vacuum bleeder from Amazon and try that.

--Since I was spinning the engine, I checked for spark, but found none. A little odd, since 12V at the coil negative terminal does seem to be toggling on and off. Yes I gapped the points before installing the distributor, but they are the original points and condenser. Could be anything; really it's way too early to be troubleshooting this.

--But most troubling is that there's no oil pressure at the car's mechanical pressure gauge on the dash. I thought, no big deal, it's just the gauge, but when I unthreaded the oil line to it and spun the engine, no oil is squirting out the oil fitting either.

Tired. Dirty. Stopping and looking at it tomorrow before I damage anything.

[search YouTube for "The Lotus Chronicles first test crank"]

(44 LIKES, 18 COMMENTS)

> **Dohn Roush** You can't have peaks without valleys. Hang in there... +4
> **TE Cole** You have a slight intake leak.
> **Justin Gerry** Forget the fancy powdercoat and get the valve cover back and just rattle can it. I've had good results with the VHT stuff. +4
> **James S Eubanks** Don't forget the option of gravity bleeding the brakes. It's shockingly effective... especially if you are killing time waiting on something else. +5
> **John McFadden** "Stopping and looking at it tomorrow before I damage anything." I don't do nearly anything as ambitious as you automotive-wise, but boy I can relate to that statement! +10
>> **Ernie Peters** I think I put a small amount of grease in the oil pump gears for a prime on my Elan rebuild. +3
>> **Scott Linn** Petroleum jelly for my TD restart. +2
> **David Kiehna** John Twist would have had this on the road already!
> **Roy Richard** One word regarding the powder coating... "Rustoleum."
> **Rob Siegel** Read on the Lotus Elan forums (all Elans are Twin-; there's more TC-related technical info there) that lack of oil pressure on rebuild or after a long sit is common. There's a procedure to prime by squirting oil into the port for the pressure gauge. Whew! +20
>> **Zoé Brady** I've had the same issue with my 944 rebuilds. Sometimes oil pressure comes right up, other times I spend a lot of time priming and getting it up.
> **Skip Wareham** Just go retrieve your valve covers. You're not priority number one obviously. +2
> **Daniel Senie** Get the head back and go for patina.

Robert Myles Sounds like you've lost or never had prime in the oil pump. As suggested above, packing it will help, as well turning it over with a drill motor the way the Chevy and Ford guys do. +1

Steve Nelson Shouldn't you also see oil in the head if/when it's actually pumping?

Rob Siegel Yes. Part of my concern. But as per above, it's a common problem on cold starts.

Tim Warner What about just having your lovely assistant pump the pedal and do a manual bleed? Is one better than another?

TLC Day 2135: Oil pressure!

APR 14, 2019, 5:39 PM

The Lotus Chronicles, Day 2135:

We have oil pressure! Thanks to Scott Linn and Ernie Peters who recommended the trick of taking the cover off the oil pump and packing the gears with Vaseline in order to help prime it. This was after using a Wizard of Oz-style oil can to prime the port on the side of the engine that supplies oil to the mechanical pressure gauge on the dash and not have that do the trick, and unscrewing the oil filter and verifying that nothing was being done pumped into the center of it. When the system finally primed, I could hear the change in the note of the engine, and the mechanical gauge on the dashboard snapped to attention, though the pressure fell by the time I snapped the photo.

111 LIKES, 34 COMMENTS

Christopher Kohler Were you listening to "Under Pressure?" +6

David Weisman Clap Clap Clap! Well written son. Love the Wizard of Oz style oil can. Damn. +1

Steven Bauer Maybe it's time to hack your own oil system primer.

Bill Snider Very happy to hear the instructional YouTube video helped you reach pressure. [link to Hanz und Franz Pump You Up!]

Chris Roberts YESSS!

Chris Roberts "nothing was being done pumped into…"? Your Texas is showing ;-). Also, in the end what will you expect the normal oil pressure to be for healthy vital signs?

Franklin Alfonso Getting there...

John Whetstone Another well served break, whew!

Franklin Alfonso [gif of Iraq oil well gusher] +6

Rob Siegel On the Lotus Elan forum, this is described as a very common issue on not only newly-rebuilt engines but also long-sitting engines. +1

Paul Forbush Charles Morris It's not an uncommon thing to have oil pumps fail to prime. As a pro, I can tell you it's standard procedure to grease a dry new or used pump before cranking it. It has saved me from several "oh shit" moments on starting a rebuilt motor.

Charles Morris I guess it depends on what you're working on... I don't dry-start engines either, but prime the pump and galleys first. Of course if the design prevents the use of a shaft and a spin-handle (or electric drill), it gets tougher :)

Dave Borasky [photo of Wizard of Oz characters with oil can] +3

Ernie Peters Alright!

Carl Witthoft Oh, look: engine temperature in Celsius. What's an American to do?

> **Rob Siegel** But the fasteners aren't metric; they're SAE. Bloody schizoid Brits. +2
>
> **Carl Witthoft** Rob Siegel I bet it only runs on Imperial gallons of gas. +4
>
> **Scott Linn** Rob Siegel My tool chest is overflowing. Tools, taps & dies, etc. SAE, Metric, BA, British Standard Whitworth, British Standard Fine, British Standard Pipe, Pozidrive +2
>
> **Paul Forbush** Carl Witthoft Given the age of the Lotus, I'm quite sure that gauge is calibrated in centigrade rather than Celsius... +2

Mike Nash "If I only had a heart" +1

> **Brad Purvis** Mike Nash Or a brain.
>
> **Mike Nash** Brad Purvis that was the scarecrow
>
> **Brad Purvis** Yeah, I know, but it IS a Lotus after all.

Kevin McGrath Holy smokes! Harkens back to the McDashboard wood McPaneling that I grew up with. I think my dad paneled that. Along with everything else that didn't move.

Dave Gerwig Under Pressure and that's a good thing!

Paul Wegweiser Couldn't you just glue a couple of ice cube trays on the top and call it done? I mean... they can't possibly leak worse than the original cover, right? +4

Roy Richard Oil pressure... we don't need no stinking oil pressure.

David Kerr Pressure Rise, or the reverse Toots Hibbert as it's called hereabouts.

Clyde Gates I recall a similar way to prime 1960 era Ford 6 cylinders

Mike Miller [gif of Fat Bastard saluting]

Brian Ach Timeline: Siegel restores entire Lotus Europa in same time I change water pump and de-A/C the 2002tii. Sounds about right. +2

TLC Day 2136: Valve cover!

APR 15, 2019, 5:17 PM

The Lotus Chronicles, Day 2136:

Finally picked up the powder-coated valve cover. It was my one frivolous expense in this entire Lotus resurrection project. I love it. It will make me happy every time I look at it.

The shop who did the powder coating, though, was a little bit sketchy. This cinder-blocked 7 Series was having its wheels plasti-dipped inside. Yeesh!

165 LIKES, 63 COMMENTS

Frank Mummolo Powder coated valve covers always make the car go faster! You won't be sorry, Rob! 😊 +1

Russell Shigeta Wait… Does that seriously say "BIG VALVE" on it?! 😂 "of emotional frustration" 😂

Rick Roudebush Looks great Rob, happy for you. +1

Bill Snider Looking Excellent Rob! I smiled when I saw the pic. +1

Benjamin Shahrabani Now that you've got shiny new valve covers, won't they make everything else look like crap by comparison? Might as well do a nut and bolt restoration now! +4

Rob Siegel Believe you me, that won't happen 😊 +5

David Weisman That looks freakin cool

Neil M. Fennessey Did they media blast the inside of the cover either accidentally or intentionally? If so, don't install it quite yet. +1

    Ernie Peters Neil M. Fennessey Because the media used can hang out in the pores of the aluminum, right? And then as the engine heats up the oil can pick it up and do bad things. +1

    Neil M. Fennessey Ernie, more like it temporarily embeds itself into the aluminum only to drop out later thanks to shock/heating/cooling cycles, and then circulate through the oil system only to potentially unleash Bad News to the crank, bearings, cams, etc. +1

    Ernie Peters Neil M. Fennessey A good explanation. I made sure the machine shop that did my Elan's head knew not to media blast it. Luckily, they were already aware, having done several for a Lotus enthusiast/restorer buddy. +2

Charles Morris I had the magnesium cover on my M50 powder-coated due to suspected casting pores… 90% of the cover is hidden under plastic bits anyway. So I don't even get to enjoy looking at it ;)

Rob Capiccioni Looks great, and good for another five horsepower! +1

Gary Beck Rob, with all this work you should drive this to the Vintage. Everyone is all psyched about it. Good job and one hell of a lot of work. +2

Scott Linn Hmm. Engine is in, suspension is in, valve cover is in hand… I see something cool possibly happening very soon. +4

Francis Dance Glad you finally got the VC back. +1

Daniel Neal That looks amazing, getting that done shortly to my M20B25 valve cover, hopefully I don't run into similar issues. +1

Jd Smith Lookin' good Motorhead Robby!!! +1

Christopher Kohler That 4-door version of the 2-door version of the 4-door thing looks like it's been there as long as your valve cover! +3

Dave Singh you're all over the internet I see you everywhere +1

Tank Şimşek They did a great job. +1

Clay Weiland There's a little part of me that wants to mill off the cam tops on the M42's valve cover and make some removable covers, for no reason other than that it would look cool. +2

> **Luther K Brefo** Clay Weiland [photoshop of Ford logo on a BMW M42 engine]

Paul Forbush Looks great, but it's amazing how much casting technology has improved over the last 40 years. +1

Delia Wolfe Just close your eyes and think about the money. +3

Jim Brannen Stay classy Rob. +2

Rupert X Pellett Rob. Love what you're doing, dig your rabid enthusiasm-press on regardless. Cheers. +2

Steve Park Looks great. Better than I would have expected, given the photo of their other handiwork! +1

Gary Rich 'Bout time! +1

George Whiteley There's NOTHING like a fine pair... of valve covers! 😊 +2

Hans Batra Powder coating is perfect for that! Makes me remember—I had an E36 M3 and a girl I was dating asked me when I might upgrade to a 7-series. I told her that a 7-series was not an "upgrade." She's not a car person so she still doesn't understand to this day. +3

> **Brian Ach** that's hilarious

Bart Collins Sexy!!!

Brian Ach "It was my one frivolous expense." No, this was completely a necessary expense. +4

Scott Aaron Brian Ach it's like he's Amish or something. "I really don't deserve this glass of tap water. I need to suffer more deeply." +3

Jake Metz I've been trying to come up with a joke about "big valve" but am at a loss. Rob "big valve" Siegel? Close, but not it. It's there somewhere... +1

> **Rob Siegel** Yes. Well. My cam cover says "big valve." Does YOURS? No? It DOESN'T? Well, then, you'll just SHUT THE FUCK RIGHT UP won't you?" +2

Jake Metz Rob Siegel ummmmmm...

+6

Rob Siegel Jake, you're fooling no one, you know... +1

Jake Metz It's a special edition. +1

Robert Myles Rob, maybe it's time to cut back on the caffeine... +1

> **Marshall Garrison** Some re-working of this Kliban classic comes to mind – [Kliban "He's the boss of the beach now that he's got BIG BALLS"]

Brad Purvis Rob Siegel I have a wrench that says King Dick.

Andrew Wilson Have to work "Big Valve" into Spinal Tap's "Big Bottom"... "Talk about valve lash, my girl's got 'em, big valve going out of my mind, how can I leave her Lowtush behind?" +3

> **Rob Siegel** Oh dear god. We're all going to hell. +1
>
> **Andrew Wilson** ...In a handbasket.

Dave Borasky No... in a Lotus +2
John Wanner why does it say Big Valve and not list the number of valves or cams or something?
Nicholas Bedworth It's hard to count beyond 5 for some people... +1
John Wanner Or is it supposed to say something else?
Rob Siegel The "Big Valve Twin Cam" label was a marketing tactic to delineate it from the earlier less powerful version of the Twin-Cam engine. +2
Dave Borasky LOL @ Spinal Tap. Rob's valves go to "11"
Chris Mahalick He died in a gardening accident. The authorities said it was best left unsolved. +1
Paul Ehrlich Oh wow... interesting way to store the car...
Doug Jacobs "A $20 shine on $3 boots!"
Nicholas Bedworth The entire project, extending over 2,136 days is not frivolous, perhaps $200,000 in labor at $5/hour, but the extra $50 for the valve cover is. I see. 😊 +2
Randy Yentsch She's a beauty. It will be like a cherry on top. +1
TE Cole Now install that thing and fire it up. +3
Scott Winfield That looks great and justifies the entire endeavor all by itself +1
Lee Highsmith I've got limited skills, but this is a T-shirt that should be made...

+1
Hans Batra Damn, industrial nihilist Bauhausian lingerie is all the rage nowadays!

This is what was parked outside the shop who powdercoated the valve cover.

Randy Yentsch Nice jack stands.
Craig Fitzgerald Well with you harassing them for your damned valve cover, they didn't couldn't afford the time and expense of $26 worth of Jack stands. +3
Craig Fitzgerald And those wheels have been off for a month.
Randy Yentsch One small step above cinder blocks. Perhaps the technicians are reverting to type.
Tom Tate That's just a security system in a tough neighborhood 😊
Chris Roberts $700 cash in hand.
TE Cole You've heard of Griots Garage, well that rear "jack stand" is from Guido's Garage.
Benjamin Shahrabani Rob Siegel that's not a 7 series!
   Rob Siegel Ha! What do I know?
   Tank Şimşek Clearly a 7! Noob! [zoom-in of trim in photo that looks like numeral "7"] +5

TLC Day 2137: New tires

APR 16, 2019, 8:13 PM

The Lotus Chronicles, Day 2137:

New tires! Achilles 122s. Correctly-sized staggered 185/70/13 and 175/70/13. $161 shipped. Second guess all you want. They fit the budget and are worlds better than what was on there.

(If I don't say "these are the old tires," some wisenheimer will say "$161 for THOSE?" So... "These are the old tires." We good?)

77 LIKES, 109 COMMENTS

Craig Fitzgerald It's amazing to think there was a time when that car drove far enough to wear a tire out. +3

John McFadden Any modern tire has to be better than the best tire available at the time that car was manufactured. Staggered 13-inch tires. OMG :-) +5

Jake Metz I feel like "Achilles" has got to be the worst name for a tire brand ever. But I'd have done the same thing. Good buy, Rob! +21

    Craig Fitzgerald Jake Metz They go with my new set of Torn ACL shocks. +11

    Jake Metz Rob Siegel if nothing else you'll have a good base for jokes with your first round of any tire trouble... +3

    Rob Siegel "Achilles? Who knew!" +4

    Craig Fitzgerald "It appears as if your tires were some kind of inherently weak link that caused a catastrophic failure. I wish there was a good name for it." +4

    Rob Siegel ["I'm having a thought... it's gone" meme]

    Craig Fitzgerald [Simpsons "Speed" movie meme] +4

Luther K Brefo At least you know it was driven! +3

Brian Ach nothing wrong with those tires. don't worry about date codes and stuff. +3

    Karel Jennings They predate date codes. It's all good .;) +5

    Dohn Roush The date codes are in Roman numerals... +6

Robert Olmedo I have a set of 13inch for free to you. Not staggered but new.

Steve Park I hope they gave you a decal you can put in the window, so you can do some proper advertising for this manufacturer.

Robert Myles "Wisenheimer"!?!? Methinks the tires aren't the only thing that is date coded... +8

Charles Morris Reminds me of that scene in "The World's Fastest Indian" where Anthony Hopkins was putting caulk and black shoe polish on his old tires to pass inspection... +7

    Alexander Wajsfelner Was just thinking that myself. +1

Bill Snider You were wise to, ah, tread carefully Rob ;-) +1

Francis Dance 205/60R13 Yokohama A006Rs were the hot upgrade from 175/70R13 back in the day. But a 6″ rim width is needed. +1

Rob Siegel and folks use 205s on Europas, but, just like on 2002, there are possible rubbing issues.

Carl Witthoft Rob Siegel "rubbing issues"—what a straight line +1

Rob Siegel uh, walked right into that one +2

Daniel Senie Well Krafted? +1

Ken Sowul Daniel Senie As a Jets fan I approve

Russell Baden Musta Francis Dance I can tell you first hand, 205/60-13s rub the rear fenders on Panasport 13x6jj rims. They do not rub on the brand lotus 13x5.5jj rims. They do not replace the 175/70-13s on the front, just the 185/70s on the rear. Just for rear, not front.

Dave Gerwig New shoes will transform your ride.

Rob Siegel Yes. I'm hoping that they will make it actually roll! +5

Dale Robert Olson It's not like you have any idea what the old tires performed like, right? +2

Rob Siegel The spare is actually untouched, so in theory I did have four drivable tires. But I did this same thing with Bertha. Once I got her to the point where she could move, one hundred-foot drive on the ancient tires was enough. The Lotus is still up on four jack stands while I'm trying to get the brakes bled. The timing was perfect to deal with tires. +3

Chris Roberts Rob Siegel Plus, new tires smell nice. +4

Ken Sowul ROLF, Rubbing, Bertha, untouched moved... I am back in Jr High.

Dohn Roush And Achilles died in the Trojan War, and no high schooler could pass that up... +2

Robert Boynton Lotus with both Lucas electrical and Achilles tires? Magic. +6

Steve Woodard Well, as always, this is very interesting, but where's the spark? When are you going to start this thing? Last I heard, the two impediments to getting on with the programme were the valve cover (done, beautifully) and the lack of spark. Yeah, I have a short attention span, when I have one at all. +2

Rob Siegel Workin' on it. Points and condenser are on the way.

Steve Nelson Rob Siegel "it's always the points"... I read that somewhere 😊 +1

Anuraj Shah So cool!! Getting closer and closer!! Love it! +1

TE Cole Just goes round and round and round. Seems like you're getting some traction. Never gets tired! +1

Doug Hitchcock You could get some sort of Hoosiers in 13-inch sizes, but probably not that skinny. Still, r-compound track tires would be perfect for that car.

Steve Nelson Doug Hitchcock FYI, Pirelli is making CN36s in 13"... that's what I just put on my 2002 👍 +1

Doug Hitchcock I just ordered a cheap Kumho 185-13 for my stock steel wheel-spare tire. $55 at Discount Tires Direct.

Rob Siegel I just don't have the budget for CNs. And this car will never see the miles I drive my 02s. +3

Rob Siegel Doug, I almost bought Kumhos. These were about $60 cheaper for the set. I try to save wherever I can. If I planned to put significant mileage on the car I might have chosen differently. +2

Doug Hitchcock I was amazed at the price you paid for a set. I'm sure you'll be fine. +1

Paul Wegweiser Steve Nelson I second the CN36 recommendation. they have truly transformed my car. No longer scared shitless of wet roads and death. They've been great tires so far. +1

Paul Wegweiser And way stickier than the Kumhos. Those Kumhos were "ok"until I got about 1/2 way through the treadwear... then they became the worst rain tire I have ever used in 35 years of 02-ing. The CN36s have nice dry grip, too, requiring a bigger swaybar last week. :) +2

Neil M. Fennessey Kumho seem to harden quicker than better tires in my experience. They're fine for a few years. +1

Paul Muskopf Hopefully they won't make you feel like a heel for buying them. +3

Jeremy Novak I want to laugh at this, but I will not. You may feel like you hit a Homer, but nobody's writing any epic poems about you. +2

Tammer Farid Boooooooo +1

Daniel Senie Messing with the tires, but won't tell us if the engine vrooms. 😊 +1

Rob Siegel workin' on it +1

Carl Witthoft It does, but only because Rob has an mp3 of engine sounds running. +1

Scott Linn FYI, most TCS people seem to just run 185's at all 4 corners to save on the complexity. When

I bought mine, he put 205 racing tires on all 4 corners. WAY too much tire. Wore out after 5k miles down to the belts. +1

Rob Siegel Scott, I didn't see the "185s all around" thing when I perused the forums. Most of the posts I read seemed to advise using the recommended staggered sizes. Plus, does a 185 even fit in the nose? I had a hell of a time getting the original 175 spare out. Damn, that's a tight fit. +1

Scott Linn Rob, yes a 185 does fit in the nose, just fine. I was worried too. It is nice to be able to do a normal tire rotation and also make use of the spare if you want. I bet that having the staggered sizes is "best" though.

Russell Baden Musta Rob Siegel no, won't fit on front either...

Scott Linn Russell not sure what you mean here. My car came with 205's all around. Had steering interference & suspension issues with low sidewalls. Didn't have a spare but it wouldn't fit in the nose anyway. Switched to 185's all around based on others' experience in the Europa forums. All working fine now and the 185 spare fits in the nose.

Russell Baden Musta Scott Linn well, more power to you Scott if the fronts aren't interfering! My 175's hit the inner well at full lock, but that's just occasional when parking. I bought mine new in 1974. It came with 185's rear and 175's front. A few years back I couldn't find 185's for the rear and someone suggested the 205/60's. At the time I had Panasports which were 6" wide vs the 5.5 brand Lotus wheels. I never had good luck balancing the Lotus wheels; why I switched to Panasports. I had 175's all around on those at first and they were nice and smooth. But the steering was strangely twitchy. Then I needed rears. I put the 205's on the Panasports and the rubbed the Sumitomo logo right off within a mile! I measured the OEM wheels vs the panas and the extra half inch was on the outside. I then put the 205's on the OEM wheels and to this day they do not rub. I had all 5 refinished to like new condition and they still do not balance well... once the OEM wheels were back on the front, the twitchy steering went away as well. I explained this experience to an engineer at Lotus and the wider wheel on from moved contact patch outward enough that made it twitchy... it tramlined horribly. Now it doesn't.

Russell Baden Musta Scott Linn the sticker in my glove box... 155/70-13 original size! I think S1's and S2's came with 155's... my spare was 175... [photo of original glovebox sticker]

Russell Baden Musta Scott Linn my spare on the floor where I keep it...

Scott Linn Russell Drove my Europa to work today. +2

Russell Baden Musta Scott Linn put about 60 miles on the Evora today, heading out shortly to meet my wife! +1

Rob Siegel You guys are just showing off. +1

Russell Baden Musta Rob Siegel soon we'll be tired of hearing where you drove your Europa to on any given day! 😊😊😊😊😊😊 +1

Scott Linn Obligatory BMW content: I drove my 98 Z3 to work yesterday...

Paul Hill I'd keep an eye on that Notre Dame fuel pump...too soon?

Steven Hage A guy I know has 205/60 R888s on his Europa, I think you need to step your game up haha +4 [photo of Europa on 205s]

Rob Siegel fat meats look awesome on ANY car +1

TE Cole That back section behind the door glass needs a half moon window like the good times vans had back in the day. +3

TE Cole And what is the white bucket for up front...is he collecting contributions? +3

Ed MacVaugh Throw your ticket in for people's choice.

Steven Hage There's already a half moon window from the tire rubbing the quarter panel lol

Jeffrey Miner Yes, for a light, good handling car, the wheel/tire combo is THE place to maximize returns. Anyone run larger wheel, lower profile tires on these cars? It's fine to have a stock set for originality, but for driving, I'd think a wider, lower profile wheel/tire package would pay HUGE dividends no a Lotus... I guess these won't fit under stock fenders though? [link to "Extra-Wide Revolution Rims" on LotusTalk]

Scott Linn Those are what was on my Europa when I bought it. Couldn't wait to get them off. They rubbed on the body, they wore VERY quickly, and they would hit the body when the suspension bottomed out, and they didn't do well in the rain (I live in Oregon). But they were VERY sticky... I've been much happier with 185s all around.

Franklin Alfonso I don't know Rob, I still saw a bit of tread on the old tires around the edges you sure they needed replacement? [gif of Tyrian Lannister doing a little dance] +3

Mike Miller I have had surprisingly good experience with Taiwanese tires in 165/80-13. +1

Rob Siegel Any particular brand, Mike? These Achilles are, I believe, Indonesian.

Mike Miller I forget the brand. Nankang maybe. I bought five for Roscoe from tires-easy.com. $200 including shipping. I figured five for 200 clams, how bad could they possibly be? Turned out three out of the five required no balancing weights! Curiously, through, a sidewall pictogram warned sternly against playing a bugle while driving on those tires. I complied. I am going to order five more for The Schwartz. +4

Mike Miller They are Nankang. Just ordered five in 165/80-13 for $230.80 to my door. +1

TE Cole The Nanking has good noodles also. +1

Rob Siegel For the Lotus, I looked at Achilles, Nankang, Kuhmo Solus, and Westlake. The Westlakes had some clear red flags in terms of quality issues. There weren't a lot of reviews of these Achilles 122s, but Achilles in general seems to have a decent reputation as an inexpensive tire; I have a set on my E39. Some folks on the Lotus forum had bought them and were happy with them. And the $161 shipped price was less than anything else except the Westlakes, and I am relentless about controlling the cost of the Lotus project. If I planned on driving the Lotus to events like The Vintage, I might have made a different choice. +1

TE Cole [link to painting of Achilles fighting against Memnon]

Mike Miller I'm taking The Schwartz to as many Hemmings Cruise-ins as possible this year, and the VT Oktoberfast. Anyone who doesn't like my Nankangs can KMA. +3

Rob Siegel Mike Miller THAT'S THE SPIRIT! +1

Mike Miller Fucks Left to Give: 0 +1

Rob Siegel (I often tell people: "Mike Miller and I are very different people who share a common value—if people don't like what we do, they can kindly shut the fuck up.") +6

TE Cole True story +2

TE Cole Old school BMWphiles are a unique bunch. Modern BMW drivers, not so much. You have to pick through dozens to find the ones with the spirit of old.

Mike Miller True, yet your remarkable transition to proper coolant and fuel plumbing indicates another common value. And for my part, I now recommend your book to vintage BMW owners favoring air conditioning. I have always recognized you as an expert in this regard. I probably remain the expert in firing all that shit into the trash can. I've offered it for free. Response? "Will you pay for the shipping, Mike?" +1

Scott Linn Rob Siegel I have the Kumho Solus tires via Walmart. So far so good. +2

Brian Ach Rob Siegel I have the Kumho Solus on Gerta and they are totally fine +2

Rob Siegel Brian Ach and Scott Linn, I have them on Louie and have had no complaints. +2

Mike Miller I have five waiting for Hans, when he needs them.

Ed MacVaugh Which wheels do you have? Lotus or another brand? I see the latest on the forum is the "Thunderer" rather than Achilles. You could get lots of humor out of that set. +3

Rob Siegel Ed, they're 13x6, I believe Cosmics. The guy who mounted and balanced the tires said they're pretty badly bent. I didn't know. I had no way to spin them, as I haven't done the brakes yet. I'm hoping his definition of "pretty badly bent" is more stringent than mine. +4

Duncan Irving So how much did mounting and balancing cost?

Rob Siegel $80 +1

Duncan Irving Ahah! Always gets me. Not bad though. Watch those in the wet.

Jeremy Shiroda Nice! I went with Achilles on the E39 Touring! No worries, mon! +1

TLC Day 2138: Spark!

Apr 17, 2019, 8:30 PM

The Lotus Chronicles, Day 2138:

WE HAVE SPARK!

131 LIKES, 32 COMMENTS

Brandon Fetch *cue Young Frankenstein* +2
    Rob Siegel almost +1
    Robert Myles Rob's a spark tease... +2
[I'm going to skip the rest of the Young Frankenstein-related references :^)]
Doug Jacobs Well, spark isn't _exactly_ life.
John Whetstone Did anything combust?
    Rob Siegel this was just plug-wire-to-ground spark +2
    John Whetstone Rob Siegel OK, I'm taking deep breaths +1
    Andrew Wilson [sparkler gif] +2
    Eric Pommerer The question still stands...
Al Larson It's always the points... +3
    Rob Siegel actually, this time it was the coil +2
Frank Mummolo Did the same with my recent BMW Airhead MC resto. Shut the lights off in the shop, grounded the plugs and looked for "light." Deep breaths, indeed! Good work, man! +1
David Shatzer Rob Siegel, master of car porn fore-spark-play... +3
Samantha Lewis The Lord of Darkness is pleased with you obviously. All hail Lucas. Now that's out of the way. Well done you, well done noble little Lotus. +2
    George Zinycz Samantha Lewis he's awakened the Beast +3
    Samantha Lewis Exxcccelllent! The Beast is pleased. +1
Gary Beck You're so dramatic. +2
Rob Koenig Great. Now start the damn thing already. +5
    Rob Siegel workin' on it +4
    Daniel Senie Rob Koenig he's going to make us all wait for days... +1
    George Zinycz Rob Siegel message me your address, I'll be there in the morning to help +1
    Rob Siegel Seriously, I've had other things going on.
David Kemether Now just add fuel. [photo of Europa on fire] +13
    Franklin Alfonso That's what he's trying to avoid lol.
Clyde Gates Same with my E34—now to check for fuel (after the storms pass)
Brandon Fetch Wow. I started something in here, didn't I? (Punned on purpose too)

TLC Day 2139: Valve cover on

APR 18, 2019, 8:34 AM

The Lotus Chronicles, Day 2139 (morning edition):

First test-fit of valve cover.

Forgive me, but this makes my dick stiff.

156 LIKES, 84 COMMENTS

Brandon Fetch Forgiven for having a dick or a stiff one? +3
Luther K Brefo [gif "UNSEE"] +4
Luther K Brefo [gif "Mental image, be gone!"]
Luther K Brefo But yea… [gif "That's hot!"] +2
Rob Capiccioni I think mine just moved a bit as well! +4
John Harvey Mine was squelched by those Strombergs +4
    Rob Siegel John Harvey, you're a… dare I say it… dick. +3
    Luther K Brefo [gif of woman holding up "this big" hands]
    John Harvey Rob, do the Lotus guys get them to perform? In my history with Volvo's, they would be replaced with HS6's or HIF6's. Now I can't remember if the bolt pattern is the same. I know racing Triumphs use them.
    Rob Siegel John Harvey, Yes, I believe that they can be replaced with SUs. As I say in both of the engineering and automotive worlds, completion before functionality, functionality before performance. I look forward to when carburetor performance is the top thing on the punch list for this car. +3
    Bruce Bubeck Now we're talking. [link to video "Lotus Elan with Weber Carburetors"]
    John Harvey Bruce, note, different head
    Brett Kay Contact Joe Curto he's the king of SUs in the US and he knows his stuff.
Andrew Wilson [link to "The Young Ones funeral"]
    Andrew Wilson Apropos of coming around to measure your stiffy. +1
Steve Nelson Gorgeous…you're forgiven 👍 +1
Brian Hemmis TMI +2
Ali Jon Yetgen [gif Nicholas Cage seconding that motion] +1
TE Cole Slow down Tonto… it needs to run
    Rob Siegel Nah… it's better this way 😊 +4
    Sean Curry Rob Siegel Right now it's Schroedinger's Engine—it both runs and does not run. +5

David Geisinger I think you need to get out more 😊. But this is awesome progress btw +1

Rick Roudebush That is a bit of an appendage straightener as the Brits might say. +2

Tom Melton So what was the final determination on the clearance of the plugged coolant port and the frame?

> **Rob Siegel** unresolved

Russell Baden Musta What is that? It's so clean!!!

> **Rob Siegel** we both know that's only because it's never been started +4
>
> **Russell Baden Musta** Rob Siegel it will be fine!!! Now I must do the same as you on my engine... was to be a winter project... is your exhaust manifold on? I can't see it...
>
> **Rob Siegel** Russell, Yes, Exhaust manifold is on. Just need to seal up the valve cover, put on the cross pipes, and connect the cables to the Strombergs. +2
>
> **Russell Baden Musta** Rob Siegel cool!!! Hylomar it to the cam cover, nothing on the head...maybe just a dab on the half moons...

John Graham Doesn't look like it fits, I don't see any oil leaking? Try to install again? It really does look gorgeous!!! +1

Roger Scilley What's with the spark plug spacing, they don't seem to correspond with the piston placement? Or do they? +1

> **John McFadden** I wondered the same. I'm guessing the spark plugs are not in the centers of the cylinder bores.

Pete Lazier ....that's more than we need to know...

Rob Anthony I dunno if I want to keep following the Lotus chronicles and see how this ends. I mean he has been telling us all along that "you can get sex out of this car"... +1

> **David Shatzer** Rob Anthony Hmm But not in it...

George Murnaghan As they say in England, keep your pecker up! +1

Dan McLaughlin Though that powder coating guy was slow as shit on a winter's day... it sure looks sharp! +1

Ernie Peters [meme: "I had a fight with an erection. This morning I beat it single handed."] +5

David Weisman Dash pots?!

Kent Carlos Everett Oooo baby.

Dave Borasky My, what big valves you have.

Marc Schatell Oh my god, fire it up already!!! +1

> **Rob Siegel** Marc, we're still at the picking-her-up-and-taking-her-out-to-dinner stage. I haven't rushed consummation of my six-year-long relationship with Lolita thus far, and I'm not going to do so now 😊 +1
>
> **Daniel Senie** We've had 6 years of ribbing Rob about the (non-) progress on the Lotus. I think he's going to drag this out for weeks to taunt us, as retribution. +6
>
> **Rob Koenig** You've been out of the loop for a while, grandpa. Everybody is pretty much all-in on the first date these days... 😊 +1

Karel Jennings There's that stressed member again. I mean on you, not the car. That does look good. But I'm biased. [photo of clean 2002 engine] +6

Samantha Lewis It always seemed a shame to put the trunk (glass fiber bin) back in, the car was supposed to look like this without it. Start up video soon, yes?

> **Rob Siegel** I hope so, but I have some other things going on.

Bob Sawtelle [gif Caddyshack "Okay. Hey, look at that!"]

Robert Boynton How many "test-fit"'s of the valve cover do you anticipate? Is there a pressure test planned too? +1

> **Rob Siegel** Seriously, there is some trimming of the gasket.
>
> **Jeremy Novak** "Trimming the gasket," you say. Mmmm-hmmm.

Philip D. Sinner *hats tipped towards the stiff dique +1

Dohn Roush Well, yes, this process has been hard on you...

James Pollaci [gif Simpsons eye bleach] +4

Michael McSweeny Schwing! +1

Bill Mann The duel carbs do it for me :) +1

Scott Aaron Hoo boy always with the classiness... +1

> **Rob Siegel** you DO know who I am, right? +2

Scott Aaron I'm aware. Oh, I'm aware.

Collin Blackburn Turn the key already!

David Kemether You may want to put one of these on before starting it for the first time. [photo whole body condom] +2

Carl Witthoft So you're saying that getting (an engine) head got you stiff?

Rob Siegel No. As Freud never said, "sometimes a valve cover is just a valve cover." +4

Robert Alan Scalla It's been a stiff learning curve for just one car. Now that you know, maybe you should look around for another. +2

Rob Siegel Oh god that's funny. In a sick twisted sort of way.

Scott Aaron Justice Scalla!

Alan Hunter Johansson You can't rush into these things.

Rob Siegel "you can't hurry love... no you'll just have to wait..." :^) +1

Bill Ecker That Better be the Supremes' tune and not the vanilla terror's version, Rob. +2

John McFadden Sheesh. 40+ replies to a 9 AM post and no reference to morning wood? +3

Rob Siegel until NOW +1

Daniel Senie John McFadden and this was just the morning edition, so we are all hanging around the virtual garage door waiting to see what comes up this evening.

Paul Wegweiser You'll have to snap a photo of this coming out of the garage under its own power, with a caption that reads: "Out... like a boner in sweatpants!" +9

Scott Linn Still waiting for "Lotus: After Dark" +1

Brad Purvis Can I call you Mr. Stiffy? +1

Doug Hitchcock Give it a brit vibe. Call him Sir Stiffy! :-) +2

TLC Day 2141: Spark, no start

APR 20, 2019, 12:21 PM

The Lotus Chronicles, Day 2141:

Valve cover buttoned up, cross-pipes attached (wanted to repaint them, but screw it). Gas in the float bowls, check (there's a plug at the base of the Strombergs you can pull; makes checking very easy). Spark, no. Had the coil wired incorrectly; low side goes through the tach and then back to the distributor. Spark, check. Did the first attempted test-start, nothing. Recheck spark at one of the plug wires with an actual plug, check. Gave two blasts of starting fluid; no difference. I could have the timing 180 degrees out or something. Or the engine itself isn't well-grounded. Or any number of other things.

But I need to put this back down for a bit. I swear to you I'm not teasing or blue-balling you. I have other non-Lotus-related events in my personal life that are far more important right now, and 15 minutes at a time is what I can muster.

(search YouTube for "The Lotus Chronicles very first attempted crank and start")

72 LIKES, 16 COMMENTS

Lance Johnson Keep going!!!!

TE Cole Well it certainly looks good.

Ed MacVaugh Take care of life, Lotus has been waiting on and off for decades :) +2

Chris Mahalick Check to ensure that the butterflies in the emissions manifold (where the cross pipes come in) are opening. Move them around a bit when trying to start the car. You are just about there. +1

Brian Lepkowski Sounds like it's ready to fire!

Russell Baden Musta Soon it will purr!

Frank Mummolo Hey, Rob, you've earned it. Back off, enjoy the holiday and come back next week with a vengeance. I don't believe any of what I just said. My balls are exploding! +5

Charles Morris [link to video "My nipples explode with delight"] +2

Brian Lowder I actually prefer the suspense. A non-running car has huge potential; once it's running, it's all about the routine maintenance. +6

Clyde Gates I agree that Maire Anne is far more important! +1

    Maire Anne Diamond Thanks, Clyde! +1

John Whetstone Damn that is a shame! Good luck with the "other stuff." +1

Bill Ennis Don't give up... gotta be a simple fix

Dale Robert Olson Having your priorities in order goes a long way to a happy and healthy life. Hang in there, all in good time. +1

Gary Beck Keep messing with it. You'll figure it out. I got mine running and it wouldn't shut off. I had the wrong 12v wire at the coil.

Lindsey Brown Condenser.

TLC Day 2141: Very close

APR 20, 2019, 2:35 PM

The Lotus Chronicles, Day 2141 (afternoon special edition):

OK, now I really AM fucking with you and trying to give you blue balls. But I found another 15 minutes and got very, very close. Besides, you folks have been along for the ride for years. If I have blue balls, why shouldn't you?

(Hopefully I'll be back later. Another 15 minutes and I might be able to, um, finish this off :-)

[search YouTube for "The Lotus Chronicles near-start with birdseed)

84 LIKES, 87 COMMENTS

> **Dohn Roush** Beware of premature ignition... +2
>
> **Layne Wylie** I assume you're aware of the ballast resistor failure issue where the engine will run while cranking, but not after you stop cranking. Might be a possibility. +1
>> **Rob Siegel** Thanks, no ballast resistor in this car. +1
>
> **Doug Jacobs** Was that prog rock in the background? Maybe you should try "Born to be Wild"? +1
>> **Rob Siegel** Whatever show was on WMBR at the time 🙂
>
> **Charles Morris** [link to video Monty Python "Drop your panties, Sir William!"]
>
> **Kai Marx** I am so excited to see it done!!!
>
> **Russell Baden Musta** ALMOST!!!
>
> **John Whetstone** I'd be holding those throttles open at that point figuring it might be getting flooded, but I don't know much about nuthin'.
>
> **John Harvey** It's been a long time since I replaced Strombergs with SUs...
>
> **Steven Bauer** Sounds like the ignition is only getting power in start...
>> **Rob Siegel** I agree with that assessment.
>>
>> **Steven Bauer** Rob Siegel do you have 12V at the coil with the key at "on"? I apologize for trying to advise the master... +1
>>
>> **Rob Siegel** The "start" setting on the key wasn't cranking the starter, which is why I am using a remote start switch. In order to clip the remote start switch on to the speed terminal on the starter, I pulled off the one from the ignition switch. I'll freely admit it's possible that there is a bit of wiring that I don't understand. +2
>>
>> **Roy Richard** Rob Siegel well hell neither did Lucas when he designed it.
>>
>> **Steven Bauer** You can always jumper coil + to 12v. +1
>>
>> **Rob Siegel** Steven Bauer, I had checked that I was getting 12 V at the coil with the key cracked to ignition and left it at that. I now can see I need to look at it a little bit deeper. +2
>
> **Tony Pascarella** Are there points in the distributor or does it have a solid-state ignition?
>> **Rob Siegel** points
>
> **Robert Myles** Rob Siegel, remember, it's English. That means positive and negative are sometimes interchangeable and there's the phantom third polarity you've got to worry about. +4
>
> **Roy Richard** Rob must make getting rid of ground loops in your sound system fun.
>
> **Markus K M Bosch** Didn't realize that you needed to pump the accelerator on SU type carburetors....... 😀😀😀😎 +3
>> **Scott Linn** Markus K M Bosch Zenith Strombergs
>>
>> **Russell Baden Musta** Markus K M Bosch my experience with my Europa you don't...
>>
>> **Markus K M Bosch** Russell Baden Musta, my thoughts exactly 😀😀😀
>>
>> [Author's note: you don't, and I was wrong to do so]
>
> **John Harvey** Plug those vac tubes on the carbs!!!
>> **John Harvey** It won't pull fuel out of the float bowls if there is a vac leak. Are those vac ports on the top by the throttle body? Is the choke pulled?
>>
>> **Rob Siegel** Vacuum ports are plugged.
>>
>> **John Harvey** Rob Siegel choke? Many SU's won't start cold without it

Rob Siegel John Harvey, not currently hooked up, but as others have pointed out, I think it's an ignition wiring issue.

Franklin Alfonso Yeah, Brit cars love to be choked... +3

Scott Linn Mine used to readily start when it was set very rich pre carb rebuild. Now I always need to use the choke when cold. But yours does seem to be missing something when the starter stops. +1

Scott Linn Is this car still running on points?

Rob Siegel Not just points—original points! (I have new P+C to install if need be) +4

Daniel Senie Rob Siegel you know how tube amplifiers from decades ago have to have their capacitors replaced due to degradation? Condenser is another term for capacitor... +2

Rob Siegel Daniel Senie, the problem is that new condensers are crap. I have one, but I would only put it in as a last resort. The car HAS spark. +2

Bob Sawtelle "Finger in my ear"... ya think you meant something else...lol. +5

Adam G. Fisher Bob Sawtelle that1 was my favorite part! +1

Rob Siegel It's a Bill Murray quote from "Meatballs" +1

Bob Sawtelle great movie! "But, the real excitement, of course, is gonna come at the end of the summer, uh, during Sexual Awareness Week. We import 200 hookers from around the world, and each camper, armed with only a thermos of coffee and $2,000 cash, tries to visit as many countries as he can and the winner, of course, is named King of Sexual Awareness Week and is allowed to pillage the neighboring towns until camp ends." +1

Bill Snider Today on BMW CCA News: "Rob Siegel was unaware of the tribute his fans now pay him; they now salute each other when approaching from opposite direction, not by waving, but by sticking a finger in their ear." +4

Skip Wareham The motor can't breathe because the exhaust is clogged with acorns. +6

David Weisman Skip Wareham and a chipmunk nest +1

Rob Siegel Possible but doubtful.

Jeff Baker Skip Wareham at least their weren't feathers coming out! +6

Karel Jennings Oh my godddddddd!!!!!! I was thinking shed fire. Good luck on the next try. +1

Brian Ach Is there gas in it? LOL +1

Rob Siegel As steely Dan said, "Is there gas in the car? Is there gas in the car?"

Dave Borasky I don't wanna do your dirty work… no more…

Steve Nelson Confetti launcher 😊

Rob Koenig Unfollowing. +4

Neil Simon Sure hope this chapter has a happy ending... +2

Franklin Alfonso Maybe he should check with Robert Kraft. He knows all about happy endings... +2

Chris Langsten That's the most high-speed carb pumping, start you bahstid, first start video I've ever seen. LOL! So close! +1

Rob Siegel Freshly charged battery 😊

Chris Langsten Rob Siegel I meant your feverish actuation of the throttle plates. Not engine rotational speed. 😊 +1

Rob Siegel Chris Langsten, to continue the sexual metaphor, I was trying to, you know, pump... +1

George Thielen I know you were raised in the Jewish faith, but you must agree that only a heathen would finally resurrect this the day before Easter Sunday. +7

Rob Siegel As Mary Chapin Carpenter said, "I take chances. Every chance I get." +2

Franklin Alfonso Maybe it would start if it was parked in a cave with a rolled stone as a door? +5

Ernie Peters Quit monkeying around and get that thing running! Seriously, do you have a good Zenith-Stromberg manual? There are a couple of things that have to be real close to spec (not perfect, but close) for it to get fuel. [photo of monkey that literally has blue balls]

Franklin Alfonso Damn didn't start and now look what Lolita did to me!

Alan Hunter Johansson Well, that was anticlimactic.

Roy Richard Rob Siegel Just don't yell or post "It's Alive" when it starts. That's when the shit really hits the fan in the Frankenstein movies. All downhill from there. +1

Kent Carlos Everett "Sometimes the best tool in your tool box is a walk around the block." That saying is from me. Well learned. +1

John Webster Nickols You are so close!!! And there's some little nuanced thing that you will find tonight.

TE Cole I called the Old Lotus Help Line for you. They have tech center in India. They asked if you put fuel in it. +3

    Rob Siegel That was after they said "Mr. Siegel, we're sorry you're having this problem…"

Scott Aaron I need to DVR this so I can fast forward through some of this stuff. 😊

Philip D. Sinner *180° out, aye. Good on ya!!! +1

    Rob Siegel it WAS!

Russell Baden Musta Was that popcorn coming out?

Guy Foster It reminds me of a 68 Mustang, I had with a 289. I was trying to start it and it wouldn't fire. I opened the hood to see if I could see anything. When trying again, the wife comes out and asks, "what's wrong." I say, "I don't think I have spark". She replied "yes you does, I saw it". It turned out to be a bad distributor cap that I took from a totaled car. 😊

Paul Hill Did you stand to attention and sing The National Anthem, God Save Our Gracious Queen? I thought not… try that and then attempt starting.

Rennie Bryant Ya know, Strombergs don't have accelerator pumps. So pump the throttle all you want. Ain't nothing gonna happen.

    Rob Siegel motherFUCKER! I actually did not know that. +2

John Jones Do you have this book? The author seemed like he knew his stuff, at least back when he wrote it. [photo of my electrical book] +3

Charles Gould Instead of pumping the throttle, try just cracking the throttle ¼ to ½ inch and it should fire right up. Resist the urge to pump the throttles as the high revs of all the fuel pumped into the chamber causes starter bendix to release before it is running. Even throttle slightly above idle is much more effective on European cars. +2

    Rob Siegel Charles, read the next post. That's exactly what I did, and it started 😊

    Scott Aaron Charles Gould you're the man

Brad Purvis I'm sorry, but my balls are made of brass.

Jeffery W Dennis Almost there!

Chapter VIII: Sploosh

TLC Day 2142: She is risen

APR 21, 2019, 8:03 AM

The Lotus Chronicles, Day 2142 (Easter morning edition):

It is particularly appropriate that on this, Resurrection Sunday, I can report:

She is risen.

It is particularly INAPPROPRIATE that I can report what's REALLY on my mind, which is:

Sploosh.

(Steven Bauer, I was convinced that you were right, that it looked like the ignition was only getting voltage while I was cranking the starter, but that wasn't the case. In the end, all it needed was the throttle cracked open, as Charles Gould suggested. In the video you can see a wrench sitting upright in the linkage. It's preventing the adjustment screw from sitting all the way down on the stop. John Harvey, you were right to ask if the choke is pulled. It's not currently hooked up. I can't find the damned cable in the engine compartment, but I see where it should be connected. Thanks to Rennie Bryant (and others) who said "pump all you want; Strombergs don't have accelerator pumps." I actually did not know that :^)

(Next tasks: Run the engine for short periods and assiduously check for coolant, oil, and fuel leaks. Figure out why trying to crank with the ignition switch doesn't work which forces me to connect a remote start switch. Then make second attempt at bleeding brakes, which is what's preventing the wheels from being mounted, the car from being taken down off jack stands, and attempting to move it under its own power. Well, that and the six years of crap sitting around and behind it (to steal Neal Neil M. Fennessey's line, that'll be the equivalent of rolling the stone away). Then, a thousand little things, as with any car that's been sitting for forty years. But this is huge. Thanks for being along on the ride. There's still much more to come.)

(Did I mention sploosh?)

[search YouTube for "The Lotus Chronicles first engine start"]

247 LIKES, 134 COMMENTS

Clifford Kelly It's been cool to watch! +2

Daniel Neal That is a most glorious sound, and in the spirit of the day (not that I am much of a practitioner of any religion), may I say, Hallelujah. +1

Jim Piette It's been fun and a privilege to look over your shoulder the past few weeks! +2

Neil M. Fennessey Congratulations! The stone is rolled away! +3

> **Rob Siegel** damn, I should've used that!
>
> **Neil M. Fennessey** Next article, be my guest!

Frank Mummolo Somehow seems appropriate, this particular day. Well done, Rob! We knew you'd get there! Now, of course, the real fun begins!! +2

Grice Mulligan Holy hell!

David Kemether [meme Patton "You magnificent bastard! I salute you!"] +8

Carl Baumeister [the guy I bought the car from] That's awesome! +1

> **Rob Siegel** Hey, Carl! I've been meaning to drop you a line! It's been a long slow slog with years-long periods of inactivity, but this winter was shit-or-get-off-the-pot time, and with a four-month push, I finished the engine assembly and got the drivetrain back in. Very satisfying! +1
>
> **Carl Baumeister** Rob Siegel I have been watching the whole process unfold, with the highs and lows but this point in the reservation is clearly a high point. Very satisfying indeed!!! I am so happy for you and for the much-needed rebirth of the Lotus! +3

Lee Highsmith Awesome! Big step! +1

Trent Weable Congratulations Rob! I think I speak for all of us when I say we feel your joy. +4

Paul Hill Praise Cheesus on chocolate egg day +3

Stéphane Grabina [gif meme Charlie Sheen "Golf clap?"]

Stan Chamallas Somewhat blasphemous if you were Christian...

> **Rob Siegel** yes, well...
>
> **Stan Chamallas** hahaha
>
> **Rob Siegel** ...worth it :^)
>
> **George Zinycz** Stan Chamallas don't forget that Christ was Jewish! +2
>
> **Stan Chamallas** George Zinycz of course. That is why one should not celebrate Easter until after Passover/

Justin Gerry I'm a believer. Finally. I remember dual SUs on my dad's old TR6. You did need choke, at least 50% or it would not start. And I do remember having to lightly crack the throttle open at the same time. +3

> **Robert Zockoff** Same thing on my 1968 Volvo P1800 and my 1972 Triumph Bonneville motorcycle.

Dohn Roush Mazel tov! +1

John Whetstone Hey that's what I said! +1

John McFadden "Period correct" (released in 1974) music to go along with the Easter theme (plus there's never a bad time for Mott the Hoople) [link to Mott the Hoople "Roll The Stone Away"]

> **Ed MacVaugh** Saw them in Scotland in '73, supported by Queen [link to video "the story behind the song Roll the Stone Away"]
>
> **TE Cole** No this is more appropriate. [link to video Eric Burden "Low Rider"]
>
> **TE Cole** or this... [link to video Santana "Oye Como Va"]

Tom Smith Huzzah! +1

Joey Hertzberg Right on!! +1

Neal Klinman Sploosh?

Rob Siegel Sploosh. +1

Charles Gould Finally. Congratulations!

Andrew Wilson Mwahaha! Winter is officially over! So happy to see this finally running! +1

Russell Baden Musta IT HAS RISEN!!! Congrats Rob!!! Now really, next time you have a guitar shipped in, cut the cardboard neatly that it came in, make it about 3' wide, about 5-6' long and lay it under the car. Oil will find its way out, but nowhere nearly as bad as mine! I cannot wait to see it!!! If you don't come to LOG, we will do a day trip to you to view it!! +1

Bo Gray Plate: è risorto +1

Brian Stauss Well done, Rob! Thanks for sharing the ride with us! +1

Dave Gerwig What made it go from kick-stumble to purrs like a kitten? Sounds great!

Rob Siegel Cracking open the throttle (after rotating the dizzy 180 :^) +1

Franklin Alfonso [link to gif of Charlton Heston as Moses parting the Red Sea] +1

Craig Fitzgerald Amazing!!

Rennie Bryant Congratulations! One of the most satisfying things in life is the first start of a freshly rebuilt engine. That simple thing justifies all the work, Pain, sweat, and time invested in it. +1

Clyde Gates I also add my congrats! Does, perchance, the starter have 2 nuts/flat spade connectors/contact places to which the small ignition wire from the switch may connect?

Greg Bare [link to gif of man, uh, polishing the suggestively-placed knob of a chair] +3

George Zinycz A Passover/Easter miracle. This shit's getting biblical! +1

Luther K Brefo [link to gif of a distinctly vaginally-shaped cave entrance with a discharge pipe at the top, meme'd as "SPLOOSH!"]

Luther K Brefo [link to gif of Amy Schumer in a little black dress getting drenched head to toe with milk]

    Rob Siegel I suppose I deserved that +1

Glenn Koerner Great news. And all done in less than one(dog) year. 🌟🌟🌟 +1

    Rob Siegel Well, after the previous five!

TE Cole So the Lotus is symbolic of the resurrection. How appropriate on Easter Sunday. It appears that the Lord Jesus Christ has favor with your car Rob. That is truly something. The holy Lotus. +3

    George Zinycz TE Cole All hail the Holy Lotus! +2

Larry Webster Wow!! Congrats!

Luther K Brefo [gif Andy Samberg masturbation face] +1

Thomas Jones [gif Mick Jagger struttin' it] +1

Thomas Jones Oh yeah, gimme some more of that throttle baby! Now, put your back into it and roll that stone away! +4

Gene Kulyk Congrats! Are you on the downslope of the punch list yet? +1

Scott Aaron Pretty exciting, Rob. Man that was a bit of a climb. It will be crazy to drive that thing—so different than an 02. +1

Paul Skelton Enjoy the first drive!

Rob Siegel That probably won't be for two months 😊 +1

Paul Skelton Have you ever driven one Rob?

    Rob Siegel Paul, only once. And I liked it a lot. Very simple and direct.

    Paul Skelton I drove Pete's around quite a bit. It was a blast but a wee bit cramped at the pedals and visibility is a terrible at the rear.

    Rob Siegel As Raul Julia said in The Cannonball Run, "First rule of Italian driving: what's behind you, it's-a not important." I know it's not Italian, but it's what I got. +2

    Philip D. Sinner Rob Siegel schoolmates' father immigrated from Italy. Sergio snatched the rearview down and tossed it in the back of his brand-new Cadillac. Same exact sentiments were forthcoming, with a wonderfully thick accent. I'll never forget. The image endures! +1

    George Zinycz Rob Siegel he said that in the Gumball Rally!!

    Rob Siegel My mistake!

Craig Fitzgerald Psalm 15:57. This is the day that Chapman has made; Lotus rejoice and be glad in it. +8

    Craig Fitzgerald Subtitle: He Is Risen. [gif for knob labeled "Pull for quick erect"] +1

    Frank Mummolo Moan... 😊

    Rob Siegel Craig Fitzgerald, truth!

Fred Blackall [gif of Minions going "yay!"]

Lenny Napier [the Amy Schumer sploosh]

    Rob Siegel Luther beat you to it

    Lenny Napier Rob Siegel Damn!

    Rob Siegel still, I never really object to a second sploosh

Steve Woodard C'mon, this must have been the plan all along; the timing is exquisite. Well done!

    Rob Siegel (actually, the video of me starting it was done last night... shhh... don't tell anyone...) +1

    Bill Ecker Steve Woodard it is so pat. The powder coater, the timing cover, the clutch pin. All gloriously synchronized to culminate on Easter. Well played sir. Well played. +2

Kent Carlos Everett Wohooo!
Martin Meissner [gif of Conan O'Brien salute]
Lenny Napier [gif of elephant with a fifth leg… no, wait a minute…]
> **Rob Siegel** oh dear god

Kevin Pennell Well done Rob Siegel, well done!!
Fred Aikins Woo freakin' hoo!!! Congrats, Rob, on the successful resurrection of the Lotus (engine) from its slumber! It is risen! +1
John Jones Huzzah! +1
Graham Thomas [gif Rick Flair "WOOOO!"]
Eric Pommerer [gif very cool pen-and-ink mechanical heart beating] +1
Brian Hart [gif Stewart and Colbert drinking coffee, on the street, looking at something, saying "WOW!"]
Franklin Alfonso Lolita has risen! [gif of Bride of Frankenstein]
> **Roy Richard** Franklin Alfonso And that's when everything starts going downhill 😷

Bruce H. Black Congratulations!!!
Franklin Alfonso [gif of Woody Allen "How'd you like it?"]
Franklin Alfonso Lolita… [gif of gorgeous Italian woman smoking a cigarette]
Jose Rosario Yeahhh.
Brad Purvis The Lotus is not yet finished teaching you, young Siegel. There is still much to learn… much to learn. +4
> **Rob Siegel** You mean "much pain to be inflicted and tolerated." +1
> **Brad Purvis** Rob Siegel Same difference. 😷
> **Russell Baden Musta** Rob Siegel yeah.

Steve Park The sound of long dormant motor springing to life is one of the joys of being a car guy who gets his hands dirty. Congrats. And a most appropriate day. +3
Doug Jacobs You make me so proud! +1
Scott Fisher "Strombergs don't have accelerator pumps," "I actually did not know that :^)" Stromberg and SU carbs don't have accelerator pumps, but there's the "accelerator pump effect" common to both. Because the dashpot rises in response to engine vacuum (after you mash the gas), there's a brief moment when the airflow into the carb is sped up, in spite of the "constant velocity" nature of the design. During this brief moment, the added speed of the intake air over the needle/jet combination causes a little more gas to be extracted, as per Signore Venturi's research. This has the same effect as an accelerator pump on, say, Holley carburetors. You can fine-tune it, if you have access to an exhaust-gas sensor of some kind (or better yet, a rolling road dyno) by changing the viscosity of the oil in the dashpot dampers. Yes, the British figured out a way to make even the carbs leak oil.

When I used my MGB as a daily driver about 30 years ago, I used to change the oil in the dashpots seasonally—heavier in summer, lighter in winter. Then something in the recirc system went wacky and gas poured out of the evaporative canister, so I bought a 45DCOE from a friend and never looked back. Between the Weber, the head work, the recurved distributor, the cam and the compression, a friend's G. Analyst said we were cranking out about 158 bhp at the wheels in that car. Big fun. +3
> **Rob Siegel** Thanks Scott! +1

Scott Linn So awesome. +1
Caleb Miller Clearly you anointed the hood with blood… +1
Brian Ach it runs too perfectly, please do something to correct this
> **Rob Siegel** It's a Lotus. that will be self-correcting. +5
> **Brian Ach** Rob Siegel genius
> **Steven Bauer** It will run fine as long as you don't drive it.
> **Rob Siegel** Simplify, then add downtime?
> **Steven Bauer** Rob Siegel Exactly. "The failure is the result of a user error. You tried to use it." +2

Bronco Barry Norman Ether is our friend! (A club for side draft carb owners!) +1
Paul Wegweiser I love how there's some part of the linkage tugging on a spark plug wire. That won't ever cause a problem. Nope. Ever. You know I love you… but it was the first thing I saw, with the sound turned off. (spark plug wire twitching with choke or throttle movement?) +1
> **Rob Siegel** I think that all that is is a twist in the just-installed plug wires getting jostled by the engine vibration. The linkage wasn't moved at all. But I'll check, thanks. +1

TLC Day 2143: Starter wired

APR 22, 2019, 7:34 PM

The Lotus Chronicles, Day 2143:

Now that the engine starts, I wanted to address the fact that I had to start it using a remote start switch because the starter didn't crank with the key. I verified that I had connected the starter exactly the way it was when I bought the car. I found that there was no voltage on the wire from the ignition to the starter when I cracked the key (obviously). I traced the wire forward from the starter and found that it and another wire weren't part of the wiring harness, and instead were suspended loosely under the car and came out behind the driver's seat. I pulled up the seat, and was stunned to find both wires underneath it, both with ring terminals on the end, and both connected to… nothing.

So then I looked at the ignition switch itself. I pulled to the back off it, and laughed when I found that there was nothing connected to the starter terminal at all. The nearest guess I have is that maybe at some point someone had installed a starter relay, but nothing like that was there now.

For now, I simply spliced in a length of wire and verified that the starter cranks when I turn the key.

Next, I connected to choke cable. The Lotus now starts up like a real car when you simply pull the choke cable and crack the key to start.

Tomorrow, hopefully I can let the engine run for a while while the car is still up on jack stands. If I don't see any egregious leaks, I can bleed the brakes, bolt on the wheels with the new tires, and for the first time in nearly 6 years, let the car down on all fours.

And THEN… I have to clean all that crap in the garage that's preventing me from moving the car.

(and fix the tach.)

(search YouTube for "The Lotus Chronicles first start with key")

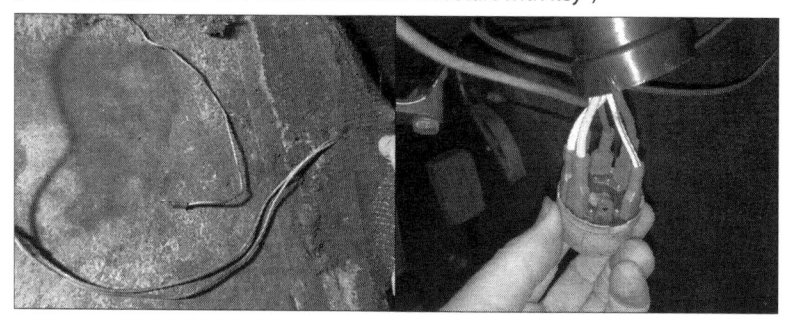

114 LIKES, 38 COMMENTS

Brandon Fetch Git'R'duN!
Bill Snider (Ahem, before the Villagers arrive) ;-)
Bob Sawtelle The great thing about a Lotus is nobody questions whether that 5-digit odometer has been around more than once. +12
Marc Touesnard It's a great feeling when things start coming together! Great work!
Tom Egan A "problem"? With Lucas electrics? Surely that's not possible! +2
John Whetstone Even cleaning the garage sounds easy in comparison! +2

Ernie Peters If you need gauge (tach) repair, I used Nisongers. +2

Rob Koenig Bitchin! Simple fixes are the best fixes. Especially those that offer a high level of convenience restored. +1

Marc Schatell This is awesome! +1

Joshua Weinstein All this work is inspiring but at the end you'll have… a… Lotus. Do you realize that? +2

    Rob Siegel fuck yeah. +9

Justin Gerry People always blame Lucas and most of the time it is a previous owner unplugging stuff. +4

    Scott Linn In reality, most problems with Lucas electrics are owners.

    Justin Gerry Yeah everyone pokes fun at them. Rightfully so. They just used weak connectors that were not weather tight. Wurth contact spray and you have on and off like normal. But nope, usually previous owner would hack, cut, splice, unplug, hide and tape everything and leave the mystery for the next guy. Lol. +1

    George Murnaghan Justin Gerry Word

Roy Richard In his best Mexican accent: Gauges, we don't need no stink'n gauges.

Lenny Napier [gif That 70s's Show "You are a god! A GOD I SAY!"]

Marshall Garrison Rob, getting almost painfully close to it moving under its own power! Considering the likely massive endorphin rush the first time you back it out the garage after 6 years immobilized and however many more decades not running, will you have someone on hand with smelling salts in case you pass out from endorphin overdose? :D +3

Dave Gerwig A real jigsaw puzzle 😊

Robert Alan Scalla Starting and running on command, a great move forward! +1

Clyde Gates Always a negative comment, Moriarty (hope I remember the name correctly from the movie), always. +2

    George Zinycz Moriarty always had the negative WAVES man… +1

John Jones a) Great news on another big milestone, and b) be careful if you don't have any leaks. That typically means it's empty. +3

    Rob Siegel right, that just means the leak is in the past tense +1

Brad Witham I empathize with the garage-cleaning part! +1

George Zinycz I wanna come over and play! +1

Paul Hill Sweet indeed. +1

Brian Hart Maybe that's a telltale tach, and that's the reason it came off the road in the first place…?

    Rob Siegel Actually the tach sprang to life this morning :^) +1

    Brian Hart Typical +1

    Bill Snider Rob Siegel Tachs' followin' your lead… as it were. ;-) +1

    Franklin Alfonso [gif Sean Hayes "OMG this is so exciting!"]

Brad Purvis The LBC gods are smiling upon thee. +1

Brian Ach buy halon fire extinguisher. put in car. thank me later.

    Rob Siegel Already one in the Lotus.

TLC Day 2144: Smoke show

APR 23, 2019, 7:38 PM

The Lotus Chronicles, Day 2144:

Another pretty significant day in Lotus Land:

--Tried again to bleed the brakes, this time making a reservoir cap that I could pressurize with my Motive pressure bleeder. Must've run two quarts of brake fluid through it. The pedal now has some feel and pumps up somewhat, but I think that, in the end, I'll have to replace the master and the rest of the brake hydraulics. After all, the car did sit for 40 years. And, as I've said, that will be a pain, as the master is NLA and there IS no bolt-in replacement; some amount of adaptation is required.

--Ran the engine for 30 seconds, then a minute, then two, then five, internally burning the Graphogen engine assembly paste off the rings, and externally burning the crud off the exhaust manifold and exhaust which had been sitting under the back porch for years. Quite a smoke show, as per the video. By the time I was done, both the tailpipe and engine compartment were burning pretty clean.

--Looked carefully for leaks. The water pump is leaking a tiny amount out the weep hole at the bottom, I think because I didn't put RTV on the O-rings of the pump cartridge as was recommended, a decision I now regret. At one point, coolant began pouring out from under the cap on the swirl pot (reservoir). Cap could conceivably be bad. Temperature gauge is running above the middle; who knows if it's accurate. Electric fan isn't turning on. Have to be VERY careful not to overheat the rare expensive Lotus Twin-Cam head. Noticed that the water pump belt I had on was waaaay too loose. Replaced it with the slightly shorter belt I'd bought. Was gratified that this is indeed possible from under the car. Still not certain if it's possible to pull the pump cartridge. If it leaks more, I guess I'll find out.

--Mounted the wheels and tires and let the car down off the jack stands where it's been for nearly six years. It'll have to go back up on stands to do the brakes, but not right now. So why'd I let it down? Because I bloody wanted to. I mean look at it.

--With all four of the Lotus' paws finally on the ground, sat in the seat and, for the first time, started it and operated the brake, clutch, and accelerator. Everything you've read about not being able to drive a Europa with shoes on is absolutely correct.

--Bounced all four corners and found that none are obviously seized or bouncy, though the front end is sitting a little higher than the rear.

--Almost tried to move the car a foot forward and a foot back just to achieve the "moved under its own power" milestone, but then my left brain said "If the clutch grabs, you're going to crash into the shelves behind you or the wall of the garage in front of you don't be a fucking idiot." I listened to it. It's a good left brain.

--I have so much crap to move before I can safely move this car under its own power.

But still... pretty damned cool, no?

(search YouTube for "The Lotus Chronicles smoke show")

180 LIKES, 102 COMMENTS

Graham Perry Smokes like burning Bovril
Ed MacVaugh You can't just get a rebuild kit for the existing brake components?
    Daniel Neal Ed MacVaugh I was going to ask this exact question

> **Ed MacVaugh** Daniel Neal Great minds, and all :) +1
>
> **Rob Siegel** You can, RD Enterprises sells it. But a) I really hate rebuilding brake components, and b) what I've read is that the bore often needs to be re-sleeved. There are companies who re-sleeve master cylinders, but I haven't looked into it in detail yet. +3
>
> **Ed MacVaugh** Rob Siegel White Post did my GT6 master cylinder for around $80—in '89. +1
>
> **Chris Reinke** I'm curious how the dual brake boosters held up.
>
> **Rob Siegel** Chris, don't know yet.
>
> **Scott Linn** Chris Reinke good question. They often don't. Most people just bypass them. I was lucky that the PO of mine had them rebuilt.
>
> **John Harvey** Rob, Apple Hydraulics. +2
>
> **George Murnaghan** Ed MacVaugh I've also used Apple Hydraulics to sleeve brake components- did a good job. +2
>
> **Chris Etridge** White Post has done right by me on several occasions +1

Charles Gould Are you familiar with White Post Restoration in White Post Virginia? They do incredible work on reboring and re-sleeping master cylinders, wheel cylinders and calipers. They bore out the original bore and install brass or stainless steel sleeves and rebuild to the original or desired new bore size. I have used them for years. They do exceptional quality work at very reasonable prices with very quick turn around times. I am sending them eight Dunlop brake calipers from a 1961 Jaguar XK150 Drop Head Coupe tomorrow morning. They will have them done in four days with disassembly, media blasting, re-bore and re-sleeve, new seals, rebuild and reassembly at $85.00 per caliper. Master cylinders are even cheaper. Tell them that I referred you. +2

> **Scott Linn** That would certainly be a great choice. I wish I had the original M/C. The PO had it but didn't give it to me with the stash of parts when I picked it up because he had stuff all over the place and couldn't find it all... Very disorganized given he had weeks of notice before I picked it up.

Russell Baden Musta The front does sit high on federal Europas Rob. With adjustable spring towers you can lower the front a bit but it works fine like this. You are just kicking butt on this! Will see you in August if not sooner!! +3

> **Rob Siegel** Thanks Russell; I didn't know that.
>
> **Scott Linn** Dang, Russel beat me to it. The front end sits high in Federal Specials unless modified (many are). +3
>
> **Russell Baden Musta** Scott Linn as mine is slightly... shorter front springs on adjustable spring tower Spax gas adjustable shocks... lower than federal TC, higher than S2. About 5 years ago tore lower left control arms off in low speed head on collision. Couldn't get OEM, insurance bought off on complete replacement tubular adjustable front suspension. Dimensionally same as OEM but caster camber tow adjustable. The car drives like it did new once again. +1
>
> **Scott Linn** Mine came with adjustable perches so it sits flatter. +2
>
> **Rob Siegel** Great. Just fucking great. So now, after everything I've gone through, after I get this thing running and driving, I NEED TO OBSESS ABOUT FRONT END RIDE HEIGHT? You guys suck. +7
>
> **Russell Baden Musta** Rob Siegel 🍺 😂😂😂😂😂😂😂😂

John Wanner Yes. Yes it is.

John McFadden Is rearward visibility nonexistent in a Europa?

> **Rob Siegel** The rear windshield is literally 4" high, so yes.
>
> **Matthew Zisk** Why would you ever need it, when nothing can keep up? +1
>
> **Rob Siegel** so you can see the tow truck when it comes +9
>
> **Scott Linn** The TCS rearward visibility is quite a bit improved from the S1/S2 models...! +1
>
> **Bill Snider** [link to video, The Gumball Rally "What's behind-a me, it's-a not important!"]

Ernie Peters Nicely done. +1 on White Post Restorations. +1

Tom Egan We need an outing with MA TCS owners, you, me and Bailey in the near future. +2

Doug Kwartler Rob I did the exact same things today. +1

> **Rob Siegel** yes, but I can't pitch-correct the Lotus' exhaust note :^) +1

Scott Linn My brake master was replaced when I got it. I can try and figure out which it is, however I'll probably have to get it on a lift to remove the panel. The replacement one was fairly drop-in. +1

Thomas Jones Oooo, ahhh... Ohmmmm. Soon Lotus driving position I see you in. +2

John Whetstone Couldn't have been a good day when you decided to install shelving behind the Lotus. +2

Rob Siegel They're wire shelves on wheels. I was smaht.

John Whetstone Rob Siegel I knew!

Chris Roberts Looks awesome, Rob ;-) Cool to see it sitting on the ground with 4 new tires. The anticipation of the first drive must be pretty satisfying +2

Eliot Miranda The front riding higher may be the adaptation required to meet federal headlight height regulations and is correctable. It can be returned to how God^H^H^HColin intended. +1

John Jones Eliot Miranda plus, cutting springs "adds lightness" +2

Greg Lewis Awesome!

Bob Jaffray Way to go, Rob. Yesterday, here in MN, winter has finally passed and it's fun to see the stored cars and cycles begin to hit the road. Easter Sunday, a gorgeous day in the 70's, I was passed by a beautiful maroon Europa, and I must say, it looked awesome!! What a cool low stance... +1

> **Rob Siegel** Yes, they are tiny, low, gossamer-like, and unlike most anything else you see on the road. It was the sighting of "The Lotus in the Wild" on the way to The Vintage two years ago that made me un-back-burner the project. I'm now FB friends with the guy who owns it, Russell Baden Musta. I was just on the phone with him for an hour :^) [

Brian Stauss Well done, Rob! Wasn't sure I'd live to see this day, but here it is! Anxious to see a video of that first trip around the block! +2

Gary Beck Way to go Rob. I drove My E9 around my neighborhood today with no hood. Made a lot of noise and made the neighbors take notice. Yes, that first movement under their own power after 3 years is pretty emotional. Longer for you. +2

Francis Macaluso This is awesome. I've loved following up on this! +1

> **Rob Siegel** Thanks! I try to be, you know, entertaining through my pain and dwindling bank account... +1
>
> **Francis Macaluso** Rob Siegel really unique stuff Rob. My favorite Lotus is the Espirit. Maybe your next project?! +1

David Kemether

+3

Scott Aaron I'm setting my alarm for 2025. Can't wait to see you driving it! 😀 +4

> **Rob Siegel** oh ye of little faith +1
>
> **Scott Aaron** Yes, that tends to be my department

Jeff Hecox Maybe try this Gravity brake bleed procedure the PMB brakes recommends. http://m.pmbperformance.com/gravity-bleed.html

> **Rob Siegel** I can't imagine that, with this car with its weird master/slave boosters in the engine compartment between the master cylinder and the calipers/wheel cylinders, gravity-bleeding is better than power-bleeding.
>
> **Jeff Hecox** Rob Siegel the key is having long tubes on the bleed bottles and suspending them above the master cylinder reservoir with the bleeders open. It's a low-cost procedure that only requires several feet of tubing.

Dohn Roush Here's one of those peaks that make you forget the valleys. Great work... +2

> **Rob Siegel** thanks Don +1

Scott Linn Mine has a Nissan F10 master cylinder. [link to article on lotus-europa.com "Converting to a Nissan F10 Master Cylinder"]

> **Rob Siegel** no longer available

Chris Lordan In the never-too-early-to-plan department, have you picked out a snow tire yet? (... and with tongue firmly in cheek, I remain, Yours, etc etc...) +2

Rob Siegel studs on all four corners, baby! +1

Brad Purvis I love the smell of unburned hydrocarbons in the morning. It smells like victory! +1

John Graham My favorite post for "The Lotus Chronicles" so far. So exciting to see it sit on its own wheels, suspension and running. Not sure how you stick with it, I would have given up a long time ago, but you just keep going. Can't wait to see the first video of you driving the car down the road. +2

> **Rob Siegel** Thanks John. In projects like this, even if you admit that you made a mistake, at some point you need to get out from under it by either a) dumping it or b) digging yourself out and trying to complete it. One of my mantras is that time will pass anyway regardless of whether or not you're doing anything on a project, so if you chip away at it, as time passes, you make progress. +2

Steve Park Damn that's a good-looking car! I'm really happy for you. +1

Paul Skelton See what I meant by a tight pedal area? LOL

> **Rob Siegel** I do now!

John Jones Very, very cool! +1

Charles V. Stancampiano Could be you have to take the head off to pull the water pump. That would be bad...

> **Rob Siegel** No, I have the Bean cartridge-style water pump kit.
>
> **Charles V. Stancampiano** Didn't know there was such a thing. Thanks!
>
> **Rob Siegel** Charles V. Stancampiano The cartridge-style pump and the front timing cover and back plate that go with that are available from both Dave Bean and Burton engineering, but only Bean has one that fits the Europa.

TLC Day 2145: Freeing Lolita

APR 24, 2019, 9:19 PM

The Lotus Chronicles, Day 2145:

Well, it's a start. At least I can see the floor.

106 LIKES, 49 COMMENTS

TE Cole I would think the Lotus had grown roots by now. +4

Brad Day That feeling of reassembling with all the parts you've been saving for years is the best. Almost as good as the feeling when you realized, "Holy shit, I've been taking up all this garage space with this crap!" +4

> **Rob Siegel** Yep. I'm nearing the point where I need to be absolutely merciless in throwing stuff out in order to clear real space and not just move around crap. +5

Paul Garnier Less itchy. :-D +3

Jeff F Hollen Can't wait to see this at Vintage. +3

Paul Garnier This pic with you in it is for the grandkids, btw +1

Chris Langsten When does the poll start on what replaces the Lotus in the garage??? I am logging formal guess of American-based wagon. +1

> **Rob Siegel** That implies that the Lotus, in addition to running and moving, is actually going somewhere. +1
>
> **Mark Thompson** Rob Siegel Rambler time. +3
>
> **Rob Siegel** if I find the right car, the right price, at the right distance, totally +3
>
> **Mark Thompson** Rob Siegel Would buy that book in a minute!
>
> **Chris Langsten** Rob Siegel

+3

Rob Siegel My wife's right. I really do have no secrets.

Robert Myles Rob, as are so many of our lovely brides, they are far wiser than we can ever give them credit for. +2

Bob Shisler I went out to work on my car in the driveway yesterday, but needed something out of the garage. Ended up cleaning out/organizing the garage for two hours. +2

> **Rob Siegel** The basic problem is that I never do that. I'm EXTREMELY focused on getting work on cars done, and just walk around piles of crap in the garage and make them higher. It's only something like this—actually not being able to move a car because it's blocked by crap—that forces me to stop and clean and organize. +7
>
> **Bob Shisler** Rob Siegel I'm working on organizing things in general. I've been cleaning and labelling with expensive special gaffer tape and special "industrial" sharpies (grease-proof both) all my sockets so I don't have to stand under a special lamp with a magnifier to read the markings. +4
>
> **Joey Hertzberg** Rob Siegel, I'm the same way. Perhaps unlike you, the projects (multiple, always) are taking 25%-50% longer to accomplish because I can't find the damn tool I need. This is self-defeating, really(!!!), and I've got to change it.. +2
>
> **Scott Linn** Usually I start working on something, pull out a tool, then have to fix the tool for some reason... +2
>
> **Rob Siegel** I wrote a piece recently about this sort of thing for Hagerty. It's very frustrating. The work is enough work without having to do extra work to do the work. +4
>
> **Ernie Peters** Mechanic school in the Army taught me: A place for everything, and everything in its place. Doesn't mean I strictly adhere to it, but it helps.
>
> **Pike Perkins** Rob Siegel [gif "Cooking with the Anal Retentive Chef"]

George Zinycz I'm not sure what the more major accomplishment is... getting the engine running or cleaning up? Jk... the cleaning is clearly the winner. You resurrect cars for breakfast, but tidying up is a real accomplishment! +2

> **Rob Siegel** truth
>
> **Dohn Roush** Cleanliness is next to impossible... +2

Rally Mono You know the nickname in the UK was squashed bread van. Good to see it running. +3

Rick Roudebush Did you find the missing bolts yet? +1

Charles Morris At least you HAVE a floor... 95% of the work on my '72 Dart was done in a dirt-floor steel barn :D

Bob Sawtelle This is now dare I say it in "The Endgame."

George Zinycz Half-time, at best

Rob Siegel Bob Sawtelle, "The Endgame" will be the title of my piece for Hagerty when I actually get the car to the point where I can drive it. +2

Tim Warner Hope the shelf rack system has wheels on it!

Rob Siegel It does! +2

Jim Strickland You are motivating more of us than you can imagine! +2

George Whiteley It is a GOOD thing that you love smaller cars and don't have a vested interest in Cadillac Eldorados! +4

Chris Roberts This reminds me of that guy that built a Ferrari in his basement. +1

Wade Brickhouse garage sale

Bill Ennis I've been working on a 1967 Nova sedan. Everything is backup and punt. I feel your pain. Endeavour to persevere! +2

Bill Snider Wait! I think I see a 10mm socket! ;-)

Rob Siegel where? where?

Bill Snider My error, just a dust mote on my computer screen.,. lol

John Graham We moved into our house a year ago. I vowed I would keep the garage organized. However, every time I need to work on a car or do something in the house or yard it is like a nightmare finding all of the things I had packed up. I think to myself "I can't stand how messy my garage is," then a second thought always comes to mind: "Obviously, not that much or you would have organized it by now." :-) +1

Sam Schultz I went to a physic this morning that told me when Rob is backing the car out of the garage he will find one bolt on the floor and then GET to disassemble and reassemble the Lotus again. +1

Francis Macaluso I know how you feel bud +1

Franklin Alfonso As a bonus, once you get it drivable you can make extra money using it as a Pet Hearse mobile. Small caskets will fit nicely in the back shelf. +1

Robert Alan Scalla I keep my garage very neat, nothing fancy, just functional neat.

TLC Day 2147: Room to roll

APR 26, 2019, 8:51 PM

The Lotus Chronicles, Day 2147:

I doubt that I WILL try to move to Lotus tomorrow, as it idles well but doesn't rev well, and as I am planning to go out to Fitchburg tomorrow and exercise the other cars, but if I wanted to, I now COULD.

Well, maybe just a few feet forward and back :-)

81 LIKES, 36 COMMENTS

Layne Wylie The garage is looking cleaner than ever
    Rob Siegel small steps +1
Jeff F Hollen Gotta make sure it'll make it to Hot Springs...
Brian Stauss Shaming me into cleaning up my shop...
Brian Lepkowski

+11

Dave Gerwig And what's behind Door #1 to the left? Recently unearthed because of a strong desire to get Lotush rolling.
    Rob Siegel The door on the left opens up under the kitchen porch. All that crap has to go somewhere. Right now, blocking it off is acceptable.
    Dave Gerwig Rob Siegel Storage Wars! +1
    George Zinycz Dave Gerwig that's the infamous garage with its own garage +3
Rob Blake How's it smell? +1
    Rob Siegel like victory :^) +2
    Dohn Roush Better than napalm... +2
    Carl Witthoft awful, just like my dog w/ his nose shot off. +1
    Rob Siegel That is NOT my dog. +2
    Ken Sowul I love the smell of unburned hydrocarbons in the morning. +1
    Rob Siegel don't we all +1
John McFadden Props on the prop rod 😀
    Rob Siegel It used to be a windshield scraper. I've upgraded. +1
    John McFadden The lumber definitely more "period correct" +1
    Garrett Briggs I think an extremely complicated system with pulleys and an old school engine crank would be perfect for that job. +1

> **Rob Siegel** you are correct sir!
>
> **John McFadden** Over-engineered like a German car, or cleverly engineered but poorly executed like a British car? 😂 +2
>
> **Neil M. Fennessey** Put a Lotus part number on it in black marker and tell everyone at the British Invasion that it's OEM. +2
>
> **Matt Schwartz** John McFadden my dad has a similar PVC pipe for the hood of his Elan and he's always called it "Lotus tool #1" +2

Russell Baden Musta Heading off to the source of the Europa Rob!!! +1

Scott Aaron Is there a Kugel/Barfaria bogo happening? Maybe ish? 😀

Robert Boynton Yes, if you wanted to you could. Me too. 😀

Lindsey Brown Condenser?

> **Rob Siegel** Probably the Strombergs. I've done very little work on them.
>
> **Lindsey Brown** 75% of carburetor problems turn out to be ignition problems. +3

George Murnaghan Carb tuning?

Fred Bersot Knowing you could is all it takes. Been following and loving every update. +2

Neal Klinman Have you been eating your way through the garage? Looks like the infamous pasta is missing from the shelves. Or maybe you moved the pasta to the space behind the door under the kitchen? Great way to feed the German and British raccoon families that live there. Congrats on the clearing. Good luck with the tuning. Ps- I wish I lived closer and could stop by to hang out in your garage. +1

Paul Garnier Omg, there's ALMOST a path! Squeee!!! +1

Chris Mahalick I just feel the need to come by and clean that garage. My OCD has kicked in big time!! +1

TLC day 2148: She moves!

APR 27, 2019, 6:56 PM

The Lotus Chronicles, Day 2148:

Another milestone in Lotus Land: Today, the car moved under its own power.

In Ran When Parked, I say that you learn a lot moving a car two feet, or trying to. This is a car with a primitive cable clutch, a very short engagement zone, and no brakes (though the handbrake does work), so I wanted to be certain I had some feeling for how the clutch worked and felt before I actually started it and put it in gear and did something stupid like putting it into the wall. So I jacked up the rear end of the car and practiced putting the car in 1st and reverse while looking behind me and seeing when the wheels spun. I had to adjust the cable several times before I got it going into reverse without crunching. Finally I felt confident that I could get it into and out of gear without drama, and that if something went wrong, I could always just pop it into neutral, pull the handbrake, or shut it off.

In the video you'll see that I eased the car forward a foot and a half, then back, but while backing up, the clutch grabbed and the car began moving quickly. I put the clutch all the way in and it disengaged.

After I shot this, I was pumped by the fact that I could hop in the car, start it, and actually move it, so I did it again. It went forward fine, but not only did the clutch grab again going in reverse, mashing the pedal didn't disengage it for some reason. Despite thinking I'd nudge it into neutral, or pull the handbrake, or shut it off, my first instinct was to hit the brake pedal, which of course didn't do anything. I accidentally backed the car into the boxes that were behind it before shutting it off. Fortunately I didn't do any damage.

But it puts the brakes on the front burner. Whether I have the existing master cylinder rebuilt (yes I've seen everyone's comments about White Post and Apple rebuilders) is going to depend on whether I choose to keep the weird brake servo power boosters or delete them. This is going to require some thought and research.

[search YouTube for "The Lotus Chronicles first move"]

144 LIKES, 54 COMMENTS

Frank Mummolo Were it not for your perseverance, Rob, this car would have been in the Crusher years ago. It owes you at this point. And better treat you with respect as you iron out its many remaining issues. Just sayin'. +3

Kevin Pennell [gif, Jeremy Clarkson thumbs up] +3

Kevin (I have to admit I never thought i'd see it together.ll Seriously Rob, well done! +1

Lincoln Benedict I'm with Frank—this car defeats all logic but appeals the soul of why we love cars. Irrational. Silly. Inanimate objects. Lumps of metal. Yet so worth it. +2

Rob Blake Congrats... let me know when you can come out and pick me up for a ride in the Catskills. +2

Brian Lepkowski

GOOD, NOW TAKE IT AROUND THE BLOCK

+7

Bob Sawtelle It would have been cooler if you had the Starry Night picture mounted on the wall in front of you when you did this... just saying. Although now that I think about it this would have been better! [photo of Munch "The Scream"] +6

Jose Rosario Congrats! I have to admit I never thought I'd see it together. +1

Matthew Rogers I was nearby today (Norwood), almost called but time was short so I scooted back north. Sorry I missed the occasion. +1

David Weisman This is awesome Rob! +1

Jake Metz This is so exciting! Congrats Rob! +1

Vince Strazzabosco Major progress! And a cleared-out garage! Congratulations! +1

Francis Dance How exciting! What drama! +1

Steve Woodard The Eagle has... well, moved. Well done! +1

Dave Gerwig Walk before you run, great to see it getting ready to breath some fresh spring air. +1

Tom Egan For whatever it's worth, I am running the stock 7/8" master cylinder and no boosters. I really have to show the brake pedal who's boss in any stoplight situation. I'm working on a plan to sleeve the M/C down to 5/8" (it is a stepped front/rear, so it's a little more complicated, but keeping it simple here) but I've kinda gotten used to stamping on the pedal with authority. A 5/8" Spitfire M/C is the usual no-boosters option, but I like the OEM option. +2

    Rob Siegel thanks Tom

David Kerr Before you know it the list of reasons for us to poke fun at the car will be down to the color. Good work! +4

    Russell Baden Musta David Kerr let me be the first... Metallic Baby Shit Brown...!!! 😂😂😂😂

    Rob Siegel I like brown cars. Fuck off. +3

Lee Highsmith Woo hoo! +1

Neil M. Fennessey Rob what make and model hydraulic lift do you have?

    Rob Siegel Bendpak MD6XP +2

    Neil M. Fennessey Thanks!

Kevin Powers First thing that came to mind
[photo, Ferrari going backwards out the window in "Ferris Buehler's Day Off"] +6

George Zinycz Way to go Rob Siegel!! +1

Bob Ball Love you're perseverance Rob. Did you, before it was parked, ever enjoy it? Autocross, commute, touring ? Or is that still an experience you get after the rescue?

    Rob Siegel I've never driven this car. It was dead with a seized engine when I bought it. I've only ever driven one Europa, owned by a guy who lived not far from me who contacted me nearly six years ago when I bought the car and began posting on Lotus forums. He had a Federal-spec '74 TCS just like mine. I wanted to know if the stock 105 hp was enough, and how it felt with the brake boosters deleted. It was all fine. I enjoyed it a lot. +3

Bob Ball Go cart. I restored a CBR1000f sport bike. Never rode. Exciting as hell. 125hp. New parts and time. Fun to build. Uncomfortable and scary to ride. +2

Rob Siegel yeah, I don't do bikes +1

Bob Ball Rob Siegel there's a reason old BMW bikes have tons of miles on them and old super bikes have none. But that engine. My brother is doing a Westfield (Lotus) 11 kit with sprite parts. I offered the bike. +2

Tim Warner Bob Ball, Peter Egan did a Westfield Kit, as I remember, and wrote about it in Road&Track many years ago. +1

Scott Lagrotteria I was more impressed with, relatively speaking, how clean the garage had become. Great job Rob.

Luther K Brefo Milestone number 3592 achieved! Fantastic work Rob! +1

Matthew Zisk Nice! +1

Harry Krix Does every bimmerhead have the "Flying-Black-CSL-poster" in their garage. I do.

Rob Siegel Yes. It is a union requirement. +1

Harry Krix Make that a Guild requirement. +1

Harry Krix Glad to see the brown bomb move on its own. Keep on keeping on... +1

Dohn Roush Harry Krix The Lollipop Guild?

Kent Carlos Everett Wahooooo!

Robert Myles 36" of movement before something went wrong. AKA a Lotus enduro... +5

Rob Siegel Lots Of Trouble, Usually Shit I Moved It Almost Three Feet Before Something Went Wrong! +2

Gordon Arnold Just don't let the freakin' smoke out. +3

Rob Siegel I used that line in the piece I submitted to Hagerty last week about the Lotus :^) +1

George Zinycz Gordon Arnold I love that line

John Jones The El Caminoto lives! +1

TLC Day 2152: Up the driveway

MAY 1, 2019, 7:35 PM

The Lotus Chronicles, Day 2152:

Freaking booyah.

History will record that, today, nearly six year after it was rolled down the inclined driveway and into my garage on June 20, 2013, the Lotus backed out under its own power and saw the sun on its smiling little fiberglass face. The first back-out, shot with the phone in the car, is shown in the second video. I gave the car a quick wash, put it back in the garage, and then, when Maire Anne came home, had her hold the phone and shoot the first video, which is much smoother and thus looks more impressive.

What enabled this momentous occasion—in addition to the nearly non-stop work since Christmas eve—was functional brakes, and what enabled THAT was bypassing the brake servos (boosters). I'd posted the question "After bleeding, the brake pedal still goes to the floor. How do you tell a bad master cylinder from bad servos?" to one of the Lotus forums, and the answer made perfect sense: Either bench-bleed the master and see if it pumps up, or bypass the servos. Each of the two servos has one goes-into and one goes-out-of connection, so I thought I could simply unscrew the in-and-out lines and connect them together with a union, but when I tried to do that, I would've had to bend the metal lines to get them to meet and I imagined them snapping. Rather than take the risk of cutting off my retreat, so to speak, I bought a couple of brake lines and bent THOSE to connect the ins to the outs. I then bled the brakes for the third time, and this time, to my surprise, the pedal came up rock-hard. I played with the brakes a little in the garage and found that they did in fact function well enough to stall the engine.

And, huge bonus, when I pulled off the vacuum line coming from the boosters to plug up the port on the intake manifold, brake fluid POURED out from the line, which answered the questions 1) How bad are the boosters (answer: fucked) and 2) Why was the engine still belching clouds of white smoke, particularly at startup (answer: it was sucking brake fluid).

It'll still be a while before a test drive, as the gauges quit working (I said nice things about Lucas a bit too quickly, and I REALLY want a functioning coolant temperature gauge before even a short test drive). Plus, it's still running like shit when I rev it (probably the Strombergs), and I really do need to concentrate on Vintage prep for Bertha.

But running like shit is still running, and backing it out of the garage under its own power, six years after we rolled it in, and seeing it there in the driveway next to the azalea bush, almost ready to drive down the street like a real car, was all kinds of awesome.

[search YouTube for "The Lotus Chronicles first back out" and "second back out"]

332 LIKES, 167 COMMENTS

TE Cole Congrats... Rob. You stuck with it and success is sweet. +1
Daniel Neal Why must the British make things so complex?
Craig Fitzgerald That is AMAZING. It's beautiful, Rob! +1
George Saylor It's always fun being upside down in a Europa working under the dash. Especially when you accidentally short something. Hope you get all the gauges working. +1
Scott Linn The REAL Lotus position. +2
    Rob Siegel was under there like that last week. not pleasant.
    Scott Linn Sometimes it helps to pull the seat out (a trivial exercise, it just slides off the bottom rails and comes right out). +2
Christopher Kohler Sounds like it's ready for BaT! +1

> **Rob Siegel** If the engine seizes... +2

James O'Brien Congrats you madman! It's like you've got a great chorus. Now to work on the verses and bridge. +2

Rob Siegel and pray it doesn't descend into a minor key +2

Brandon Vitello Sweeeet! +1

Josh Wyte "Runs and drives!" "Future classic!" "$40k for quick sale"

> **Rob Siegel** you offering? SOLD! +2
>
> **Wil Birch** Future (?!?) classic?
>
> **Scott Linn** $40k... I would have to think about that even though it's 2.5x what I paid for mine. I REALLY like driving it.
>
> **Josh Wyte** Rob Siegel ha! No!! I want nothing to do with English cars!

John Harvey F'n A man!

Scott Hopkinson I think we should all start looking for a TR3B or MGC (6 cyl. from AH 3000) for Rob so we can put him over the edge... Congratulations on a hard fought and won campaign. +1

Brandon Vitello Rob can I share this in our little car enthusiast group, Born Classic? I invited you.

> **Rob Siegel** sure!
>
> **Brandon Vitello** Rob Siegel cool!

Wil Birch Fantastic! Guess I'd better get going on my MGB restoration so I have an excuse to see the Lotus at the Larz British Car Day 😊 +2

Steve Nelson Woo hoo 🎌🎉 +1

Lenny Napier [gif "like a boss!"]

Chris Mahalick Awesome job!!! +1

John McFadden So are you saying my size 13 feet preclude me from autocrossing one of these? +1

> **Rob Siegel** yes, that is what I am saying. +1
>
> **John McFadden** E39 less suited to autocrossing, but more suited to my 25% German feet. +1
>
> **Rob Siegel** have your feet had a DNA test? +1
>
> **John McFadden** Well not officially, they just seem to prefer these socks, so I just assumed... [photo German flag-colored socks] +3
>
> **Robert Boynton** John McFadden Perfect for Flintstone car brakes! +1

Faith Senie WOOHOO! +1

Daniel Senie A car that requires bare (or sock) feet. Brilliant British engineering! 😊 +1

Cameron Parkhurst You have to feel great about today! So cool to see it move under its own power. +1

Chris Mahalick You may as well just take out the brake boosters. I removed mine and the braking is fine. I just may go with a more aggressive front pad. +1

> **Rob Siegel** I'm reading up on all this. I'm thrilled to have been able to bypass them as a test, and have the test be so definitive ("yup, they're junk--responsible for a soft pedal AND passing brake fluid into the intake")
>
> **Daniel Senie** Chris Mahalick Lucas made brake boosters?
>
> **Rob Siegel** Daniel Senie, Girling
>
> **Rob Siegel** I expect that, even though I have functional brakes, I'll still wind up replacing the master, the calipers, and the wheel cylinders before the car is truly road-worthy.
>
> **Chris Mahalick** Rob Siegel I did a lot of reading on this as well. A million different approaches. I hope you are having many beers to celebrate the great thing you have done. As of this week, there are now TWO more Europas moving about under their own power. Oh, I just received your book today. I'll be reading all night, LOL. +1
>
> **Daniel Senie** Rob Siegel Princess of darkness?
>
> **Chris Mahalick** Rob Siegel I installed all new soft lines, rebuilt the master, rebuilt the front calipers and put in two new wheel cylinders on the rears. Brakes are something I will not half-ass. These cars have zero safety features whatsoever. My e39 530i feels like a tank after I have driven the Lotus.

Paul Forbush Yours may be the only driveway on the planet that contains a Europa, a Rialta, and an e39 BMW. +3

> **Rob Siegel** you forgot Bertha in the garage, and the rotted faux tii at the end of the driveway.
>
> **Paul Forbush** Didn't forget them. I just figured it was esoteric enough as I described it. +3
>
> **Rob Siegel** For a while, I wanted a Vixen because not only does it have a BMW M21 motor, it, like

the Europa has a Renault transaxle. I loved the idea of having the two most different vehicles with a Renault transaxle. I can be an eccentric guy. +2

Paul Forbush Ya think? +1

Layne Wylie Should have M21 swapped the Lotus. +1

Rob Siegel In retrospect, that would've been easier. And cheaper. And think of the torque! +1

Layne Wylie And you could put an exhaust stack out the top!

Rob Siegel useful for fording bodies of water

Dave Gerwig I hear there's been a sighting in W Newton: a Classic on the //Move! Way to go! +1

David Kemether [gif Simpsons clown car]

James Pollaci From this day forward the exclamation 'Eureka!' Shall now go by 'Europa!' Congrats!!! +3

Thomas Diamond The Lotus Chronicles: This is the story of the hack mechanic, who boldly goes where no other sane mechanic would go. His mission is to seek out and repair dead vintage cars and bring them back to life. Cue the Star Trek music. +2

Jake Metz Hooray!

Scott Sturdy Tara King approves! [photo of Tara King and Europa from The Avengers] +5

Noam Levine Mazel tov!

Alan Hunter Johansson Why does a 1600-lb car need power brakes?!? Our three and a quarter ton HEARSE had non-power brakes, and was daily-driven by my mother! +1

Rob Siegel Alan, it doesn't. The earlier Europas didn't have boosters; only the Twin Cam cars.

Charles Morris What kind of lap times did your mom turn in the hearse, though... :D

Alan Hunter Johansson Charles Morris I don't know. My brother and his roommate at Davidson had a circuit laid out on public roads outside the campus called "The Road Course", over which my brother could put said '48 Cadillac only 10% off his roommate's speed in his '68 350/350 Corvette with the gymkhana option. On the other hand, I watched from the passenger seat of a Lotus Elan Sprint being driven by another of his friends as my brother walked away from us over those same roads in a '62 Checker, and later over different roads in a '52 Plymouth with seven people crammed in it, no shocks, and daylight showing under the inside rear in corners. So some of that may have been skill.

Paul Wegweiser Ya' know... I wanted to come up with some smartass comment, but you know what? I'm going to let you have this moment of glory. Well done, my friend! Really looking forward to seeing you at The Vintage. +7

Scott Sturdy [gif of "Yep!"]

Rob Siegel Paul, your restraint is admirable. Why do I think I'm going to pay for it later? +3

Franklin Alfonso Feather be careful around Paul... +1

John Whetstone Paul Wegweiser WHO ARE YOU AND WHAT HAVE YOU DONE WITH PAUL?! +4

Franklin Alfonso [gif of Star Trek "It's a trap!"]

Gary Beck Sunshine, me too. If I could get insurance straightened out. PIA! [photo of E9 seeing light of day] (8)

Rob Capiccioni This is YUUUGE!!! 🍑🍆 +1

Levent Erhamza Better love story than Twilight. +1

Rob Siegel Anything is better than Twilight. Jerking off with Lava is better than Twilight. +2

Rick Roudebush Happy for you Rob. It's been a long and winding road. +1

Rob Siegel The Lotus should like the "winding" part. The "long part," well, we'll see about that :^) +2

Kevin Pennell [gif Snoopy happy dance] +1

Scott Lagrotteria Nice accomplishment Rob, shows what perseverance can accomplish +1

Rob Siegel ain't that the truth

Brian Stauss Well done Rob! +1

Rob Halsey I have loved following this story. Thank you for sharing the adventure with me. +1

Rob Siegel It was my pleasure. Except when I wanted death to anyone who designed certain parts of the car.

Rob Halsey Rob Siegel those were some of the best parts. As Fredrick Douglass once said, 'there is no progress without struggle.' I am sure he was also referring to Europas... +1

Franklin Alfonso Rob Siegel That's all part of Lolita's charm... 😊

Harry Krix Mazda Balls! (oh, it's Mazal Tov!) So proud of your achievement; you are an inspiration to us all! ...and he's applauding

+2

Rob Siegel you honor me, sir! +1

Kevin Pennell Well done sir, congrats on your excellent endeavor! +1

Kimberly Cox I remember the first push!!! +1

> **Rob Siegel** YES!
>
> **Binki Witherby De Collibus** Kimberly Cox are you referring to Tyler's birth? +1
>
> **Kimberly Cox** Binki Witherby De Collibus lol. No, I helped push the Lotus into Rob's driveway! 😊
>
> **Binki Witherby De Collibus** Kimberly Cox I have done my share of pushing car projects.
>
> **Rob Siegel** I believe it was Kim, Maire Anne, and my neighbor Dave Role and his son Jake pushing, with me steering. It's a 1600-lb car, it wasn't too difficult for them to push it around the block and me to make three right turns into the driveway. 'Course I realized it had no brakes, and my driveway slopes down into the garage, so that was interesting... +1

Lee Perrault Wows a, you are one hot shit, Rob 😊 +1

Jim Strickland Beautiful sounds and action. Congrats Rob!

John Turner Love the odometer gears tee. Strange looking El Camino though +1

Peter Gleeson Well done and hearty congrats from a sympathetic soul +2

Tom Egan Major congrats, bra! See you on the highway! +1

Robert Spina Jr Congratulations my friend. I've really enjoyed the progress, can't wait to see some rolling shots! +1

Patricia Rice Please put up a photo of what it looked like going in... thanks.

> **Rob Siegel** Patricia, it looked about the same when I rolled it into the garage, maybe even slightly better. But the engine was seized. I removed it within the first week that I had the car, but then ran into a quagmire that lasted years.

Seth Connelly even a nice way to end my day!! :-) +1

Django Vaz Bitchin' +1

Bob Sawtelle Congrats! I could not help but smile when you drove her to the edge of the driveway, and wow Dat Ass! those Europas look so damn good from that rear 3/4 shot. Many a person has been seduced by that rear end. And Maire Anne congratulating you was so sweet! She truly is your soulmate. +1

Jim Denker Great! Still strongly prefer your BMW rescues... +1

> **Rob Siegel** me too +1

Charles Gould They are very sweet cars to drive. Lots of power for such a light weight skate, and very nimble handling too. I hope to take the maiden voyage in the Jag this weekend. +1

Wade Brickhouse Yay! +1

David Weisman That reminds me of when as I was driving to school my senior year at LHS my Volvo p1800 was belching white smoke like a mosquito patrol. Same thing. It had power brakes and was sucking brake fluid into intake manifold due to a bad O ring in the master cylinder. +2

> **Rob Siegel** It's fairly common, especially on cars that sit. The brake fluid absorbs water, the bores in the hydraulics corrode, and the seals harden and crack. +1
>
> **David Weisman** Rob Siegel sounds like my arteries. +2

Aashish Dalal You've accomplished the unthinkable. You're my idol! +1

Derek Barnes Rob Siegel please fix your "O"!!! +1

Paul Hill Rob Siegel note to self. Buy some slip-on driving shoes. This continues to be a fantastic read and another proper car resurrection adventure. Thank you so much for sharing. I found myself whooping with excitement at the video but I have now controlled my emotions and have made tea and marmite on toast in celebration. I raise my tea cup to you Congratulations Rob, cheers and bottoms up! +1

John Whetstone That thing growls!

    Rob Siegel that thing has holes in the muffler :^) +1

    John Whetstone Rob Siegel Just another case of things being seen through the "Facebook lens", or maybe microphone in this case!

Russell Baden Musta This is freakin AWESOME! I noticed you have the original wiper blades! +1

John R Brandl Awesome Rob! Hilarious with "remove the right shoe." This is standard operating procedure when I drive my 67 Beetle or the MG kit car on the 74 Beetle Pan! I have found that Academy Sports sells a gray Skechers shoe that is very thin, almost like a driving shoe, that I can wear when driving these 2 cars. Not expensive either. Congrats again! +2

    Lee Highsmith If Rob has to remove his shoe, I'd probably require amputation to drive the car...

    Lee Highsmith Marie Anne has a steady hand. Congrats on getting the car this far. It looks WAY better in the sun! ;) +1

    Rob Siegel Don't we all. Well, maybe most of us.

    Dohn Roush I look my best in the dark...

Russell Baden Musta First big trip!!! To end of the driveway!! Pack an overnight kit and go for end of the block today!! +1

Bill Howard Fiberglass? Can the British make fiberglass rust? Short out? +1

    Rob Siegel No, but they CAN make it leak oil. +1

Larry Wilson Very cool. +1

Robert Zockoff I'm tired from watching all the work that went into getting to this point. After all your work, no one could possibly pay you enough! +1

    Rob Siegel There's a chapter in my first book called "Restoration and why it makes no freaking sense." And here, I'm not "restoring" this is a mechanical resurrection. And it STILL makes no freaking sense.

    George Murnaghan Rob Siegel I just read that chapter an hour ago.

    Garrett Briggs does doing anything with old cars "make sense"? (he says after 4 hours trying to un-seize one screw) (^me^ not Rob) +1

    Rob Siegel Garrett, me too, just not today

Mark Abare Happy for you Rob! +1

Robert Boynton Think you must be thinking of Fiat for rust +1

Lisa Binder Wow! Nice :) +1

Brad Purvis Brown is beautiful. +1

Colin Tennessen Europa is my absolute bucket list car. Love it +1

Go ahead and call it a bread van NOW, motherfuckers.

56 LIKES, 13 COMMENTS

David Holzman What's this? It's got a TV license plate from Illinois!

    Rob Siegel That's when it was last on the road—1979. +1

George Saylor I just can't get over that view... Soooo cool.

    Rob Siegel Yeah... people obsess about the "bread van" thing and don't realize how cool the car looks from other angles, especially this one. +3

Karel Jennings Rob Siegel It's a beauty. So much so you can forgive the square box. +1

Ken Doolittle [meme "the face you make when you hear the motor fire up"] +1

Scott Lagrotteria Europas are very good-looking cars; I like the John Player Special black and gold best. (3)

    Rob Siegel Nearly everyone would agree, but the "Twin Cam Special" (my car) is simply a John Player Special in a different color. +3

    Tom McCarthy Guitar player special? +6

    Doug Jacobs Scott Lagrotteria oh not me—I absolutely prefer the metallic infant shit version our man has here. It makes me all loose in the bowels, which can be a good thing at my age. +1

    Scott Lagrotteria I remember going up to Canada around 1978 and excitedly finding John Player Special cigarettes, which I bought (although I don't smoke). I might still have the pack in one of my memento boxes. +2

Clyde Gates Rob wins! +1

Hans Batra What a beautiful, sincere and happy face! It's sad that modern car designers can only design aggressive/angry faces now. +1

The El Camino / Breadvan view everyone makes fun of.

140 LIKES, 55 COMMENTS

Rob Blake I'm no mechanic. But the way this car looks... wow! I had the Matchbox version and even the HO size (remember them?) version for my race car track in the basement when I was in 5th grade... I hope to at least see the car (and you) someday. +1

Rob Siegel It is a uniquely-shaped car. This is probably the most polarizing view. It looks more unified from either the front or the back. I think it looks totally badass from the front. +1

    Rob Blake Rob Siegel sorta reminds me of the Jag Hearse from the movie Harold & Maude. +1

    Rob Siegel trust me, that's worth a lot more money :^)

    George Zinycz Rob Siegel it looks like a Frankencar from the side view. Both ends seem like they belong to very intriguing designs... two completely unrelated intriguing designs though 😂😂 +1

    George Zinycz Rob Siegel And I like brown cars too...

    Rob Siegel yeah, oddly, in that way, it's not unlike the Z3 M Coupe +3

    George Zinycz Rob Siegel oooo I'm going to want to see them together now... +1

    Rob Siegel George Zinycz, I'm sure that happened at one point in the last six years in my garage... +1

    Rob Siegel In the side view, it accentuates the "cut-down sail panels" that, in the earlier S1 and S2 cars, continue at the level of the roof line toward the back of the car. That looks a little more unified, but still a bit odd. +1

    Charles Gould Rob Siegel The $64,000 question is whether a man of your age and stature can get in and out of it gracefully!

    Rob Siegel "Can get in and out," yes. "Gracefully," well, that's in the eye of the beholder :^) +1

> **Dohn Roush** Rob Siegel Or beer holder… +1
>
> **Rob Siegel** Damn it, Dohn, quit being funnier than me! I thought I'd been clear on this point! +1
>
> **Dohn Roush** Rob Siegel As I've said before, you've got mechanical skills, musical talent, and hair. You gotta leave me something… +1
>
> **Rob Siegel** WELL YOU'RE TAKING TOO GODDAMED MUCH! +1
>
> **Dohn Roush** Rob Siegel Ok, I'll take the hair…
>
> **Joey Hertzberg** Rob Siegel seems to evoke similar judgment as a 914, which I always loved. I do like the styling of the Europa, but what I really like about it is how low it is.

Corey Dalba Wish it had a pickup bed. That would be so cool.

> **Rob Siegel** Meh. Then someone would just want to hang truck nuts from under the bumper. And that'd probably be enough to bend it. +1
>
> **Corey Dalba** The visualization of that is absolutely hilarious. Given (like you said in the video) it's a 1600-lb car. And we both know your statement about the truck nuts is true. Esp here in Mississippi! +1
>
> **Jeffrey Miner** Corey Dalba it's beggin' for a gooseneck or a fifth-wheel hitch set up… it just needs a tiny fifth wheel camper to go with it 🙂 +1
>
> **Rob Siegel** STEP AWAY FROM THE LOTUS +3
>
> **Jeffrey Miner** A couple box flares, a set of wheel studs, and you could run duallies on that bitch 🙂 +2

Drew Lagravinese It kinda looks like the designer was not sure if he was designing a sports car or a pickup truck.

Rob Sass Digging the Tawny Gold. Great 70s Lotus color.

> **Robert Myles** Can Tawny Kitaen be far behind? [link to video of Whitesnake "Here I Go Again"]
>
> **Rob Siegel** David fucking Coverdale shows up on a Lotus post. Life is still full of surprises. +1

Mark Thompson Pretty sure I had the "Matchbox" of this…Loved the XKE they made… Lesney! +1

Rally Mono I always thought with this car and the TR7 that the rear end design was done on Friday afternoon. At least now you can deliver bread in your area. Haha. Joking aside congrats on getting on the road! +1

Nick Wood I think you need to get a 1/4 scale RC helicopter to land on the back. +4

Carl Baumeister It looks good Rob! +1

Jeff Lutes You should have parked it the other direction - with the rhododendron poking up from the back. It would look like a shrunken funeral flower car… +2

Jeff Lutes Just kidding! Very impressed that it's running/driving(?) again! +1

Brad Purvis Goes well with the siding. The Azaleas ad a nice touch of colour. +1

Ned Gray WOW!!!!! That motor sounds really sweet. A massive undertaking… props to you! Can't wait to see it up close, maybe even get a ride… +1

> **Rob Siegel** Ned, let me get it up and down the street, then up and down two exits of 95, before I think about Amherst :^)
>
> **Ned Gray** That's more than fair. You should be proud! +1

Paul Hill I think it rocks! The way it looks, colour and the resurrection story. I can't wait to hear how it drives and read about the epic adventures ahead. There is a bright yellow one near me that goes past me in the summer months every now and again. There is one notable constant. The owner has a permanent smile at the wheel. I'm sure your cheeks will be aching very shortly and not just due to an uncomfortable seat. +1

Frederick Tutman It sorts of remind me of an Italian version of an El Camino. +1

Rob Siegel yeah… the better view is nose-on

Robert Burchett What model is this?

> **Rob Siegel** Lotus Europa Twin Cam Special
>
> **Robert Burchett** Rob Siegel that's a sweet car! +1
>
> **Thomas Jones** Life-sized Matchbox

Chris Lordan That Illinois plate issued during the Carter administration is the very definition of cachet. It might be cachet squared. +2

Tim Warner So happy for you. Thanks for bringing us along! +1

Rupert X Pellett Magnificent Wheelbarrow she is. +1

Matt Pelikan Shit, man. It has flowers growing out of the hood! +1

Steve Teager All that work to get it running and you turned it into a planter?! +1
Tom Smith A triumphant profile!

53 LIKES, 51 COMMENTS

John Russell great color... +3
- **Rob Siegel** I like it
- **TE Cole** 70s brown. +1
- **Rob Siegel** Hey! I LIKE brown cars! (photo of my dear departed 911SC)
- **Douglas Wittkowski** Rob Siegel, so do I. 😊
- **John McFadden** The brochure called it "Mink" 😊

Mark Larson Wet? From being in the rain?
Rob Siegel no, I just washed it +2
Ed MacVaugh Rob Siegel Paint correction comes next, right? +2
- **Rob Siegel** HAHAHAHAHAHA oh god that's funny
- **Mark Larson** Rob Siegel Clay bar at least!
- **John Whetstone** Mark Larson Clay bar might be even funnier!

Charlie Barzola I like how you kept the Illinois plates! I have so far kept the California plates on my car! Great to see the Lotus running again, congrats!

Carl Graving I can't tell if that is the ugliest pretty car, or the prettiest ugly car. +4
- **Rob Siegel** As I said elsewhere, straight on the nose is its best view. Even I think it looks weird straight on the side. +1
- **Carl Graving** Keeping the brown? +1
- **Rob Siegel** Keeping everything about the exterior. Except I'll straighten that "O" on the trunk lid :^) +2
- **Carl Graving** Looking forward to the drive by video hearing the engine + Doppler at high speed!
- **Rob Siegel** You and me both!

Robert Alan Scalla Congratulations! It's a nice-looking car from most views. I think the side view is, like maybe a pickup truck profile of some sort. Then again, I know what it's haulin' back there. +2

Bob Coffey I like the custom "Reclining Buddha" nameplate. +1
- **Rob Siegel** I liked it too, but I rotated it back into compliance.
- **Bob Coffey** Rob Siegel propah, yes.

Scott Aaron Looks like rear visibility will be top-notch. +1
- **Rob Siegel** yeah, I had to open the door while I was backing up +1

Scott Aaron Cool car though. No doubt. Good job.

Rob Siegel Wait. Where's the snark? +1

Scott Aaron I've been doing PPT and Excel, so I'm in a work zone where I'm relatively pleasant.

Rob Siegel So I only get abused from your home environment? +1

Dohn Roush When I use PPT and Excel I always end up saying a bad Word or two...

Paul Wegweiser Rob Siegel See? Even Scott Aaron shares my compassion for you at this stage of the project. Don't worry... it'll pass. 🐜 💨 (farting ant is my favorite emoji) +1

Russell Baden Musta It really is good rear vis, they cut the sails down on the rear on the TC version to improve 3/4 vision, but you look up to everyone! The stock side mirrors are not great on driver side and useless to the driver on the passenger side... +1

Dohn Roush Russell Baden Musta So I guess the best rear view mirror would be a sunroof... +1

Russell Baden Musta Dohn Roush if you're waaaay tooo tall perhaps... but don't ruin a Europa with a sunroof!!! I have Vitaloni mirrors on mine... I bought new clams a couple years ago for bringing mine back to original... put driver dude on first, couldn't see enough, took it off.

Carl Witthoft Hey, it's a Lotus. Nobody's supposed to be able to catch up to you. +2

Steven Bauer Is the two-tone paint job stock or patina?

Rob Siegel patina +1

Paul R. Elliott Is the back bumper supposed to go on top of the trunk (boot)?

Rob Siegel it's correct where it is

Paul R. Elliott I like the way it looks but wonder if it would be effective so high. Then again, with so many high-riding pickups and SUVs, it's probably perfect. +1

Rob Siegel Paul, it's a fiberglass-bodied car with a rear bumper the thickness of a woman's bracelet. If a pickup or SUV even LOOKS at it with impact on its mind, the car will crack into a million pieces. +8

Scott Linn He's not lying... +2

Scott Lagrotteria Paul R. Elliott It's not that the bumper is high, it's just that the Lotus is so low.

James Pollaci When you want to get started and don't know when you'll stop, Europa. From the makers of Rialta. +3

Blair Meiser The fiberglass bumpers on my Lotus Elan SE were good for about 0.1 mph hit! I believe I had to reglass, sand and repaint them at least twice each year! Is your bumper metal?

Rob Siegel yes, metal +1

John Thomas Blair Meiser so, was yours a lightweight model? Didn't they do fiberglass bumpers on the e9 csl?

Blair Meiser John, mine was a red over black 1967 Lotus Elan SE roadster, Twin-Cam engine, close ratio 4 speed, 1490 lbs wet. Sold it to a guy from CT to get enough down payment for our first home. We lived in an apartment and the old ladies parallel parked by bumping into the cars in front of and behind their parking spaces! +2

Derek Barnes Nice contrast with the Rialta. Lotus=little in back. Rialta= bubble butt 😂 +1

Ken Battan cool man +1

[second video: search YouTube for "The Lotus Chronicles second back out]

105 LIKES, 63 COMMENTS

Marilyn Rea Beyer You're wonderfully goofy and brilliant all at one time. Miss you, ole buddy! +1

Samantha Lewis Huzzah! <3 Congratulations! +1

Francis Macaluso Hahah remove the right shoe! +1

Rob Siegel truth!

Samantha Lewis The Europa's peddle box is a wee tight, although as I recall, less offset than the Elan, yeah, it's small down there.

Rob Siegel I kept on my left shoe so I'd have lots of force on the clutch, but took off the right one so I could be certain to hit the brake pedal and not impinge on anything else. +1

TE Cole Jeff Caplan says he appreciates the plug. [the "Odometer Gears" t-shirt]

Rob Siegel I knew he would. Jeff, you gave me enough shirts that I was highly statistically likely to wear one in a landmark video. That was your plan all along, WASN'T IT???

Ernie Peters I might have shed a tear of joy just now. So very, very cool. Well done, Rob. +1

Rob Siegel When it was in the driveway, I couldn't stop just walking around it going holy shit, it's outside, and it got there under its own steam. +1

> **Ernie Peters** Rob Siegel You did it. Dream, plan, execute, enjoy. A good example for all. +1
>
> **Ernie Peters** FYI, this is the book I used to go through the ZS's on my Elan. If you need some of its info let me know. 😊 [photo Haynes Weber carburetor book]
>
> **David Holzman** About that rough running, you haven't said anything in your Lotus Chronicles Odyssey (Odysseus visited the land of the Lotus Eaters) about anodizing the muffler bearings. That's probably at least part of the problem. +1
>
>> **Rob Siegel** They have no bearing on the problem :^) (see what I did there?) +1
>
> **Robert Myles** See? He takes his shoes off one at a time just like the rest of us! +1
>
>> **Rob Siegel** Robert, I only do that because it makes walking on water easier :^) +1
>
>> **Robert Myles** I prefer to use the other old family trick and just part the water instead. It keeps my sneakers dryer. +2
>
> **Jackie Jouret** I'm really proud of you! +1
>
>> **Rob Siegel** Aw, thanks Jackie. What I lack in common sense, I make up for in brutal stubbornness :^) +1
>
>> **Jackie Jouret** Let's call it "perseverance." Whatever it is, you deserve a freakin' medal for not hurling that thing into Boston Harbor.
>
>> **Rob Siegel** Jackie, being mostly fiberglass, I couldn't be certain it would sink.
>
> **Jeff F Hollen** Rob Siegel I'm serious. You commit to bringing this to Vintage and I'll bring my Lotus powered machine as well. David Kiehna will too... +1
>
>> **Rob Siegel** Jeff, no; Bertha and I have a date :^) +1
>
> **Larry Webster** OMG, I'm so happy for you. And oddly proud too, even though I've only watched, LOL. You even have the exhaust on, a detail I usually skip. I hope that Lotus brings miles of smiles. +1
>
>> **Rob Siegel** Thanks Larry!
>
> **Christopher P. Koch** Phew, glad I'm not a passenger with that odor wafting around... +1
>
> **Greg Bare** Seeing in next to the E39 makes me realize how small that car is. Absolutely Beautiful. 👏👏👏👏👏👏👏👏👏 +1
>
> **Caleb Miller** What is that surface upon which the Europa is parked? Is that the mythical Siegel garage floor? I think that's almost more impressive than the Lotus motivating under power! 😊 +1
>
> **Joey Hertzberg** Man, I want to get at that thing with a buffer and some shiny-making-liquids +2
>
> **Fred Bersot** Congrats and glad the moment was digitized, posted, and immortalized. That car suits you. Suits your look and what I perceive to be your personality, based on over 20 years of reading your column in Roundel. +1
>
> **Bill Gau** Rob, I've been following this post since 2017 just so I could post a photo of this stray camping up the hill at Monterey above turn 10. Brian Lowder spotted it first and the legendary Seth Malcolm can authenticate the sighting. I'd post other photos, but this speaks volumes. Anyway, congrats—who doesn't love Lotus?? +2
>
>> **Rob Siegel** Thanks Bill!
>
> **Brian Ach** Fucking A. +1
>
> **Alexander Wajsfelner** Cudos to Maire Anne Diamond for filming and putting up with your obsessions +1
>
>> **Maire Anne Diamond** Someone has to keep an eye on him. No telling what sort of trouble he'd get into otherwise. +2
>
>> **Alexander Wajsfelner** The trouble was buying the Lotus in the first place 😊
>
>> **Rob Siegel** Alex, you DO know that every horizontal surface (and several vertical ones) of our bedroom are now occupied by quilting supplies, and the sewing machines are threatening to outnumber the cars, right? +4
>
> **Bil Smith** WOW! They said it couldn't be done! +1
>
>> **Rob Siegel** Who? I'll show him what for! +1
>
> **Larry Johnson** Mad respect to you, sir. Congrats, enjoy. +1
>
> **Greg Lewis** Epic! +1
>
> **Scott Aaron** Try to look happier +1
>
> **Tom Harvey** Love the smile on your face. Clearly, a great moment for you and the car. +1
>
> **Steve Rapson** Yes, pretty darn cool.
>
>> **Rob Siegel** thanks Steve!
>
> **Scott Winfield** Wow!!!!
>
> **Doug Jacobs** That entry and exit brings a modern dance to my mind.

Mark Manasas Glorious Day!! Don't change the orientation of the "O" in Lotus on the tail! +1

[first video: search YouTube for "The Lotus Chronicles first back out"]

17 LIKES, 9 COMMENTS

    David Holzman Love that sound!

        Rob Siegel holes in the muffler :^)

Neal Agran That looks an awful lot like a Blaupunkt Reno SQR46 head unit. If you don't know what that is, you're in for a pleasant surprise if it still works.

    Rob Siegel Yup, that's what it is! Haven't even turned it on yet.

    Neal Agran That looks like it's in remarkably good shape. I didn't know that Porsche ever offered those as a remanufactured part. So... you know that this is probably worth $1,000 or so, right?? That is the OE stereo for for 944's, 911's, and 928's from 1987 to 1989. +2

Nicholas Bedworth Mrs Robby should her an award for tolerating, possibly abetting, 6 six of total insanity. And that was before the car was even operating. Now what happens? There's a lot of evidence accumulating that would support involuntary confinement. But congratulations... What a job, well done. +1

Interjection: The passing of a car guy

MAY 2, 2019, 10:55 AM

Yesterday, in the middle of writing up my triumphant first moving of the Lotus under its own power, I received awful news that a member of our Boston musician community, Rob Ayres, had suddenly passed away. I was especially stunned because I'd seen that Rob had posted not 20 hours earlier about attending an open mic.

I first knew Rob as my friend Tom Smith's brother-in-law, but we grew to be friends. In addition to being an avid musician who was in the process of recording his first CD, Rob was a car guy. He'd owned a BMW 2002tii like mine back in the day, and he used to drive his Porsche Boxster to open mics. We'd joke that, if you're a musician, you're supposed to drive a 15-year-old Subaru, and how we were usually the two guys with wildly inappropriate cars in the parking lot.

A month ago, at Tom's 70th birthday party, Rob told me the following story. He said that, years back, when he was selling his 2002tii, he took it for one last drive. He went up Rt 3 and wound it up to a hundred. Unfortunately, blue lights came on behind him. He honestly spilled the beans to the officer, explaining that he was selling the car, the new owner was coming to pick it up in the morning, and this was his last drive. He apologized for his exuberance. The officer remonstrated him, but was sympathetic, and wrote him up for a warning. Rob, telling the story, said "Then I figured I'd push my luck. I asked him if there was any further police presence up ahead. "I can't tell you that!" the officer protested. He walked back to his cruiser and I got back in the 2002. Then I saw him walking back to me. With a wink, he said "It's clear up to Lowell." " With Rob's warm, guile-less manner, I could easily imagine the whole thing happening.

Although it was a trivial thing in the face of such devastating loss, I couldn't help but think how much I would've enjoyed running into Rob at the next open mic and telling him I'd gotten the Lotus running. I mentioned this to Tom, who said "As a ten-year-old, all he wanted to do was design cars. He relished his conversations with you."

Tears go out to Tom and Margo and the Ayres family.

TLC Day 2154: In front of the house

MAY 4, 2019, 1:11 PM

The Lotus Chronicles, Day 2154:

At this point, you might think that I'm just fucking with you, driving it 40 feet back and forth in the dead-end street in front of my house instead of around the block. But in my defense, I don't drive unregistered cars lightly. The brakes are functional but poor. I do these things in small deliberate steps.

And no, my comment that "I keep waiting for it to catch fire" wasn't prophetic, even if Maire Anne then said, in all seriousness, "I smell something burning…"

[search YouTube for "The Lotus Chronicles first drive on pavement"]

124 LIKES, 49 COMMENTS

    Ed MacVaugh Back up lights work! +7
        Russell Baden Musta Ed MacVaugh took the words out of my mouth!! THAT is awesome! I know! +2
        Rob Siegel I know, right? +2
        Russell Baden Musta Rob Siegel fantastic!
        George Zinycz Ed MacVaugh THAT time anyway +2
Roy Richard Forwarded this to the Mass RMV 😀 +4
Jeffrey Miner No brake lights. +1
Doug Jacobs YOU'RE fuckin with US? You're the one with a moldy Lotus in your garage. The joke's most definitely on YOU. +3
    Doug Jacobs Or maybe we should call it "The Scrotus." +4
    Susan Brown Strickland Doug Jacobs how easy you boys digress +1
    Doug Jacobs Susan Brown Strickland oh, if it's not shopping or shoes, it's a digression? 😈 😀 +1
    Rob Siegel Looks like I have trouble bussed in from multiple states… +2
    Susan Brown Strickland Doug Jacobs don't forget manicures and musicals! Boys topics for digression (including any variation of said word) are listed but not limited to the following words: anus, penis, dick, ball, shaft, scrotum, wood, jewel, tool, pecker, thrust, pound, nail, blow, suck and stroke. +1
    John Jones Susan Brown Strickland That escalated quickly +1
George Saylor If it's any comfort, I *think* the rear brakes are from a GT6 😀 +1
Phil Morgan It looks better every time I see it. +1
    Rob Siegel my wife used to say that about me, but it's been a while +6
    Susan Brown Strickland Rob Siegel as I just pointed out to Doug Jacobs…you boys digress so easily +1
    Rob Siegel back off, Stag girl…
    Susan Brown Strickland Rob Siegel you already have one foot in the grave…do you really want to go there with me??? [Note: Susan is a mystery writer who says she is going to kill me in print. I actually DO live in fear of her.]
John Wanner "It's probably just the clutch." Man I hope so.
Chris Roberts In my next life, I aspire to find a woman who (at least) humors my automotive passion. +1
    Rob Siegel Good luck with that. Maire Anne is taken! +2
Josh Wyte Better insure it now! Just in case it, ya know, spontaneously combusts. +1
    Ali Jon Yetgen Josh Wyte don't use such terms! They are scary… use "thermal event" please. ;) +1
    Rob Siegel Oh, it's been insured since the day I bought it +1
Rick Roudebush Those that have never held a piston in their hands, dipped it in oil, compressed the rings, inserted it into a block, wiped all the bearings with assembly lube, carefully torqued all fasteners, double and triple checked cam timing and valve clearance, etc, etc, will never know the unmitigated joy one receives when that engine spins to life and breaths on its own and then eventually propels the vehicle. Well done sir. May your chest be all puffy and stuff. +12
    Rob Siegel Truth right there. +3
Scott Winfield Very cool. +1

Frank Mummolo One small step for a man... +1
Seth Connelly Isn't the parachute supposed to fly out when you get to the end of the road? +1
    Rob Siegel That, or the passenger when you hit the eject button. It IS a British spy car. Those things just come standard. +2
Francis Macaluso HAHAHAHAH Remove the right shoe. You get me with that one every time. +2
Marc Schatell I'm ready to drive it +1
John McFadden I'm tempted to paint a tunnel entrance on that fence, a 'la Wile E. Coyote! +2
Bob Sawtelle She is such a sexy beast! +1
Rob Koenig I'm beginning to think that The Lotus Chronicles is a Monty Python tale... +1
Brad Purvis If you value your home either leave it parked outside tonight or disconnect the battery. I'm SURE you are way ahead of me on this, but sometimes the father/officer/captain shit takes over and I can't help but say something. +2
Dave Gerwig Roll on brother! +2
Franklin Alfonso Smelling something burning is standard on British cars. Goes along with the smoke and intermittent switches 😊 ... +1
Bill Mann It sounds like an aeroplane but is still "smashing" :) +1
Gary Beck Sounds good Rob. Get a tag and go for a ride. You find some things and hopefully make it home again. I almost didn't with failing fuel pump. +1
Stan Chamallas I want to see a burn out! +2
    Rob Siegel That's what we used to yell in high school: "LAY A PATCH!" +1

TLC Day 2154: First drive, live

May 4, 2019, 3:55 PM

The Lotus Chronicles, Day 2154, live edition:

(I did a Facebook Live session on the car's first drive on a public road, and as I began to drive down the road, I let out a cackle of laughter, and the live feed terminated, leaving everyone to wonder what the hell had happened.)

[search YouTube for "The Lotus Chronicles first short drive"]

    Rob Siegel · 0:00 I have no idea why the live feed terminated. I made it back. That's one short drive for man... +9
        Erik Wegweiser · 0:00 Interference from the flux capacitor. +4
        Rob Koenig · 0:00 "No idea", ehh? We heard that cackle. +1
    Eric Pommerer · 0:00 Dropped transitioning from Home WiFi to cellular data? +1
    Stan Chamallas · 0:03 Lucas jumped? +1
    Tim Warner · 1:09 Yay!
    Jim Strickland · 0:41 Awesome!
    Joey Hertzberg · 0:25 Yessss!!!!
    Joey Hertzberg · 0:08 Too short!
    Joey Hertzberg · 1:02 I'm going wide-screen
    George Saylor · 0:00 Pointy toe cowboy boots work well.
    Adam G. Fisher · 0:22 Go go go!
    Scott Aaron · 0:00 I assume there was an explosion of some sort during this drive, and I'm super bummed you were blown up or otherwise seriously killed. Totally worth it, though, to spend all that time and money on it. Even though you were probably immolated at least to a certain extent. +1
        Franklin Alfonso · 0:31 Nothing worse than being seriously killed. That's almost as bad as being killed dead... +1
    Bill Snider · 0:00 Mad Scientist Maniacal Laughter replete - and due. Well done Rob
    Brad Purvis · 0:00 It was the Dr Frankensteen laugh. +2
    Dohn Roush · 0:02 One giant leap for a kind man... +2

Marc Touesnard · 0:58 Fill me in on the "Remove the right shoe" thing
Adam G. Fisher · 1:07 I think it's because the pedals are so close to each other.
Brian Stauss · 0:00 Sweet!
Peter Wright · 1:08 That cackle of delight!
Erik Wegweiser · 1:08 Gas pedal sticks up to high? Rob's right leg is longer than his left leg?

TLC Day 2155: First drive, replay

May 5, 2019, 8:29 AM

The Lotus Chronicles, Day 2155:

As many of you saw, yesterday was a landmark day in Lotus Land, as I consummated my six-year-long relationship with the car by driving it.

--First I posted the video of me driving the car the 50 or so feet back and forth in front of my house and testing the brakes. Shortly after that, I relieved my case of blue balls and actually drove it down the street. I did this live on Facebook, but the feed terminated just as I accelerated and let out a cackle of joy (as someone else noted, possibly as my phone was switching from WiFi to cellular). In the video below, I recreated that first drive, going down to the first side street and turning around. It was enough to get the car into second gear.

--The take-away data was that the brakes are functional but poor (more later), and that, as soon as the car is moving enough to push air through the ventilation system, there's a very strong smell of rodent, so it's likely the heater box is going to need to come out and be cleaned.

--After that, I took it out three more times, going a little further each time, the last time actually driving it around the block and getting it into third gear. It's running surprisingly well. Although I gave the carbs a cursory synchronization, my friend Lindsey Brown is coming over this morning to play with the Strombergs and see if he can get them to idle more smoothly.

--With the car back in the garage, I wiped the interior down with a few cleaning cloths, and was astounded at what the absence of dust and a little sheen did.

--While wiping down the seats, I discovered something hilarious. As I wiped in the crevice between the thigh bolster cushion and the seat bottom, I felt something. I spread the cushions and found a toggle switch. Then I realized: This was a starter kill switch. The two wires I described a few weeks ago that had ring terminals on them and were lying under the seat cushion connected to nothing were for this switch. It is still, however, a mystery why nothing was connected to the start terminal on the ignition switch.

--When I came back in for dinner, I said to Maire Anne "I've spent years saying "I just want to get the car to the point where I can move it in and out of the garage." I'm now at that point. I'm not sure what to do with the rest of my life." I kid; obviously, I still have loads to do on the Lotus. The brakes work enough to pilot the car around the block, but they need everything to drive it safely at speed. I'm still running on the original calipers, discs, pads, drums, shoes, and wheel cylinders. With the booster servos bypassed, while they're functional, it's advised that you should use a master with a smaller bore if the servos are bypassed. And, to be inspectable, the car needs a windshield, and they're rare and expensive. So lots of work is still ahead in Lotus Land, even before a real drive above 20mph.

But it is incredibly satisfying to have reached this point. When I look in the garage and see both the Lotus and Bertha, both awoken from long slumbers and resurrected, and both at the point where you can jump in them, turn the key, and drive, I've just got to smile. Thanks for coming along for the ride.

And now my attention REALLY DOES have to turn to Bertha, as she and I have a date to leave for The Vintage in ten days, and she's been very tolerant of this little British interloper. I think if I ignore Bertha any longer, I'll find a wrench in the side of my head. But at least it'll be a metric wrench. (Maire Anne, don't get any ideas...)

131 LIKES, 60 COMMENTS

Larry Filippelli so cool to see it running. the story has been great to follow! +3

Hunter William West We had quite a time getting the brakes to work correctly on our Europa. Are the s2's still rear drum? +2

Rob Siegel This is a TCS, and yes, still rear drum.

Hunter William West Rob Siegel is the pedal just not there? Or is it just stiff?

> **Rob Siegel** Hunter, the car sat for 40 years. The pedal initially went to the floor despite repeated bleedings. I bypassed the booster servos and re-bled and then the pedal firmed up. It's perfectly firm now and the brakes are functional. But none of the brake components except the rubber hoses have been changed; calipers, rotors, pads, wheel cylinders, shoes, and drums are all 40 years old. It's all going to need to be gone through.
>
> **Hunter William West** Rob Siegel ah, nvm
>
> **Hunter William West** I only really got to drive ours for a short period of time before it got sold, was fun, tho can't imagine how quick the tcs is.
>
> **Rob Siegel** This is a stock Federal-spec TCS, which, despite the "big valve" hype, I think is spec'd at 105 hp, which, in a 1600-lb car, is more than it sounds. Before I rebuilt it, a guy was advising me to upgrade the engine to European Sprint specifications. I drove another bone-stock Federal-spec TCS and decided it was plenty quick for me. +1
>
> **Hunter William West** Rob Siegel yeah, we had a weber on ours and I think the Renault engine was like 90 hp and the car was 1300-1400
>
> **Jeffrey Miner** Rob Siegel I "revived" a totaled E30 for my son last year, and it had sat for three years, and I was amazed at how bad the brakes were from simple rust corrosion. I thought they would wear and bed back in. It was very hard to get the ABS to kick on, so I replaced the pads and rotors. So that's probably a lot of what you're feeling in the pedal now. But to prevent a catastrophic hydraulic failure, the rubber caliper seals on each piston will need to be replaced also. And I can't image the rubber seals in the master cylinder are any safer to rely on.
>
> **Russell Baden Musta** Hunter William West All Europas are rear drum unless it's aftermarket such as a Spyder chassis, or a Mark 47 GT... those have rear discs.

Bob Sawtelle The long and winding road in Lotus Land. But "functional but poor" isn't that the tag line for an erectile dysfunction commercial? +1

Scott Aaron Jeez Lindsey Brown can just swing by and fix up your carbs? How great is that? That's like Richard Thompson swinging by and showing you how to play the beginning of "Calvary Cross" +2

> **Rob Siegel** As Joan Cusack said in "School of Rock," "Yes, it's cool. It's very, very cool." +1

Ernie Peters Ain't it great when you uncover something like the "kill switch", and then realize how it is related to a prior mystery? And, you are getting into the head of a prior owner who was probably a little risk averse and wanted to immobilize it to prevent theft. Bloody cool that the bloke Rob aced it, as a Brit might say. +2

Mario Langsten Dare you to drive it to The Vintage 😀 Great to see you drive it! +3

Tom Egan I'm driving booster-less with the stock MC, and yes, you have to show that pedal who's boss to stop quickly. I'm considering a simple alternative to the reduced-bore MC; simply drill a new clevis pin hole in the brake pedal closer to the fulcrum and block the pedal assembly up by the same amount to keep the forces inline with the MC. 3/8 inch ought to do it. [insert pic: clevis pin illustration.jpg]

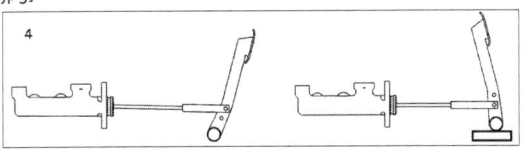

+2

Rob Siegel Tom, I may try new rotors and pads with the existing MC before I flood my garage with any more brake fluid. +3

Tom Egan Rob Siegel Prevailing wisdom says "Green Stuff" pads (whatever they are) make a big difference, but presumably you have to replace them more often. +1

Rob Siegel I may try some ten-dollar RockAuto Centric pads first. Anything would be so much better than the 40-year-old consumables that are on the car. +1

Tom Egan Rob Siegel BTW, the above fix requires no blood, er, brake fluid to be spilled. +1

Jeffrey Miner Tom Egan I don't know if they updated the compound, but EBS pads evaporated for me in every application I tried. +1

Jeffrey Miner Rob Siegel no, no. Don't use crappy pads. That will only destroy your pedal feel. I strongly recommend Hawk HP Plus. They will help offset your non-boosted system with more rotor torque and give you great pedal feel. They may actually solve more than half your hydraulic boost problem with increased rotor torque.

Rob Siegel Jeffrey Miner, got it, thanks. Just FYI, the recommendations for Green Stuff pads for the unboosted Europas and Elans are pretty pervasive across several forums. +2

Chris Mahalick I am also running the stock MC on mine. All it really needs is a more aggressive pad. +2

Jeffrey Miner Rob Siegel well maybe they changed the compound, or maybe they're so light the pads don't "evaporate" like they did on my 318is, or maybe they don't drive their Loti like they're running a qualifying like I do my cars 😀

Tom Smith Wahoo indeed!

Rob Siegel Tom, you know that I thought of Rob [Tom's brother-in-law who'd passed away just a few days prior] and how much we both would've enjoyed my telling him this :^(

Tom Smith Thinking exactly that. He would have loved that engine sound, and of course the smell of rodent and brake fluid. 😀 Carry on! +1

Charles Morris "I just want to get the car to the point where I can move it in and out of the garage." I'm now at that point. I'm not sure what to do with the rest of my life."... I understand the post-coital depression feeling ;) It took 21.5 years (so far) to get my big-block '72 Dart to this point - and now I have just been walking past it to work on other projects! +1

Chris Roberts ⭐⭐⭐⭐⭐ +1

Paul Hill Right shoe off!! Little spin round the block… right shoe on!! +3

Rob Siegel yes, the Lotus version of the Hokey Pokey +3

Paul Hill I feel a T-shirt coming on Paul Wegweiser +3

Eliot Miranda the guy I bought my TC from used ballet shoes, which worked okish for me until a friend gave me his puma Ferrari driving shoes 😀 +2

Paul Wegweiser Paul Hill in the spirit of Chapman, I'll mail Rob Siegel a spool of thread and call it done. +4

Doug Jacobs Have you inspected the motor for… leaking?

Rob Siegel Yes. So far so good. +2

Doug Jacobs So by that you mean it is leaking, which every proper running British car should be. +2

Sean Curry Does Mass require a perfect windshield for inspection? I know many states just say it can't be completely cracked or impede the driver's field of view.

> **Rob Siegel** I don't know what the exact laws are in MA, but a top-to-bottom crack down the center (which is what it has) is certain to fail. +1
>
> **Sean Curry** Rob Siegel Ah oh yeah, I guess that's a pretty standard gotcha. Glue a length of chrome trim over it and pass it off as a two-piece! :D +3
>
> **Scott Lagrotteria** Sean Curry Massachusetts used to require that no cracks are in the sweep of the wiper blades, although the past few years they have become more stringent, so it currently might be tighter. +1
>
> **Francis Dance** Rob Siegel what I recall as a former Mass RMV inspector, any crack in the path of the windshield wipers was cause for rejection. So you are right.
>
> **Lindsey Brown** Rob and I are seasoned veterans at windshield replacement. Well, at least we did one once. Well, twice. +1

Frank Mummolo Kind of reminds me of the Wright brothers initial flights. You and I could unzip and pi$$ further than they went of flight #1, but by the 3rd flight, as I recall, they got to about 1000 feet. They say if you refuse to quit, you narrow all your possible outcomes to just one: winning. Which is what you did. Well done, Rob!

Bailey Taylor Congrats! Drive it down to my neck of the woods for a real shakedown. +2

Jamie Eno The freedom that you get when cars move on their own! Congrats! +1

Paul Forbush Congrats, Rob! Whodathunkit would ever come to pass?! +1

Chris Mahalick So you found a cut-off switch under your seat. We found a half smoked joint from the 80s in mine, LOL. +1

Thomas Siegel Gonna be a tough act to follow!! +1

Russell Dejulio Great car only wish it was in Lotus yellow or BRG

Brad Purvis Nothing better than the blue balls of success. +1

Paul Garnier Oil the door hinge

Bill Howard Always dubious about car livery that can be mistaken for primer. +1

> **Rob Siegel** "Always dubious about car livery that can be mistaken for actual liver."

(video, 37 likes, 20 comments)

Russell Baden Musta Interior looks very good Rob!!! +1

Alan Hunter Johansson Woot! as the kids say. +1

Tim Warner A level side pic that has the 5 series in the background would show off the crazy height difference! So happy it's alive! +1

Franklin Alfonso Now you've upset the mice... 🙂

> **Rob Siegel** Fuck the mice. Living rent-free in my Lotus. Socialism never works! Where the hell do they think they are, Canada? :^) +1
>
> **Franklin Alfonso** In my house the mice get recycled as bird food...

Don Moy You left the key to the frunk in the bonnet.

> **Rob Siegel** I know. Otherwise I need to either lock it, or leave it unlocked, and then the latch sits on the fiberglass.
>
> **Tom Egan** If you leave it unlocked it will rise up to about eye level at 40 MPH bringing much needed parity to forward and rearward visibility. +3
>
> **Rob Siegel** Yes. The perfect physical manifestation of the Bernoulli Effect. Happens on the hood of my 2002 as well.
>
> **Tom Egan** BTW, regarding your license plate, isn't cable TV channel 2680 deep into the Verizon porn region?
>
> **Rob Siegel** um, I wouldn't know :^) +1

Clifford Kelly You have to be grinning from ear to ear. +1

> **Rob Siegel** oh yeah +1

Christian Mulero Need a ride

Andrew Wilson [gif of Cinderella transforming for the ball]

Bill Howard First time I saw it, I have the same thought as now: World's smallest pickup truck, but it has a tonneau cover. +1

John McFadden I think you need to sell me the E39 to make room for more Loti. 🙂 +1

Trisha Knudsen YAY!

Kent Carlos Everett Ahhh olor de rod'ent y brake flu'id. Makes my eyes water just thinking of it. But sounds great, shifts smoothly... you have taken a British pos and made it into a car. Well done sir. I'll enjoy watching you beat the brakes into submission. +1

Caleb Miller Still reminds me of the illicit love child of an Elan and an El Camino.

Jeff F Hollen Was your right shoe already off?

This interior is rapidly growing on me, and I don't mean with mold.

35 LIKES, 10 COMMENTS

Scott Aaron Wowza

    Rob Siegel I know, right? Just a few Armor All cleaning cloths to wipe the dust off. +3

Russell Baden Musta The switch is a pain in the butt I'll bet!! +3

Doug Jacobs It's looks like possums don't live there anymore. Or they're really tidy possums I guess. +1

Scott Linn Nice to have the ash tray. Those are becoming unobtanium. +1

Dirk Rasmussen Man that looks inviting. You have to hand it to the Brits, they really know how to make up for all of the electrical nightmares, and oil hemorrhaging motors. Too bad most of the gauges never work, but who cares? It's a splendid place to sit while waiting for the tow truck. +3

    Dohn Roush You don't need a tow truck; just a couple of friends to carry it to the repair shop... +1

Adam G. Fisher Just...WOW! +1

Pete Waldeck Wow! I've never been a fan of the Europa styling but that interior makes up for the body. Pure Euro whips and leather sex. +2

Keith Roth Ok from the inside making a little more sense.

Interjection: The stages in sorting out a car

In *Ran When Parked* and *Resurrecting Bertha*, I talk not only about the details of sorting out both of those BMW 2002s, but about the general stages of resurrection of long-dead cars. I'm not sure I've ever laid it out this finely, but I think of it like this. Each stage requires the success of all of the preceding ones.

Engine starts	Requires freely-turning engine, compression, clean oil and fuel, correctly-timed spark, charged battery, functional starter.
Engine runs and idles for several minutes	Requires leak-free fuel and cooling systems and oil leaks to be small.
Engine idles indefinitely without overheating	Requires functional water pump, belt, thermostat, and radiator.
Car moves under its own power two feet	Requires functional clutch and unseized brakes.
Car can be driven back and forth 20 feet	Requires functional brakes, shift linkage, and transmission.
Car can be driven around the block	Tests how well the word "functional" applies where it's used above.
Car can be driven safely on short low-speed trips on public roads	Requires that the brakes actually perform well as opposed to merely being "functional," that front end components aren't so worn as to be dangerous, that drivetrain giubo (flex disc), center support bearing, and universal joints aren't badly degraded, that it isn't leaking any brake fluid, and that other safety systems such as seat belts and brake lights work.
Car can be driven on the highway at traffic speed	Requires decent tires, wheels are not badly bent, well-balanced tires and wheels, front end and driveline components without any obvious wear or play.
Car can pass state inspection	Requires all lights and wipers work, handbrake works, no holes in the exhaust, no cracks in the windshield, no worn front end components.
Car can be used and driven "like a normal car"	Requires all gauges work, charging system tested and functional.
Car is comfortable	Requires that suspension isn't ox-cart-like, no strong odors, seats aren't so worn that springs are sticking in your ass.
Car can be driven on a long multi-day trips	Requires tires are very good to excellent; no ominous rumbling or whining from transmission, differential, wheel bearings, CV joints, and idler pulleys; verification that rubber fuel lines and coolant hoses aren't soft or cracked; cooling system perform like new or better to handle high ambient temperatures when car is stuck in traffic; no major transmission, differential, or power steering leaks.
Car is a joy to drive	Car is largely free of thunks, clunks, rattles and squeaks; engine and suspension perform as desired; you're satisfied with exterior and interior condition.

When you begin sorting out a dead car, you may have no idea where on this list it's going to land—whether, with an oil change, fresh fuel, and inflated tires, it's going to rocket straight to the top, or require hundreds of hours and thousands of dollars at every step.

In the post above, I say "I've spent years saying that I just want to get to the point where I can move [the Lotus] in and out of the garage." That's true, but it's disingenuous.

Unless you're a collector with a warehouse that can swallow so many cars that you lose count, a car really needs to be *for* something. Maybe that's just Sunday morning trips to Cars and Coffee. If the car runs poorly and overheats in traffic and the brakes are marginal but the C&C is just a mile away and you get enjoyment out of using it that way, that's great. But most of us want a higher level of functionality and performance from a car. We may not admit it, but we do. If all a car can do is be limped around the block, its functionality is little more than a parlor trick. You can't do much with it. I'm reminded of when I bought back Bertha, the '75 2002 I sold to my friend Alex in 1988 and then sat in his neighbor's garage for 26 years. Because the garage was behind the house and at the bottom of a hill with no paved driveway to get a ramp truck down, and the only practical way to get the car out was to drive it out, I spent a solid week in that garage un-seizing the brakes and shift linkage and getting the car to run on three cylinders so it could make the 100-ish foot climb up the hill, but once it was on the street and taken by ramp truck back to my house, it's not like I was going to continue to drive it that way. It still needed everything. I may say that I like cars with patina, cars that are closer to ratty runners than to concours, and that's true, but it belies the fact that even cars like that have to be fairly close to fully functional, and within that functionality have areas where they perform well, in order for us to be happy with them. This is especially true with performance cars. After all, they're not old pickup trucks that are supposed to lumber around slowly. You're supposed to get on them—nail the accelerator, throw them into curves—and they're supposed to light your fire when you do. If they don't, they just drive you nuts—you wish they were something more, something else, something better—and sell them. And in my case, this wasn't just a "performance car." This was a fucking Lotus.

Yes, I'd just limped the Lotus around the block. It was a huge milestone. I was very excited about it. I felt an incredible sense of accomplishment at having dispelled the stench of abandonment and failure that had surrounded the car for six years. And, from a practical standpoint, it was great that I could pull it in and out of the garage. But with the path from barely-functional to well-performing yet to be driven, what the Lotus' actual future was still completely unclear.

TLC Day 2155: Paying it forward

MAY 5, 2019, 5:17 PM

The Lotus Chronicles, Day 2155 (afternoon massive pay-it-forward edition):

My friend and pro Lindsey Brown came over and tweaked the Strombergs in the Lotus, until a second episode of the car's overheating and blowing coolant out the cap while idling caused him to stop (yes the electric fan was running). I'll try replacing the radiator and electric fan with something that isn't 45 years old. But probably not anytime soon. Thanks so much, Lindsey!

Then, a guy I know (Dezi) came by. He found me a few years ago online. I sometimes help him with the pretty Verona 2002 he owns. He's a little eccentric. I liked him immediately. Over the phone he described a problem that sounded like shift platform bushings so worn (or disconnected) that the shifter was laying down on the driveshaft. I thought it'd be 15 minutes in and out. But when I put the car up on the lift, I found that I was wrong. The problem was that a) the giubo was in seven pieces and b) one side of the transmission support bracket was unbolted.

"Damn," I said, "I don't think I can do it right now. I'm not sure I have a spare giubo here in the garage."

"I have a giubo," Dezi said.

"You DO?"

"Yes," he said. "You made me buy it after you looked at the car two years ago.

To paraphrase Gandalf, I have no memory of this giubo.

I replaced the giubo and reattached the transmission support bracket. Obviously it was more than 15 minutes. "You're a lifesaver," he said. "What do I owe you?"

I then explained how, having just been the recipient of a free faux 2002tii not-quite-a-parts-car, and free transportation for the free car from my friend Tom Samuelson, and free tuning of the carburetors of the Lotus, that if I accepted money for replacing his giubo, much less replacing it with the one that I'd told him to buy but had no memory of, the automotive universe would smite me.

My universe is a kind, generous, and interesting place.

124 LIKES, 49 COMMENTS

Brad Purvis Yes my friend, and you are the center of that universe.
Tim Warner Ye olde slide.
Luther K Brefo Lindsey!

Paul Wegweiser This is the "bend over… I need to take a sample" pose. +2

Scott Linn After SUs, it took me a while to get used to the spring-loaded needles in these ZS carbs.

> **Rob Siegel** I know nothing about either.

Steven Bauer If they're maintained, they can work quite well. I've delved into the SU world via the carbs on my 240z. I don't think these carbs were designed with ethanol in mind attention needs to be paid in that area.

Joey Hertzberg I had a properly functioning set of SU carbs on my Datsun roadster and they were excellent carbs. Fired right up, instant acceleration.

John Harvey Is the tune of the carbs any better?

> **Rob Siegel** I think so, but I haven't run it since I shut it off after it dumped coolant.
>
> **John Harvey** Ugh. I absolutely adore SU's. I know the Strombergs can be made to perform I just don't have the same experience with them. Do the Lotus guys stick with the Strombergs or do they go to a pair of HIF6's?
>
> **Rob Siegel** Some do.
>
> **John Harvey** The design concept is excellent.
>
> **Scott Linn** Some go with SUs, HIF6's, Weber side-draft, a few downdraft, etc. I've driven mine with the ZS's "as rebuilt" by "experts" when I bought it (including rocks in the carb to hold open the cold running valve), a Weber 32/36 DGV and then my ZS rebuilt by me. The Weber actually felt at least as good as the rich ZS', and pretty similar to the rebuilt versions. I really didn't see the downside to running a DGV. I know there are people who think it's blasphemy but it seemed to work just fine for me in the short time I ran it. But I'm back to the ZS carbs and they seem to work okay and now get much better mileage after my rebuild (20mpg vs. 31mpg).

Robert Myles The universe doth already smote you; you've got a Lotus… +5

Rennie Bryant I've got an idea for your overheating. Maybe you still have air trapped in the gerbil track of pipes that go forward and aft in the cooling system. Try one of the vacuum bleeder systems that we use on Minis. They also use them on water cooled Porsche for the same reason. +5

> **David Ibsen** Rennie Bryant bingo
>
> **Rob Siegel** I would be surprised if that's necessary. I don't see it reference on the forums, and I don't see a lot of chatter about the Twin Cam as being difficult to bleed. There's a bleed port on the radiator, and the highest point in the cooling system (the top of the metal pipe) has a bleed port that vents to the coolant reservoir. Of course I'm wrong about many things. +2

TE Cole Rob It's an easy procedure to do. And you would kick yourself later if it were that easy.

Brad Purvis SUs are a great carburettors (correct spelling & pronunciation). They are excruciatingly simple, but people make way too much of them. Once set, they are a joy forever, and rarely need adjustment unless you fuck something up with the timing or valve adjustment.

> **Scott Linn** I ran SUs on my Midget for the first 23 years I owned it. They were great when set, however you had to top up the damper oil and tune them every 6 months or so. PCV valve needed to be cleaned every so often too. I switched to a Weber DGV about 20 years ago and I put a drop of oil on the throttle shaft and choke pivot points once every 6-12 months. That's it in 20 years... Much less maintenance. And the mpg between the two is nearly identical.

Scott Fisher A modern multi-core radiator with shrouded fans is probably the right choice, but the last car I tried to update that way was my Shark. I ended up putting back the engine-driven fan (plus a new water pump and t-stat and hoses) because clearance issues with e-fans prevent using the stock shroud. And I don't know how much room the Europa has around the radiator.

> **Rob Siegel** Scott, this is a mid-engine car with a front-mounted radiator. There's no mechanical cooling fan, thus no shroud. +1
>
> **Scott Fisher** Oh, right, with those looooong coolant tubes from engine compartment to radiator. My only mid-engined experience was with a car that was DESIGNED to blow its coolant all over the place. So your comment, "I'll try replacing the radiator and electric fan with something that isn't 45 years old," is probably right on. It's been my experience, though, screwing around with progressively less ancient automobiles, is that fan shrouds are a Good Thing, with or without electric fans (but with them, whenever feasible). Also that one difference between hopelessly shot old British sports cars and BMWs is that you can probably cobble something together using parts from the hardware store that is superior to the factory cooling system, while if your BMW is not running cool enough, the best thing is to put it back to the factory configuration with all new parts. ;-) +1

Andrew Wilson Rob, this post is exactly why I love the BMW community. +1

John Whetstone Our friend Bruce would be happy with the proper spelling (and hopefully pronunciation) of giubo! +1

Lindsey Brown While I rarely opt to spend any of my all too infrequent time away from my day job working on even more cars, working on cars with Rob is more like visiting an amusement park. +8

David Weisman I have a Cuica?!

Brent Fauntleroy Rob that last line says it all; life is joy.

David Kiehna Bleeding the coolant system in a Europa is fun! I know from experience 😊

Jeff Dorgay From now on you shall be called St. Rob: The patron saint of automotive patience... :) I'm serious. +2

Rob Siegel Trust me, I can be as much of a dick as the next guy. It's just that the episodes are fewer and further between the older I get... +4

John McFadden Rob Siegel maybe it should be automotive patients?

Scott Aaron Rob Siegel 😊

Brian Stauss I sure wish you didn't live 1000 miles from me. I need a fellow wrencher like you nearby! Well done, Rob, paying it forward! +3

Brian Ach mensch

Lindsey Brown After leaving Rob to his afternoon of adventures in 2002 repair, I met someone who told me about a nest with a pair of Great Horned Owlets in it, and was kind enough to lead me all the way there, thus completing the cycle.

+6

Doug Hitchcock I love this story!

David Ibsen It's spelled guibo +1

> **Rob Siegel** no, it's spelled "giubo," but by agreement with Satch Carlson, writers for Roundel magazine will continue to pronounce it "GWEE-bo" because, well, it's funnier [Wikipedia link to "giubo"]
>
> **Robert Myles** Ahh, Uncle Satch, the man who gave us, amongst other things, "The Incredible Awful." Just remember, he was a SAAB guy first. I've got an autographed edition of his book.
>
> **Rob Siegel** Robert Myles, that's a rare book these days!
>
> **Carl Witthoft** Rob Siegel So, pronounce like "geoduck" +1
>
> **David Ibsen** A clam? How about driveshaft kupplung? Leave it at that? +1
>
> **Gordon Arnold** Rob Siegel I'm sure you meant giubo, Rob.
>
> **Sam Pellegrino** It's so Italian! You would say "giunto" as GEE-OON-TOE. Then, the contraction of "GIU-BO" would be GEE-OOO-BOW. But, it's been Anglicized/Americanized. Back in my yoot, we had a family friend, Edgardo Giunchi. Everyone said GWIN-CHEE. My grandfather always reminded us that it was GEE-OON-KEE. I did not realize all of that until we got on the Uncle Giubo thread. THANK YOU!!
>
> **Lindsey Brown** Giubo is a contraction of the Italian words giunto ("joint" or "coupling") and boschi (the surname of the Italian engineer who designed and patented the first flex disc, Antonio Boschi (1896–1988)).[1] Boschi later founded GIUBO SpA, a company solely dedicated to manufacturing flex discs. +2
>
> **David Ibsen** I'm never going to hear the end of this... oh well, it won't be the first time. I'll let myself out...
>
> **Thomas Jones** Lindsey Brown I learn something new every day, at least one, and this is by far the coolest I've learned in a while. I've been calling them giubos forever and not only did I just learn that I've been spelling it wrong all these years, but to learn the background of the name is so cool. Thank You Sir. +1
>
> **Lindsey Brown** Chevy Novas sold well in Spanish-speaking countries, a general did not utter

"Jeepers creepers!" after riding in a Jeep prototype, and Guido Botticelli did not invent the rubber driveshaft coupling.

Brad Purvis Rob Siegel Come in Rob. You were an Alfa guy once. Once. It's Gubio. Satch still thinks it was Bimmer in the 80's when the rest of the WORLD (Okay, Yuppies) called them Beamers.

TLC day 2179: New radiator

MAY 28, 2019, 9:08 PM

The Lotus Chronicles, Day 2179:

With 5 out of 6 of my articles in, and the 6th just a paragraph away from completion, I stole some moments with Lolita in the garage and unpacked the $280 shipped-from-China aluminum radiator and fan I'd ordered before I left for The Vintage. I ordered the full-width (86mm) one with the fan on it so I'd be sure the fan holes would align with the welded-on bracket. Looks about what I'd expected. Yes, several folks I trust advised that the overheating problem I experienced when Lindsey Brown was tuning the carbs might be due to air in the line, and a vacuum bleeder might solve it, but that would've been nearly $100. I instead bet on replacing the sat-for-40-years radiator, which, together with the cartridge-style water pump and hoses, gives me an entirely new cooling system stem-to-stern. I'll know more in a few days.

(74 LIKES, 33 COMMENTS)

Eric Pommerer [gif, Indiana Jones, "You have chosen wisely"] +3
Rob Siegel we should have the GIF ready for «he chose poorly» :^) +2
Eric Pommerer [gif, "he chose poorly"] +2
Steven Bauer Shiny new toys are always good.
Rob Siegel As much as I didn't want to drop nearly three hundred bucks on a new radiator and fan, I tried the "well the old one doesn't LOOK corroded" path. +1
Steven Bauer Rob Siegel Did you check temps at different points in the system when it was running hot?
    Rob Siegel No. It blew the radiator cap. Twice. +1
    Steven Bauer like a baby spitting out a pacifier??
    Rob Siegel Like a baby hurling with the pacifier still screwed into its face. +7
    Daniel Neal Rob Siegel as I am holding my son getting him to sleep that is not an image I wanted popping into my brain. +4
Clay Weiland Damn, that thing is thick.
    Rob Siegel that's what SHE said. BOOM! +12
    Dave Gerwig Clay Weiland Rob Siegel Lolita needed a thick one. This creates another opportunity to blow out the plumbing from engine bay to the front cross-member. Have you bypassed the

heater core or is that fresh? Maybe you've already covered that topic, forgive me if I missed that Chronicle. Great progress on all fronts.

Rob Siegel Heater core has been flushed in situ, that's all. I haven't seen it leak so far, but when the speed crept up during the around-the-block drive, there was a lot of rodent smell coming from under the dash. I'd wager it's the heater box.

Clay Weiland A clean, freshened up box will make the rides more enjoyable. +7

Rob Siegel I'm dyin' +1

Dave Gerwig Rob Siegel Clay Weiland its needs some Wessonality don't ya think? +2

Rob Siegel Wow, haven't thought about THAT one in 40 years... +2

John McFadden Randomly triggered Looney Tune memory from "I'm dyin'" post [link]

Chris Mahalick I don't think these cars have a heater core. But I could be wrong.

Daren Stone Chris Mahalick Europas have a wide, thin heater core clipped in a sheet metal box located directly beneath the center of the windscreen. If you open the front cover and look at the screened opening on the back wall of the storage area you are actually looking into the center of their heater box. +2

Chris Mahalick Daren Stone You rock! I am going out to look now.

Daren Stone Chris Mahalick glad to help! I've had my S2 since 1994 and have been one of the Dave Bean boys off & on since 2004 so most anything vintage Lotus, Ken or I can help with.

Matthew Zisk 👍

Bob Shisler [gif, John Maloney from Frasier, "thic"]

Scott Fisher I was going to type some of this in, but then I remembered the time I interviewed a half-dozen or so of the top cooling system engineers in the country about how to make sure your radiator was big enough. [link to Scott's April 2018 article in PRI magazine]

Franklin Alfonso But now you have Chinese stuff in a British car??? Is that even Kosher? What will Lucas think? +1

Franklin Alfonso Where is the screw top to add coolant in that thing?

    Rob Siegel Franklin Alfonso, the fill reservoir is in the engine compartment.

    Franklin Alfonso They have to do it differently don't they...

Daren Stone Assuming that's a bleeder on the RH tank shown I'd suggest getting the front end up as high as you can when it comes time to bleed the cooling system.

TLC Day 2180: Beginning the brake job

MAY 29, 2019, 5:41 PM

The Lotus Chronicles, Day 2180:

Quite a bit, actually. Got the replacement radiator installed, but not before cursing the fact that what I thought were studs that held the fan onto the bracket were instead Phillips-head screws with no way to access the backs because they were put through the bracket and then the bracket was welded on. Finally figured out to lock two nuts together at the tip of the screw to hold the screw in place to tighten down the nut beneath them. But it's done, installed, filled, and wired.

Then I went for extra innings and started doing the front brakes. Pulled the wheels and yanked off the calipers. I have new rotors and pads, and had said to myself that I'd buy rebuilt Centric or Cardone calipers if they became available again (they're Spitfire calipers, and only like $30 each when RockAuto or CarID has them in stock). Checked online, and CarID had them. Neither of mine were obviously seized, but the left one had a torn dust boot. And one pad was completely detached from its backing plate. Glad I didn't drive much over 15mph. Ordered the calipers. I already replaced the flexible hoses. It'll feel good knowing every bit of the front brakes are new.

The hubs on these are like a 2002; the rotor is on the inside of the hub. So you have to take the hubs off the spindles to change the rotors. So if there's anything wonky about the wheel bearings, it makes sense to change them as part of the rotor replacement. Sure enough, the passenger side one is ticking. Looking at replacements right now. These also appear to cross-reference to Spitfire parts. It's nice that, after all that expensive insanity with the engine, I'm in the territory of ubiquitous inexpensive parts I can get on RockAuto or Amazon.

Unclear how long the delivery is on the calipers. I may just slap the new pads in it and put it back together for now and see how it feels.

44 LIKES, 21 COMMENTS

John McFadden I think it's been said before—15 lb weight savings on radiator and fan = 15 whp? :-) +1

Scott Winfield Go Rob, go! +1

Colin Tennessen Europa has always been top of my "if only" bucket list cars. It's nice to see someone else going through all this trouble and just pretending it's me doing it +2

    Daniel Senie There's your potential buyer, Rob :) +2

    Colin Tennessen Daniel Senie give me 5-6 years and yes!

    Colin Tennessen It's even brownish.

Franklin Alfonso Well, moving along even if it seems nothing is every easy with Lolita.

Steven Bauer The rotor/bearing connection was common with older cars. My 240z is the same. +1

Rick Roudebush Gravity is not your friend when it comes to this coolant system with rad in front and engine behind, but you can make it your bitch by filing the rad while the front is as high as possible. This was my method (after a few bad encounters with air pockets) when doing coolant flush on the Pontiac Fiero we had years ago. Good luck as you go forward. +4

    Rob Siegel yup, that's what I did +3

    Rick Roudebush Then I would let it run until the thermostat cycled a few times with the cap off before letting it back down. I could see the coolant, and sometimes air, rush by the fill neck as it cycled. That always gave me comfort before I let my wife drive it on her hour long commute the next day. It's been fun to be a fly on the wall as you repeatedly get right back on the bull at Rob's Rodeo Palace.

Bob Sawtelle I'll bet this is the only time in the history of automotive writing that someone wrote "it cross-references with a Spitfire so that's a nice thing." Meanwhile somewhere in a small dark garage in the English countryside a mechanic is cursing at a Triumph Spitfire as I write this. +7

Jeff F Hollen Part of the reason I ditched my Range Rover Classic: rust. The other determining factor: There was absolutely zero captive nuts. Everything was bolt thru... it's ridiculous. Also, doncha love parts bin cars? My Jensen is basically a Triumph underneath, with a 2002 dogleg trans (literally ordered the shifter parts from BMW), and Lotus engine. +3

    David Kiehna Jeff F Hollen there's also Vauxhall mixed in there too +1

    Brian Aftanas Thank goodness all low-production cars have parts used in other higher production cars. When I pulled out the electric mirror switch on my Ferrari 308 I was astonished to see the BMW Roundel cast on the back; Ferrari co-opted the one from the E28. My fuel injectors are the same as a VW Rabbit, the windshield washer pump is the same as a Jeep Cherokee… and the oil filter is also used in some kinda forklift. It would be crazy-expensive if every part were custom designed. Not sure Bosch would even be willing to make, say, a specialized fuel injector in low quantities since they are pretty generic things. +2

Robert Myles I wish you better luck with Rock Auto hydraulic parts than I've had over the years. Every caliper I've ever bought from them has either leaked, had mismatched hardware, been filled with sand (presumably from cleaning) or otherwise been unusable.

Rob Siegel Huh. I've never had those problems, and I've used many on BMWs.

Robert Myles Glad you've ducked that bullet. When I was still young and optimistic I encountered all of those and more across a variety of cars, including Mazda, Pontiac, Toyota, Volkswagen, and a Ford. Hard pass for me now, I'll spend the big bucks for the OE stuff if I'm not rebuilding them. It's worth it to it have to do the job twice. Don't even get me started on their return policy.

Brandon Fetch *mashes the gas twice, reaches for the choke control and keys*

TLC Day 2190: Happy anniversary

JUN 9, 2019, 8:37 AM

The Lotus Chronicles, Day 2190:

According to my email, it was exactly six years ago Friday that I saw the ad for the Lotus on Craigslist in Chicago, and six years ago yesterday that Ben Thongsai looked at the Lotus for me in Chicago and sent me these photos. The car was basically an accent piece, a coffee table in the lobby of a European car repair shop. Ben said "It's actually in very complete unrestored original shape. If this is what you want, you could do a lot worse." I authorized him to negotiate on my behalf, and about half an hour later, he called me up and uttered the following immortal words:

"Congratulations. You own a little brown thing."

I'm not sure I've ever forgiven him.

Happy birthday, you sexy troublesome little brown thing. Soon as I get the bearings, hubs, rotors, and calipers back on you, you'll be whizzing around the block again.

155 LIKES, 54 COMMENTS

Brooklyn Taylor Little Brown Fun Machine!
    Patrick Hayes Brooklyn Taylor I see what you did there
    Alan Hunter Johansson Brooklyn Taylor I bet that was your nickname in junior high.
Richard Shouse I for one would like to see a video of you getting in and out of that fiberglass sardine can without making any, ahem… older person noises. Or farting. 😂 +5
    Adam G. Fisher Ditto!
    Rob Siegel Right here, dudes :^) [link to video of first back-out]
    Adam G. Fisher Rob Siegel I think I heard something. 😂 +1

Wade Brickhouse Adam G. Fisher that was the Lucas mice at work. :-) +1

Rob Siegel Wade Brickhouse, Lucas mice don't piss, they leak oil 🙂 +1

Wade Brickhouse Rob Siegel positive earth thing I guess? +1

Richard Shouse All kidding aside, congrats on getting this project so far along!

Bailey Bishop Jr. Same here. Mine was a piece of art in the shop's lobby for years. I couldn't resist. [photo of Bailey's car in the lobby] +3

Trent Weable Happy anniversary to you two Rob, I had totally forgotten that the Lotus came from these parts and that Ben had negotiated the deal. I can totally hear him saying that line about "little brown thing." +2

    Martin Meissner me too

Brian Aftanas John Deere makes the perfect colors for a Lotus. +3

John Graham The only thing the Lotus Chronicles makes me sad about is how I gave up on a 79 Mercedes I had purchased. It sat for 2+ years and I just ended up practically giving it away. I sort of wish I had still tried to make the car work. The issue though? I hated the thing. The moment I drove it, after I bought it, I hated it and from there things went downhill. The Lotus looks great. Although, we are all still going to make British and Lucas jokes, even after you win Concours at Pebble Beach! :-) +2

    Rob Siegel If you hated it, then you made the right choice. +6

    John Graham I learned a lesson that day: Only buy a car that you are willing to keep because you may just have to! +2

Brian Hemmis [gif champagne glass]

Stuart Moulton Whizzing fluids +1

Caleb Scully Strangely enough, I was selling a car to a guy in the backwoods in Maryland, and I saw one of these sitting on his property, no idea if it was real or not (if they had kits), but the only reason I knew what it was was because of your posts! +1

    Rob Siegel Caleb, no kits; they all were produced by Lotus. The later cars like mine with Lotus-Ford Twin-Cam engine and the cut-down rear quarter panels are generally worth more than the older ones with the full-height rear quarter panels and the Renault engine. +2

    Caleb Scully It definitely looked like yours but I didn't inspect it closely, unfortunately it will probably sit in that man's yard and continue rotting away

    Rob Siegel Since they're fiberglass-bodied cars on a steel backbone, it's really only the backbone that rots away, but, yes, any car without a six-figure restored value, when left outside, becomes nearly worthless. +1

    Ed MacVaugh If it was blue, I have driven it. Out on the way to the Hagerstown Penitentiary.

    Caleb Scully Ed MacVaugh it was black and yellow, in between mount airy and new market +1

Walter Eschelbach Always more to do, re do, and have a break down out in nowhere. Good luck with Lucas. +1

Bob Sawtelle I cannot wait to see this bad boy blasting around! Are you shooting to make it to British Car Day at Larz Anderson?

    Rob Siegel Not impossible, but unlikely. The car isn't insured or registered yet. And it has a cracked windshield, and replacements are expensive, so I will only have a one-week grace period after registering it before I need to quit driving it and replace the windshield.

    Richard Shouse Rob Siegel watch the Saab episode of Wheeler Dealers regarding your windshield.

John McFadden Spotted at the local annual British car show last weekend. Drove in under its own power. 🙂 [photo] +3

David Kemether Turd brown is just so appropriate. +1

    Rob Siegel Hey!

    David Kemether I struck a nerve. It was certainly a turd when you bought it. Less turdy each day. +1

    John McFadden Franklin Alfonso I thought it was "shorts when you realize the brakes are inoperable while underway" brown?

    Rick Ramsey Franklin Alfonso more like a skid mark!

Ernie Peters I would have been inspired to purchase the Europa based on the photos. At least yours had four wheels and tires, and all the left front suspension. My Elan was worth the effort, though, as I am sure your Europa will be. [photos of Ernie's Elan] +3

Hugh Forrest Mason Well, you could not have had a better negotiator on your side. Ben's great guy. +1

Josh Boone Such a weird, but cool car. +1

Matt Schwartz That sounds exactly like Ben. Congrats on 6 years of Lots of Trouble, Usually Serious. +2

Larry Webster Jamie Kitman

Scott Chamberlain This is why there should be a Surgeon General's warning on Craigslist "The Surgeon General has determined that this activity is addictive, and can lead to divorce, job loss, banishment from decent neighborhoods, and being disowned by your family" +2

Hugh Forrest Mason Scott Chamberlain preach!

Scott Aaron Oh God is this Europa thing still going? Let me right this ship via a pic I found [photo of my Bavaria] +2

Brian Ach I love these cars +2

Marc Bridge Got to admit it would make a sexy coffee table!

TLC Day 2194: Removing the wheel bearings

JUN 12, 2019, 6:29 PM

The Lotus Chronicles, Day 2194:

While replacing the front rotors, it made sense to also do the wheel bearings, since a) the right front wheel bearing made a slight tick-ticking sound, and b) the rotors are on the back of the hubs, so you have to pull the hubs off to replace them, and when you do, the outer wheel bearing literally drops in your hand. I make no apologies for trying to contain cost, so when I found that RockAuto had good quality SKF wheel bearings at a great price, that's where I ordered from. Unfortunately, I tried to install them last night and found that a) neither the inner nor the outer ones fit--like they were WAAAAY off-- and b) the original inner seal is felt (very British), the new replacements don't fit correctly, and replacing it with something modern led me down some forum rabbit holes. Two phone calls later, one to RD Enterprises and the other to Dave Bean Engineering (not everything in this world is click-and-buy), I had it resolved, and the correct parts on the way, though at three times the price. Sometimes, as they say, you just need to pay the man, but I am grateful for these two reputable Lotus parts houses.

60 LIKES, 24 COMMENTS

Ed MacVaugh Aren't those Spitfire hubs? +1

> **Rob Siegel** They are. Oddly (or correctly), RockAuto does not list these SKF bearings for the Spitfire. +2
>
> **Layne Wylie** Triumph, or Supermarine? +2

Paul Nerbonne Apple has now dropped Car & Driver into MY newsfeed. You should be proud. 😊 +1

Ernie Peters Sometimes it's better to incur the expense of purchasing through one of the "meisters" to get the right parts. +1

Joe Eaton If it's a part from another British car that Lotus used, I've found the prices tend to be best at Moss or British Parts Northwest. Also be sure to check the suppliers in England. Often cheaper, and delivery is often faster as well, oddly enough. +1

Russell Baden Musta I use RD, pretty much next day delivery if I order early... not next day prices... but to be fair I'm also in PA. Pay the Man!

> **Rob Siegel** I use RD for most things, but I put this order through Bean because they had the Nilos seals that replace the felt seals, as well as to share the love. +3
>
> **Russell Baden Musta** Rob Siegel gotcha! Wanna hear or see pic of these seals...
>
> **Rob Siegel** they're basically just thin metal dust caps: https://www.waikatobearings.co.nz/service/nilos-rings
>
> **Russell Baden Musta** Rob Siegel what were your crank seals like when you reassembled the engine?
>
> **Rob Siegel** They were the regular kind, not the rope kind. I don't remember the condition they were in.
>
> **Russell Baden Musta** Rob Siegel mine are rope... I think that's why my engine is soaked... and clutch chatters occasionally.

Richard Shouse My $6.00 Autozone Chevy C1500 bearings worked very well after being submerged after Hurricane Harvey. Replacing them anyway, along with everything else but still... nice to see them still working.

Scott Fisher It is a truth universally acknowledged, as Jane Austen might say (or her niece, Jane Austin-Healey), that a vintage-car owner must be in want of a reliable parts source. More than one, whenever possible. I further find that it is both necessary and proper to share the love among all parts sources, lest one awaken some morning to discover that one's favorite source for parts to fit one's chosen marque has struck its tents and vanished into the night for want of custom. If I ever get all my cars back to where I can drive them, I will endeavor to follow my own advice. Meanwhile, I'm driving a Sixties-era Fiat Spider because the Shark won't start. +3

Al Larson Same rotor behind the hub setup in my Pathfinder. It had a bearing noise, so did it all at once. This is the lovely that came out. No wonder it was noisy. [photo of grungy bearing]

Alonzo Graham Taylor III Are those GT6 or spitfire brakes?

> **Rob Siegel** Spit

Chris Roberts That whole "very British" thing would drive me crazy.

Brian Stauss I am looking for a project car when I retire soon. I'm thinking a Europa. Heck, all I have to do is save off all these chronicles, buy your presumably forthcoming "How to Resurrect a Dead Lotus" book, and I should be good. What could possibly go wrong?? +2

Jeffrey Miner Wow, funny what doesn't change—my mother's Sunbeam sat for weeks while the wrong wheel bearing parts kept arriving from Britain. That was in 1971. Don't leave a pregnant woman with two kids trapped in a NJ suburban home... thus the 1972 Chevy wagon that replaced it. And oddly, to turn off the Chevy, I immediately noticed she no longer had to blow the horn. How odd these American cars I thought at the time. (I was 5). +1

Jeffrey Miner My '87 911 requires complete wheel hub disassembly to replace the floating rotors, so whenever I do the rotors, I put the wheel hubs in the oven, heat them to 350 degrees and fill the house with the horrible aroma of baking wheel bearing grease to swap all the bearings and races. But then,
it's a track car. Curiously, dropped wheels and curb impacts usually bend the wheel bearing cages anyway. +1

Scott Linn Yup. RD & Bean are great. +1

TLC Day 2198: Installing the wheel bearings

JUN 16, 2019, 4:28 PM

The Lotus Chronicles, Day 2198:

Made an executive decision this morning to skip a rainy German Car Day at Larz Anderson and instead be completely self-indulgent and give myself a different Father's Day present and concentrate on getting the wheel bearings and the new rotors installed on the Lotus. Done (and yes I know the dust caps are still off, so STFU). Rebuilt calipers should arrive next week. Then I'll need to clean out all the crap that's crept back in and surrounded the car since I put it back in the rear spot in the garage a few weeks ago. Honestly, it's like the Red Sea after Charlton Heston says to god "Okay, you can drown the Egyptians now."

(And yes the Lucas wheel bearings come with the smoke still in them :^)

64 LIKES, 25 COMMENTS

Franklin Alfonso Sweet. Never be this clean again...

Lindsey Brown But do Lucas wheel bearings still turn in the rain? +2

    Rob Siegel Yeah, they probably don't, but the hood latch probably does instead.

    Don McMahon Rob Siegel drive it hard and cook-off the moisture…

Daniel Neal I like how the motto is "Trust Lucas."

    Rob Siegel Right, they have to say that, because otherwise, you wouldn't, you know, trust Lucas. +3

    Tim Warner Daniel Neal "Trust Lucas to steal months of time and lots of money to keep their old cars running!"

Ed MacVaugh Did you get your fancy metal seal plates for the back side?

    Rob Siegel I DID! Ordered them and the wheel bearings from Bean so I'd be sure they'd fit. I was a little surprised that the wheel bearings where the same Lucas ones that are available cheaper on eBay, but no harm no foul. You can sort of see the thin Nilos seal behind the bearing in this pic.

+4

Andrew Wilson Time well spent. +1

George Zinycz Don't let the smoke out!!! I don't know much, but I know you're not supposed to do that.

Christopher Kohler What kind of grease are you using?

>**Rob Siegel** I used the Exxon blue synthetic grease that came with the wheel bearings.

Tom Tate We missed you today at Larz Anderson. '84 M635 CSI took best BMW.

David Holzman You picked a good day to miss the lawn event. +1

Dohn Roush Never give a sucker an even brake… +1

Rob Blake Looks FANCY

Chris Roberts Man. You know it's bad when Rob tries to de-snark his own post before the snarks can even get started. Give the guy a "brake"… Aw, hell. They just sort of write themselves, I guess. +1

>**Rob Siegel** Chris, when my first book was nearing publication and we were doing the cover design, Bentley came back to me with a mock-up cover that used the photo of Alex and I standing in the garage in front of his Volkswagen Passat whose engine we had just replaced. It was an appealing photo, as we both were greasy from head to toe and smiling like idiots, but it was wrong for the book and its title. I had to push back hard against Bentley and literally say "do you have ANY idea how much shit we will take for having a book who subtitle is "how fixing broken BMWs made me whole" and having a photo of a fucking Volkswagen on the cover?" So that was my trying to "de-snark" my own book cover 🙂 +2

Lee Highsmith Rob Siegel Stay true Rob, stay true… ;)

David Schwartz British Car day at the Larz Anderson Auto Museum is this coming Sunday, June 23rd. Will the Lotus be ready in time?

>**Rob Siegel** Unfortunately not.

>**David Schwartz** My 1950 Morris Minor spent the last two years on jack stands in my garage. My goal was to drive it to British Car Day this year, and I got the MM back on the road with a week to spare. +1

TLC Day 2203: Calipers and rotors

JUN 21, 2019, 5:41 PM

The Lotus Chronicles, Day 2203:

Front brakes done; just one final wheel bearing check and bleeding left. Getting my moments of zen when I can.

105 LIKES, 19 COMMENTS

    TE Cole Who did you use to rebuild all the brake stuff?

        Rob Siegel Click-and-buy of Centric rebuilt calipers from RockAuto. $29.95 each after core return. Ya gotta love Spitfire-compatible parts. +3

        TE Cole Yeah I have purchased a few of those. I recently had an E24 M6 in with brakes locked up all the way around. Instead of dicking with the shitty little Brembos they used on that model we did an upgrade to 840 Brembo fronts... I had a couple used sets and the best match for those on this car is the E34 540 rear vented. I got a second hand set of calipers that looked like dookie. So I ordered a set of Cardone rebuilds and sent the used ones in as my cores. Still saved a mint over new. Incredible how much better the brakes on the e24 M6 are with this set up. That car was a hippo to stop with the OE setup. +1

    Luther K Brefo [gif of Spongebob "Stai Tranquillo!"] +3

    Russell Baden Musta Beautiful! Mine looked like that once!!

    Dohn Roush Bleeding left? You schedule your bleeding, too? +1

        Rob Siegel Yes. I have to be in the mood to deal with that volume of brake fluid. There's a cleansing ritual involved both before and after. +3

        Dohn Roush Rob Siegel Brake fluid? I thought you were talking flesh wounds... +1

        John Whetstone Rob Siegel Agree! Diff fluid a close second. +1

        Paul Wegweiser Rob Siegel Ewww... like... you know... a COLON cleanse? #browncarasmetaphor +4

        Scott Aaron Ew

    George Zinycz Shiny... pretty...

    Maris Mangulis Europa at my local car cruise tonight. MAN these cars are LOW. I was kneeling to take this photo. [photo of Europa] +6

    Josh Boone Baby steps!

    Thomas Jones Such Girly calipers have so much flabby air in them. We're gonna pump them up... [gif Hanz und Franz "Pump YOU up!"] +1

    Neil Bradbury New rotor porn +1

TLC Day 2204: Adjusting the wheel bearings

JUN 22, 2019, 9:34 AM

The Lotus Chronicles, Day 2204:

Mom seems stable for now; trying to slip in garage/therapy time whenever I can.

Finished up the brakes and wheel bearings on the Lotus this morning. Despite my having adjusted the new wheel bearings yesterday and triple-checked them, after I banged on the dust caps and put the wheels on, the passenger-side bearings felt like they clearly had a little too much play. Had to pull the cap off and tighten the castellated nut by one notch and re-pin it. These idiotic Spitfire hubs and caps have no seating lip where you can pry the cap; it's very difficult to pull the cap back off without mangling it. But it's done, and the Lotus is back sitting with all four feet on the ground.

The way that junk in the garage crawls back in to surround the Lotus, despite my having cleared a path to the car just a few weeks ago, is like something out of Narnia where vines instantly close in behind you to overgrow a path that was just there. I guess that's what I get for doing this and the a/c repair in the E9 at the same time.

69 LIKES, 15 COMMENTS

Rob Koenig Garage entropy is a thing for sure.

Tim Warner I'm about to "cut back" my vines surrounding my 11-month Toyota Tundra V8 project and get it out of my garage. A little afraid to, since it was outside before, when a mouse climbed up into the timing belt area, built a nest, and made the damn thing skip time! Filled up every tooth of crank belt gear... crappy design! But nothing like your long-suffering Lotus project. +1

Ben Greisler Considering that bearings are all standard sizes I'll be a trip to Tractor Supply or the equivalent will find you some trailer bearing caps of the right size with the lips. We had to do the same thing with our 2002 rally car up at an event in Minnesota once. Way cheaper than OE parts too! +2

Mike Clifford You can pop a divot or two in the side of the cap as a place to work from, unless that counts as mangling. Also please let me drive that CS. +2

Andrew Wilson Let's coin the term "Garagia" for the mess that comes from nowhere & nothing yet suddenly appears. +8

John Whetstone Dulled chisel (I had a cap like that on something years ago) to get it started, but I'm sure you've already figured out a way that's as good or better. +1

Ed MacVaugh I recall the Spitfire technique was single use :) you drill a hole in the end and screw in a sheetmetal screw to remove the cap. +1

    Mike Clifford That'll do it 😄

Paul Forbush I have occasionally had success by grabbing them with channel lock pliers wrapped with a rag, and gently wiggling them off. +1

Ray King Garagia. That explains it. Just at the shit that accumulated on my car while I was doing major surgery on a couple of motorcycles. [photo parts on BMW 2002 hood] +1

Kevin McCurdy Rob, it appears that your vehicles are in the garage sideways, unless you have an extra garage door that does not appear in the photo. +1

Don McMahon On the VW bug there was a fat washer under the castellated nut which you could move with a screw driver pry if the nut was tightened to the correct pressure on the bearings.

Don McMahon Get cap with BF Pliers

Rob Siegel It's sloped. They slide off.

Tom McCarthy Every time I go to work on my E46 more things have accumulated around it. +1

TLC Day 2206: Baby steps

JUN 24, 2019, 7:29 PM

The Lotus Chronicles, Day 2206:

Took the first baby steps out of the garage since installing the new radiator and fan and redoing the front brakes. It's leaking a fair amount of antifreeze under the cap and out the drain tube. There's no catch bottle, so it's both jarring (pun intended) seeing that antifreeze dripping out, and a little embarrassing. Brakes definitely feel better, but I decided to go no further until I replace the 45-year-old radiator cap and install a catch bottle.

But on the plus side, I figured out that the car is much easier to get in and out of the garage if I back it in, although of course the nonexistent rear visibility makes this challenging. And I placed it next to the E9, which is a little higher so I get the two-level effect with a little path running down the middle. Bring me... ANOTHER SHRUBBERY!

145 LIKES, 39 COMMENTS

C.R. Krieger Fifty bucks gets you a backup camera you can mount as a plate frame. +10

Bob Shisler But you need a screen for it too.

C.R. Krieger Bob Shisler it clips onto the mirror.

Adam G. Fisher iStrong Backup Camera Wireless 5" Monitor Kit for Car/SUV/Minivan/Pickups/Smaller Truck Waterproof License Plate Rear View /Front View Camera 7 White Light LED Night Vision Guide Lines ON/OFF [link on Amazon]

Kim Dais I also need a catch can for my W108. Been looking at the ones Canton sells (I have one in my race car). Just trying to figure out what size and where to mount it.

Robert Myles NEET! +1

George Zinycz Robert Myles I see what you did there... +1

Lindsey Brown "Take this thing away, and bring me another Lotus!" +2

Jonathan Poole I've been running a Heinz catch-up bottle for years on the 02, because it was empty and in the recycling bin and the tube fit the top and it has a vent already.

+13

    Rob Siegel I LOVE IT!

    Mike Tunney I used an empty seltzer bottle. I drilled a hole in the cap and ran the tube through it.

    Rob Siegel I just verified that one of my six 2002 windshield washer bottles will fit. EAT VDO, LUCAS! +4

    Justin Gerry Jonathan Poole I used a Gatorade bottle in YaleR. I think it's still there. +3

    Lee Highsmith @jonathan poole - that's very Hack :) +1

John Thomas Hang a tennis ball so you'll know where to stop. +2

Chris Mahalick Just put on a new radiator cap and call it a day, LOL. +1

Chris Lordan Rob, what is your Top Speed (Observed) in the Europa at this point?

Rob Siegel Maybe 15 mph, at which point the rodent smell from air flowing through the (unheated) heater core becomes pretty damned strong. +12

    Dohn Roush Rob Siegel Hey, they're running as fast as they can. Personal hygiene takes a back seat to getting you up to 15 mph... +2

    Franklin Alfonso If ever a car needed a back-up camera this is it! +2

Bob Sawtelle [gif of Python "Ni! Ni! Ni!"] +7

    Andrew Wilson Icky Icky Icky Ni Wom Zewing!

    David Weisman Andrew Wilson 😂😂 oh no you di int!

    Daniel Neal For Rob's next challenge he must cut down a tree with a herring +3

    George Zinycz Daniel Neal the mightiest tree in the forest... +2

Daniel Senie Garmin makes Bluetooth backup cameras that communicate with their dashboard GPS units. Though I'm sure all that would be anathema in a Lucas-wired car. +2

Russell Baden Musta Daniel Senie backing up is very easy in a Europa! Think bell star helmet, you turn around and you're against the window, and everyone is higher! Except kids or adults lying on the ground...

David Holzman The E9 is gorgeous!

Hugh Forrest Mason So nice to see her out in the sun!!! +1

John McFadden Vintage Lucas rear view camera in order. Of course all you'll see is electrical smoke. +2

Robert Alan Scalla That little brown front end is pretty good lookin'. +1

Russell Baden Musta Don't need a catch Rob! Mine only goes out when overheated... last time was on Talladega after running 8 laps up to 100 mph...

Steve Park Great to see it outside. It's been quite a journey. +2

John Jones https://www.hyndsightvision.com/cameras/journey

Brad Purvis [link to video of Python "Roger the shrubber"]

    Rob Siegel I am Rob(er) the shrubber. +1

Bailey Taylor I saw a rat rod at a car show that had a whiskey bottle for the overflow. +1

Scott Chamberlain I love the Europa, but then I've always had a fetish for pretty bad girls with nasty habits... +2

Bernard Sharpe A PATH! A PATH! +1

TLC Day 2216: Buttoning up the engine compartment

JUL 4, 2019, 11:25 AM

The Lotus Chronicles, Day 2216:

Installed the stock air cleaner with a new filter. And to do that, I had to reinstall the engine compartment luggage bin, since that's what the air cleaner housing attaches to. It'll probably need to come back out since it's in the way of other parts I'll probably need to screw with, but for right now, one could almost mistake this for a whole intact car :-)

73 COMMENTS, 53 LIKES

Курт Вин Excellent way to keep food warm too. +2

> Daniel Senie I was thinking he could replace that with a stainless steel bin and have it filled with a casserole. Cook on the way to a meetup. +2

Andrew Wilson A magnificent mistake. +1

Eliot Miranda Put some K&N filters on and a catch tank and throw away that junk airbox. The engine breathes better, one can take the luggage box in and out easily and the car is a couple of pounds lighter. [photo of this setup in Eliot's car] +1

> George Zinycz Eliot Miranda simpler AND lighter! +1

Tom Egan Is the cracked windshield the only thing keeping you off the street?

> Rob Siegel Tom, I was about to go to the registry yesterday morning and get plates, but when I pulled out the title that came with the car and looked at it for the first time in six years, I saw that it wasn't from the guy who I had bought it from, but from the person HE had bought it from, and was dated 2012. I'm sure I knew this six years ago, but had completely forgotten about it. This will cause problems with the Massachusetts registry, as they would assess me for back sales taxes and penalties for seven years. I'm trying the "Vermont loophole" where you register it by mail in the state of Vermont. I sent the paperwork off yesterday. I'll know in a couple of weeks whether it was successful. +5

> Mike Savage Which issue causes the back taxes? The 2012 date or the fact that the name of the seller? How would RMV even know the second issue? A bill of sale? I am not sure I even needed a bill of sale the last few times. You know more about RMV than I do, but I am kind of surprised they're going to charge you back taxes on an old title. I guess the sale date is noted on the title?

> Rob Siegel Mike, it's the date. Google "Boston Globe RMV Triumph tax mistake" for a nightmare scenario. They DO check the date, and seven years IS likely to be a big problem.

> Mike Savage your basic premise is correct, but I googled that story and the $11k initially quoted as sales tax / penalties from 1982 was later found to be incorrectly calculated. His final bill was about $1,000... still an unpleasant surprise when expecting to pay nothing.

> Roy Richard Damn it Rob Siegel this is the second time I've had to call the RMV about you in the last 3 months. +1

> Rob Siegel Mike Savage, the point is that a title dated seven years ago is almost sure to be flagged, and no one gets the "incorrectly calculated we are sorry" free spin card without intervention from someone like a reporter at the Boston Globe.

> Benjamin Shahrabani Rob Siegel would it be back taxes on the declared or assessed value of the car? What would those even be?

> Rob Siegel Benjamin, it's penalties and interest on the state sales tax for the car, which would be based on the NADA low value. Part of the problem is that when you go to the registry, THEY

can't tell you what the formula is; they tell you the formula comes from the Mass Department of Revenue, they don't know what it is, it's part of their computer application process, and they have to process your registration application in order to tell you what the number is. I've looked online and can't find the formula anywhere. I've been afraid of a car's VIN somehow getting flagged if I have it partially processed and tell them "no." That's probably just paranoia. The story in the Boston Globe of the guy with the Triumph is a highly unusual case of a) the Globe reporter intervening on his behalf and the registry backtracking (this basically NEVER happens) and b) the registry saying that in 1982 they used actual paid value not assessed value. The chances of either of those things happening if I show up with a title with a seven-year-old date on it are less than zero.

George Thielen Perhaps, it's just the British getting revenge for our insurrection. +2

Roy Richard Rob Siegel sell the car to my son who lives in NH then he will sell it back to you for a small profit 😊

Rob Siegel Roy, that's what I did with Louie the 2002tii in "Ran When Parked"—sold to a friend of mine in Maine, who sold it back to me.

Roy Richard Oh boy third infraction to report to the RMV.

Josh Wyte Rob Siegel this is one spot where the DMV in California does it right. You can register a car there as "non-operable" and work on it as long as you want. When it's road worthy you go back to the DMV and change the status. Here in Colorado they fine you but I think the maximum fine is only $100. +3

Scott Linn Mine came with a Napa cone filter that clamps onto the stock airbox, so a catch tank wasn't necessary. Still allows easy removal of the trunk. I compared driving using the stock air cleaner and the cone and didn't notice any difference so I've been using the cone for convenience. +1

    Eliot Miranda Scott Linn can you post a picture?

    Scott Linn Eliot Miranda not right now. The Lotus is up on a lift right now. I'll try to remember next time the Midget underneath is out.

Rob Siegel I'll probably switch over to some aftermarket setup as soon as the stock one becomes inconvenient, but right now I'm digging the fact that the stock setup is back in the car and functional. FYI I had to order the stock air filter element through RD Enterprises. I tried three times through eBay and Amazon, using part numbers that cross-referenced to the Ford number stamped into the housing. They were all too big. It wasn't outrageously expensive through RD, but I try to save money wherever I can. +3

Russell Baden Musta I am so freakin' proud of you Sir!!!

Russell Baden Musta You can get 3 bags of mulch in a TC and one in the bonnet!! Just so you know what a workhorse that car is!

+9

    Rob Siegel that's awesome

Brian Ach COMPLETELY ABLE TO JUSTIFY TO WIFE NOW +6

Russell Baden Musta Brian Ach it is totally practical car!!

Scott Linn Even better, it has a hot trunk and a cold trunk. Perfect for picnics with wine. +3

Russell Baden Musta Scott Linn absolutely correct!!

Dohn Roush Russell Baden Musta So you're ready for the next time it breaks down and you turn it into a planter? +1

Russell Baden Musta Dohn Roush breaks down!!!? What are you talking about!? Actually... that's a great idea for a planter!!! +1

Rob Siegel #ChiaEuropa +2

Russell Baden Musta Rob Siegel ohhhh Gawwwd! Ha ha ha ha! +1

Paul Wegweiser Meanwhile… I'm pouring a blend of ATF and acetone into each cylinder of this thing… still thinking I can get it roadworthy in less time than that Lotus thing. Status Update: engine does not rotate with 36" breaker bar / 100lb/ft +/- on crank nut. [photo Paul's new 2002 project] +6

Rob Siegel yeah, that's not good :^)

Paul Wegweiser Rob Siegel My Mantra: "…at least it's not brown." +11

Justin Gerry Paul Wegweiser top oil and zen patience. Await the results. +2

Duncan Irving And dogleg +1

Collin Blackburn You need a bigger breaker bar, comrade +2

Carl Witthoft Some things were just not meant to be. +1

Rob Siegel When I bought the Lotus, knowing it had a seized engine, and tried to un-seize it, and couldn't, and yanked it out and tore it down, and saw what the cylinders were like, I thought thank heaven I couldn't un-seize and try to start it. It might have put a rod through the side of the block, and I really needed to re-use the block. +4

Don McMahon Get plumbing pipe and extend breaker bar to 4 feet. Try both directions!! +1

Don McMahon worked to get pulley nut off M30 engine. Had 200 pounder doing chip-ups on it while a small tap with hammer it let go. +1

Julia Cuestión Mark Do you and fellow Lotus enthusiast Roger Metcalf know one another?

Rob Siegel We do not!

Roger Metcalf Thanks for the intro Julia Cuestión Mark! What kind of Lotus do you have Rob? Looks like an Elan or Europa perhaps from the engine bay photos here? I've got an 07 Elise :)

Rob Siegel Roger Metcalf, '74 Europa Twin Cam Special. It's a long story. If you google "Rob Siegel Hagerty Lotus Europa," you'll find six or so pieces I've written about the car.

Scott Aaron I love that it has a "luggage bin." Like a luggage dumpster or something.

Interjection: The passing of my mother

On July 12th, my mother, Bernice Siegel, the best, kindest, wisest, fairest person I will ever know in my life, passed away at age 89 after a brief illness. Both my eulogy for her at her service and the car-centric eulogy I wrote as a Roundel Magazine column can be found on my blog, thehackmechanic.blogspot.com. Re-reading these Facebook posts during preparation of this book and seeing the veiled references to "personal issues" and "other things going on" was tough. I could say that it's remarkable that I got as much done as I did, but I was trying to grab my moments of Zen wherever I could.

Chapter IX: Barely Legal

TLC Day 2233: Plated in Vermont

JUL 23, 2019, 9:09 AM

The Lotus Chronicles, Day 2233:

Thanks so much, everyone, for the birthday wishes. As you might imagine, this was a hard one. I'm a generally a relentlessly cheerful fellow, but with my mother's passing... I'm just not accustomed to this level of sadness and melancholy.

On the positive side, my birthday present arrived, which is that, according to my online bank statement, after a two-week wait, the state of Vermont finally cashed my check for my mail-in registration on the Lotus (as per prior post and this article [google "Rob Siegel Vermont loophole"], there were reasons why I took this path). With the check cashed, I'm expecting the VT plates to arrive any day, allowing me to finally legally drive the car.

110 LIKES, 41 COMMENTS

> **Richard Shouse** Why $601.00? Forgive me if I've missed the explanation before.
> **Rob Siegel** registration fee plus sales tax +1
> **Richard Shouse** Rob Siegel Texas comes up with its own "value" no matter what you paid for the car. I bought a 95 530i for $250 and paid $350 in taxes and registration... +1
> **Philip D. Sinner** Do they refer to that tax as "AD VALOREM" by chance? I recall GA as having that system in place. Doesn't matter what you r purchase price is, they assess the tax amount by their own "book."
> **Richard Shouse** Philip D. Sinner yep that's what they do +1
> **Caleb Scully** Richard Shouse in MD anything over seven years old just has to pay the minimum tax which is like $37 👍 +2
> **Paul Wegweiser** Richard Shouse It's $500 in Whitworth. $634 in British Standard and 7 in Sterling. +1
> **Philip D. Sinner** Notice! Everything I own is over 7yrs old. I'm moving to MD, apparently +1
> **Stuart Moulton** Caleb Scully I have never in my 35 years of registering cars in MD only paid min tax. Maybe my stuff costs too much?
> **Caleb Scully** Stuart Moulton you're doing it wrong 😂
> **Caleb Scully** There's some gray area about putting what you actually paid for a car on the title, but every time I've gone to a tag and title service they've just written in '$500' which is the minimum taxable amount. If the car is less than 7 years old they have to go buy the book value of the vehicle, not the purchase price.
> **Brad Witham** Having spent a few of my own best years waiting in line at the DMV in Montpelier, I

empathize, Rob. +2

Rob Siegel This was all mail-in. I checked my email and on-line bank statement daily for either a cashed check or a notice of rejection, but for two weeks, nothing. I tried calling the Montpelier DMV to ask for status, but never got through. Two weeks passed. Finally, today, the cancelled on-line check. +1

Brad Witham Their PBX systems are set to "permanent busy signal," I swear. They finally set up satellite offices around the state because Montpelier hotels were overflowing year-round with unrequited DMV supplicants. +1

Philip D. Sinner Reasons being of sound logic and legit process, I'm sure. Vermont has had a long-standing openness about registering vehicles without the need for... hoop jumping a.k.a. verification of certain things required in one's home state (most all other states). Many a motorcycle barn found, abandoned, etc. without title has found its way back in use and on the road legally through this legal loophole. +1

Stuart Moulton Ed MacVaugh. See, this is how I learned the process. From others. +1

Philip D. Sinner Excellent news, ayup. The faithful old Green Mountain end around. I'm glad to see it being enjoyed/employed in such manner.

    Rob Siegel As I called it in the Hagerty piece, "The Vermont Loophole."

    Rob Blake So I could re-register my NY State vehicle in Vermont?

    Philip D. Sinner They'll gladly take anyone's dollars, in effect

Rob Blake PS - What's with your penmanship? I thought you were a product of Lexington Schools. +4

    Rob Siegel HA! Yes! And I took typing!

Jack Fletcher WAIT... we buy these cars to actually DRIVE them?

Rupert X Pellett Yes, if you go to many old motorcycle shows, many are adorned, oddly enough, with Vermont tags. +2

    Rob Siegel Yes, "the Vermont loophole" is referred to on several motorcycle and scooter forums because of the combination of a title not being needed, not needing to be a VT resident, and being able to do it all by mail. +1

    Anne B Barnard My recently discovered "Vermont Loophole" is the fact that they allow you to buy high VOC stains in gallon sizes rather than quarts. Cuts the price a lot. Fortunately I have a nephew living in Montpelier.

Brandon Fetch Guess I no longer need to pay for an autograph on one of your books. 😀😀

Bob Sawtelle Ah, "The Registry/" I have actually just dealt with both of the issues in your article. I recently purchased a 97 Jeep Wrangler with an open title for $1000 and went to "The Registry" to get it registered and titled and to my great dismay they based the sales tax on this 22-year-old "Project Cheap Jeep" on a book value of $4400 for a base 4 cyl model resulting in a cost of over $400 for title reg and sales tax. Also my son is now currently in the process of getting a title for the 94 Toyota pickup he purchased from my father who recently passed that had been sitting for several years up in NH with nothing but a registration and a bill of sale. Purchased by my son from his grandfather for a dollar prior to his passing but my son hadn't bothered to get it registered as it is also a project vehicle. +2

    Rob Siegel MA isn't the only state that assess on book value. VT uses the same NADA low book value that MA does.

    Alan Hunter Johansson Rob Siegel I protested the valuation of my 122S wagon in North Carolina, and won. I'd really like to see a legal requirement that the state stand ready to purchase any vehicle for 90% of their assessed value. +1

    Rob Siegel Protesting was possible in Massachusetts many years ago. Now, you can fill out the form at the registry for abatement, but it comes back denied. They give you a numeric code for the reason. When you look up the code, it says "abatement is not granted for this type of tax." It's a sham. +1

Dohn Roush Now all you need is a Maltese passport and you're good to go anywhere... +3

Chris Roberts I don't think I've had actual paper bank checks in almost 20 years. This may be the first time I've seen a practical use for them in all that time. +1

Jeff Lavery I haven't been as excited to see money go out of my account then when the great state of Vermont registered the Mercedes 190E 2.3-16 Cosworth I bought out of a Pennsylvania junkyard. It's a beautiful thing, those Green Mountain folks. +3

TLC Day 2240: Seized rebuilt caliper

JUL 28, 2019, 8:15 AM

The Lotus Chronicles, Day 2240:

Three weeks after I sent in the application, the Vermont plates and registration finally arrived. I was all set to channel Etta James and quote her song "At Last," but of course nothing with the Lotus is that easy.

I slapped on the plates and was all set to take the car for its first real drive. After two preliminary runs around the block, the first thing I noticed was that the heat was on. Combined with the afternoon heat, I was getting baked. A quick trip back to the garage revealed that I'd zip-tied the heater valve open instead of closed.

With that repaired, I picked up speed, jammed on the new brakes (rebuilt calipers, new rotors and pads) harder than before, and heard an obvious grinding sound from the left. I stopped and could see through the wheel that something was wearing the outer edge of the new rotor.

I pulled back into the garage, jacked up the car, yanked the wheel off, and tried to figure out what the cause was. I was initially stumped. It seemed that the caliper itself had to be was hitting the rotor, but that meant that either it was moving or the rotor was, and the caliper seemed perfectly secure on the strut. I noticed that the rotor WAS just a little loose. Since I'd replaced the wheel bearings as part of the brake job, those WERE a new variable, so I tightened the bearing nut and drove it again, and initially it seemed better, but when I got on the brakes hard, the same thing happened. Back to the garage, back up on the jack.

It took me a while, but I figured it out: My newly-rebuilt caliper has a stuck outer piston (see last photo). When I stand on the brakes, the pistons aren't squeezing the rotor; instead, the inner piston is taking advantage of the wheel bearing and shoving the rotor into the caliper. Other people have told me that they've had problems with these inexpensive RockAuto-sourced Centric rebuilt calipers. This is my first bad experience.

Unfortunately, Centric-rebuilt Spitfire calipers are completely out of stock (as they were when I tried to buy them; I pounced on them when RockAuto had a pair). And the irony is that the original calipers weren't seized; they just had torn boots. I thought I'd do the won't-have-to-worry-about-the-front-brakes thing and buy rebuilt calipers. But NNNnnnnnooooo.

I looked on eBay and found a guy in upstate NY selling used Spitfire calipers, $50 or best offer for either left or right. I messaged him yesterday evening, asking him to verify that the pistons weren't seized and the boots weren't torn, then thought "no vendor checks his eBay messages on Saturday night." But in ten minutes he responded and confirmed that his calipers have freely-moving pistons and intact boots, and that he could send it out first thing Monday. I was so thrilled that I didn't dicker and met his asking price.

So, hopefully mid-week, we'll try this again.

As Ringo said, it don't come easy. Ringo probably owned a Lotus.

(112 LIKES, 50 COMMENTS)

Mark D'Sylva Don't you think that Ringo owned a Jaguar E-type? 😊
    Rob Siegel George Harrison certainly did [photo of George and the record player in his E-Type] +2
    Alex Lipowich Rob Siegel George also had a McLaren F1
Tom Egan Girling will rebuild, or swap for rebuilt, your old calipers I'm told.
Chris Mahalick I am surprised you didn't just get the stainless pistons and some rebuild kits from RD. These are so easy to rebuild. +2
    Rob Siegel I dislike rebuilding brake hydraulics. I haven't had good success with it over the years. I prefer to buy rebuilt calipers and install them. I've never had trouble until now. +2
    Chris Mahalick Rob Siegel And your stories always make me laugh, so you need some good material, LOL. You will soon be driving your trouble-free beast. Just wait until you throw it into a corner. It will all be worth it. +3
    Russell Baden Musta Chris Mahalick indeed, good material! Send me your OEMs Rob and I'll rebuild them! For postage!
    Rob Siegel Russell Baden Musta, I appreciate it, but part of the irony is that they're gone; I had to submit them as cores to buy the Centric rebuilds from RockAuto 😊
    Russell Baden Musta Rob Siegel ohhhhhh... fair enough!
John Harvey Why not just carefully pull that caliper apart and free up the piston? I mean, how hard could that be? +4
    Chris Mahalick Exactly! +1
    Rob Siegel John, I bought it as a newly-rebuilt caliper by Centric from RockAuto. It should work. The piston didn't free up when I stood on the brakes repeatedly. I'm not going to take it apart. I'm going to return it. If I wanted to take apart a caliper, I would've rebuilt the ones that were originally in the car :^) +2
Frank Mummolo As Kurt Vonnegut used to say, "and so it goes!" +2
    Bill Ecker He says that when someone dies. Not applicable here,... we hope. +2
    Frank Mummolo Bill Ecker I had forgotten that. Thanks for the info. Indeed, not wishing harm anywhere.
Collin Blackburn Sounds not unfamiliar from the 2002. Was getting ready to go for a drive yesterday and the brake light was on, engine shudders when the brake pedal is pressed. I think the master

cylinder seal blew and it's pooling in the booster. Yayyyy!

Andre Brown The joys of an English car.

Philip D. Sinner Bravo for the verde mont plates!

Alan Hunter Johansson Ringo's cars all have drum brakes. +6

> **Dohn Roush** Alan Hunter Johansson But Rob did get snared... +4
>
> **Alan Hunter Johansson** Dohn Roush The shoe is on the other foot, now!
>
> **Rob Siegel** Alan, Dohn's just pissed you beat him to a punchline that seems like a natural softball for him to have smacked into my face :^)
>
> **Dohn Roush** Rob Siegel Alan gets full credit. That was great. +2
>
> **Roy Richard** I think they were Ludwigs

David Schwartz Ringo also owned a 1967 Mini https://silodrome.com/ringo-starr-mini-cooper-s/

> **Franklin Alfonso** I would think Ringo was more of a Rolls Royce kind of guy...
>
> **Brad Purvis** Franklin Alfonso I don't think they allow Scousers to buy Rolls.

Jeffrey Miner If you can clamp the right piston and hit the brake pedal can you free the stuck piston? Insert a wood block to keep them from striking each other. I'm stumped as to how they jammed a piston like that on a rebuild though... if the rubber piston seal is jammed, it would be leaking fluid. I'd force it—either it frees and works or it tears the rubber seal and leaks. Either way, you're no worse off, but you might have a working car sooner if it works. +1

> **Rob Siegel** Not going to do that. It didn't free up with my jamming on the brake pedal as hard as I could, which is effectively clamping the other piston with the brake rotor (subject to it tipping slightly on the bearing, which it did). Bought it as a newly rebuilt caliper. It should work. Returning it for a refund. +1

Andre Brown I've used my air compressor to unfreeze a stuck caliper piston before. May be worth a try?

> **Rob Siegel** Andre, the brake hydraulics themselves wouldn't push it out.

Neil M. Fennessey FWIW, their "rebuilt" Bosch starts are crap too. +1

Bill Mack Caliper rebuilds are easy if you can get the &#$=%/ pistons out. At least they are not rare.

Neil M. Fennessey Brake lines have the same size as a grease Zerk fitting. Screw one onto each side of the caliper. Fill a grease gun with cheap engine oil and have at it. If one side moves and the other doesn't clamp the moving side with a c-clamp. It's both effective and safe versus the idiocy of using compressed air.

> **Rob Siegel** Neil, I tried to move the piston in with big channel lock pliers. Nothing. I tried to move it out by standing on the brakes. That's almost like clamping the inner piston with a C-clamp. The inner piston moved enough to shove the rotor into the caliper. That's a lot of pressure. The outer piston didn't budge. I'm done with this "rebuilt" caliper. +1
>
> **Neil M. Fennessey** Rob, the grease gun most certainly exerts more force that big channel locks or your foot on the brake pedal. As you described, the bearing "gave" when you stood on the brakes. A grease gun delivers up to 15,000 psi which why this works. If one doesn't split the caliper into its halves, removing grease is impossible which is why I use engine oil. Brake fluid could splash on the paint ruining both the car finish and your day.

Lindsey Brown "I build my cars to go, not to stop!"—Ettore Bugatti.

Robert Myles I'm one of those people who warned you about Rock Auto. I don't want to say "I told you so" buuuut... never trust something as important as the brakes to underage third world slave labor, m'kay?

> **Rob Siegel** Yes you did. This is the first problem I've had, out of close to ten sets of Cardone or Centric rebuilt calipers bought through RockAuto.

Brad Purvis [link to Ringo Starr "In My Car"]

Jeff Ireland Cardone is sometimes hit and miss. Had several Tii calipers, I guess newly manufactured, that did not have the passage machined between the two calipers 😂😂😂 AKA there was no fluid pressure to 2 of the 4 pistons. Don't recall if from RockAuto or not. Either way always a good idea to thoroughly check these before slapping them on you car and giving it a whirl. Most of them, I'd say 95% work perfect no issues. +1

> **TE Cole** Yes Cardone is about 50/50.

TLC Day 2243: Stuck used caliper

JUL 31, 2019, 6:45 PM

The Lotus Chronicles, Day 2243:

Today's shitstorm involves calipers. Again.

If you recall, several months back, I decided that the front calipers—WHICH BOTH WORKED FINE but had torn boots—should be replaced with rebuilt ones. I hate rebuilding brake hydraulics, I'm not great at it, and the Centric / Cardone rebuilds are cheap on RockAuto, like $30 per side after core charge. I liked the idea of rebuilt calipers, new rotors, new pads, new wheel bearings, boom, done. But the rebuilt calipers were out of stock. I was about to just re-use my old ones when finally rebuilds came back in stock on RockAuto and I grabbed a pair.

Then, if you remember, last week, when I finally got around to driving the car with the new front brake setup on it, there was a grinding sound that turned out to be caused by the driver's side caliper having a seized outer piston, resulting in the inner piston shoving the rotor into the caliper.

So, just wanting to get the damned thing done, I found a guy in upstate NY on eBay selling used Spitfire calipers. I messaged him on Sunday night asking him to verify that the pistons moved and the boots weren't torn, he did and let me know within 10 minutes, I met his $50 asking price for the caliper, done.

Well, funny story...

So the thing arrived today, I ran out to the garage and installed it, but after what I went through last time, I thought "I'm testing that both pistons work BEFORE I lower this off the jack."

Guess what? The INNER piston on this one is seized. I tried standing on the brake pedal with the other piston clamped. Also tried compressed air. The inner piston moves about 1/16", then stops.

I just chatted with the guy. He'll refund my money. He's a good guy. He postulated that I'm not meant to drive the car this week, and implied that it being marooned in the garage might be saving me from some unsavory fate. He might be right.

But hokum smokum, what the ever-loving fuck is up with calipers?

Also tried compressed air on the stuck piston on the rebuilt caliper. That one won't budge either. And yes, thanks for the suggestions on freeing the piston up with a grease gun. I read up on it, watched the videos, and checked to see if I can do it. The grease gun that I have doesn't fit over the bleed nipple on these calipers, and doesn't appear to be the kind where I can thread the gun directly into the caliper. I'd need to rig something up with adapters. Maybe next time.

If I had the old calipers, I'd use the left one, but I don't, because I submitted them as cores for the rebuilt ones. D'oh!

I just ordered a new caliper for $89. Hopefully THIS time it will be boom, done.

Lolita does not give up her secrets easily.

52 LIKES, 68 COMMENTS

Frank Mummolo Lolita was a Nabakov novel in the 60's. Different premise. Similar outcomes. Just sayin'! +1

    Rob Siegel Oh, I know :^) +2

George Zinycz When it's time, she will roll... +1

Dave Gerwig Todd Howerton can you help Rob Siegel figure out his caliper dilemma?

Rob Siegel Nothing really to figure out... I got bad rebuild with a seized outer piston, then I got a bad used caliper with a seized inner piston, and now I'm buying a new caliper! +1

    Dave Gerwig Rob Siegel I gathered that. Todd at Outlaw Brakes had all the resources for future challenges

    Todd Howerton Yes: I can unstick most any piston. I won't say all because; well we all know. Anyway; there is a good chance that I can unstick the rebuild if you decide to go that route. I can't image there is a huge demand for this and it is likely that they have been sitting on a shelf for many years. Also depending on which caliper this is: I might have a billet aluminum 4 piston caliper available as a replacement. Many of the British cars used a Girling caliper (r-16) which I make a replacement for. +3

    Rob Siegel Thanks Todd. I have a new caliper on the way. If something falls through with THAT, I'll let you know. +1

Rob Blake NAW... You just forgot to implement the Jedi Mind Trick. Works every time. Step up to the car. look it squarely in the headlights with a glint in your eye with a grim look, and say... "These aren't the calipers we're looking for." And Hocus-Pokus, you're done. +2

    Rob Siegel damn, must've been off my Force game

Mike Savage I assume you tried it, but the grease gun idea suggested installing a grease Zerk fitting in a threaded brake line inlet (same size apparently) and using the grease gun filled with motor oil. It was not supposed to go over a bleed nipple. +1

Marshall Garrison If this keeps up, you'll find the next RockAuto calipers that become in-stock and available that you buy are the same ones you sent in for cores and got refurbished! +3

    Rob Siegel That is in fact likely!

Bill Snider Given that you can't brake, surely the song for the moment is The Police's "Roxanne" (You don't have to put on your red light tonight) 😀 +1

    Wade Brickhouse I don't know, that would involve Lucas! +1

Steve Kirkup It's British... boom, done. +1

Steven Bauer Better Lolita than Lola… +1

Steven Bauer Is the bleeder nipple on the opposite half of the caliper from the fluid inlet?

    Rob Siegel no, same side

Eric Pommerer I'd argue that the grease gun theory is bunk. Unless a grease gun is designed to deliver pressure anywhere near that of a compressor?

Charles Morris Most of our air compressors are either 120 psi or 175 psi (some two-stagers). Grease guns easily deliver several thousand psi at a limited volume, and some can do 15,000…

Eric Pommerer Wow! OK. Grease gun it is!

Rob Siegel Eric, I was highly suspicious of the grease gun theory until I looked into it. I think the 15,000 psi estimate is high, but there's a YouTube video of a guy measuring the pressure delivered by a small $20 grease gun at about 1700 psi. The thing I don't understand is why it apparently works better than the brake master cylinder itself. After all, both are just cylinders moving hydraulic fluid under pressure. +1

Charles Morris But that isn't logical... the psi is what's acting on the caliper piston. 1700 psi from your leg via master cylinder exerts the exact same amount of force on the stuck piston as 1700 psi from a grease gun. The reason you can get more pressure is because the grease gun "master cylinder" is substantially SMALLER than the brake MC bore, and there is a longer lever advantage. (The pressure generating part is that little thingy sticking out the side that's attached to the handle. Not the big bore where the grease reservoir is...)

Rob Siegel Charles Morris, I agree that I don't understand it :^)

Charles Morris I also think that better quality guns should easily put out 3000 psi or more. Or maybe the YouTube guy just had wimpy arms ;)

Bill Mack Rob Siegel he's got it right.

Rob Siegel I'm not saying he doesn't. I'm saying I haven't taken the time to completely understand it.

Bill Mack Takes some thought.

Jeffrey Miner Aside from the PSI, is the grease being introduced helping to "unstick" the "stuck" part as well as the add'l pressure?

Eric Pommerer Is this going to be on the final exam?

Jonathan Selig Rob Siegel, when you say you tried compressed air, did you use an airline directly on the believe nipple? I had success in this situation by brazing a quick disconnect air hose fitting on to a spare bleed nipple and then giving it 120 psi.

Rob Siegel Compressed air into the screw-in fitting where the brake hose threads in, sealed not perfectly with the rubber tip of the air wand. Not as good as what you're talking about.

Bill Mack Compressed air is nowhere near the line pressure of the brake fluid. And despite what I read here, a grease gun produces very high pressure on the order of 10 times what you see in an air line. Look at the parts and do the math...

Neil M. Fennessey The grease gun isn't supposed to fit over the bleed nipple. Remove the bleed nipple and screw in a grease Zerk. The nipple and the Zerk have the same NPT threads. Mike Savage has this right. Use oil in the gun and not grease. +2

Bill Mack I would fix those for you if this comes up in the future. I use stainless pistons. Works best for longest. No less of pia to get apart... not stainless at that point. +1

Isabel Zisk So, it's either third time's the charm OR you're better off building one yourself with bubble gum, duct tape and a wire hanger?!?! +1

Rob Siegel Bubble gum, duct tape, and a wire hanger could probably stop the 1600-lb Lotus quite well! +3

Bill Mack Rob Siegel I think you could stop it with a 2 by 3 bolted to the side of the car like a soapbox derby car.

Jacob Johnson Somehow I never had the front calipers off the Spitfire I had. Now the clutch slave cylinder... +1

Bill Mack To do the grease gun thing you need to remove the tip and find/ make weird adaptors. Not that fun but doable. +1

Dohn Roush I'm thinking a cutting torch and a good pair of shoes might be the easiest braking solution...

Rick Roudebush I had similar "fun" with my 88 M6. I rebuilt my stock front calipers, 4 pots, and during honing process there was a slight pit, I mean slight pit in one cylinder, that just didn't want to be removed. I honed for 20 more minutes, and after thorough cleaning proceeded to install the rebuild kit and O-rings. The car worked great for a year, but I don't drive it that much. One hot summer day we were to lead a club drive to a Wine and Shine event at a local winery. The battery was dead from sitting so I jumped it at home, bought a battery along the way and set it in the trunk. The 30-year-old AC hose burned in two about ten miles from home, too close to the exhaust. At the rendezvous point I noticed the car seemed to be sluggish while circling the parking lot. The right front caliper was sticking. I slid under the car with the tire iron and pried the piston back in, vowing not to use the brakes much on the way to the winery. The car wouldn't re-start, so I hooked the jumper cables from the new battery to the old battery, MacGyver style, and proceeded to lead the drive to the winery, 30 miles or so, slowly in the 100 degree heat with no AC in an E24 greenhouse, no fun for anyone!! Limped the car home, 60 miles or so, at the end of the event and ended up buying remanufactured calipers for a later model M3 and adapter brackets kit from Turner Motorsports. Larger single pot has more breaking force then the little 600$ each 4 pots and a huge selection of pads to choose from. Two reman calipers, pads and the adapter kit under two hundred bucks. The ones I rebuilt are hanging on my wall of shame. Perhaps there is a similar upgrade for your Egyptian water lily. Good luck... tiny little pit, barely see it, grrrr!! +2

Charles Morris :D [gif "Buy an old car, they said… It'll be fun, they said…"] +1

Thomas Jones I've never liked the idea of having to clean all the grease out of a multi-piston caliper after using the grease gun method.

    Al Bell Thomas Jones that's why you put oil or maybe even brake fluid in the gun

Thomas Jones This guy has built a really ingenious setup of pumping brake fluid in with a modified bottle jack. https://youtu.be/qc4zIGaJgoA

    Neil M. Fennessey Kent is an interesting character. The original tool he markets is called a "pop tester" and is used to test the onset opening pressure and spray pattern of a Mercedes diesel fuel injector.

    Rob Siegel yup, watched that one

    Neil M. Fennessey A grease gun full of cheap oil is lots less expensive than his tool. +1

Clyde Gates So Lolita is being a total bitch?

    Rob Siegel yes!

    Clyde Gates Total is the problematic word! +1

Ken Carson I have run into problems with Cardone rebuilt products. Generally I avoid Cardone.

    Rob Siegel Others have said the same thing. I've bought maybe a dozen sets of calipers. This is the first one I've had trouble with. And actually, I misstated… they're Centrics.

Gregory Bradbury BTW, I've looked at this photo more than ten times. Why don't I see a brake disc in the middle of the "brake pad sandwich"?

    Rob Siegel Because the goal was to try and un-seize the seized piston in the caliper by stomping on the brakes without the rotor between the pads. If the pads are completely removed, the un-seized piston can pop completely out. Hence, trying it with the caliper unbolted from the strut but the pads in place. I also tried it with the pads removed and the un-seized piston held in place with a C-clamp.

    Gregory Bradbury Rob Siegel, of course. I'm asking as a friend who tried to do brakes (years ago) came to me with a bigger problem, car was pulling to one side after pads and rotors. He'd forgotten a rotor on one side. 😀 Back to you, a shame on the Cardone. Great value as many know, I've schlepped BMW ones transatlantic in a suitcase. ~$50 exchange vs. ~250€, this is an easy decision. +2

    Bill Mack Rob Siegel pretty much did all I would have done except the cursing and stomping around +1

William Jeffers Off the car, it's hard. Put it back on the car, pull push the free side back, Put a C-clamp on it, then use the brake pedal to push the stuck side out.

    Rob Siegel tried that

    William Jeffers Wow!

David Holzman Isn't the word, "caliper" derived from Caligula? Wait, I'm sorry, "caliper" is derived from "Caliban." +1

TLC Day 2246: Live drive with working brakes

AUG 3, 2019, 9:38 AM

The Lotus Chronicles, Day 2246, live version:

[live Facebook video of first drive in car to the package store with decently-working brakes]

[search YouTube for "The Lotus Chronicles first real drive with decently working brakes"]

156 LIKES, 16 COMMENTS

Dave Gerwig · 0:00 Happy for you! +1
Greg Calvimontes · 0:00 Sam Adams, right? +1
Neil M. Fennessey · 0:00 Congrats on kicking that caliper's butt and taking a drive! I thought for a moment you were heading to 128 but was relieved to see that instead it was the packie. +1
Tom Schuch · 0:08 Dude, I'm sure Scott Sislane would love to lend you his helmet cam! ;) And congrats on 'going mobile'!
Brian Ach · 0:15 I'm watching this twice with the sound up really loud because I need some positivity today
Kristi Bodin · 0:38 what the what?
Robert Heitz · 0:43 Seatbelts?
Michael Foertsch · 0:53 You sound like Richard Dreyfuss in this video. And that's not a bad thing.
Doug Jacobs · 1:01 Dude, get a GoPro or something so you don't have to hold your camera all the time.
Marc Bridge · 1:05 This gets better every time!
Jack Fletcher · 2:22 I tell you... go to the bowling alley and steal a pair of bowling shoes!
Bill Snider · 2:26 Filmed in SiegelVision (ala Blair Witch Project)! Still, Great that the Louts is finally rolling & stopping on all four.
Andrew Rich · 2:29 I wish I could see the road
Russell Baden Musta · 2:32 Yours is like mine, idles at 4800 rpm!
Larry Wilson · 2:33 Good work Rob!

TLC Day 2247: The first short legal drives

Aug 4, 2019, 9:38 AM

The Lotus Chronicles, Day 2247:

As some of you saw from Saturday's video, yesterday was a watershed day here in Lotus Land, possibly the single most auspicious day since I got the engine running and first drove the car around the block.

In the morning, the new front caliper arrived. I installed it and checked that both pistons were working. Initially, the inner piston did NOT seem to be functioning, which, since two previous calipers had a stuck piston, made me utter a loud WTF in the garage, but it turned out to simply need a more thorough bleeding. I briefly ran the car around the block, and found that the brakes worked very well in their revived configuration with new calipers and rotors, soft pads, bypassed boosters, but the original master cylinder.

With that, and with the car wearing its legal Vermont plates, it was finally time for a real drive. I first took the car in a wider circle on larger roads where I could hit 3rd gear.

Initial impressions: It's uncomfortable. It's weird. It's buzzy. It still smells of rodent.

I couldn't get enough of it. I drove it five times.

And, even in its heavily-patina'd monkey-poop-brown clothing, this car got more attention than anything I've ever driven, anywhere. A guy on a moped at a light. A woman walking her baby in a stroller. A man and a kid at the hardware store. Of course the guy in the 240Z and I thumbs-up'd each other.

--The first drive was to the gas station to fill the twin tanks with high test. It was her Pinocchio "Look! I'm a real car!" moment.

--The business about the pedals being so close together that I have to take off my right shoe to separately actuate the brake and gas is only the beginning. I need to wear a left shoe because the cable-operated clutch requires quite a bit of effort. Further, you need to hit the clutch pedal with the tip of your foot, because there's not really enough clearance around the steering column and the bottom of the footwell to simply mash it with your whole foot. Further still, there's really no dead pedal room anywhere for your left foot, and it's easy to find it placed under the clutch pedal where it snags when you need to lift it up and use the clutch. So you need to be aware of where your feet are.

--The seats are slightly reclined and their angle does not adjust. It takes some getting used to.

--The rear visibility through that 4"-high rear windshield is as bad as you think. It was compounded by the fact that both the driver's side and the interior mirrors were missing. I found them both in the glovebox. My second drive (the first in which I ventured more than a mile from the garage) was to the hardware store for Gorilla double-sided sticky tape.

--With something resembling rear visibility, I drove the car a bit further and harder. The cable clutch is very grabby, and the shifter is tough and notchy (the linkage runs the length of the car), so it's not an easy car to drive. I did get it into 4th gear, which was very satisfying.

--The steering is unbelievably light. Even while parked, you can easily rotate the wheel. If you can do that, you can imagine how flick-able it is while in motion.

--The suspension is a little creaky and squeaky but basically functional. You can bounce the car and feel that the shocks appear to neither be blown nor seized, but it feels soft and squeaky. I bought trunnions, tie rods, and ball joints, since the rubber boots in the front end are clearly deteriorated, but I haven't installed them. But even in this condition, it doesn't feel loose or wander-y or obviously out of alignment.

--The main issue is that, even with every component in the cooling system replaced, it's still running 3/4 of the way up the gauge, it blew coolant out the brand-new radiator cap and into my jury-rigged overflow bottle, and sucked it all back in while it was cooling. I bled it again, but it didn't seem to make a difference. I need to watch this very carefully before I take it on the highway at speed.

--After I let it cool down, bled it, and verified the cooling system wasn't actively leaking, the third drive was a celebratory beer run (the video I posted yesterday).

--After I secured a rattling passenger seat, the fourth drive was the longest yet, but still around local Newton and Waltham roads.

--On the last drive, one of the front sway bar links, which was broken when I bought the car and which I'd repaired with my Neanderthal welding skills, broke again, so I need to deal with that.

In short, there is nothing relaxing about driving Lolita, but I can't wait to do it again. She still needs a shit ton of work before I can do the nail-and-wail-on-the-entrance-ramp thing, but, as Carly Simon said, "Anticipation..."

As I wrote in my book "Resurrecting Bertha," not everything is 2000-mile gonzo road trips to The Vintage; part of the joy of owning a vintage car is in simply being able to hop in it on a Sunday morning and drive it to Trader Joe's for cereal. Lolita has, at last, crossed that Rubicon. And THAT is massively freaking cool.

261 LIKES, 63 COMMENTS

Doug Jacobs Man, you are an ENTHUSIAST. And I mean that as the highest complement. Many, many, many people would have pulled the plug before now. YOU, my friend, are the champions. +4

    Rob Siegel Doug, by getting it running, I have the opportunity to lose FAR more money than if I'd bailed... +6

    Wade Brickhouse Rob Siegel there might be a way to minimize your losses. 😀 +2

    Doug Jacobs Wade Brickhouse, are you referring to a friction fire? That's where your title rubs against your insurance policy... +2

George Zinycz Congratulations of the Highest Order are in order!!! +2

Henry Noble If there is an obscure award for saving brown sports cars, you deserve it! +1

Daniel Senie Did you need a shower and change of clothing afterward for the rodent smell?

    Rob Siegel I needed a shower and a change of clothing afterward for MY smell! +5

Bob Sawtelle This put a smile on my face, just the thought of this cool little thing banging around Newton's ridiculously potholed streets and you dodging and weaving around them makes me laugh. +2

Paul Hill I can't wait to hear about further adventures. +1

Sean Leuba So very happy for you Rob Siegel. It's been a tremendous fun to read along as you saved the car. Congrats and here's to many years of enjoyment. +2

Marc Bridge Love the write up, as usual witty and funny! When's the Resurrecting Lolita book coming out? +1

    Rob Siegel It'll be called "The Lotus Chronicles" and I hope to start working on it after "Resurrecting Bertha" is published in the fall. Thanks for asking! +1

    Daniel Senie Rob Siegel hope you sell enough copies to pay off a bit of the parts expense. +4

Chris Mahalick Very well done! And yep, mine does not like this hot weather either. +1

Don Stevenson I hope no Lotus Elans show up on Craigslist soon, you have got to be out of space by now.. +3

Alan Hunter Johansson 💜💜💜💜💜

Adam G. Fisher Awesome!

Mark Rowntree We'll done Rob, I'm with you all the way! Currently resurrecting my '72 Twin Cam. +2

    Russell Baden Musta Mark Rowntree keep at it!

Daniel Neal This has always been the highlight of your posts to me, so glad you got to take her for some longer and longer trips. +2

Peter Morton Great job. I've had my Twin Cam off the road for 20 years, moved it to two different houses and still have never driven it. +1

    Russell Baden Musta Peter Morton get it going!!!

Michael Hüby Congrats! Make sure your radiator cap has the correct height and pressure rating. Mine did the same, only to find out the PO had installed a too shallow radiator cap that did not seat correctly on the swirl pot. Blew coolant prematurely.

Rob Siegel Pretty sure the replacement cap is correct. Ot seems to pass coolant only when the temperature creeps up past 3/4 of the way up, which, for a 7psi cap, seems reasonable.

Michael Hüby Rob Siegel there are 2 different heights of caps that both fit. It's easy to miss. Did you check the temp with a thermometer, at the radiator? Just to make sure the thermostat opens correctly. Sometimes it helps to jack up the front of the car to release any air bubbles from the system. And be sure that the heater circuit is properly vented, too.

David Weisman This is great! I've been following the whole story. +1

Lee Levey A great accomplishment to get the Lotus on the road. The next challenge will be a 500 mile road trip. +1

Russell Baden Musta See You at LOG39! 👏👏👏👏

Jeff Cowan Fabulous! Well earned.

Matthew Rogers I'm going to have to swing by and see it now that it's up and running.

Ray King Sounds like you should have done an automatic transmission swap while you were in there. Now THAT would make a book... 😉

Carl Witthoft So what's the eBay value now?

Daren Stone Re: cooling woes, ensure the front compartment is as air tight as possible so all air coming in the nose goes through the radiator. That, and get the front end up as high as possible to bleed the system. If you do both these things, you should have a totally marginal cooling system. +4

> **Rob Siegel** Thanks Daren :^)
>
> **Daren Stone** Rob Siegel anytime!

Tom Egan Ain't they a fecking hoot to drive! I have no way near found the cornering limits on my car, and I have been trying my best within sanity boundaries on public roads. Regarding the shoe thing, I spent the morning grinding the bunion rubber off the sides of a nice pair of Piloti driving shoes my long-suffering wife gave me to narrow them by about 1/4 inch. +1

Charles Morris Congrats on getting it running! (y) Bee-double E-double R-U-N, Beer run! [link to Todd Snyder "Beer Run"]

Daren Stone I've also made a couple of mods to my Europa to make the clutch and shifter more user-friendly if you're interested. +1

Brian Stauss I must be out of my mind to want a Lolita. But I do. +1

Jim Piette Congrats! I drove one for a while as a college student, but my dad and I had a "discussion" (tuition or Europa?) that put a stop to ownership. Always wanted another one, but those flashbacks are kicking in - I'm not 20 anymore, maybe I'll hang on to my M Coupe. Thanks for sharing the adventure! +1

> **Rob Siegel** M Coupe is a MUCH easier car to drive :^) +1

Scott Fisher Reading your description of the cooling system and having owned a couple dozen hopelessly shot old British sports cars, I find myself asking, as Jack Nicholson once did, "What if this is as good as it gets?" +2

> **Rob Siegel** He was perilously close to sleeping with Helen Hunt by the end of that movie, so when he asked the question, he was wrong :^) +3
>
> **Scott Fisher** Rob Siegel point taken. +1

Steve Park What a great feeling it must be after the long "journey" we've been following. Your perseverance is inspiring. Maybe there's hope for some of my projects. Congratulations. +1

Dennis Fenimore Europa is a vehicle whose happy place is intense within its narrow band. Pretty much everything has to be right. My journey proved that to me. Mine came to me looking pretty but it was mechanically full of "wrong". Once it is all corrected the reward is substantial and very real. +1

John Jones Thanks for the great write up. Wishing you all the best in working through the list and not getting stranded. +1

Greg Bare I'm at this exact same stage with my Bavaria. Frustration and joy all mixed into small trips to the store and dinner. +3

Andrew Wilson Rob, thank you for the Pinocchio reference. I celebrate everything about this return to Lotus Land!

John Graham Did you have it towed to the gas station? 😉 Seriously, it is really neat to see it out on the road. I will always remember where I was when I saw this picture. I might not remember you, but I will remember the picture! +1

Bradley Brownell This is massive, dude! Congrats! +1

Christopher Kohler So, details on the cereal. 😉

Mike Tunney Met a guy at a car show yesterday who told me he once had one and used to get it up over 160mph. That's a lot of speed for a very light car. You'll get there 😊

Der Alte Classics Congrats! Re: the light steering. I'll be more worried about massive understeer on low speed (especially in the wet) due to no weight on the front wheels. With higher speeds I think the front generates enough "downforce" to be grippy. Don't get caught out is what I'm saying. Is there any reference on this on Lotus fora?

Lincoln Benedict Great for TJ cereal runs... if you don't mind a faint whiff of rodent 😊 Congrats on getting it on the road!!! +1

Dave Singh rob that is a super cool car! never seen one and I can imagine how awesomely terrible some of it may be haha. hope you figure out the cooling system, I'm sure you know how to diag that so I won't say anything. +1

Chris Lordan It does not matter how slowly you go as long as you do not stop. –Confucius

TLC Day 2247: The first windy road

AUG 4, 2019, 5:23 PM

The Lotus Chronicles, Day 2247 continued:

It's still running 3/4 of the way up the temperature gauge, and I don't know why, but it seems to stop there, and it hasn't puked coolant since I re-bled the cooling system, so I took it for a MUCH longer drive along a proper New England windy road on which I exercise the other cars. I'm still being gentle with both the engine and suspension, but this thing is bloody addictive. It's much torquier than I expected a little 1558-cc four-cylinder twin-cam engine to be. Once I stand on it, I imagine it'll be all over.

Say it with me like Sean Connery: I like the Lotush. I don't understand it, but I like it.

Oh, remind me never to patch the hole in this muffler.

[search YouTube for "The Lotus Chronicles first drive on windy road"]

125 LIKES, 67 COMMENTS

Scott Winfield Bloody cool!

David Kiehna Ring it out Rob. Quit pussy footing around. I had my 85hp Renault powered Europa S2 at a floaty 110mph once. Never installed seat belts either! +3

    Rob Siegel Small steps.

Justin Sz Arkowicz Could you measure the resistance on the sensor of the warm engine to see if it's a faulty sensor/gauge combo? Probably doesn't explain the puking coolant, though. +1

Russell Baden Musta Now to see you in the "wild" on I81!!! +1

    Rob Siegel Not there yet, but closer. +2

Drew Lagravinese Too were awfully quiet for a guy who was having so much fun.

    Rob Siegel I was quiet when I used to drive my 911SC as well. It too took concentration. +4

Stan Chamallas Yes and he preferred the Beretta 418 over the Walther PPK, but it jammed... +4

Christopher Kohler The teaser photo for the video looked more velocitous. +1

Sam Schultz Have you manually checked the coolant temp? Eliminate the simple items first, make sure it's really running warm and then diagnose between the gauge and the sensor.

    Rob Siegel I have not, and I should, but the fact that it blew coolant made me believe it.

    Bill Mack Rob Siegel use one of those temperature guns.

Bailey Bishop Jr. I hope to get mine on the road like you did. +1

Samantha Lewis Isn't it just the best thing ever! There's nothing like a Europa that's running anywhere near right. A car that you wear, that really connects you as viscerally as all get out. Addictive, oh yes sir! Enjoy <3 Oh, BTW, that .85G lateral stick, it comes unstuck pretty suddenly after it's exceeded, explore limits with caution, but no worries, the oil pressure usually goes before that anyway, it's all good. +4

Kevin Pennell Awesome! (and you really need a go pro 😊) +3

Bill Mack That's undershtand...

Scott Aaron Pretty sure I heard the frame crack when you hit that bump 😉 +1

Robert Alan Scalla Don't patch the hole. +1

Lance Johnson The crappy road surface in the video looks like it was a good suspension test. How about SCDA HPDE at Thompson Speedway Motorsports Park on 8/24 or BMW CCA CT Valley & Patroon Chapter 2 day Driver School at Palmer Motorsports Park on Sept 14th & 15th. It wants to be free on the track!

> **George Zinycz** Lance Johnson I have a standing offer to Rob for a free private track day as my guest at Palmer but he said he's too old (or something like that 😉). Maybe the Lotush will reinvigorate him enough to take me up on it? +2

Dave Gerwig It looks glued to the road, and drivers are molded into Europa's once seated. All in unison doing a dance with the highway. Looks like fun with a big heap of satisfaction. Congrats! +1

Paul Forbush I just spent a relaxing vacation week in southern NH. I had almost forgotten how much fun New England back roads are... pot holes, frost heaves and all! +1

Ed MacVaugh Did you ever check for the 10V after the voltage regulator for the instruments? +2

> **Rob Siegel** I have not checked that yet, no. +1
>
> **Scott Linn** Bad ground on the regulator (err, stabilizer) can also cause issues. +1
>
> **George Zinycz** Scott Linn Err, earth :^)

Paul Forbush I noticed how 35 seems killer fast at less than 3 ft above the pavement...passing Audi's just loom over you... +4

Joshua Weinstein Is the thermostat new? If the gauge is correctly reporting the temp, you may want to ensure the thermostat is opening, assuming you can't live with it at 3/4.

> **Rob Siegel** thermostat is new from RD Enterprises, and definitely opening

Tim Warner Thanks for holding the phone higher! Nice to look out the window.

Jacob Johnson Will somebody sent Rob a windshield mount for his phone or something? Congrats on the drive though! +2

> **George Zinycz** Jacob Johnson +1

Tony Pascarella Check the temperature of the engine with an infrared temp gun.

> **Rob Siegel** A number of people have said this. I HAVE an infrared temp gun. The problem is that no infrared gun reaches into the coolant. You CAN use a gun to get a measurement of how hot the engine, or the hoses, or the radiator tanks, or the outside of the temperature sensor are, but it's not the same thing as the probe sitting in the coolant. +1
>
> **Bill Mack** Check anyway. Don't overthink. +2

Jean Laffite I don't know why I expected this to end in a Dukes of Hazard style jump, but I did. +2

Brian Ach Welcome to Europ(a) +1

Chris Mahalick Nice!! +1

William Davis Why does the sound of the engine remind me of the scene from "A Clockwork Orange"?... ;) Probe 16 (AB/4)/Durango 95 +2

Franklin Alfonso Keep in mind while fun to drive at speed I don't think it's a very crashworthy cage in terms of safety... 😉

> **John Wanner** Franklin Alfonso buzzkill. +1

Ken Carson That was frickin' cool! Thanks for posting, Rob. And congrats!

Kent Carlos Everett One gratifying drive in the country.

Wade Brickhouse Are you relying on a gauge that is tied into the Lucas system? Asking for a friend 😉 +1

Rob Blake Face it... Lolita is one HOT car... or maybe someone dumped some of that leaky radiator leak fix goop into the system long before you ever laid eyes on her? +1

Mark Abare I can be seen waving a thumbs up in this video at the 2:28 mark +1

Bill Mack There are quite a few British sending units. Those will give you calibration issues.

Brad Purvis It's just toying with you, like a cat with a mouse. Eventually, its native instincts will take over and it will rise up and smite you yet another time. +1

Isabel Zisk Go with the simple answer: perhaps she's as hot for you as you are for her. +1

> **Rob Siegel** I like you. :^) +1
>
> **Isabel Zisk** I like you too! P.S. It's not too late to some to Princeton for the house concert on the 10th!

Lee Levey Now you have gone and made me miss my old Elan. +1

Robert Verhelle Is it running too lean?

Scott Fisher For peace of mind, look into a calibrated temperature gauge with a bulb that will fit inside one of the hoses, so you can find the temperature of the coolant itself. It COULD be that the gauge/bulb/sending unit is suffering some kind of mismatch. Is it a mechanical or an electric gauge, for example? I wonder if this chart could help. It lists the voltage, current, and the resistance and voltage at the sender. You might be able to do some DMM work on the whole temp gauge circuit. After a little research, it could be poor voltage stabilization at the instrument panel. The footnote on the "Sender Resistance" column head says: "If, for example, the voltage is allowed to rise to 14v the water temperature gauge will indicate 120 degrees instead of the actual 90 degrees. It is quite possible for the voltage to rise into the 14 volt range above 3,000 rpm."

Andrew Wilson "Europa my old friend..." [link to Thomas Dolby "Europa and the Pirate Twins"]

> **Rob Siegel** so 80s
>
> **Andrew Wilson** Rob, you have a Pirate Twin Cam! +1
>
> **Dimitri Seretakis** Looks like Weston or something.
>
> > **Rob Siegel** That's exactly where it is—Newton Street, heading south toward Comm Ave. +1
> >
> > **Dimitri Seretakis** Rob Siegel Ha!

Jack Fletcher A proper day for it.

Jose Rosario Europa Earth's cry, Heaven's Smile. +1

TLC Day 2248: The first highway drive

AUG 5, 2019, 8:53 PM

The Lotus Chronicles, Day 2248:

Couldn't stand it anymore and took Lolita up onto the highway.

And...

Not great. At about 50 mph, the front wheels started vibrating badly enough that they felt like they were going to fly off. Took her back to the garage, jacked her up, spun them, and they're... not so bad. The Cosmic wheels aren't obviously bent. The el-cheapo Achilles tires are just a little out of round. It may just be a not-so-great balancing job when the tires were mounted.

Took her back to the highway, pushed through 50 up to 60, and it smoothed out a little, but still wasn't confidence-inspiring.

The hot-running issue isn't resolved, but this was an evening run, it was cool out, and it ran just above center, so whatever's going on isn't an integral issue of heat generation.

Still, I've got nearly 70 miles on the car, and that in and of itself is astonishing.

(Forgive the "unique automotive butts and noses I've known and loved: pic. Life's been good to me so far :^)

181 LIKES, 43 COMMENTS

Collin Blackburn [gif "Not great, not terrible"]

Stuart Moulton ///Mmmmmmmm butt +1

Russell Baden Musta Rob, I've rarely heard others with Europas complain, but my OEM Lotus wheels which you have, have ALWAYS been a nightmare to balance. When you saw me in the wild, I was experiencing same as you. I used Panasports for a few years and they were spot on. They were 6" wide and I changed back to OEM because I wanted correct look and the 205/60-13 on the rear rubbed the right side fender. I have zero advice as to making that better... it sometimes seems like a harmonic that goes from front to rear and back, then I'll find a sweet spot at say 71 mph where it's smooth for a while. BUT, once you are on a windy back road and tossing it back and forth, it won't shake. I may have missed but did you put new thermostat in yours? Without seeing my temp gage, I forget where mine runs... but, it does get to 100 at times and I overheated it on the way to Asheville that trip going up a long grade in W VA... to the point it lost power and downshifting to 3rd solved it... then it cooled to normal after cresting and ran normal rest of the trip! With that new radiator can't imaging you having trouble with that... +1

> **Rob Siegel** I checked my receipt from RD, and new thermostat is 165 degrees. Top of the thermostat says "TOP" on it, so I know I actually installed it right-side up :^) [photo of thermostat installed correctly] +3

> **Russell Baden Musta** And know what, that's the one I use and I overheated it twice... the one I described in WVa and the other time was on Talladega speedway two LOGs ago!! It soaked the left rear tire in the pits... then hour later did 8 more laps and no trouble... up to 110mph. +1

Ali Jon Yetgen Goodness, that car is small!

Bill Snider Has to be shared [link to Joe Walsh "Life's Been Good To Me So Far"] +5

> **Rob Siegel** vividly remember when this came out the summer of '78... +1

> **Andrew Wilson** This song parallels my life to some degree, and I quote it often.

Keith Martin I've been patiently watching and reading as your saga has unfolded. One of my favorite cars was a white/black 1973 Europa Twin Cam. I owned it probably 30 years ago, let's say in 1980 so. So it was barely ten years old. I bought it for $6,500 and drove it around for a year without doing anything to it except removing some extra smog-related throttle bits from the Strombergs. I think it had about 50,000 miles on it. Original paint and upholstery. A little crazing to the wood dash. I still recall thinking I had my own miniature GT-40 each time I got behind the wheel. And I marveled each time I took out the trunk liner, and looked down at the suspension, the transaxle and the road beneath. It was a totally cool car. The only issue I ever had with it was when I tried to put larger front tires on it. I never got them to balance properly. So stick with stock.

It's funny to think that cars that are now completely worn-out were once just used cars, with nothing totally decrepit about them. Thanks for bring this car back to life. The Europa was a remarkable singular expression of one man's vision of what a sports car should be, and we are all the better for it.

I will say in closing, your story makes me appreciate my 2006 Lotus Elise (20,000 miles on it). It's a car I sold and bought back two years later because my daughter Alex was so angry with me for letting it go. Four airbags, a/c, Toyota engine—as pure a driving experience as you can buy these days. +8

Isabel Zisk She's cuter from the front. +1

Dave Gerwig I know every car adopts a name along the course of ownership, with that green plate and titleship route, might I suggest Varmit. They are small quick rascals and that mink brown color, well it sort of fits. Just a suggestion 😊 +2

Neil Bradbury You're my [not feeling so] guilty pleasure Rob +1

Marshall Garrison Rob, apparently a giant stepped on your brown clown shoe on the right so hard it squished the grille all the way off of it! ;) +1

John Harvey "Can't complain but sometimes I still do." Smaller and lighter the wheel the harder it is to balance.

Jon Allen Love the JW reference. Not quite sure how one rides in the back. What doors will you lock in case you're attacked? It's brown, it's Malaise, and it's British. "Warts and all", as Joe Strummer said. Love it. +1

Thomas Jones I had to have the tires for my '91 318i road force balanced to deaden a bad vibration at 65...

Garry Summers I have a friend who vintage-races one

Tanya Athanus So what is it that you like about a Lotus... You certainly know good engineering from your German cars. Tell us... one thing that they do well.

> **Rob Siegel** 1600-pound fiberglass-bodied mid-engine twin-cam road-going little race car; what's not to like? +4

Pete Lazier Rob Siegel ….the colour!

Paul Fredendall Tanya Athanus the term «Handles like it is on rails» must have been coined to describe a Lotus. They are amazing to drive-when they are drivable.

Bill Mack I'm tellin ya get the temp gun out and shoot it around under the hood and around the radiator. I just became aware of a Lotus group event at Thompson at end of month. +1

Robert Boynton How about Road Force wheel balancing? +1

Bob Shisler My old (indie Mercedes shop) boss used to say "You can balance a square, but that ain't gonna make it roll." (referring to shitty tire brands.) +1

Josh Boone Such a cool little car

Scott Barnes The Europa is one of the few cars that makes the M coupe look big.

Tom Egan Right? I thought it was a horrible X5 until I took a closer look. +1

Martin Bullen I parked my Z3 next to a Fiero GT a couple of years back. I was very surprised at how petite the Fiero was compared to the Z.

Ernie Peters My Z3 and my former Midget for comparison. Midget is much shorter, and much narrower in the cockpit.

Wade Brickhouse You took a picture of the Lotus with your piggy bank. 😊 +1

Cherche Buddy Vuhmahnt plates FTW +1

Scott Aaron Pretty cool man

Bob Sawtelle Rob I so love this little brown thing... I think I told you before, I saw a picture of a JPS in Road and Track back in the day and went and got the model kit for it. In all my years of going to race tracks I think I've only seen a handful of these usually with the thicker sail panel (not as good looking as yours) and yours is the only malaise brown I have ever seen and is still so cool as it sits with the patina. But gotta change the lyrics to "my Lotus only does 85."

Clyde Gates Quick, sell it while nothing is broken. Claim all of that is normal («they all do that…») +1

Wolfgang Kovac My friend just sent me an $11k listing on a brown Europa

Tim Warner Persevere… enjoy! Love the comparison to a small GT40. +1

Stephanie Fox Smithwick It's smaller than an m coupe?!?! My goodness! Both are wonderfully weird. And SO awesome.

Brian Epro Rob, I'm not flirting with you or anything, but that driveway would give me a boner. +3

Chris Mahalick Mine did the same thing, but in my case, the tires were about forty years old. I put on a set of Kumho Solus and balanced. For a little over $200 my vibration was solved. Also, did you measure center to center, front to back to do a basic alignment test?

Franklin Alfonso Man that car is low…

TLC Day 2249: Driving the toys to a gig

AUG 6, 2019, 7:24 PM

The Lotus Chronicles, Day 2248:

My small frivolous toys and I are out on the town :-)

[Note: The destination for this drive was one of the open mics I regularly played at. At this one, the featured performer was supposed to be Rob Ayres, my friend I wrote about on March 2nd in "The Passing of a Car Guy." Instead, his kids played. I performed a song I'd written about my friendship with Rob called "We'd Talk About Cars." What I didn't say either on Facebook or in the song was that the car that I describe Rob having sold, taking it out for one last drive, winding it up to 100 mph, getting bagged, and talking his way out of the ticket, was a BMW 2002tii. The video of me playing the song at this event can be seen by searching YouTube for "Rob Ayers tribute Rob Siegel talk about cars." The Lotus was right on the hairy edge of being ready for the drive—this was the furthest I'd driven it, and certainly the first time I'd used it like it was a real car, with an actual destination at which I needed to arrive as opposed to it being a test drive—but given my friendship and history with Rob, I couldn't not take it. I'm not one who feels that departed people are looking down and smiling on such things, but I know he would've heartily approved, and he would've been chomping at the bit to see the car during intermission.]

151 LIKES, 32 COMMENTS

Bill Mack Nice!!
Ed MacVaugh Did I miss an article on the windshield replacement? +1
    Chris Lordan... or magic tape?
Daniel Senie You managed to fit a guitar in there. I'm impressed! +2
    Rob Siegel "It's a Mini in a Lotus!" +8
Jack Fletcher Bowling shoes?
Kevin McLaughlin Bond... James Bond... +1
Russell Baden Musta Spoiler alert, I can get my Gibson Firebird and my Les PAUL in my Europa in their cases! +3
Lance Johnson How does it compare to a GT40? Height-wise of course...
    Rob Siegel Two inches taller +3
George Zinycz Sooooooo loooooow...
Steven Anthony Smith Isn't that Natick center? +1
    Rob Siegel It IS! I was at the Center for Arts in Natick. +1
Noam Levine Not too many Loti in Natick...
David Holzman Chubby Checker would have liked this car.
    Rob Siegel From afar. He never would be able to get in it. +1
Dohn Roush Do you have to pay full price for a parking space? You're only using half of it... +2
    Rob Siegel Are you talking vertically?
    Dohn Roush Vertically, longitudinally, laterally, literally... +2
Richard Shouse Door handle high to the car next to you! +1
Doug Jacobs It looks a little like a brown Gremlin.
Daniel Senie We can fit 6 instruments and a sound system in our Tesla Model 3. And we appear to be the only Tesla in town. +2
David Kemether You're no Barry Gibb.

+6

George Zinycz Waaaaaay too much crotch +1

Russell Baden Musta David Kemether love those early or Euro S2's… get him off the car!! +1

Maire Anne Diamond You made it, yay!! +3

Tom Egan Ironically there is no room for a guitar AND a passenger in the TCS. Groupie or guitar, not both. +1

Franklin Alfonso Easy choice. Groupie can take the bus…

Duncan Irving I don't park next to vans in my car if I can help it. Too hard to back out. Can't see. But this… +1

Chapter X:
A Mini Road Trip?

TLC Day 2250: LOG? Maybe.

AUG 7, 2019, 8:53 PM

The Lotus Chronicles, Day 2250:

So.

The national Lotus Owner's Gathering (LOG) is in Sturbridge MA the weekend after this one. If you'd asked me in the spring, I would've said no freaking way. And now... it's not inconceivable. It's only 50 miles.

I've put 135 miles on the car since its revival.

My "Big Seven" issues likely to cause a vintage car to leave you stranded are ignition, fuel delivery, cooling, charging, belts, ball joints, and clutch hydraulics. Yes, I've had a hot-running issue, but it doesn't appear critical so long as it's not hot out. I've had no ignition issues, no fuel delivery issues, no charging system issues, although those systems aren't new, just tested. The belts are new. There are no clutch hydraulics, just a cable. The front end does need to be rebuilt, but that's because the boots are dried up and pulled back, not because things are loose and banging. The car's general status is twist-and-go.

This is pretty amazing, really.

To see how I feel about the idea of highway-driving to Sturbridge, I just took Lolita on the longest highway drive yet, up I95 to RT 2 and back. Twice. I'm still not nailing and wailing, but I am, shall we say, goosing and juicing. But swapping wheels front-to-back hasn't changed the fact that, above 50, there's an anxiety-inducing amount of vibration. As I think I said a few days ago, the Cosmic wheels are surprisingly straight, but the Achilles tires were cheap, and all of them do show some visible out-of-round. I have a road-force balancing scheduled for next week. It smooths out a little at 65, but at that point, a feeling of lateral instability, like the car can get blown off the road like a leaf, likely caused by the car needing both front and rear alignment, begins to get scary. Really what I should do is put the car back in the garage, tear apart the front end, install the trunnions and tie rods I've already bought, and do the alignment and wheel balancing together. I'm just not sure I have the time.

Lolita is somewhere between Louie, who on his first drive on pavement, jumped out of his own tii body and said to me "YOU FOOL! DRIVE ME BACK TO BOSTON RIGHT NOW!" and Bertha, who, when I tried to push too hard to get her ready to go to Pittsburgh, mumbled "ya damned kid you woke me up too soon from my nap you got any warm milk and Old Mr. Boston that doctor who did my knee is a BUM."

So?

I'll treat her right, listen to her, and we'll make a decision together. We have time to make up our mind.

(Not looking for anyone to talk me into or out of it. Not that I'd listen anyway :^)

138 LIKES, 74 COMMENTS

Steven Hage I love these little cars.

Doug Jacobs You should do the work you mentioned and then go to the LOG. Bask in the glory my brother! +2

Samantha Lewis <3

Marc Bridge You are taking your Lolita for a spin and the best you can do is Sturbridge?

Bob Sawtelle That picture should be the cover of a Boston Detective series where the grizzled detective moonlights as a folk singer by night and solves crimes and mysteries involving the hoi polloi of the city by day. +3

    Rob Siegel "It was raining in the city. No, wait; that's antifreeze..." +6

Jim Strickland Well, she's pretty in black and white, or in color

Isabel Zisk When Alexandra was a 2lb 9oz preemie in the NICU, the doctors and nurses always used to say. "She's telling us XYZ." And I kinda thought they were crazy. But they knew how to read the signs. So: BE THE LOLITA WHISPERER... +1

Tom Egan What was your engine break in routine after the rebuild? Maybe take it easy until the break in period is up?

    Rob Siegel Haven't really done anything yet; no oil change, no valve adjustment. I HAVE been taking it pretty easy.

    Tom Egan Also, car looks awesome in B&W.

Justin Sz Arkowicz That's awesome! I drive by there every day for work. I'll have to check it out.

Alan Hunter Johansson Chris dragged a '41 Studebaker Champion out of the weeds at Truett Ray's (yes, Shadetree Garage Truett Ray) in 1976 or so when he was living in Creedmoor and studying engineering at NC State. When he got it going, the joke was that he would drive it to a nearby SDC meet, and then say, "Well, it got to Goldsboro and back, I should be able to get to (2xdistance between Creedmoor and Goldsboro." So he'd go to a meet about that far away. By 1978, we drove it to the annual meet in South Bend, Indiana, and in 1982, to the national meet in Seattle-Tacoma. The car was famous in the SDC community for forcing a rule that no car could win "Most Needs Improvement" more than twice, since it was intended to encourage people to work on their cars, but Chris seemed to take it as a badge of honor. +2

> **Rob Siegel** "I shall take that crown and wear it proudly."
> "Uh, it's not a crown."
> "I SAID I SHALL TAKE THAT CROWN…" +1

John Whetstone Are the headlights on?

> **John Whetstone** Seriously, congratulations. Huge project to us laymen.
>
> **Rob Siegel** Are the Lotus' headlights on? Not in this photo, no. +1
>
> **Alan Hunter Johansson** John Whetstone Always hard to tell on British cars. +4
>
> **Rob Siegel** As I say in Resurrecting Bertha, "it's only cruel because it's true."
>
> **Tom Schuch** John Whetstone "A gentleman doesn't motor about after dark." —John Lucas. 😊 +1

Eric King good photo

> **Rob Siegel** Just my phone, converted to B&W. I took a series of shots as a car drove by, thinking the lights would look cool if I converted to B&W. It wasn't entirely an accident. +1
>
> **Eric King** Rob Siegel Just needs some water on the street and a bit of fog in the background; femme fatale indeed. +1
>
> **Rob Siegel** you're hired :^)

Dohn Roush On your list of things that could leave you stranded, you left off swollen feet... +1

> **Rob Siegel** ya know...

Thomas R. Wilson Rob, love the lotus noir photo. Have you checked the sending unit for temp problem? After trying everything else on my MG checked engine temp with infrared only to discover it was fine.

Rob Siegel Thomas R. Wilson, a number of people have asked me this. The infrared readings aren't terribly helpful, at least not to me. When it's reading 3/4 of the way up the gauge, the input radiator neck is 193, the output 187 (remember that the radiator is in the nose and it's a mid-engine car), the threaded sensor port in the radiator for the fan 178, the thermostat rubber hose and thermostat cover 240, and the threaded temperature sensor itself 350 to 500, depending on where I shoot it. Is that "fine" or is that "a little hot" or is that "hot?" I can't really tell with any certainty. It shows that the trip to and from the radiator is cooling the coolant. Beyond that, I can't say.

> **Thomas R. Wilson** Rob Siegel Ah yes I forget if simple you would have found the solution already. Well good luck, it is good to see Lolita on the road again.
>
> **Bill Ecker** All of those temperatures sound perfectly reasonable. Do you have any reference of what other Loti run their temps in those locations?

Sam Schultz Nice proper B&W photo of her.

Thomas Jones Fifty miles!? Only fifty miles! I do that each way to work... And, I've even been to Sturbridge! Though it was thirty five years ago. +2

> **John Whetstone** Thomas Jones Almost to the day!
>
> **Thomas Jones** Yup! [photo of badge for 1984 BMW CCA Oktoberfest] +1
>
> **John Whetstone** Direct from Mike Cinnamon, Oktoberfest Chairman- August 13-17 1984!
>
> **John Whetstone** Thomas Jones Did you get the furthest traveled award?
>
> **Thomas Jones** John Whetstone Nope, it went to a couple who drove their freakin' Porsche from Marin county. About fifteen more miles… My mom still feels wronged, she and my sister drove the tii out. Sister wasn't quite 16 and was on her learner's permit. I was with them, but was only nine years old. Dad had to fly out due to having to stand in the picket lines of the AT&T strike. Dad had to change a guibo in the hotel parking lot. Thankfully Michel Potheau hosted a group drive to CTC so folks could buy some much needed parts! +1
>
> **John Whetstone** Thomas Jones That stinks, I hope that they have since changed the criteria to the members who drives the furthest specifically in a BMW! I hope it was at least a member in the Porsche.

Bill Howard 50 miles? No major hills. You should make it. Go in daylight. Don't tax the headlamps. +1

> **Rob Siegel** Really, my biggest concern is that I'll get my ass run over if I drive 50 on the Mass Pike. +1
>
> **Rob Siegel** I could just head straight out Rt. 9. It's actually not such a bad idea. +3

Bill Howard Rob Siegel For God's sakes don't go up to Bose. That's a freakin' mountain. Or The Mountain as they call it.

Neil M. Fennessey I think that Rt. 20 might be less busy than Rt. 9. Plus, with Spag's gone, Rt. 9 is hardly worth the bother. +2

Adam G. Fisher Rob Siegel Rt 9 to Rt 20 and you can get some beer at Treehouse on your way by. 🍺 +1

George Zinycz Neil M. Fennessey Long live Spags!

George Zinycz Rob Siegel Yeah seriously Rob... the only problem is your assumption that you have to take the Mass Pike to Sturbridge. Set Waze to "avoid freeways" and go there the old school way... +1

Paul Skelton Just go. You'll regret it if you don't.

Rob Siegel Paul, it's less about regrets and more about what's reasonable and feels right. +1

Lee Highsmith #listentolololita +1

Daren Stone Your "Big Seven" vintage car issues are obviously borne out of experiences and are spot-on, except we're talking about old Lotuses here so there's a random factor that no matter how well you prep and check, the door may still fall off or the brand new alternator may seize. (How's that for a run-on sentence?) So prep & check, but by all means take Lolita to LOG. If you make it w/o incident it'll be the first event milestone, and if something craps out, it'll be another chapter in the life of Lolita. +1

Brian Stauss As we used to say around the frat house in college: "No guts, no glory."

John Whetstone If nothing else you're bound find a much more sympathetic crowd there, surely well worth the ride!

Tom Samuelson If the worst happens, just call a buddy with a trailer. +4

Rob Siegel I love you, man...

Randy Yentsch Great photo. Have you run a deodorizer through the vents, blower etc? Also, an ozone treatment at an auto detailer should help with the odor.

Rob Siegel Randy, the more I've driven the car at speed at all ventilation configuration, the more the smell dissipates. +1

Jack Fletcher "I love the smell of rodent pee in the morning" +1

Tim Warner Damp rid might take some smell out too

Mike Miller I love the color! +1

Mike Miller Seriously, if that color is sprayed properly I'll bet it would be gorgeous. Although... there is GOLF. +1

David Weisman OoooOOOOoooo sexy

Doug Jacobs "Whinging? I'm not whinging." "You're worried about driving 50 miles and your mouth is moving, that means you're whinging."

Franklin Alfonso Maybe you could swap tires with one of the other cars? If they are out of round and see if that helps.

Rob Siegel different size

Franklin Alfonso Wow, nothing in the arsenal rides on that size, eh? Bloody Brits!

David Holzman Trunnions and bunions and tie rods, oh my! Oops! forgot the onions! +1

Jack Fletcher Put some Water Wetter in with the coolant

Andrew Wilson Have fun!

Paul Forbush It's 50 miles on the fucking turnpike! Just do it! You're going soft in your old age, Rob.

Brian Ach "install the trunnions and tie rods." this.

Paul Onboost Rob, although your Lotus tale has been inspiring, sadly I may have to part with my 74 TC. With the numerous other projects in play, I just don't think I'm gonna have time to really get into this one. :-(If ya know anyone...)

Mike Kovacs Is it true that Lotus Europa owners never need to pay for parking in a garage because they can drive under the gate? +1

Scott Winfield Don't worry Rob. I can follow you in my '70 Sedan DeVille. Any problems and we can just put the Lotus in my trunk. +1

Anne B Barnard As a frequent driver from Hartford to Boston area, you could hop over to route 9 or 20 if the pike makes her run hot. +1

TLC Day 2251: Lighter socket and USB

AUG 8, 2019, 8:15 PM

The Lotus Chronicles, Day 2251:

I decided that the most important thing to do today on Lolita was to install a cigarette lighter socket. That's after looking for a socket to charge my phone, not finding one, reading online, learning that they actually don't have one, and thinking "This is a little strange, isn't it, especially considering that the most desirable version of the Europa is the John Player Special, AND THEY'RE A FUCKING CIGARETTE COMPANY, NOW, AREN'T THEY?"

$10.99 on Amazon. One lighter and two USB sockets. Nice little package with fuses and connectors. Just about invisible unless you know it's there.

And yes, it felt great to use the right-angle drill attachment for the second time ever, and even better to have been correct about where I'd put it and able to lay my hands on it.

93 LIKES, 58 COMMENTS

Bob Sawtelle That is fancy but the purists will howl about you hacking up that pristine glove/tea service area. +2
    Rob Siegel The left side of "that pristine glove/tea service area" is literally a piece of cardboard with black paint. +4
    Russell Baden Musta Rob Siegel you put a ground wire under it right? There is a metal strap along the bottom edge of the wood dash all the way across. I fabricated an aluminum bracket and riveted it to the bottom of the dash… that worked as ground.
    Rob Siegel Russell, I didn't know about the metal strap along the bottom of the dash. The wires that were part of the kit were plenty long to run over to the driver's footwell and tap into the metal structure above your feet that appears very well grounded.
    Bob Sawtelle Rob Siegel I too am amazed this thing doesn't have like three ashtrays and two lighters already as it is a proper 70's British Sports car… maybe they figured the owner could light his cigar off of the flaming embers of the molten fiberglass when it bursts into flames from an electrical gremlin on the way home from the dealership anyway… +10
    Russell Baden Musta Rob Siegel perfect! Mine is right 3" from the fuse box… I don't know what they're hot-wired to…
    Rob Siegel Bob Sawtelle, IT HAS AN ASHTRAY BUT NOT A CIGARETTE LIGHTER!! (I shit you not.) +4
    Russell Baden Musta Rob Siegel True! The car CAN be a source of fire! +2
    Dohn Roush Silly, you light your cigarette with your pearl-sided Ronson. Very continental… +1
    Bill Snider Rob Siegel "I shit you not" just one of Robert Wuhl's great quips from "Assume the Position " show on HBO. [link to video]
Paul Garnier Heathen!
    Paul Garnier he probably used these too! (oh the humanity!) [photo of wire splicing clips] +4
    Rob Siegel nope, wired straight into the fuse box with a Y-spade connector like a civilized Hack Mechanic +4

> **Rob Siegel** (I hate those fucking things in the picture) +2

Russell Baden Musta I did that in 1974 to mine! Updated a couple years ago to accessory socket... same thing, no lighter but a cap! +1

Marc Dobin As I was reading the beginning, I was thinking "USB." Genius.

> **Rob Siegel** I'm not sure I'd want a naked uncovered USB socket permanently staring me in the face while driving a vintage car, but this one is very nicely hidden. +1
>
> **Jacob Johnson** Rob, you get used to it. I've got one in the 2002 that will be hidden if I ever redo the center console. Doesn't really bother me where it is though.
>
> **Rob Siegel** Jacob, it's a personal preference, not a moral judgement. Yes, I realize that it's almost silly when we routinely slap phones on windshield mounts and drive with them for hours, but the older I get, the more sensitive I am to seeing things mounted in or on the dash of vintage cars that look like they teleported in from the wrong millennium. +6
>
> **Jacob Johnson** Rob, I totally get that. It is a little bit anachronistic. It will get hidden, along with a hidden modern stereo as well. Someday.

Wade Brickhouse Keep fooling around and you will be in the dark! +5

> **Rob Siegel** there IS that risk

Rob Theriaque I can only imagine the endless entertainment somebody in your position could create by pointing drill bits at various things, posting the picture, and causing forehead veins to explode. +4

Bronco Barry Norman Invisible for the driver anyway. Interesting little spare cubbyhole in the dash. Must have been put there in anticipation of future tech!

> **Samantha Lewis** Actually it was sized perfectly for stacks of 8 track tapes. My '73 had a Lear 8 track under the dash (original owner's decision, I swear I had nothing to do with it). The true horror of the 70's the 8 track. +2
>
> **Bronco Barry Norman** I also had an 8 track (in my Spitfire). One friend made the smart choice with cassette player in his Beetle.
>
> **Rob Siegel** My GT6 had an 8 track AND a cassette player! +3

Brian Aftanas A USB PORT???!!?? WHAT'S NEXT???? A CUP HOLDER???!!!??? WHAT HAS THIS WORLD COME TO???!!!??? +5

Daniel Senie You're going to trust Lucas electrics to charge your phone? 🙄 +5

> **Rob Siegel** Hahahahaha!

Jeremy Shiroda I understand. [photo of USB in BMW E30 console]

Andrew Wilson You are officially an honorary English cabinet maker. +2

Mike Miller It's part of the lunacy. It wouldn't be a Lotus without the lunacy. +1

Neil M. Fennessey I have a right-angle drill that has only been use once too! Congrats to you and yours for finding a proper second application! +1

Franklin Alfonso So now we know where the source of the fire will originate from. +1

> **Rob Siegel** oh ye of little faith
>
> **Franklin Alfonso** Go balance your tires lol.

Dohn Roush Add lighter nest?

John Jones Sooooooo cocky now. You get Lolita up to freeway speed ONCE and you decide to modify the ELECTRICAL SYSTEM?!?! Hubris. You are a well-read man. What do you think that gets you? +2

> **Rob Siegel** a charged phone to use when it dies? +5
>
> **John Jones** Rob Siegel ...if you get out in time +2

John Jones "Burned to the Ground When Parked: How I Resurrected a Decades-Dead Sh$& Colored Lotus, and Watched It Go Up In Flames Because I Spit In Lucas's Eye" +9

Franklin Alfonso Yup, He's messing with the Lucas Electrics Demon...BEWARE! +1

> **Kevin Pennell** Only those of us who have owned and worked on anything Lucas can truly appreciate this dialogue +1
>
> **Kevin Pennell** [gif Antman Lewis "Oh we can handle it we ARE professionals"]

Scott Aaron Nice hole in the dash. I mean dash-hole... +1

Jack Fletcher Nice place to put a screen for a backup camera. +1

> **Franklin Alfonso** Actually, a backup camera on all the time could be handy in place of the rearview mirror.

Bill Mack Dang. I gotta find my zippo with the lotus emblem on it.

Bill Mack You have produced a disturbance in the force

Bailey Taylor I'm not sure that adding more sources of electrical resistance heat to a British car is a great idea, but that's just me. +2

Rob Siegel Bailey, actually, as I say in my electrical book, intended resistance is a GOOD thing. What you don't want to add are: 1) Unintended low-resistance circuits. Those draw a lot of current and burn wires. That's what a short circuit is. ...and... 2) Unintended resistance to existing circuits in the form of corroded connections. You could certainly argue that a Chinese-made USB socket could be poorly engineered and thus be an unintended low-resistance circuit, but "this thing is junk and almost burned up my car" usually shows up in Amazon reviews. +2

> **Jacob Johnson** Yes, but Amazon reviews tend to be for Toyotas and such. I'm doubtful too many people are installing anything into ancient British cars.

Gary York Is nothing scared?

Lance Johnson Surely, the Black & Gold version had a cigarette lighter?

Tom McCarthy I would assume it came with a carton of smokes in the glove box.

Marc Bridge Very nice! What in the hell did they do with their phones back in the 70's? 😎 +2

> **Rob Siegel** um, long cords and driving around the block in circles? +3
>
> **Tim Warner** Marc Bridge they used smoke signals- if they didn't do like Rob said! +1
>
> **Marc Bridge** Tim Warner there are times that I wonder how we functioned in the pre mobile phone age (aka Stone Age according to my kids) +1
>
> **Rob Siegel** Marc, I don't have a clue. "You mean... we used to READ MAPS? And not be able to call people and tell them we'd be LATE??" +2
>
> **David Schwartz** Maxwell Smart never plugged in his shoe phone. Must had had great batteries. +1
>
> **Bob Crownfield** they filled the trunk!

Kevin Pennell Well done, Rob. Very nice and clean looking. +1

> **Rob Siegel** I thought so! +1

Drew Lagravinese And it works?

> **Rob Siegel** Oh, yeah, I guess I didn't say that. Yes it works :^) +2

Clint Carroll Run the same rig on my 02 tucked under the radio; you can see the lightning pigtails here, but if I pull them it's basically invisible. [photo of USB in 2002 console] +5

> **Nicholas Mav** Clint Carroll that console is the cleanest, most well put together '02 console I've ever seen. That radio is perfect. +3
>
> **Clint Carroll** Thanks Mav Nicholas!
>
> **Christopher Heine** Very nice!

Eric Pommerer By Jove....!

Russell Baden Musta Nicer than mine!

TLC Day 2254: Committed to LOG

AUG 11, 2019, 9:10 AM

The Lotus Chronicles, Day 2254:

Well, I did not think this would be possible, but I am now committed, or I should be: I sent in my check to attend the Lotus Owners Gathering (LOG) in Sturbridge MA on the 24th. Nothing like money on the table to, uh, commit yourself :^)

Replaced the shifter bushings to tighten up the notoriously sloppy linkage (mid-engine car; long linkage) and to help me get the car into 5th gear, which I've been unable to do. Actually, I shouldn't say "replaced," as the bushings were completely missing. What a PITA. Much of the linkage is in a tube under the car. To access the bushings, you need to undo the entire linkage and slide it backwards out of the tube. Got it done, most of the slop is now gone, and shifting it is now much more enjoyable.

But I still can't find 5th gear. Not that it really matters right now, as the wheels still shake when I drive over 50.

But, more important, when I drove the car, it exhibited electrical issues: The alternator light came on and off, including when I hit the brake pedal. WTF? My immediate thought was that you folks who razzed me for tempting the Lucas gods by installing the cigarette lighter socket and USB port might have been right. Plugged in my cigarette lighter voltmeter (which I can now do, given that I, you know, have the socket), and I could see that it wasn't charging. Turned out to simply be that the negative terminal had pushed out of the back of the plug on the alternator. Completely coincidental. Suck it, oh ye of little faith.

Continued on my test drive. 160 miles now on Lolita. The novelty of driving this weird fragile little thing with my butt four inches off the pavement hasn't even close to worn off.

144 LIKES, 53 COMMENTS

Frank Mummolo If this were "1001 Arabian Nights", you'd never run out of stories! I hope you're collecting these for you next book. Or should I say chronicles! +3

    Bill Snider Well imagined, Frank. +1

Russell Baden Musta Well done Rob!! I have been thinking for years to make oilite bushings for mine to really snug it up... make sure the roll pin near that joint by the bell housing is safety wired of any sort so it doesn't fall out! I'm looking into new wheels for mine for the same reason... harmonic shaking at various speed... keep you posted! +1

Carl Witthoft Can't help but see a dangerous parallel between this Lolita and a certain Christine. To be on the safe side, Rob, make sure there's always something left to repair +1

Doug Park [link to video of Original Log Commercial from Ren & Stimpy]

Brian Aftanas I don't understand why your expectation is that the shifter will be sloppy due to the long shifter shaft. If the bushings are tight, I would expect that the shifter would also be tight. Wait, I just looked at internet photos of the shifter linkage routing. Never mind. Eww. +1

Collin Blackburn I guess if you can do all that to replace your shifter bushings then I can definitely do the ones in my Volvo that involve popping off 4 clips. Still too lazy.

Tim Warner Lolita makes an original Mini seem very robust!

Jim Denker She's a temperamental thing, isn't she?

Jim Strickland If anyone can do the 51 Lolita miles to LOG, it's you. The Lama easily made the 700+ mile round trip to The Vintage. +2

Dohn Roush Lolita is a nice name, but I still think Sisyphus would be more accurate... +6

    Rob Siegel Yes, but a stone is actually heavier. +4

    Steven Bauer Dohn Roush the Aegean Stables may be a better mythological analogy... +4

    Dohn Roush Steven Bauer Not sure that would fit on a vanity plate... +1

    Steven Bauer Dohn Roush "Augean" would. I checked on the correct spelling. [Note: I had to look this one up. "Augean. Relating to Augeas. Of a task or problem requiring so much effort as to seem impossible."] +1

Lee Highsmith But Lolita is a nod to the "you can get sex from it" lore... +2

Dohn Roush Just so you don't try to get sex IN it... +2

    Rob Siegel ON it? +1

Dohn Roush Rob Siegel Redefining the Lotus position... +1
Drew Lagravinese Rob Siegel that would work on that nice flat rear end.
Steven Bauer Rob Siegel that engine deck may be the right height...
Lee Highsmith Dohn Roush on it more likely than in it, especially at our age...

Wade Brickhouse You're assuming there is a fifth gear in there. 😊 +6

Rob Siegel I'm virtually certain that I verified it by transmission model number, physical appearance, and shifting it through all gears while it was out of the car, but, hey, I've made stupider mistakes.

Wade Brickhouse Maybe it's physically missing, It's British after all. 😊 +1

Rob Siegel Right. It COULD have leaked out. +9

Wade Brickhouse 😊

Rick Roudebush Or caught fire and melted into the bottom of the gear box. ;-) +2

Martin Bullen Except the gearbox is French. +1

Rob Siegel Martin, yes, but it is connected to the shifter via a "tunnel." Coincidence? I think NOT! +2

Franklin Alfonso If it's French it may just refuse to work with a British Car... +3

Martin Bullen More likely just a French part being stroppy for having to do the bidding of the British. +1

Robert Boynton Martin Bullen If it's French, then intuitive logic opposite of expected (like finding the high beams on a Peugeot 504). (5th gear probably located in reverse position, but only above 50 mph and with headlight switch on.) +1

Dave Gerwig Maybe you own an extremely rare 4 speed Varmit and the prior owner screwed on a 5 spd knob to look cool 😎 +5

Steven Bauer Dave Gerwig or just to mess with the head of the next owner... +2

Dave Borasky Day 2254: Rob is voluntarily committed. The Lotus drove him crazy. +1

Paul Aspden Keep up the posts Rob, good to hear your exploits.

Mark Rowntree Keep posting, Rob, this is very interesting. Drove mine for the first time in 31 years just a few weeks ago and I'm still grinning from ear to ear! +1

Chris Mahalick Regarding fifth gear. You have to really slap the lever hard to the right to engage fifth. It took me a few tries, but I finally found the winning formula. +1

Franklin Alfonso Hmm... Go for it, what could go wrong?

Bo Hellberg No problem changing the bushings from within the car. Done it several times. With new bushings the change is greatly improved, negative is that they don't last very long. Strongly suggest changing to oilite(?) bushings. Play here is multiplied before reaching the rear. +1

Rob Siegel Just my being a newbie then!

Bo Hellberg Rob Siegel, been there, done that!

Daren Stone Is the clutch effort still excessive or have you gotten used to it?

Rob Siegel Daren, it's not so much that the clutch effort is excessive (it's slightly stiff but not monstrously so); it's that all of the action occurs in a fairly short amount of motion.

Dave Borasky that's what she said +2

Rob Siegel BOOM!

Tom Egan Ever see the movie "Freaks"? "One of us... One of us..." +1

Clyde Gates Remember why the British drink warm beer...

Marc Bridge Rock on Lolita! 👍 Sounds like she is Sturbridge bound. +1

Geoff Bartley Well, I don't know anything about shifter bushings or linkage but it's very cool that you do, and cool you have the car! +1

Rob Siegel Thanks Geoff!

TLC Day 2256: I am a big doofus, again

AUG 13, 2019, 8:58 PM

The Lotus Chronicles, Day 2256:

While driving Lolita yesterday, I noticed that, when turning hard, like backing out of the driveway and angling the car to take off down the street, or pulling into a parking spot, I heard the front tires rubbing. This was odd, because it was a new development.

Then, I looked at the space between the front wheel and the back of the fender arch, and was alarmed to see that it was virtually non-existent, like a quarter of an inch. I'm aware that the car could use front end work, as after sitting for 40 years the rubber boots are completely missing on the ball joints and tie rods, but I've checked and haven't found any play that could cause such a shift in the front end.

I posted the photo below to one of the Lotus forums, and folks immediately jumped in, saying that the body must've slid on the frame, or the double-wishbones must be mashed back, or the sleeves in the bushings in the wishbones must've slipped. I ran down to the garage as the responses came in to check on people's suggestions. No, everything appeared okay, but still, this clearance had become alarmingly close over the span of a few days. I wondered if my trip to the Lotus Owners Gathering in Sturbridge would be in jeopardy.

Then, someone commented "I've only ever seen that when someone used 185/70/13 tires on the front." I sat there and looked at the comment, and I realized what had happened. The car has staggered tires—185/70/13s in the rear, 175/70/13s in the front—but in troubleshooting the problem with the car shaking at over 50mph, at some point last week I swapped the wheels back-to-front, completely forgetting that the tire sizes were staggered.

So, yes, this episode of The Lotus Chronicles is actually an episode in a longer-running series: I Am A Big Doofus.

169 LIKES, 57 COMMENTS

    Drew Lagravinese Live and learn. We learn from our mistakes my mother used to say. +2
        Randy Yentsch Drew Lagravinese with the advent of forums, Facebook and the like, it is easier to learn from others' mistakes and experiences. Thank you for the education and entertainment Rob. +1
    Frank Mummolo With cars of this vintage (and origin)... yes, a BIG doofus! Imagine what folks from the next generation of older car nuts are going to face!! Bill Gates is going to be Joe Lucas!!

David Kemether I bought a 12-year-old Chevy pick up to use as a work vehicle. When I purchased it I knew it desperately needed shocks and immediately replaced them. The ride improved but was still jittery. I jacked it up and everything seemed tight. I thought maybe I should have bought pricier shocks. After driving it for several months I finally thought to check the air pressure. They were all 10lbs over-inflated. D'oh! +2

> **George Zinycz** David Kemether so funny because I've seen that too. First thing I do when buying a car now is check the tire pressures. The first elective service after that is an alignment. Gotta make sure that stuff is right as a baseline.

Shad Essex Tape those wheel weights before you lose them. +1

Doug Jacobs That taint nachrul.

Neil Bradbury A free lesson. No harm, no foul. +1

> **Rob Siegel** The best kind. +3

Paul Onboost Dooh!!

Wade Brickhouse Attention! Attention!! The Hack is a mere mortal!!! 😊 +2

Layne Wylie Never thought I'd see "staggered" and "185" in the same sentence. +10

> **Rob Siegel** More like "limped" +2
>
> **George Zinycz** Layne Wylie and with 13" wheels! +1

Steven Bauer Glad the solution was simple.

David Kiehna Could they have put any more weights on those rims? I'd take them to the recycler and get a set of good Minilite replicas.

Glenn Koerner That much lead open to the environment is probably illegal in some states. +3

> **George Zinycz** Glenn Koerner especially Massachusetts! +1

Chris Roberts Did the same thing with directional tires on the 02. Took a week or so to figure out why the front end suddenly had a bad shimmy at about 50MPH. I was too embarrassed to actually tell anyone tho. 😊 +2

Russell Baden Musta That's funny… I'm laughing with you, not at you! You saw yesterday I hit a new pothole on a local road and my rear-view mirror fell off! Tonight I took a contractor friend for a ride in the Europa and hit the sway bar (just a scuff) on the table top speed bumps. always a little something. +1

Rusty Fretsman Welp—beats the alternative :| [photo of insanely small tire in wheel well] +1

Dohn Roush That would make one circumspect…

Brian Aftanas Out of curiosity, I checked Tire Rack website to see what / if tires are actually available. Surprised there were 5 options including snow tires… all with UTQG numbers bigger than I've ever seen. Nice. +1

Guy Foster I HATE it when that happens!

Bill Hiddle Glad I'm not the only one!

Robert Myles I've had a starring role in that same show as well… +1

Scott Linn Wait, aren't they the same diameter? The PO of mine put 205's on. They did sometimes hit the body. The 185's at all four corners almost never hit the body. I thought it was fixed for good but found some circumstances where they hit. Not a big enough issue to go to 165's or staggered.

> **Bill Ecker** I think it's more a function of the aspect ratio than the tread width, based on the solution.
>
> **Brian Aftanas** This one, on auction at Mecum's next weekend in Monterey, is wearing 15" wheels. Interesting. [link to Mecum's]

Hans Batra FaceBay: Great storyteller, humorous, witty and honest. Would do business again, highly recommended. A+++++ +1

David Weisman 😊

Aaron Cranshaw That'll learn ya

Paul Hill Phew, I was thinking… WHAT!!

Tom McCarthy Personally I like easy fixes. That didn't happen to stop the shaking when you swapped them back did it?

> **Rob Siegel** Nope, no such luck. +1

Rob Siegel I'm taking the car and its wheels and tires in for a proper road force balancing this morning. We'll see if it helps. +5

> **Justin Gerry** Rob Siegel I wonder if these are like some BMWs that are super sensitive to slightly bent wheels. +1

Dan McLaughlin I've done similar things... had switched to a staggered setup on a Z4 years ago... I had one side switched front to rear and the other side correct. Still won the autoX challenge!

Clyde Gates Where can you find 13 inch tires these days?

> **Rob Siegel** Tire Rack has several. eBay and Amazon have more.
>
> **Brian Ach** Kumho Solus, the go-to, cheap, not that bad tire for lightweight vintage cars
>
> **Rob Siegel** I should've spent the eighty bucks or whatever it was more, maybe not even that much, and bought the Kumhos, but I don't know that tires are the problem.

Josh Wyte That's a lot of balancing weight on that wheel. Despite them not appearing to be bent, I bet they are. +2

Brad Purvis As much as I know you would like to be a Borg, it's being fallible that make us human. +1

> **Dohn Roush** Persistence is futile... +1

Rick Roudebush That is a shit ton of weight on that wheel as Josh ^ has said. Maybe something is wrong with that tire or wheel that it "needs" that much weight. I wonder if you have been pranked. Someone stick those weights on while you were doing an errand as a joke? I never add weights to the outside of a wheel for appearance. If there is a heavy spot in the tire that needs that much weight, that could be what is causing the shimmy. You might try breaking the tire down and spin it 90 degrees on the wheel, re-inflate it, remove all the weights and balance it again, placing the weights on the backside of the wheel so I don't feel so itchy when I look at it next time. ;-) +1

> **Rob Siegel** See upcoming post. Just had the wheels and tires road-forced balanced. +1

Bill Mack How did force balance go?

> **Rob Siegel** It went well. I'll write a post on it tonight.
>
> **Bill Snider** Tonight? Tonight? We have to wait `till TONIGHT? Our collective minds will explode ;-) +3
>
> **Rob Siegel** mwaHAHAHAHAHA!

Brian Ach Don't you love it when a huge problem is made small by your own moron-osity? I wrote the book on it. +1

Geoff Davis Hellaflush

Wade Brickhouse It must be night somewhere.

JT Burkard Amazing how the wheel well is so tight to the tire that an increase in one size up causes your tire to rub. That is some tight clearances.

TLC Day 2257: Road force balancing

AUG 14, 2019, 6:29 PM

The Lotus Chronicles, Day 2257:

Took Lolita to Lindsey Brown at The Little Foreign Car Garage in Waltham for a road-force balancing. This is where, for each wheel and tire, the springiness of the tire and the roundness of the wheel are measured, and the machine tells you the optimal "indexing" (rotation) of the tire relative to the wheel that will produce the best roundness and balance prior to adding the weights. All four vintage Cosmic wheels were remarkably straight. The inexpensive Achilles tires I'd bought weren't all that bad, and many of the ridiculous number of weights that several people commented on in the photo in yesterday's post were shed once the pairs were properly indexed. Only the right rear wheel and tire could not be brought into spec as well as we would've liked.

When I took the car on the highway to test it, I found that while can still feel vibration, probably telegraphing in from the right rear, there's no question that it's much better—no longer scary when I cross 50 mph.

So, $211 well spent. Thank Lindsey and TLFCG!

130 LIKES, 18 COMMENTS

Ed MacVaugh Did they put the big tires on the rear? +7
Scott Linn Out of my 5 original alloys, one required a bunch of weights. That's the spare... +1
Frank Mummolo Good judgement to know when you need an expert "expert"! Well done! Now you can unlock the next level of mystery in this car! 😀! +1
Brian Aftanas Are you going to roll the dice by buying a fifth Achilles tire for the right rear?
Jim Gerock Now it "Just needs a buff."
Daniel Neal RoadForce Balancing is the only way to go +2
Thomas Jones Alignment? +1
> **Rob Siegel** It needs that, no question, but it's not obviously pulling or crabbing. Rear trailing arms were out of the car and re-bushed (by me). Front end is tight but tie rods and ball joints have no boots left on them. I have the parts. I'll probably do the front-end work over the winter, decide if I want to also do new shocks at the same time ($$$), and get it aligned in the spring. +2

David Kiehna Rob Siegel what was the MG Lotus dealer in Waltham in the mid 70's?
> **Jonathan Selig** New England Lotus owned by Jeff Brown?
>
> **Lindsey Brown** Jeff Brown owned Ultima, which was a Lotus franchise. High Performance Cars was a British Leyland, Alfa, Maserati dealer.
>
> **David Kiehna** I bought locally here in Germantown, TN a '74 MGB from the original owner and also a few years later her late husbands Europa.
>
> **Jonathan Selig** Lindsey Brown ... did you happen to know Joe Sampson, their chief mechanic at the time.? They rebuilt my '71 2002 after it was recovered following theft from Allston in '75
>
> **Lindsey Brown** I used to work with Joe. Sadly, he's long gone.

Lindsey Brown Colin Chapman was the Prince of Lightness. +1
Paul Fredendall So, I learned from that. Tks. +1
Paul Hill Getting close to not needing that spare pair of underpants. Close but not quite yet...

This is how you know you've arrived at the right repair shop.

75 LIKES, 30 COMMENTS

Rally Mono I love that place. +1

Richard Shouse I'm impressed! Must be a good shop. All that Lucas-ness sitting there and no smoke at all! +3

Scott Aaron Love those MGB GTs

Dohn Roush Hmmm. Why are there so many of them?

Jon van Woerden Love that MGB GT. Had a 69 MGC GT back in the early 80s. +1

> **Rob Siegel** MGBGTs are still pretty cheap. The mini-shooting-brake lines have held up very well. +2
>
> **Jon van Woerden** Rob Siegel Yes but I think I would like another MGC GT but then again I'm looking for a Dutch 1802 Touring. +3
>
> **Wade Brickhouse** I had an MGC GT too! +1
>
> **Jon van Woerden** Wade Brickhouse Awesome speed but didn't turn great at speed.

Dan Tackett Is it Morris' Garage?

> **Rob Siegel** The Little Foreign Car Garage in Waltham MA
>
> **Dan Tackett** "The company had its roots in the Morris Garages (MG) company begun in the 1920s." [link to Wikipedia entry on MG Cars]
>
> **Charles Morris** Definitely not my garage :D I had a '72 MGB for one year in '88 and got really tired of having to fix it every weekend before I could have fun. Mostly Lucas electrical, of course.

Rick Roudebush Electrical gremlin convention? +5

Trent Weable I had a 73 Midget back in the 80's and there was a shop that I took it to that sold bumper stickers that said "The parts falling off this car are genuine British Leyland." +1

Kent Carlos Everett Comforting to know that even you need a little help occasionally. I am currently stumped by a problem with my roadster. Humbling.

Chris Raia Twin of my '62 MGA MK II! Except less rodent goo, I imagine.

Rupert X Pellett Brittle.

Rally Mono I rebuilt a 76 Midget about 17 years ago.

Ed MacVaugh Everyone of them properly backed into place. Either working clutches and reverse gears, or strong muscled folks working here. +1

> **Rob Siegel** They'd have to be REALLY strong-muscled. It's uphill from the roll-up door to this spot. +1
>
> **Sean Curry** Rob Siegel Well, they each only weigh as much as a big sandwich. :D

Cameron Parkhurst Was driving to Portland, Maine from Boston a few years ago and stayed off the highway and came across a place called Brits Bits. Old English car heaven. +1

> **Robert Myles** Cameron Parkhurst Brit Bits is a really cool place. I stopped in once to ask about a TR6 they had for sale and the owner gave me a solid 30 minutes of his time educating me about the breed. +2

Cameron Parkhurst [photos of Brit Bits]

JohnElder Robison The Rolls-Royce and Bentley version of that scene [photo of JE Robison Service]

Tim Dennison British auto repair = job security 😀 +1

TLC Day 2261: Fun things

AUG 18, 2019, 8:46 PM

The Lotus Chronicles, Day 2261:

Fun car thing #1: I was moving around cars so I could drive the 3.0CSi after it's been sitting for months in the garage having its a/c rebuilt. I had the Lotus parked in the driveway, practically hanging its tuchus out into the street. I went up to the 3rd floor to take a shower and change into clean clothes before parking MY tuchus in the 3.0's beige Recaros. The 3rd floor bathroom window faces the street. I'd just stepped out of the shower when I hear, in a strong Jamaican accent, "Oh mon, dat's a LOTUS!" I look outside to find an Amazon delivery van stopped in front of my house and a guy checking out Lolita. I wrap myself in a towel and stick my head out the bathroom window and have a brief conversation with the guy. "Eye'll com back, ya know, some other time, eef dat's okay?" he asks. "Anytime," I said, "I'm usually around."

Fun car thing #2: I bought an MG tachometer on eBay to use while I try to fix the unique one that's in the Lotus. I swapped messages with the guy from whom I bought it. Turns out he's also a BMW guy, races an M3 at NHIS and Thompson Speedway. He says he has some other Lotus engine parts including valve shims, which as it turns out, I need, as I should readjust the valves after a few hundred miles. He says to call him. I do, and I, the native Long Islander, can tell a New York accent in the first five seconds. I call him on it. Turns out we're both nice Jewish boys from New York with a predilection for German metal and Brit bits. He says he'll gladly swap valve shims for a signed book.

What could be better to round out a weekend?

235 LIKES, 24 COMMENTS

Tom McCarthy Are they gazing longingly at each other? +1
    Rob Siegel I think something resembling grudging respect. +4
    Joshua Weinstein The coupe respects Lotita? I find that hard to believe. +1
Tom Melton Jamaican not know it's "Lotush"? +1
Russell Baden Musta Bring all your released CDs to LOG, COD Sir! Need music for the Evora! +1
Rob Siegel Yes sir! +1
Russell Baden Musta Rob Siegel I'm serious! Glad I remembered to ask! +1
Cameron Parkhurst 3.0CSi is staring down the interloper that has emerged from the depths of your garage. +1
David Kemether I can only assume that everyone did the Jamaican accent in their head when reading that. I know I did. +9
    Rob Siegel The best Jamaican encounter was when I'd parked my silver Euro 635CSi, came back to find two guys admiring it, and one said to the other "Dat's OLD school Beemah, mon!" +7
    Lenny Napier Yup, I did lol
    Eric King Rob Siegel Stir It Up! [link to article on Bob Marley and his BMWs] +2
Bo Gray Jah +1
Robert Krantz Now, that's a good weekend! +1
Robert Krantz Also, so nice to see beautiful, elegant chrome bumpers facing each other. +2
Rob Blake A Klezmer band concert maybe? +1

Jeff Dorgay Saw a nice green Elan driving around today and thought of you instantly! +1
Gary York Cool running +2
Gary York Many English cars found their way to Jamaica (former English colony)
Lindsey Brown No park im in dem bumbleclot driveway deh, mon. +1
Mel Green Can you say Splif?
Christopher P. Koch So, you've been banished to the third floor for showers?
Brad Purvis You need an Alfa back in your life. +2
Roy Levine EVERYONE needs an Alfa back in their life. +2

TLC Day 2266: Heading to LOG

AUG 24, 2019, 7:01 AM

The Lotus Chronicles, Day 2266:

Do you know that classic scene in "King of the Hill" where Hank Hill rents a backhoe, drives on to his front lawn to dig up his septic system, and says "my entire life has been leading up to this moment?" Yeah, it's kind of like that.

Lolita is about to attend her official coming out after being off the road for 40 years. We are headed for the Lotus Owners Gathering (LOG). Hopefully the next photo you'll see will be Lotii (Lotuses? Lowtushes?) in the parking lot of the event hotel, and the one after that, the concours at Thompson Motor Speedway in Connecticut, and not Lolita on a flatbed.

I'd say wish us luck, but I know you are :-)

315 LIKES, 55 COMMENTS

Gail Stewart Good luck!

Carl Witthoft What's the Lotus equivalent of "break a leg"? +1
    Craig Fitzgerald Carl Witthoft Throw a rod +11
    Wade Brickhouse Blow a fuse
Frank Mummolo Fair winds and following seas…
Larry Filippelli It's been fantastic to follow this story over the years! Good luck! +2
Scott Winfield Woo hoo!!!
Russell Baden Musta Love it!! Welcome Rob!!! Photos to follow!!!
John Turner It's "LOG"! by Blammo! +1
Phil Morgan Awesome! Enjoy it my friend! +1
Levent Erhamza Safe travels!
Lee Shumway Congrats Rob 👍 🏎 +1
John Whetstone This is great, but we'd better not lose you to the other team! +2
C.R. Krieger You got this, pal, whatever that ridiculous tub throws at you. Well, you and a VISA card, anyway… +1
Dave Gerwig Let the good times roll! Congrats! +1
Harry Bonkosky Have fun!
Dave Borasky Are you sticking a banana in it for good luck? +2
Chris Roberts I know it will be a fun adventure come what may, but with Lolita in such great hands you will take good care of each other +2
Gary Apps Are you just doing this for an excuse to write books?
Norma Coates I saw one of these on the Mass Pike yesterday and thought of you. Enjoy! +2
Marc Schatell I'm excited, so I know you must be!
Hugh Bernard Love the King of the Hill reference +1
Jake Metz Wahooooooo! Enjoy the journey, go fast, safety third, etc +1
Richard Shouse Enjoy life two inches off the pavement!
Susan Rand So exciting! Congratulations!
Doug Jacobs Ride fast, take chances!!!
Andrew Wilson Long may you run. +2
Dohn Roush As Chapman never said, "Add lightheartedness!" +1
Scott Lagrotteria You got it down Rob; have a great time showing her off & schmoozing +1
Richard Shouse and try not to bump your knees on the roof… +1
Larry Wilson Excellent job, Lotus daddy :-)
Gary York Live long and prosper
Chris Pahud Onward! [painting of… I'm not sure how to describe this… a waifish blonde girl, in a white dress, with a snowy owl perched on her forearm, riding, an elephant, clutching a lotus blossom in its trunk]
Joseph Hower Good luck!
Philip D. Sinner The British debutant presented! Big fun!
Franklin Alfonso Did you ever find the 5th gear? +2
Kevin Pennell "Luck is not a factor," but sheer determination, well that's just a standard quality we all know and love about you, Rob +3
David Weisman Love King of the Hill.
David Weisman Rob have you ever been to Epping speedway?
Bob Sawtelle I have no doubt you will arrive with a smile on your face and a bare foot.
Chris Lordan Congrats Rob! Well-earned… well-earned for sure. Do the more recent cars—the Elise, the Evora, etc—get any love from the Lotus folks? Or is it like the BMW people who sniff and any M3 that's not S14 powered?
Marc Bridge You got this, Lolita! Be a good girl. 👍 +7
Bob Shisler Count the number of guys who offer to do a cut n buff on it, or at least bring it up. (That's one.)
Chip Chandler Get it on! Good luck!
David Holzman Drive fun!
Daren Stone Eagerly awaiting updates.

Steven Bauer Enjoy the rewards of your efforts.

Rob Capiccioni Awesome!

Lance Johnson Hey Rob I'm driving at Thompson today. I've seen several Lotuses (Loti?). But no Lolita :-) +2

Jeffery W Dennis Great fun.

Maire Anne Diamond Don't forget your other shoe—and I'm not referring to the M Coupe. +2

Rob Halsey That is a shit eatin' grin if there ever was one. Hope you have a great time. +1

Francis Dance I was instructing today at Thompson with SCDA. Sorry I missed you. Enjoy the event!

Geoff Davis It's ten thurtee and already the coolant ain't raht. +1

Sean Magee Doesn't make me feel so bad about the 924 we've spent 3 years trying to get running and safe, it's only been off the road for 12 years! +1

TLC day 2266: Arrived!

AUG 24, 2019, 8:07 AM

The Lotus Chronicles, Day 2266 (continued):

In the words of a great Englishman, yeah baby.

282 LIKES, 62 COMMENTS

Russell Baden Musta Awww!!! We are right round the corner eating!! Back shortly!!!

Jim Strickland Beautiful success!

Kenneth Yee Made it!!! Have a good time!

Mark Thompson Lotus Land...

C.R. Krieger Yay! Half way ... 😊

Kevin McLaughlin [link to Austin Powers "Yeah, baby, yeah!"]

John Graham Where are the tow trucks? Hahahaha! Had to make one joke. Looks really great sitting with its fellow expats! +4

    George Zinycz They wait until the afternoon to arrive so as not to ruin the vibe of the meet. +2

Bob Coffey Yay!

JT Burkard [gif "Congratulations! You survived!"]

Doug Jacobs THEYS A HO LOTTA LEAKIN GOIN ON!!! +5

Rupert X Pellett "I'm all sixes and sevens, Basil..." +1

Dohn Roush That's a whole lotta low...

Christian Mulero Thought of you [photo of beat-up Europa]

Tom Melton Look! Lotushes!

    Franklin Alfonso That's Low-tuches pronounced with a Sean Connery accent... +2

Brian Aftanas Wow! Three Europas in one place. I am not sure if I have ever seen one "in the wild." +1

    Wade Brickhouse A gathering of parts cars?

    Jeffery W Dennis Brian there's actually 4 in the top photo it's like where's Waldo.

David Weisman Lotii

Eric Gerber Oh, lord. Stuck in Lotii again! +4

Kieran Gobey Awesome

Hilton Joseph Sexy cars.

Matthew Rogers You made it! Hallelujah!

Andrew Wilson That's your bag, baby, yeah! +3

Jeffery W Dennis Is this a club rally and track day? There was a meet a few years ago at Barber Motorsport very nice. I always wanted a Seven or Europa or the Elan 130 S2, ended up with ALFAS instead.

Jackie Jouret So now that you're in the groove, please inform of the proper plural of "Lotus." Is it Lotii? Lotae? Lotuses? Or simply Lotus, using the same form for singular and plural like "deer"? +3

    Rob Siegel "Lotii" is both joked and actually used.

    Jackie Jouret I like it.

    Rob Siegel The president of Lotus USA spoke at the dinner last night and said "Lotii." I call that more than just tacit approval. +3

    Nick Wood Jackie Jouret I like Lotususes. +1

    Jack Fletcher Rob Siegel but he's probably a Yank...

    Clint Carroll Jackie for the plural of "Lucas" electrics you simply add an "s"... +1

    Dohn Roush Clint Carroll Actually it's Lucasssss, like the sound of an electrical short... +1

    Don Bower and the collective noun: Lotta

    Greg Aplin Rob, I've always said Lotii and I also say Lexii. Too much Latin beat into my head in college choir. Our choir director wanted us to understand what we were singing as well as pronounce it properly.

    Sean Curry BOTH! :) [link to grammar-monster "plural of Lotus"]

    Matt Schwartz Jackie Jouret I used to be very active in the Lotus community before I got my 02 (we have an Elan). Back then, when this question was asked, one answer that always came up was "the plural of Lotus is just Lotus."

    Charles Morris That only works if it's a fourth declension noun... if third, Lotes would be correct :D

    Sam Pellegrino Jackie Jouret it's almost as In-Crowd-like as "Bimmer, Beemer and Beamer." We got our own way of walkin, we got our own way of talkin'...

    Rick Roudebush I have it on good authority a gathering of Loti is called a Leaks. ;-)

    Brad Purvis Sam Pellegrino Let's not go there...

Henry Noble Gives me the chills like fever and impending doom.

Joseph Oneil You win a prize!!! It's in the mail!

Eric Evarts Congrats, you made it!

Tim Warner Yay!

Duncan Irving There's something about that Cortina. I guess enough with the square cars though. Awesome back story and presentation Rob! +1

Blair Easthom [photo, t-shirt, "Lucas, Prince of Darkness: A gentleman does not motor about after dark"]

Marc Bridge So how Lolita stack up?

> **Rob Siegel** Easily the rattiest car there, but well-appreciated
>
> **Jack Fletcher** Rob Siegel a single step...
>
> **Franklin Alfonso** Oh, shut up Iron Butt! lol...

TLC Day 2266: Lolita pukes at the prom

AUG 24, 2019, 10:37 AM

The Lotus Chronicles, Day 2266 (continued):

Well, I made it to the concours at Thompson Motor Speedway in CT, and the drive here on local twisty road was spectacular, but it took so long waiting in line to get into the park that the car overheated and overfilled the coolant overflow bottle, resulting in a steaming engine hatch, a big puddle, and much pointing. We'll see if it's okay after it cools down.

149 LIKES, 34 COMMENTS

Wade Brickhouse I see potential parts cars around you! 😄 +1

Neil Bradbury I spy a front wheel drive...

Stephen Timko There's a guy near me with two of those under tarps in his driveway. +2

Kenneth Yee At least she is among her peers now!

Dave Gerwig Lolita just got a little excited at the Coming Out Party in her honor 😄 +4

Tom Melton Premature eja... +3

John Whetstone Tom Melton Seems to have gone the distance!

Franklin Alfonso So, A normal British arrival then? +5

Ed MacVaugh You look to be parked next to John Collier, the guru, in the dark blue Europa. +2

Rupert X Pellett Lotus pray. +2

Jim Tennessen Rob leaves a mark

Frank Mummolo So you have managed to resurrect this car without compromising its core (you'll pardon the pun) character. Brilliant, Rob! You rock! +1

Thomas Jones Is there a chance that water pump just doesn't flow enough at low speeds? Or, do you have enough electric fan?

Jeffrey Miner I can't believe that radiator can't do the job. How many bottles of Water Wetter have you added? I put that crap in everything...

Paul Fredendall They weren't pointing, they were saluting on of their own!

Eric Pommerer "Functioning within normal parameters." +1

    Robert Myles Eric Pommerer aka "Inherent in this design."

Dohn Roush AKA, "They all do that..." +1

Doug Jacobs Man, I just want to say again, Bravo. With the time and energy you put into this car you could have whipped out 11 e30's, but all those e30's wouldn't have meant as much as this Europa. I know I like to give you a hard time to balance out all the fan boy praise you get, but this time I really mean it from the heart— you put it all into this car and I'm very impressed and happy for you for it. Congrats! +11

    Rob Siegel Thanks Doug! +1

Bob Shisler At least the trail looks oil-free. ;) +3

Gary York Were you the only Grease Monkey?

Rick Roudebush After all these years of captivity it is just a little excited to be among others and is just marking its territory. +4

Mike Hopkins Love to see it out on the road!

Andrew Wilson Sounds wonderful and normal. +1

Henry Noble Her entrance was well-noted no doubt.

    Rob Siegel No doubt!

David Holzman I hope you have no trouble on the way home

Jack Fletcher Citric acid flush... Check for compression leaks and then add Water Wetter

Josh Lyons A good car has to leave its mark!

Ken Doolittle Just like any old guy after a long drive... you just gotta go.

TLC Day 2266: We make it home and I put her to bed

AUG 24, 2019, 7:36 PM

The Lotus Chronicles, Day 2266 (addendum):

Well.

I've put 400 miles on Lolita since her resurrection in early summer. 150 of those were today—out the Pike to the Lotus Owner's Gathering (LOG) event hotel in Sturbridge, then sinewy back roads down to Thompson Motor Speedway and back, and then home on the Pike. Initially I hugged the right lane like it would save me from something, but by the time I was heading home, I was blasting down the left lane at 70, then checked the speed on my phone and found out that the car's speedometer reads 5 mph too slow, THEN realized that, in addition, the car's speedometer seems to stick at 70 and found my phone reporting that I was going nearly 80. Previously, the furthest I'd driven the car in one spurt was maybe 30 miles while I afraid to take it over 50 mph due to wheel balancing issues. So, yes, this was a very significant trip, and Lolita and I can both claim to be total badasses.

Lolita, EASILY the rattiest most patina-laden car there, was warmly received. A surprising number of people came up saying they'd been following the story in the weekly pieces I write for Hagerty or here on FB, and how cool it was that I'd rescued the car, gotten her running, and brought her. While the fact that this was her first real road trip IN FORTY YEARS was a great story, any number of people told me about similarly long lay-ups of Elans and Europas. Passion seems to be highly communicable with these cars.

The Lotus people at the event were great. You never know when you come to a new one of these things whether you will find "your people," but these totally were. Engaged. Passionate. Scrappy. Funny. Friendly. And you know what else? I NEVER HEARD ANYONE TALKING ABOUT HOW MUCH THE CARS WERE OR WERE NOT WORTH. God I fucking loved that.

The business of Lolita puking coolant while in traffic in hot weather does not appear to have harmed the car, but it's troubling. Many people chimed in with theories. The best one came from Tom Radcliffe, coincidentally the husband of a woman I was a fellow physics major at UMass with (Elizabeth Brackett). He verified that the cooling fan was spinning the right direction, and that the fan blades were oriented the correct way, but noticed that a lot of air was bouncing back and not being transmitted through to the other side of the radiator, which is another way of saying it's a cheap shit Chinese-made fan and not a high-quality fan like a Spal. So, homework and probably purchase of a decent fan.

Lolita has had my focus since New Years, and I have a never-ending list of things to do with the other cars, so I may run her out to Fitchburg soon and swap her for one of the other vehicles for a few weeks. The fact that she has reached the point where I can hop in her, twist the key, pull her out of the garage, and do that is simply astonishing. So, although I hadn't given this a moment's thought until now, the trip out to LOG and back really might mark a significant closing of a chapter with Lolita.

As Randy Newman said, "My life... is GOOD."

[search YouTube for "The Lotus Chronicles driving to LOG"]

172 LIKES, 34 COMMENTS

Marc Schatell This seems very satisfying +1

Scott Winfield Mazel tov! +2

Thomas Jones So much awesome-sauce! +1

Dohn Roush Looks like a fun road. Congrats on another successful resurrection! +1

Tom Smith I think I am going to start describing my complexion as "patina-laden." Sounds better than "liver marks" or "old man skin." I can see the personal advert already, "vintage man, excellent patina." +5

>**Dave Borasky** Tom Smith pukes coolant? +2
>
>**Tom Smith** Dave, stopping short of that. +1

Dave Gerwig Quite a milestone. Glad all went well. +1

Susan Rand What a wonderful reward for your dogged persistence!

John Wanner Inspiration. Through and through.

Carl Witthoft That's odd about the fan. I guess ICE cars do tend to have fans forcing air into the heat exchanger? We mech engineers know fans are more efficient when dumping into zero resistance, so ideally a fan would pull air thru the exchanger. That's how roof vent fans, for example, work. Personally, if the airflow sucks, I'd blame the exchanger fin structure rather than the fan, but I'm certainly not a car-fan-guy. +2

George Murnaghan Congrats, Rob, this whole journey has been very inspiring! +1

William Jeffers If it overheats in traffic, the fan is the likely culprit. +2

John Thomas Had similar coolant barf from the 635 early in the summer. Wound up being a resistor on the a/c fan. +1

Rennie Bryant Doing what God and Colin Chapman intended. +1

Bob Sawtelle "My Lolita Twin Cam does maybe 85, I registered her in Vermont, lost my license now I can't drive" +2

Scott Aaron Huh. Huh-huh. You said "Log" +1

Andrew Wilson Awesome, Rob! +1

Francis Dance Congrats 🍾 on another milestone

Russell Baden Musta Rob, you being here just completely made my day, So very happy you made it!!! So I've now seen YOUR Europa in the wild, Welcome to the group!!! +1

> **Rob Siegel** Thanks so much for your friendship and advice, Russell. I'm not sure I would've come without you gently egging me on, and I'm so glad I did! +2

Franklin Alfonso Rob Siegel did you finally find the fifth gear on the shifter?

> **Rob Siegel** Franklin, Yes. I was doing the lean-against-the-detent-until-it-pops that most folks do with reverse and 5th, and that works with reverse on this car. Someone commented that, for 5th, you really need to do more of a wind-up-your-wrist-and-smack-it-against-the-detent motion. Sure enough, that works reliably, though it's not pretty.

> **Russell Baden Musta** Rob Siegel appears you made some friends too so that just was icing on the cake! Just thank you coming!!! +1

> **Rob Siegel** Russell, Yes, totally. The people were awesome. Clearly "my crowd." +1

> **Russell Baden Musta** Rob Siegel I thought you'd fit right in!! Great group as I'm certain your people are as well! +1

Tom Egan That car ahead of you seems exceptionally well driven. Handsome driver too! +1

> **Rob Siegel** Yes. The driving was so precise, the risks so well considered, that I swore that Niki Lauda himself was at the wheel. [obviously that's Tom's car] +1

Henry Noble I loved following this Lotus adventure.

> **Rob Siegel** Thanks Henry!

Daren Stone Following along as Lolita became a driver and you became a Lotus person has been one of the most enjoyable stories I've read in a long time. Thank you and welcome to the Lotus family. +1

Bill Snider Thanks Rob. A truly great tale shared expertly. +1

Mark DeSorbo Can't wait for the book about the Lotus, Rob, even though I've been reading every post.

Lee Shumway Nothing like the visceral experience of older cars. Windows (or top in my last experience) down, engine revs freely, and that huge smile and feeling of freedom… and then there are people who have never experienced it, and they have no clue… poor souls. +1

Chapter XI: Life After LOG

TLC Day 2267: To Fitchburg

AUG 25, 2019, 7:34 PM

The Lotus Chronicles, Day 2267:

Eventful day. Drove Lolita out to Fitchburg and put her away for a bit, making it the first time she'll sleep under a different roof since she came to live here in June 2013. I don't want to make too much of the idea that, after the car's lengthy resurrection, the burst of activity in the spring and summer, and yesterday's successful 150-mile drive out to the Lotus Owner's Gathering in Sturbridge, driving her out to Fitchburg (and BEING ABLE to drive her out to Fitchburg) and putting her away represents the closing of a chapter. I'm only doing it because if she's here, I'll keep working on her, and I need to turn to other projects. She'll be back soon. I miss the batshit crazy girl already.

Speaking of other projects, drove Kugel back to Newton. Why? Stay tuned...

107 LIKES, 24 COMMENTS

Phil Morgan Had the pleasure of following you in Kugel out of Hot Springs 3 years ago in my e28 with Molly Morgan riding shotgun. That is a sweet car. +2

Brian Stauss Something about that Europa backed in the garage and looking back out that I love about it... +2

Neal Klinman Kugel? Did she come with that plate?!? I love kugel. Or is it kiggle?

   Rob Siegel Neal, cars in Massachusetts don't come with plates; a new plate accompanies a new registration. I got the vanity plate KUGEL for her. +1

Robert Myles It could be "KEGEL" but this is a family show. +4

David Weisman Neal Klinman 😂😹😹 +1

Neal Klinman Rob Siegel I just didn't see you as a vanity plate kind of guy. Now I'm curious about the origins: does it handle like wet noodles, a reference to a previous owner, pale in color, or? +1

Brian Hemmis I love kugel. Your car is cool too. +1

Alexander Wajsfelner Neal Klinman: https://en.wikipedia.org/wiki/Kugelfischer

Kevin Fernandez Bring the clownshoe home...

Rob Siegel Oh, the shoe has been in the driveway for a few weeks. I've been driving the shit out of it. +6

Dohn Roush Rob Siegel Wouldn't it be easier to just vacuum it out? +2

Wade Brickhouse Rob Siegel don't break anything! 😀

Dimitri Seretakis Kugel looks a lot cleaner, Did you buff out the paint?

Rob Siegel Had the entire car dipped in carbon tetrachloride. Highly effective :^)

Dimitri Seretakis Rob Siegel So did you grow a third hand?

Rob Siegel (seriously... yes, I did clean it, but I didn't buff it; just washed it)

Dimitri Seretakis Rob Siegel Paint looks shiny!

Dave Gerwig Looks like Fitchburg may have had some much-needed roof repairs this summer?

Rob Siegel yup, that's all done +3

Richard Baptiste White tiis damn things get under your skin... +1

Kent Carlos Everett In the time I have been chasing down a vacuum leak, you got that thing on the road. Congrats! +1

Scott Linn The siren song of the Europa is strong... +1

Scott Aaron I know. Even though I don't know. I think I know. What's going on w Kugel, that is. +1

Ed MacVaugh I like garage space that comes with wheel chocks. +1

Dave Parenti She looks sad and lonely +1

Brian Ach Awesome. Well done. Great car.

TLC Day 2298: Retrieving Lolita

SEP 24, 2019, 6:00 PM

The Lotus Chronicles, Day 2298:

"Rob Siegel, you've just pushed 125 personally-inscribed copies of your new book out the door, filling the backlog of pre-orders from your loyal readers. What are you going to do next?"

"Well, it's the sweet spot of fall and I should be driving my vintage cars, and at the house, I have the 3.0CSi I can't drive because I still haven't put the console back in it after the air conditioning work this summer and wires are still hanging out everywhere, the '72 2002tii I can't drive because I cleaned it all nice and pretty and then sold it, and the '73 2002 with 48,000 miles I can't drive because I haven't registered it yet. So..."

"I'M GOIN' TO FITCHBURG!"

118 LIKES, 46 COMMENTS

Scott Linn Drove mine to work the last two days. Always a fun ride. +1
John McFadden I should have such problems 😊 +1
Benjamin Shahrabani Can I still get a personally inscribed copy?
    Rob Siegel Absolutely; just go to www.robsiegel.com and click on the books you want.
Henry Noble [gif Homer Simpson "Oh Lordy!"] +3
Jake Metz Looooootus +1
Maire Anne Diamond I have insider information that he didn't get there. Bertha threw a fit at being put back in the box and all. +16
    Maire Anne Diamond He's back, she's sulking in the driveway after being towed home. +3
    Maire Anne Diamond The real irony here is that, before he left, I said "Don't break down, I've had a long day and I'm tired." Only joking of course. But feeling like I jinxed the trip! +5
    Rob Siegel Outed by my own wife. Boy, talk about having no secrets :^) +3
    John Whetstone Rob Siegel Wouldn't have been a secret for long, Soon would have been a column headline +5
    Susan Brown Strickland [gif Stephen Colbert "I TOLD you so!"]
    Susan Brown Strickland Maire Anne Diamond your grace under fire inspires me. Sadly, I'm not there yet! +1
    David Holzman Maire Anne Diamond Poor Bertha. She probably wants someone to wash her.
    David Holzman Rob Siegel You are very VERY lucky to be married to her.
    Rob Siegel To Bertha? :^) +3
Rob Siegel Lindsey Brown, remember when you described prophylactically replacing an ancient electric fuel pump with a new electric fuel pump because it seemed to be the smart thing to do, and the new one died shortly after, and you put the old one back in, and I had JUST done EXACTLY the same prophylactic replacement of a 30+ year old electric fuel pump in Bertha with a new one, and I read your post, and THOUGHT about putting the old one back, but DIDN'T? Well, funny story... +5
    Paul Wegweiser Not sure I'd trade installing a tii starter (TWICE) for that experience. Note to self: even modern batteries showing 13 volts should have the water/acid level checked periodically. I am a sucker. +3
    Rob Siegel Yes, Paul, today, we are both big doofuses. +1
    Paul Wegweiser Rob Siegel I guess the good news is: I discovered a cracked brake fluid elbow on my master cylinder during all this tomfoolery and no longer have brake fluid dripping on my 3-year-old fresh frame rail.
    Rob Siegel so, as Carl Spackler said, you got that goin' for ya +2
    Paul Wegweiser Rob Siegel [gif of strange sad-eyed creature "I need cuddles"]
    Lindsey Brown Works with condensers as well. +1

Franklin Alfonso Feel so sorry for your first world problems Rob… 😂 Life is a bitch when you own all those cars and don't have enough garage space to keep them all at home and handy… sad. Maybe if you did some bodywork on that rusty hood she would be happier 😂.

Bob Sawtelle Hmm the day the book comes in the mail she throws a hissy fit because you were banishing her to the hinterlands of Fitchburg…Bertha wanted to bask in the limelight a little longer me thinks. +1

Rob Siegel I think this is exactly true +1

Rob Siegel In my defense I had her at the house all week and have been driving her around quite a bit +1

George Zinycz Rob Siegel they get used to being treated well +1

Russell Baden Musta Take the Roper!!!! Had mine out for a drive tonight!! +1

Rob Siegel that was the plan, man +1

Russell Baden Musta I just don't get tired of this view! Still pinching myself!

+2

Paul Wegweiser With a Boston accent, it's pronounced "Ropah" anyway, right? +3

Russell Baden Musta Paul Wegweiser but Ropa in UK tends to come out Roper! +1

Scott Linn Russell Baden Musta at work yesterday (and today): [photo, Scott's Europa in the parking lot] +2

Russell Baden Musta Scott Linn I think that one was at LOG… that's the original color of mine, and taking it back to that soon as I can appropriate funds!

Scott Linn Russell Baden Musta that's mine and it didn't go to LOG. I'm in Oregon. Regency Red. +1

Russell Baden Musta Scott Linn mine was regency red too..I repainted to this red in 1985… seriously thinking of changing back, putting the sills back on… and stripes… looks rich!! You'll laugh… I put new front tires on last week on oem wheels… local small garage spun balanced them and they are better than they've ever been!! Front is glass smooth now!! Unbelievable!

Scott Linn Nice to have new tires correctly balanced for sure. The pin striping was pretty easy to put on. When I bought it the PO didn't like the stripes, but I did. I couldn't find the right width pinstripe, or it was super expensive. I found one the right width and it was only $6/roll (2 rolls required for the car). When it arrived it wasn't a single strip as advertised, but two thin ones that together were the right width. I decided to put it on anyway, again because it was so cheap. I think it looks good even 8 years later. Some of the best $12 I've ever spent on a car for sure. +1

Russell Baden Musta Scott Linn never can beat that on an old Lotus! I'm original owner of mine… love the car. I'm very lucky to have 2… never thought I'd do that!

Russell Baden Musta I know the guy who does all the striping for Classic Lotus… I'm thinking of bribing him to hand paint mine!! But more likely tape…

Miguel Francis Rob you and your first 🌍 problems 😂 🤦 +2

Clay Weiland "Off the [Tow] Hook: Adventures in Breaking Down" I smell a new book in the works. +1

David Holzman Rob Siegel my mechanic had a beater second generation Taurus wagon. That car was (quite appropriately) called Bertha. I'm not sure how you can use that name on a 2002. +1

TLC Day 2299: Lolita gets LOUD

SEP 25, 2019, 6:40 PM

The Lotus Chronicles, Day 2299:

Made attempt #2 at driving Bertha out to Fitchburg and swapping her for Lolita, since, during attempt #1 last night, Bertha threw a hissy fit and withheld fuel pump functionality on me. Removed the four-month-old electric fuel pump and replaced it with the one that'd been in there over 30 years until I replaced it before the drive down to The Vintage this spring. Works just fine. So much for prophylactic maintenance.

Swapped cars. Drove home in Lolita, windows down, cool fall weather, livin' the dream. Then... was this exhaust ALWAYS this loud? Then... NASCAR LOUD. Did my chimney flashing and hose clamp muffler repair get overwhelmed? Did the headpipe crack off from the muffler? No. The new old stock asbestos gasket between the manifold and headpipe failed. Got home, went on eBay, and bought the only other one in existence so I can continue driving the addictive temperamental little brat through the fall. #HackLife

110 LIKES, 27 COMMENTS

Luther K Brefo Make a pattern of the replacement before using it. ;) +4

Russell Baden Musta Love it!!!

Paul Wegweiser You HONESTLY didn't believe I was going to let "temperamental little brat" go unnoticed, did you? Oh look! A BROWN ONE!

+26

    Rob Siegel you put that and Lolita together and you get two cars with normal-looking rear ends +3

    Christopher Kohler Two cars with the spare tires up front. +2

    Collin Blackburn Paul Wegweiser I need that for reasons +2

    George Zinycz Paul Wegweiser it looks mad as shit

    Bill Gau Now THAT is a creampuff!! +2

Bryan Lepley I can make you more, non-asbestos gaskets if needed. Just need the dimensions... On the house... +2

    Rob Siegel Bryan, I may well take you up on that, thanks.

    Bryan Lepley Rob Siegel please do! Your help from your books have more than paid these gaskets. I run a waterjet company, so no problems... +1

Neal Klinman Sounds like you should just make some and sell them! Someone's gonna need them, and Rob just got the last one. +1

Chris Mahalick Well, you put on their asbestos you can. +5

Rick Roudebush I had constant header gasket failure on an old 351 Ford stroker motor I built years ago. Got tired of replacing gaskets and ended up using orange high temp silicone gasket maker on both sides of new gasket, let it sit for a day before starting it and voila, no more leaks after that. +1

Isabel Zisk Is Bertha's license plate attached with zip ties? If so, she deserves better! :) +1

Rob Siegel Yes it is, and yes she does. A product of being about to get her inspected out in Fitchburg last month, realizing that I had to put the front plate on, looking in the trunks of all of the cars for what I could use to find the plate, and, voila! Zip ties to the rescue! +4

Bob Sawtelle Rob Siegel apparently they forget the "Hack" part of The Hack Mechanic... +4

George Zinycz Rob Siegel put the fn zip ties on the fn car...

Richard Shouse I hate it when prophylactics fail. +5

George Zinycz NOS... accent on Old

Andrew McGowan Those old school asbestos exhausted gaskets are such garbage. My memories of high school era maintenance on my TR4 is chock full of trying to deal with these.

George Milonogiannis Amazing description of my life

Mel Green Geez, when my Dinky Toys got rusty and temperamental, I "got rid of them." +1

Brad Purvis [link to National Mesothelioma Claims Center] +1

TLC Day 2300: Stripped studs

SEP 26, 2019, 1:24 PM

The Lotus Chronicles, Day 2300:

I just wanted Lolita back at the house for a few weeks during this glorious fall, but as I wrote yesterday, the asbestos exhaust gasket between the headpipe and the exhaust manifold shredded on the drive home. They're long since NLA. I put the only NOS one available on eBay order last night. Today I thought I'd drop the exhaust, pull the broken piece of the gasket out, and at least snug the two halves up close so I can move the car without it sounding obnoxious.

I snapped off one of the three studs on the exhaust manifold.

Well. Fine. This exhaust is a goner anyway, with the muffler having multiple holes patched using chimney flashing and hose clamps. The muffler is available and not terribly expensive, but the problem is that the intermediate pipe connecting the muffler to the manifold is NLA, and it and the muffler are rust-welded together. The solution is a new set of headers that bolt directly to a new muffler, but that adds up. The headers, as they always do, replace the exhaust manifold and the intermediate pipe. I hate to spend the money, but clearly the time has come to, as they say, pay the man. So I call Ray at RD Enterprises, order the headers, muffler, clamps, and while I'll be living under the back of the car, I also order the seals for the transaxle and the tool to change them, thinking that I've just mapped out several weeks of a winter project. Ray says that I've just, in fact, bought the last muffler he has, and that he'll have to tell the next person who calls that they're now on backorder. It's a non-trivial bill, but I'm glad to be THIS guy and not THAT guy.

Again, all I want to do is get a few weeks out fall pleasure out of Lolita, so I run down to Autozone and buy a tube of J-B Weld Exhaust cement, thinking that two of the three studs should be plenty to pull the two exhaust flanges to within sealing range. I lay down a big ugly bead, fit the headpipe's flange over the two remaining studs, and start to tighten the nuts down.

The threads of one of the studs strip out.

Maybe this new exhaust won't be a winter project after all.

I do look forward to both a new tight exhaust and a transaxle that doesn't mark my driveway.

#TheJoysOf45YearOldCars

39 LIKES, 27 COMMENTS

Ernie Peters The new beginning of another never-ending story. +4

Russell Shigeta She's definitely a fighter… just not fighting *for* you… 😊 +4

Frank Mummolo I have no doubt the worst is yet to come…which is why I am a retired wire wheel owner of many years.

Bill Ecker Don't you, as SOP, deep-soak any exhaust hardware with SiliKroil for at least 24 hours before touching them?

    Rob Siegel Often, yes, but this was a flange that I had reinstalled just in the spring.

Steven Hussein Bernstein Ironically, this meme was posted by another car friend, appearing RIGHT after this thread… Cosmic. :) [meme: "Not everyone will understand your journey. That's okay. You're here to live your life, not to make everyone understand." +9

    Frank Mummolo Steven Hussein Bernstein so true. On many levels!

    Bill Ecker Add: You're not the idiot whisperer. +1

    Bill Howard I just want to understand if left blinker means left turn, or grandpa changed lanes 20 minutes ago. Freakin' John Glenn, second time around, made three complete orbits with the turn signal on. +1

Wade Brickhouse That's what happens when you use those knock off Elbonian "Snap Off" tools! 😊 +1

Jeff Dorgay I've got an E30 with an exhaust situation similar to yours, so winter project it is as well. +1

Richard Shouse Be Ant Anstead and just whip up your own header and exhaust. Only took him 10 minutes or so on tv… 😎 +4

Bob Crownfield And they say that a boat is a hole in the water into which you throw money. They told me never do a spreadsheet, but I did. It cost $38 to put the keys in the ignition. If you turned it on, it went up. +2

Daren Stone Been there, broken that. +4

Brian Aftanas To provide a bit of encouragement and to state the obvious, cars CAN be sorted out. It is the application of logic, identifying those places were weakness existed in the original design and construction and addressing them. The vicarious pleasure of reading about your work is how you apply this logic. I just returned from a 660 mile trip on my 1983 Ferrari 308 GTS QV. Fifty miles of that was at 6,500+ rpm racing up the Virginia City Hill Climb. It is a car not with a good reputation (although immune from Lucas jokes). I fix the car myself, it is now well sorted and it ran flawlessly. It annoys me how peoples' maintenance expectations regarding my car are so different from the realty. I admire the good spirit of British car owners as they seem to enjoy the bad reputation as much as they loathe it. Enjoy your awesome little buggy. +1

    Franklin Alfonso Everything on cars can be sorted with a high limit Mastercard... +2

Chris Mahalick The header will be totally easy. I hope you ordered the stainless clamps. Also, start hitting the manifold studs with PB Blaster now, so the nuts will come right off. How about you hammer through the exhaust for a weekend, and save the trans seals as your winter project?

Rob Koenig Just imagine how boring your life would be if any of this was easy. +2

Kenn Sparks Splendid story Rob! Knowing what's ahead, I am—only for the moment—delighted to be me and not you! 👋👋 +2

Thomas Jones I'm not seeing the copper anti-seize paste I'd expect you to have applied the last time you assembled those exhaust nuts... +3

    Rob Siegel yeah. no. not this time.

David Holzman the patches were hosed

Brad Purvis I find it interesting that at the beginning of fall things seem to start breaking at an exponential rate just so we have enough (too much?) to keep us occupied until spring. +1

David Kiehna Should have sent the exhaust parts out to be blasted and jet hot coated while the motor was out. A simple $100 universal muffler could have been used to recreate an OEM like exhaust from the header back. Would have been a great time to upgrade the flanges to a modern clamp or v band style connection.

Chris Longo You know... whenever a stock exhaust manifold fails on me I just order a turbo manifold... but I'm not exactly "right" in the head. 😊

Bob Shisler Every 10-minute job is a broken bolt away from a three-day ordeal. +5

    Steven Hage Bleeding the brakes on my Tii turned into a month-long ordeal lol +2

TLC Day 2301: Muffler cement

SEP 27, 2019, 9:30 AM

The Lotus Chronicles, Day 2301:

My J-B Weld Muffer Cement kluge on Lolita's exhaust is holding, which allows me to keep driving the car through this glorious fall weather. As I say in my book Resurrecting Bertha, "One of the main ways I enjoy my cars is to have them at the house so I can be able to hop into one of them and run out to pick up a gallon of milk at Trader Joe's on a Sunday morning. It's the automotive equivalent of smiling across the breakfast table at your spouse. Not everything is mind-bending back-arching nail-clawing sex in a Paris hotel room with a view of the Eiffel Tower, nor does it need to be."

88 LIKES, 44 COMMENTS

Rob Capiccioni Great analogy! +1

Der Alte Classics Very true. Try to take a classic whenever weather permits. From getting groceries to going on vacation. Nothing beats the satisfaction of pulling into your driveway after returning from a two-week 2000 km vacation (and visiting that Paris hotel room while you're at it) with a classic car. +3

Al Navarro To springboard off of your analogy... while I love my Seven, I can only really drive it when I'm in the mood. I have to be okay with feeling a little beat up afterwards. Which does not make it the best car to commute to work in. +1

Pam Loeb I used to work for a doctor that had a car collection. He had an old Ferrari (this was around 1989, and the car was older then) that he would drive whenever he was on call and had to go out in the middle of the night... In good weather, he also drove it on the Saturdays that he had office hours (it was a group, everyone took turns for Saturdays). +1

Jim Strickland That's living well!

Rob Blake Hey Man ... Here's an offer you might NOT want to pass up... Drive out here to scenic Germantown, NY [167 miles from Newton], (10 minutes from Bard College, 15 minutes from Hudson, Just across the river from Catskill) I'll feed and house you and direct you on the most amazing Catskill Mountain foliage drives at the height of the fall colors...all I ask in return is the shotgun seat in Lolita. +3

    Rob Siegel Rob, appreciate the offer. Not sure Lolita is ready for a drive that long yet. +2

    Rob Blake Well... you can take that "Community Chest" card put it in your back pocket and cash it in anytime when you're ready. +4

    Rob Siegel Thanks Rob! +1

Josh Lyons I am in complete agreement, though, to be honest, it may be partly maturity setting in! I think I've been able to drive mine everyday this week! +1 [photo of Jaguar]

    Andrew Wilson Careful with that... [link to SNL "Hitchhiker"]

    David Schwartz Nice vanity plate on one of my favourite jags. [plate says "ROAR"]

Kevin Pennell Always enjoy your insights, Rob. Looking forward to our copy, soon to arrive!

John Thomas Not a bad grocery getter... +1

David Kiehna Their corn puffs are pretty good.

Susan Brown Strickland Stop being a bad influence on Jim Strickland...now he will never take me back to Paris. He'll just pull out random quotes from our collection of your books! +5

    Jim Strickland Susan Brown Strickland if I'm influenced by Rob's songs, I'll keep loving you 'til I get it right +2

Paul Hill Shoe on. Shoe off. Trousers on. Trousers off... +2

    Paul Wegweiser Paul Hill ...rubber chicken codpiece on... lollipops in the ears...and bullfighting music. +3

    Paul Hill Paul Wegweiser ah the olde chicken up ya fandango.. +3

Dohn Roush So that's why they called it the breadvan... +1

Marc Bridge Wow! Needed a cigarette after that Paris hotel reference! 😊 +3

Andrew Wilson Your post reminds me of this song... [link to Bow Wow Wow "Sexy Eiffel Tower"] +3

    Susan Brown Strickland Andrew Wilson Bahahaha...I'm so glad I wasn't the only one who thought of this! +1

    Andrew Wilson Me too!

> **Rob Siegel** Just listened. I've never heard this! Thank you for making my morning! I shall think of this whenever I drive Lolita!

Bob Sawtelle Hmm remember something in Memoirs about British cars: "The Triumph was a miserable piece of garbage. I never bought another British car (proving that men, I fact can learn)" +4

Doug Jacobs Well, it _could be_. If you'd buy that TVR.

> **David Schwartz** Listed in the October edition of the British Marque which came out today. FOR SALE: 1988 TVR 350i. Two owners, LHD, very rare. CT registration. Yellow / black, V8, 5-speed, convertible. Great condition. $17,500.
>
> **Doug Jacobs** David Schwartz No, shit no. He needs an old, hoary TVR. Either that, or a thigh belt with spikes. +2

Mel Green YOU paint an interesting and provocative picture… +1

David Schwartz Seeing my 1968 Mini or 1950 Morris Minor in the parking lot at work always puts a smile on my face. +2

> **Rob Siegel** as well it should! +1

Geoff Davis Bust a Nutella +1

Dave Gerwig Oh Maire Anne Diamond. Just need one more grocery getter, I see the strategy now! +1

Maire Anne Diamond Dave Gerwig I'd send him to Costco but the Lotus is space-challenged. +2

Dave Gerwig Maire Anne Diamond space no problem, will just have to make a few more trips! 😀 +1

JT Burkard I prefer waking up pantsless in a $40 hostel in Chatou overlooking a polluted section of the Seine River. The story becomes so much more interesting. +1

David Weisman Ahhh suburbia. I miss that.

Rennie Bryant Charged up the battery on my Midget today for this exact reason.

Scott Chamberlain At this point in my life, I'd trade ALL my cars for even ONE night of back arching nail clawing sex, based on what I remember of such things. +1

Scott Chamberlain Hell, I'd settle for one night in Bangkok. [link to video "One Night in Bankcock"]

> **Rob Siegel** Hadn't thought about THAT one in a while!
>
> **Scott Chamberlain** Thinking about a night in Bangkok is no longer something one does in polite society.

TLC Day 2330: Retrieving Lolita, again

OCT 26, 2019, 4:29 PM

The Lotus Chronicles, Day 2330:

After I got back from cars and coffee at Larz Anderson, I made a snap decision, ran Hampton out to Fitchburg, and picked up Lolita. Why? 1) Because Hampton now can do that, and 2) Let's face it, because I CAN.

92 LIKES, 11 COMMENTS

Thomas Jones Best part, because Lolita can! +2
Paul Wegweiser Ooooohhhhh...so THAT'S what a 2002 is like below 5000 rpm! I had no idea! +12
    Rob Siegel you're such a dick +4
    Paul Wegweiser Rob Siegel A dick having every bit as much fun as you... and for the same reason! Lily is an absolutely sublime and civilized car! I may or may not have just driven it illegally to get groceries. Around town, I may prefer it to the F Bomb! +6
Franklin Alfonso Lolita summoned you eh... bewitched you are lol.
    William Jeffers Did they ever fix the roof?
    Rob Siegel Yes! +2
Scott Aaron Will start referring to you as "Nigel"... you are starting to identify as British.
    Jim Denker Scott Aaron : Clive sounds good too.

Tom Egan Good choice! +1

Franklin Alfonso Lord Robert to all you plebeians...

TLC Day 2344: Fuel level and heat

NOV 10, 2019, 5:43 PM

The Lotus Chronicles, Day 2344:

With other automotive events holding center stage, Lolita has been out of the spotlight the past few weeks, but I've been driving her every chance I can get, and doing some very satisfying small repairs.

I got the fuel level sensor working, which is great because it helps eliminate the possibility of running out of gas and feeling like an idiot.

And I got the heat working. It has a heater valve internally sprung to be on unless the cable shuts it off. Initially I just zip-tied it off. Then when the temperature dropped, I removed the zip tie, figuring that I'd just leave the heat on and moderate it by opening up the windows. Bad idea. MASSIVE amounts of heat. So I bought the little barrel connector I needed to secure the cable to the valve at a hardware store. The friction of the cable wasn't quite enough to overcome the internal spring in the valve, so I added a little helper spring to pull it closed. Voila: controllable heat. It's funny how, with the car's exotic looks, slot-car handling, and WAAAAAAAAH waaaaaaaah acceleration noises, I'm so excited about it doing normal car things like having a working gas gauge and controllable heat, but that's the way it goes.

I hope today's drive out to Natick for my friend Mark Thompson's 60th birthday wasn't the car's last for the season, but it may well be, as I need to lay it up on the lift for a while to do the front end, adjust the valves, and tweak the cooling system, and as snow can fall any day, I need to start getting the cars where they need to be to overwinter.

As my mother would say, a happy problem.

168 LIKES, 39 COMMENTS

Sean T. Going Great to see you and finally meet this girl today. Thanks for the nickel tour!
    Rob Siegel You too, Sean! +1
Marc Schatell And pretty satisfying, I'm thinking
    Rob Siegel Oh, hugely.
Russell Baden Musta Put about 80 miles on the Evora today... should have taken the Europa! +1
Franklin Alfonso Don't worry Rob, I'm sure there are other ways you can feel like a fool besides running out of gas lol.
    Rob Siegel truth!
    Franklin Alfonso That's a nice photo of the mini hearse.
Josh Isenbarger Man love that color. It was nice here today too. Took my e46 out for spin.
Franklin Alfonso That's British Racing brown man... +4
    Rob Siegel I describe it as "monkey shit brown," but I say it with love +11
    Franklin Alfonso At least it's not pink or mustard...
    Brian Hemmis You've never seen the Mary Kay Europa? +1
    Carl Baumeister Actually, "MSB" as it is affectionately called looks quite nice on the car when shiny! +2
    Greg Calvimontes It looks great in that light. Overcast OTOH... 😀
    Stan Chamallas Rob Siegel brings new meaning to Captain's log! 😀 +3
    Peter Potthoff Rob Siegel Clearly haven't spent much time with monkeys. It's more of a putrid green... +1
    Rob Siegel "The more you know..."
    Peter Potthoff Spoken as the caretaker of an animal colony during my Ph.D. studies...
    Rob Siegel I shall in no way contest your monkey shit color analysis skills! +2
    David Weisman Stan Chamallas 😀
Garrett Briggs How much time do you spend calculating the date on the Lotus Chronicles? And does anyone check your math? Hmmm... +1
    Rob Siegel https://www.timeanddate.com/date/duration.html
    Garrett Briggs That's cheating. +2
David Schwartz The greater Boston weather forecast for Tuesday night is 16 degrees with a 50% chance of snow. This was a good weekend to play with the fun cars. I wore a ski hat and gloves so I could drive my Morris Minor with the top down. Sadly the destination was a gas station to top off the tank and add fuel stabilizer before putting it away for road salt season. +3
Mark Thompson Great to see you...Love the Lotus! +1
    Rob Siegel Woof!
    Mark Thompson Rob Siegel Tuna!
Frank Mummolo Could not be more NOT BMW. Absolutely love it! +1
Alex Lipowich I recall my '87 Range Rover... no heat for a while... then found a weird vacuum hose loose. Plugged it into what seemed the closest available nipple. BAM... it was a vacuum-operated valve to direct something in the right direction. Truck long gone, but many lessons learned! +2
Richard Shouse Leave him alone... it's Steve McQueen brown. [photo of Steve McQueen's brown Ferrari] +2
Scott Linn Ha ha! Correct on the heater blowing you out of the car. Massive heating ability. And I always thought my Midget was good... +2
Paul Wegweiser I totally feel this. 6 hours ago Lily became a member of the working speedometer and gas gauge club, too! [photo of Lily] +6
Steve Park Few cars can pull off that color. I think the Europa does quite well. +2
Robert Alan Scalla I put the snow tires on the 318ti today. +1
Brian Aftanas Wow! It's really sparkling in the sun. Did you polish it ... or is that just the shiny corner?
    Rob Siegel It's almost just an optical illusion. It's really not a shiny car.
    Mike Miller I can fix that.

Chapter XII:
The Winter Projects

In the interests of trying to bring this book in at sub-*Gravity's Rainbow* length, I'm going to omit the posts from November 18th 2019 through March 22nd 2020 and instead summarize the activity.

As I headed into fall 2019, Lolita was starting to feel almost like a normal car I could simply hop into and drive. There were, however, five projects I'd planned for the winter: Rebuilding the front end, upgrading the cooling fan, replacing the transaxle seal, adjusting the valves, and replacing the cracked windshield, which was the one I didn't get to (it's still cracked). Collectively, these comprised a pretty substantial second round of sorting out. Much of this work was written up for Hagerty and posted online in great detail and with color photographs, so if you want more of a step-by-step how-to (as opposed to the original Facebook posts which were "holy crap how much else can go wrong here), my weekly Hagerty pieces can be found at https://www.hagerty.com/media/author/rsiegel/.

Front end rebuild

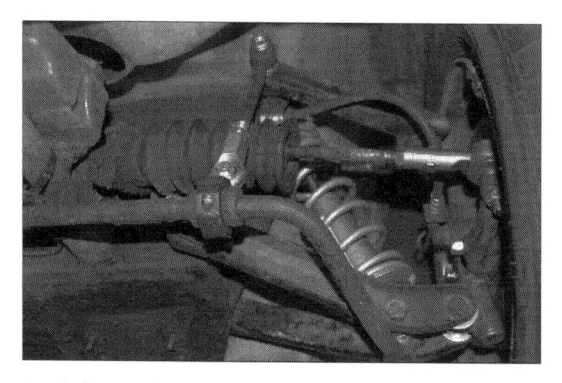

Fresh front end components.

The front end rebuild was fun and enormously satisfying. There were four separate goals: A general refreshment to replace any worn parts which might be contributing to the car's continued front-end shimmy at speed; lowering the car's nose from its silly boat-on-plane height caused by federal headlight height specifications; replacing its original shocks which were so soft that they were barely functional, and replacing a broken front sway bar link.

I completely disassembled the front end, replaced the ball joints and tie rods whose rubber boots had deteriorated out of existence, and rebuilt the trunnions. I planned to replace the bushings in the upper and lower wishbones, but I could not press them out without beginning to bend the no-longer-made thin metal wishbones. I read that the required action was to burn out the rubber, then use a hacksaw blade to cut through the metal sleeves. There are four wishbones on each side, with bushings at each end, for a total of 16 bushings. On close inspection, none of the rubber looked or felt deteriorated, and rather than facing endless hours tediously sawing something out that wasn't obviously bad, I left them alone. My car, my decisions. As I said in *Memoirs*, if you go into the cave, look the beast in the eye, and are not prepared to battle it to the death, sometimes the thing to do is to back slowly out of the cave.

I looked at the price of a new front and rear adjustable suspension setup (shocks with adjustable damping plus lowering springs on adjustable perches), but the $1300 was too, um, stiff for my taste. I found a used set of probably 35-year-old Spax shocks with adjustable damping but fixed spring perches on eBay for $225 shipped, and bought them with the understanding that I could return them if the shocks were blown, seized, or the adjusters didn't work. I had to build a spring compressor to disassemble them (the coils on the springs were so close together that no spring compressor I could find would work), then manually verified that the shocks themselves appeared to be fine. I did a lot of reading on the Europa forums and learned that 10-inch front springs were the prevaling choice for dropping the nose to a visually-appealing height. The Europa is a very light car, and the engine is in the middle as opposed to directly over the nose, so the replacement springs don't need to be very stiff. I decided to try a set of QA1 125 lb/in springs—not a brand generally used in the Lotus world, but available for $80/pair on Amazon. I reused the stock rear springs on the newly-procured used adjustable Spax shocks.

The lowered front end necessitated shortening the sway bar links. Since one link was already broken, I sawed them both, threaded them each to a Heim joint, and fashioned a clamp around the sway bar.

Cooling fan

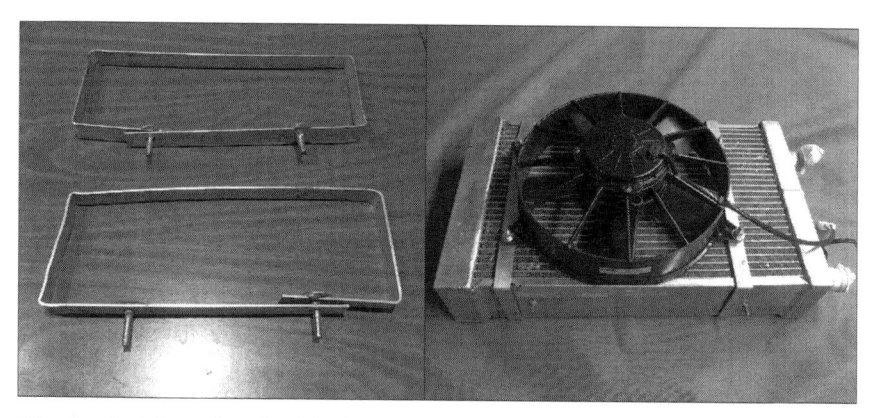

The new Spal fan and my hand-built mounting brackets.

To replace the anemic Chinese-made fan that came on the aluminum Chinese-made radiator I'd bought, I sourced a proper Italian Spal fan, the most powerful one available in a 10-inch size. Mounting it, though, was challenging. The mounting brackets for the original fan were welded to the radiator, and didn't fit the new fan. I cut them off and fashioned a pair of rectangular straps that went all the way around the radiator to distribute the weight load. Once installed, I wired the fan with a relay to accommodate its larger current load.

Transaxle seal

The finned transaxle nut that cracked when I foolishly tried to press the seal out the wrong end.

The right-side seal of the car's Renault 5-speed transaxle had a pretty good-sized leak. I bought the seals and the $80 tool to remove the finned nut into which the seal is pressed, but unfortunately, when I was pressing out the seal, I tried to push it out from the wrong end, and cracked the finned nut. They're not carried by RD Enterprises or Dave Bean, but a web search revealed that two European sites had them. I ordered it from Renault16shop.com in Denmark. However, there was a complication. The tightness of these finned side nuts sets the play on the pinion and the pre-load on the bearing, and unless you're going to go through an arcane process of measurement, the procedure is to mark the finned nut and the case, count the number of turns required to take the nut off, and tighten it back to exactly where it was when you removed it. If you use a different finned nut, all that goes out the window owing to nut-to-nut variations in the amount of space on the face of the nut before the threads start to bite. I read a lot, settled on installing the nut finger-tight plus just a skosh, and waited to see if it whined or howled on acceleration or deceleration.

Valve adjustment

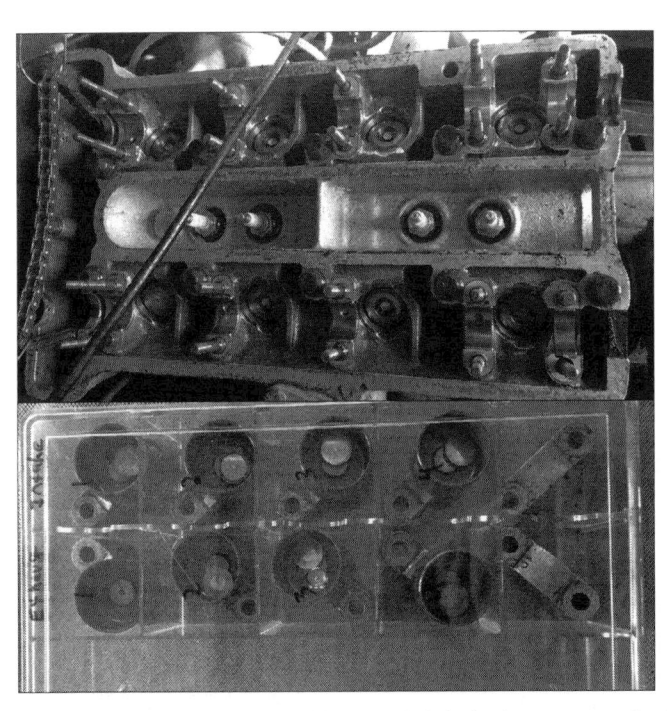

The cams removed from the twink engine, and the bucket tappets and valve shims carefully placed in an organizing container.

The Twin-Cam engine uses shims below the bucket tappets to set the valve clearances. Thus, valve adjustment consists of these steps:

- With the engine dead cold, carefully measure the valve clearances with feeler gauges, and record the clearances in a table.
- Rotate the engine to top dead center (TDC) and remove the camshafts.
- Remove the bucket tappets and valve shims, putting them into a labeled organizing container so you're absolutely certain which one goes where.
- Measure the thicknesses of the valve shims and record their sizes in a table as well.
- Calculate the size valve shims you need to obtain the desired clearances and order them.
- Place the new shims over the valve stems.
- Reinstall the bucket tappets and camshafts.
- Measure the resulting clearances to see if you got it right.
- Lather, rinse, repeat if you didn't.

I'd never done a shimmed valve adjustment before, so I did it very carefully (there are three pieces about it on Hagerty online). Soup to nuts, it took me three weeks, at the end of which, the car refused to start. This turned out to be because the jackshaft—the in-block camshaft left over from when the Ford block was used in a pushrod engine—turns the distributor, and although the distributor itself hadn't been moved, the gear on the front of the jackshaft is driven by the same timing chain that spins the cam gears, and in the process of removing and reinstalling the cams, the tugging on the chain had moved the jackshaft gear.

There were also odd events regarding oil and coolant. I checked the oil level prior to draining it pre-valve-adjustment. Although it appeared correct on the dipstick, the volume of oil in the drain pan seemed small. I poured it into containers to measure it, and it was almost two quarts low. The oil also appeared really thick and black. I sent a sample into Blackstone Labs, who said that the consistency was probably due to the large amounts of Graphogen assembly paste I'd used. After the valve adjustment, I filled the oil filter and installed it, then carefully measured the correct 4.75 quarts that should go into the pan, ran the engine, checked the dipstick, and it was pretty close to the original mark. I never really figured out the root cause of the discrepancy, but sometimes, like in *Shakespeare In Love*, you just need shrug and say "it's a mystery."

The coolant issue was jarring. When I started the car, there was a cloud of sweet-smelling white smoke. I shut it off and saw visible green coolant in the tailpipe. This seemed to indicate a badly-cracked head or blown head gasket. However, I also saw a small puddle of antifreeze on the floor under the exhaust side of the engine. I bought a Stant cooling system pressure tester, pumped up the cooling system to 7 psi, and found several leaks, including ones from the hose and heater valve into the thermostat housing that's a cast part of the head. As unlikely as it seemed, it appeared that the coolant was running on the underside of this housing, down the side of the head, and weeping past the gaskets on the exhaust flange to get into the headers. I eliminated the leaks, ran the engine to bake out the coolant, and the problem never recurred.

Whew!

Chapter XIII: Emerging Into Covid-19

TLC Day 2478: Nose looks good

MAR 23, 2020, 8:04 AM

The Lotus Chronicles, Day 2478:

Yesterday, after I finally fixed Lolita's no-start problem, I let it warm up, then got it out of the garage, pulling it off the mid-rise lift in the back where it's been for three months for its three major winter upgrades (front end rebuild with lowering springs and adjustable shocks all 'round, bigger radiator fan, and valve adjustment). It swapped places with the E9, making the Lotus the car that's now directly behind the garage's single roll-up door (although having the E9 there allowed me to take some delightfully unexpected winter drives in it). The nose of the car does look settled down from the Federal-spec boat-on-plane tilt it had before. I'd hoped to begin running around the block to dial in the alignment, but today looks cold, with rain and snow moving in. I'm thrilled at all the progress over the winter months, but still, it's probably a poor choice of vehicle if Maire Anne, Ethan, and I need to escape the viral zombie apocalypse. [Note: I later wrote a piece for Hagerty about this; search for "Rob Siegel Hagerty vehicles apocalypse"]

[search YouTube for "The Lotus Chronicles engine start after warm up"]

100 LIKES, 32 COMMENTS

George Zinycz Yes, a shitty choice indeed. The brown should clue you in on that lol

Scott Winfield But really; wouldn't you rather be eaten in a Lotus?

> **Rob Siegel** It should be their new marketing slogan! +3
>
> **Richard Stanford** Oh myyyyy!

Richard Memmel I wanted to take the 02 out this morning only to wake up to snow 🎃🎃 +1

Clay Weiland From the video, it starts and idles better than any carb'd car I've ever owned. +2

> **Thomas Jones** 'Cause you and carburetors don't much agree. 😬
>
> **Clay Weiland** Thomas Jones correct. That's why all my cars are on Atkins. I do appreciate and envy the learned, old-school practice of laying your hands on mechanical devices and maintaining their operation through sensory examination and articulate manipulation of adjustment hardware, but EFI provides a shortcut to reliability and efficiency. My current resources lean me in that direction.
>
> Rob Siegel Atkins, the diet? I thought that shit went out with the new millennium.
>
> **Clay Weiland** Rob Siegel limiting carbs still works for me, at least in the garage. +7
>
> **Rob Siegel** Okay, believe it or not, I just got that :^) +7

Bill Ennis Purrrrfect!

George Murnaghan But now aren't the vestigial jackcamshaft lobes improperly timed for their theoretical valve openings? This could upset some delicate force causing unknown consequences. Because British, you know. +3

> **Rob Siegel** yes, the phantom pushrods are all fucked up +1

Daniel Senie Perfect timing having the Lotus as the accessible vehicle. It's going to snow this evening.

> **Rob Siegel** DONUTS! +2

Rob Blake Are you KIDDING!!! It's the perfect car for the GREAT ESCAPE... doff long purple scarves flowing out the window while wearing gold silk N-95 masks... and some strings of tin cans dragging behind you as well, just for the hell of it. +1

Wade Brickhouse With all the brains they eat you would think zombies would be smarter than that! :-)

Chris Mahalick When I did the front end on mine, I had the same deal. The lowering springs and shocks don't lower it as much as you would expect. So that tells me you did it right. +1

Russell Baden Musta I'm so fucking proud of You!!!! Car looks GREAT!! Love the stance!! Once you start driving it checking alignment, curious how self center is working.. once I went tubular (after my shunt) made all the difference being able to adjust the caster... but can be done stock. +2

> **Rob Siegel** I can't wait to drive it. It had the original 46 year old shocks in it, which, bouncing each fender of the car, felt functional but very soft. When I pulled them out. one rear shock was borderline non-functional. Now it has those used vintage Spax adjustable shocks (adjustable damping, fixed spring perches). For starters I set them all to the middle setting. I can now push down each corner, and I feel it give, but I also feel the fiberglass in the body give! +1
>
> **Russell Baden** Musta Rob Siegel yeah, don't sit on the front fenders!!!! Ya don't need it too stiff... it's softly sprung... +1

Brian Stauss Well done, Rob! And thanks for all the updates!

Paul Wegweiser Last night, just to get out of the house, Wendy Burtner and I went for a 20 minute drive... to nowhere... in A SUBARU FORESTER... WTF is wrong with me when I "choose" to take an SUV on a relaxing drive instead of a 2002? I'm hoping sunny warm weather will re-program my brain... maybe next week. +3

> **Rob Siegel** Clearly you're infected and must be put down. +2
>
> **Paul Wegweiser** Rob Siegel Or sent to some kind of "re-education" camp. +1
>
> **Dave Borasky** I once visited a re-education camp in Viet Nam. They didn't have any cool cars.

Charles Gould So what was causing the "NO Start" issue?

> **Rob Siegel** The timing had moved because the distributor is driven by the jack shaft (the internal cam shaft left over from it being a Ford pushrod Block), and my jostling of the timing chain had moved it.
>
> **Charles Gould** Ah, that makes sense. So, I was on the right track when I suggested that the distributor may have stripped, shifted or rotated.
>
> **Rob Siegel** You totally were. I wasn't against resetting or re-checking the timing; I just didn't understand why it would've moved. The Europa factory manual doesn't say a thing about resetting the timing as part of the valve adjustment procedure, but the procedure references the procedure for removing/reinstalling the cams, and in there there's one line that says "Check ignition timing." I found it because there was a single post on the Elan forum from one guy who

had it happen. I supported the timing chain on two pieces of a coat hanger, but I guess one of them drooped enough that when I pulled it tight, it spun the jackshaft gear.

Charles Gould Rob Siegel That makes sense. Let me clarify. What happened makes sense. The design doesn't!

TLC Day 2479: First short drive after winter projects

MAR 24, 2020, 11:40 AM

The Lotus Chronicles, Day 2479:

First short drive in Lolita after a winter's worth of work:

--Rebuilding the front end, including new front lowering springs, ball joints, tie rods, trunnions, and adjustable shocks at all four corners

--Replacing a leaky side axle seal in the differential

--Replacing the radiator cooling fan

--Adjusting the shimmed valves, including solving the vexing no-start problem caused by the internal jackshaft having moved the distributor timing

--An in-the-driveway alignment this morning to center the steering wheel and get the toe-in good enough to drive

She's BAAAACK!

[search YouTube for "The Lotus Chronicles first short drive after winter work"]

146 LIKES, 74 COMMENTS

Harold Simpkins Sounds wonderful! +1
Dave Gerwig It's puuurrrrrring again. Fantastic! I'm happy for you!
Rob Anthony Does "get the toe in good enough to drive" refer to alignment or physically being able to get both feet on the pedals? 😎 +7
    Rob Siegel The former is "toe in," the latter is "foot in" :^) +3
Brian Aftanas Congrats. Enjoy. That is one cool little car you have. +1
Linda Bernstein Congratulations! Enjoy!
Jim Strickland Sounds and looks great! What dashboard mount do you have? Very steady video.
Jim Denker Your OCD is paying dividends today!
Paul Wegweiser I love the fact that it sounds like you have a pissed-off house fly attached to your camera. Perhaps glued to a piece of thread? Also: I'll have to make a similar video of me in my freshened-up Subaru Forester. It will be thrilling for my viewers, I'm sure. +9
    Rob Siegel Position the camera over the head gaskets so we can see them leaking in real time. +11
    Rob Siegel (*ducks*) +3
    Ed MacVaugh Rob Siegel That is a myth perpetuated by Porsche enthusiasts who don't want anyone Japanese swapping their precious cars. +4
    Paul Wegweiser Ed MacVaugh Nope. It's true. ALL of them fail... until you replace them with higher quality gaskets. Mine seeps oil for sure. +1
    John Turner Paul yes, please
    Lee Highsmith Rob Siegel [gif of two mudskippers fighting] +1
    Marc Touesnard Paul Wegweiser, almost every tourist I see in my area of Nova Scotia drive around in their Subaru is wearing a Tilley hat and has roof racks on it for the kayak they never carry.
    Tim Warner Paul Wegweiser I loved the housefly sounds...
Scott Aaron Sounds good. It looks great, too, with the nose lowered. Nice nice nice. +1
Michael McSteen Well done. She sounds great. What is Lolita's redline?
Rob Siegel I believe that the redline on the Europa TCS with the "Big Valve Twin Cam" engine is 6750 rpm.

Jeff Cowan Okay, okay... but what's the driving experience like?

Rob Siegel The Europa is a weird, crude, loud, buzzy little 1600-pound bit of a thing. It's challenging to get into and out of because it's so low. The pedals are so close together that I literally need to remove my right shoe every time I drive it because otherwise when I hit the accelerator pedal I also catch the brake. It's a cable clutch with a grabby disc that makes take-off a little challenging, especially when it's not fully warmed up. The seats are permanently tilted back. It's un-BMW-like in every way possible. And mine is pretty ratty. It sat for 40 years, from 1979 'till last summer when, after the 6-year slumber in my garage and engine rebuild, I finally finished the engine rebuild and got it running and sorted enough to begin driving it. It still needs a thousand things.

I love it.

Driving it is raw and direct. Saying "it handles like a go-cart" is such a cliché, but… it handles like a go-cart. Sneeze and you'll pull it into a phone pole. And couple that with its 42-inch-high roof line, it's just more fun than anything involving wearing pants. And—not that I care—but it generates attention like nothing else I've ever owned. From EVERYONE. Young mothers with strollers. Other guys my age. Of course every teen-age boy. You'd have to land in the space shuttle to make more of a scene.

I'd love to concentrate on de-buzzing it. This is the layers-of-the-onion nature of vintage car resurrection. This round of winter repairs peeled several big layers. But driving it is an absolute hoot. It's addictive. I could say things that will get me in trouble, so instead I'll say see the chapter in Memoirs of a Hack Mechanic titled "So Why DO Men Love Cars Anyway?" +18

Jeff Cowan Brilliantly said. What great fun!

Benjamin Shahrabani Rob Siegel have you thought about modifying the pedals to avoid your issue with catching the brake?

Rob Siegel Benjamin, yeah, people bend the clutch and brake pedals to the left. It's on my list. +2

Robert Myles Rob if you think a Europa handles like a "go kart" you really, REALLY don't want to drive an Elise. Think "atomic go kart on steroids." Every time I drive one I (particularly on track) am immediately both sad and grateful that I don't fit in one comfortably. +2

Bob Sawtelle Damn thought we were going to see Newton Center... too short video! +1

John Morris How's the oil level?

George Zinycz John Morris lol too soon… give the man an hour of triumph, at least! +2

Rob Siegel I've done a carefully-calibrated fill with the exact 4.75 quarts that should be in the pan. I've mentally made a new mark on the dipstick, which wasn't where the old one was. I don't have an explanation for it. I got an oil analysis back from Blackstone, which I'll post one of these days, but nothing earth-shattering; it's not made up of blue goo from The Expanse or anything. +2

William Jeffers Rob Siegel love the Expanse series.

Andrew Wilson Wonderful!

George Zinycz Must feel good to feel all those positive changes at work!

Doug Jacobs "Yeah bitches she's baack." Could you be any whiter? +3

Rob Siegel 'cause I'm a dirty white boy... +1

Doug Jacobs The prosecution rests. +1

Robert Myles [link to The Tubes "White Punks On Dope"]

Dave Borasky "I have removed my right shoe" sounds like Fredrich Frankenstein starting an experiment! +1

Rob Siegel [gif "Put… the candle… BACK!"] +2

Paul Skelton Great work Rob! I miss driving Pete's! +1

Daniel Senie The phone mount helps, but could use vibration damping. She's a rumbler for sure.

Gary York Wait a minute there Rob Siegel, isn't this really Day 2480?

Franklin Alfonso So glad you got Lolita running again.

Steven Bauer Will she be able to get you away from the zombie apocalypse?

Barry Wolk I have two Europa memories. I had a friend named Corbin whose father worked for Continental Teledyne and was able to obtain a Renault engine that was a higher horsepower than stock. He called one day and said he needed some help. He had unbolted the sub-frame and had us lift the car by the rear wheel openings as he dragged the drive train out from under it and set up jack stands for us to set the body back down on. Is my memory correct? I also spent several weeks working at a shop in Dearborn that had 3 Europas that had dash fires. I was hired to install new wiring harnesses and electrical components. I'm a giant so I had to remove the seats and the windshield so

I could work inside the car. Another memory test is I'm thinking that I traced the problems back to bad grounds. There were many makeshift repairs I found with grounding screws passing through the fiberglass to the backbone steel chassis below. Is this a correct memory? +1

> **Rob Siegel** Barry, I dropped and re-installed my engine from under the car as well. When I pulled it out, I first removed the head, which let me slide the block out from under. To reinstall the whole engine with the head on, I did need to lift the back of the car up, which I did with a winch attached to a lag bolt in the ceiling. Yes, you're correct that, since it's a fiberglass car with a steel backbone, the grounding points are unusual as compared with most conventional cars. I've jury-rigged a few new ground wires myself that run to the backbone. +2
>
> **Barry Wolk** If I recall correctly I used a hole saw to make a 7/8" hole in the body and drilled and tapped a 10-32 hole and made a stud that extended through to the inside of the car and sealed the floor with a piece of aluminum double-nutted to the stud.

Joey Hertzberg Yahoooo

Scott Linn Whenever I get in the Lotus after a while it's always "damn this is fun!" +2

> **Rob Siegel** Yes!

Fred Aikins Does she still puke coolant, Rob? +1

> **Rob Siegel** She only did that when idling in direct sun in hot weather. It's 50 degrees out. I won't know for months whether the bigger beefier Spal fan makes a difference. +2

John Harvey New ride height looks spot on old chap. +2

> **Rob Siegel** Thanks. So far I'm pretty happy with it. +2

Neil M. Fennessey Maybe you could buy just one? [link to article "Best driving shoes 2020: Heel and toe in style"]

Clay Weiland [gif "And we're back!"]

Lee Highsmith You're not a large man, so I'm guessing my 12D version of "I have removed the right shoe!" would be more like "I have amputated half of my right foot!"... +2

Brian Ach And? How was it?

Luther K Brefo Do you have one of these? [link to foot-shaped gas pedal with toes]

> **Rob Siegel** HAHAHA yeah no

Luther K Brefo So you want one is what you're saying.

Paul Hill SHOE OFF!!!!

Kenneth Yee Good job! Now time to make a run for beer and Cabot's ice cream!!! +1

Russell Baden Musta It sits nice Rob!!! Beautiful!! +1

> **Rob Siegel** Yeah, I'm really pleased with that, considering I was doing it the low-budget way I do most things. It's the QA1 10" front springs I bought ($80 for the pair on Amazon :^), the original rear springs, and the vintage set of Spax adjustable shocks. It really doesn't look like I lowered it at all; it just looks like the nose is where it's supposed to be, which was what I was going for. Put about 10 miles on the car today. It feels pretty good with the shocks at the middle settings. I think there are like 10 clicks on the adjuster. I'll probably dial it way soft, then way firm, just to see what those feel like. +3
>
> **Russell Baden Musta** Rob Siegel I have Spax with adjustable towers on the front and without in rear. I have the same springs from Ray. Not really lowered as well but after my shunt we balanced the car with the spring towers, it works well. If it is real twitchy with crowned or truck rutted roads and slightly on good flat pavement, a tad of caster can help. We can Facetime sometime and I can show you how it was done on mine with OEM control arms. Oh... I believe I have my fronts set softer than the rear, but probably 1-3 clicks only...

Phil Sullivan Smooooooth!

Chris Lordan Yay!

Scott Winfield Such a sharp looking car!

Tim Warner My Europa memory is of my friend having the body of his in his living room- to watch movies while sitting in it! It was cool. It had to leave when he got married - after he was in his 60's!

Francois Doremieux Fun sound. Nice video, thanks. Ok, time for preventative maintenance, after such a long run!

TLC Day 2480: Squash that buzz!

Mar 25, 2020, 6:08 PM

The Lotus Chronicles, Day 2480:

Yesterday I mentioned that, with several big winter projects now completed, I'd love to put some time into peeling some of the layers of the onion of Lolita's rattles and buzzes. A particularly snotty one had developed over the winter that, in yesterday's short drive, really bugged me.

I just took the car for a longer drive in an attempt to find it. It seemed to be emanating from in or around the glovebox, but I had the hardest time isolating it. Turned out it was coming from a thin black piece of plastic trim that's supposed to line the lower corner of the windshield and is almost completely detached. Simply moving it an inch stopped it for now. I'll probably just yank it completely out. The last big project is locating and installing a new windshield anyway.

But it's amazing how much more relaxing it is to drive the car now that I'm not waiting for the rattle to reappear at 3500 rpm and feel obligated to stick my right hand all over the right side of the dashboard, grabbing and pressing down on things when it does.

42 LIKES, 24 COMMENTS

Francois Doremieux Sourcing a windshield and the relevant gaskets may be a lot of fun?
    Rob Siegel available but expensive
    Francois Doremieux Rob Siegel the gift that keeps on giving!
    Scott Linn Rob Siegel they become available, then not available, then available. The PO of mine had run a wiper arm w/o blade across and really scratched the hell out of it. Otherwise it was in great shape. Spent 6 months sourcing one and eventually had to find a NOS one (with delam bubbles in a corner). No one near me would install it, so I had to do it myself. A year later, they became available again (for a short time). If you find one, you should buy it, because it might not be there in the future when you really want it. +2
    Scott Linn TCS didn't use any gaskets. It was originally held in with butyl rubber, but nowadays people glue them in with urethane adhesive. +2
Neil Bradbury Learned the pressing and occasional dash slap technique from my dad. Squeak troubleshooting step 1. +1
Russell Baden Musta Good job! When you're done, you can come sort mine! +1
    Rob Siegel you... want... me to come over and stick my hands all over your dashboard? +2
    Russell Baden Musta Rob Siegel yeah! If you can isolate the rattles!! After this thing is done of course! But you have to drive Lolita here so I can do a remake of the photos of me that you took a few years back!
Rob Koenig Squeaks and rattles will be the death of me. I have zero tolerance.
Scott Linn Apparently the person who re-did the frame on mine used some sort of foam chassis pad

vs. something like mohair/original. It squeaks, and there isn't anything I can do unless I remove the body...

Samantha Lewis It once took me three months to find a rattle in an old Mercedes coupe. Finally turned out to be a Bic pen that had fallen down the defogger vent. I almost went mad tracking that one down. +1

    Rob Siegel That's perfect. +1

Sean F. Gaines At one point I had removed the entire dashboard of my original Saab (my silver 99EMS) and before I reassembled everything, I grabbed a caulk gun and sealed up everything and glued at will. Worked very well; that car ran without creaks or squeaks for decades—at least from the dashboard. +1

John Harvey Nice that the right side of the dash is only an inch or two farther than the shifter. +1

Paul Wegweiser You'll laugh at this: My gray 69 '02 had the craziest rattle when I'd drive it around this winter. After about 15 minutes it would go away... turned out it was water that had leaked behind the dashboard that had frozen into what I imagine was a 6" long ice cube... and once it melted, the noise/rattle went away. #shakennotstirred #icemachine +5

    Francois Doremieux How did you actually find out?

    Paul Wegweiser Francois Doremieux The car leaks MASSIVE amounts of water when it rains... new window seals are on my "to do" list. The glove box contained a 20cm x 5cm block of ice several times this winter... but the rattle was from some other chunk of ice behind the glovebox. +1

    Francois Doremieux Paul Wegweiser I also have a leaky car. There's going to be a big time bullet to bite... +1

Dohn Roush Wait a few years. The creaking of your knees and shoulders will drown out the car noise... +3

Bob Sawtelle Ok so I'm not the only one who drives one handed beating the crap out of a dashboard trying to find a rattle... +1

    Rob Siegel hell no +1

Paul Wegweiser I have to figure out if the rpm-related rattle in the gray 69 is:
a) loose oil pump chain
b) alternator internals
c) loose rocker arm pads
d) rusty fenders flapping near the driver's door
e) under the dashboard somewhere
... this quest haunts me every time I drive it. I HATE rattles. +1

Christopher P. Koch I once rented a 9 month old Allegro class A Motorhome. There was nothing on the entire vehicle that DIDN'T rattle! Unbelievable cacophony of noise!

TLC Day 2481: Longer drive after winter projects

MAR 26, 2020, 1:05 PM

The Lotus Chronicles, Day 2481:

Put about 20 miles on Lolita in a combination of back roads and highway driving as part of a fairly thorough test-out of the completed winter projects (the front end rebuild and lowering, adjustable damping on the shocks, transaxle reseal, valve adjustment; won't be able to really test the upgraded cooling system until summer).

--The lowered nose feels perfect. The shocks feel great until I hit some of Newton's patched pavement. I think I'll dial them a click or two softer and see how it feels.

--Lowering notwithstanding, one of the big reasons to redo the front end was that, even though nothing felt loose to the touch, the car shuddered pretty strongly (steering wheel shimmy) at about 65, even after having the wheels and tires road-force balanced. The rubber boots on the tie rods and ball joints were completely gone from 40 years of sitting, and the trunnions had zero oil in them, but I didn't know if any of those things were the cause. The shudder/shimmy isn't gone, but it's better. There's certainly more blood flow into my knuckles while I'm driving it :^)

--My seat-of-the-pants driveway alignment is pretty damned close. I don't feel it scrubbing or wandering.

--The transaxle isn't howling, something I was concerned about since replacing the side seal requires tightening its finned nut onto the bearing, and there are consequences if it's too loose or too tight.

--I'm still driving her pretty gently, not putting my foot fully in because it's still only about 800 miles since the rebuild, and cornering very gently because, frankly, the vaunted Europa handling notwithstanding, she's a fragile little fiberglass bit of a thing with all the crash-worthiness of an egg and I'm scared of getting surprised, swapping ends or something, and putting her and me into the guardrail.

--At the end of the drive, parked, the only leakage I see is from the valve cover, on which I intentionally left the old gasket with no new sealant in case I need to pull the cover back off.

Not bad.

#IGotYerSocialDistancingRightHerePal

[search YouTube for "The Lotus Chronicles Weston" and ""nailing"]

161 LIKES, 67 COMMENTS

Andrew Wilson Sounds great, no rattles?
   Rob Siegel Well I wouldn't say "no," but I squelched the biggest snottiest one yesterday. +1
Neil M. Fennessey I second that!
Brian Ach Sounds like it runs super. Get a damn phone camera mount!
   Rob Siegel I have one, and it is used in the other video, but in the first one I wanted to show the instrument cluster at least a little. +1
   Brian Ach I'm just busting your carbs +1
Brian Epro Oh man, Rob, she sounds great.
Scott Aaron Wow that engine sounds great—a little throaty and it sounds really smooth.
Hilton Joseph Cooool.
Patrick King Fun! And I even felt a twinge of nostalgia for potholes!
Bob Wright Nice to see you enjoying the day... can't wait to see you open her up... 4 or 5 speed? Make sure your horn works and is loud... think like you're on a bike... would you ever auto-x it? +1
   Rob Siegel It's a Twin Cam Special, nearly all of which are 5-speeds. I don't do HPDEs. I don't have the concentration for it or the inner ear anymore. +2
   Russell Baden Musta Rob Siegel my TCS is a 4 speed. They used to say the 5s weren't as reliable, but I don't know... I only ever heard of one failing and that was looong time ago... unbelievable, that car was one I was supposed to buy then I got laid off from work... it got sold and about 7 months later I got mine. They actually parted that car out which really pissed me off back then!
Dave Borasky Brown Barchetta
Mark Thompson Route 128 when it's cold outside, with the RADIO ON!!! +3
   Rob Siegel Except I've got my radio OUT! 🙂
Jim Denker Very impressive. It is purring nicely! +1
George Zinycz It sounds GREAT in the video! +1
Daniel Senie Makes me tempted to do a video like this in my Model 3. Except there wouldn't be much audio... +1
Bill Howard Next time, give us a video at 2X with engine sounds at 1X. We'd believe it. +1
Wynne Smith Weston?
   Rob Siegel Exactly. Heading west on Route 30 and then north up into Weston on Newton Street. +3
   Wynne Smith Love it!
Jim Brannen I like it!!

Frank Mummolo You should be really proud of yourself, man. I bet there are professional mechanics out there who would have thrown in the towel long ago. Plus, if you ever decide to sell it, it can be featured as a "previously owned and rebuilt by the Hack Mechanic" which will launch the price way beyond a "normal" one...provenance and all that! Well done, man! +7

    Rob Siegel Thanks Frank! +1

    Jim Strickland Frank Mummolo agree! I have benefited from Rob's great work twice, with the Lama and Kugel. Highly recommended seller 😊 +2

    Rob Siegel There's really nothing special about me as a mechanic. Anyone who slings a wrench and works on their own car can do what I do. But, like I was for nearly 35 years as an engineer, I'm very persistent and really don't like throwing in the towel and giving up on projects when nearly all of the problems actually do have solutions that can be implemented, often at a cost I can swallow and a timeframe that's manageable. The turning point for the Lotus came a few years back when I had the rebuilt head sitting in my basement, the bottom end was still at the machine shop, and the car was occupying precious garage space, and I looked at the whole situation and thought "the time has come, NOW, to nut up or shut up." A bit of work, daily if possible, over time, creates forward motion, which eventually turns into literal forward motion of the vehicle. The fact that I happen to like ratty patina-laden cars certainly helps. But I really had no idea whether, once I got her up and running, then driving, then sorted, I'd like her or not. The fact that I just can't get enough of the quirky little thing is something I could not have foreseen. +5

    Frank Mummolo Rob Siegel yeah. What you said. I am a fellow engineer somewhat lacking in your persistence at times. Hats off! +1

    Rob Siegel Jim Strickland, I am totally not selling you a Lotus :^) +2

    Jim Strickland Rob Siegel I don't think Susan Brown Strickland would allow 😊 +1

    Susan Brown Strickland Jim Strickland definitely a firm, hard NO! +1

    Michael McSteen Well done. Book to follow? Because I want to read about the 2481 days up to this point!! Thank you for keeping us entertained.

    Rob Siegel Michael McSteen, yes, it is in the works!

    Dave Borasky if course if you follow on Facebook it's akin to being the people who read Dickens' Pickwick Papers via serial publishing.

Chris Raia Side note: I love and miss that road...used it as my escape west when cycling +1

Rob Siegel My go-to drive is up Newton Street to Weston center, then continuing on the backroads through Lincoln, coming out at Walden Pond, then Rt 2 east to 95 and back down south to Newton. I did the abbreviated version this morning, getting onto Rt 20 and popping out in Waltham. +1

Chris Raia You probably passed me dozens of times on the way to Walden if so! +1

Daniel Whynot Rob Siegel I used to live and work in Weston. Loved driving those exact roads! I knew every turn and pothole on Conant road, along with Concord and Merriam streets! +2

George Murnaghan I prefer jumping on 2A and going through Minuteman Park before getting back on 128 to stay on the 2-lane as much as possible. But I read somewhere that you and Rt 2 have a history.

Sam Schultz Ahh the sound of a vintage car. That's the good stuff... [gif When Harry Met Sally "Oh God!"] +4

    George Whiteley I'll drive what HE'S driving! 😊 +2

    Rob Siegel See the chapter in Memoirs of a Hack Mechanic on "So, Why DO Men Love Cars, REALLY?" +1

    Sam Schultz [gif Wedding Crashers "Ma! THE MEAT LOAF!"] +1

Kent Carlos Everett Nice job. Well-deserved big smile. +1

Chris Mahalick If that bumpy road didn't dislodge anything, you should be good to go. +1

Marc Touesnard Caution: Upside-down pedestrian crosswalk. [photo of street sign with upside-down pedestrian]

Francois Doremieux Congrats. I am glad Lolita is under the temporary delusion she is a Toyota and giving you a break. +1

Rupert X Pellett Smiths gauges accurate as ever? The Tachometer telling Greenwich mean time -and the voltmeter saying the petrol-tank is under pressure? [photo of British is it a car or a house?] +1

Steve Park Very cool. If anyone deserved a good test drive it was you. You went through hell and back with this car. Congrats.

Hans-Jürgen Krieger I'll never put my 1,94m in a Europa! So I decided to wait until someone gives me the money for this... play loud 🔊 [link to Jensen CV8 Rally Tarmac] +1

Brian Stauss I feel like I've driven down these same roads before with you, Rob! +1

Jim Tennessen Great results.

Mike Crane Lovely day for a poot around the neighborhood. 👍 +1

Robert Myles You really owe it to yourself and that car to try it out on track. You don't need to invest a whole day in the process—the SCCA runs a neat program called TNIA, Track Night In America. It's a great, low pressure, controlled way to have #funwithcars, particularly if you're new to the track experience. Give it a try! A couple of hours time on a weekday afternoon and you'll get an hour + of track time. There are several tracks not far from you - Thompson, Palmer and Lime Rock. Bonus points is I know the novice coaches for those tracks, and I coach at NJMP. +1

> **Rob Siegel** Robert, I used to do driver's schools through the BMW CCA, but it's been 30 years. I tried to get back on a track about six years ago, but I had to accept that simply don't have the concentration or the inner ear for it. A few burst of adrenaline on entrance ramps and some drives on twisty roads where I'm not worried about "the line" and people aren't trying to cram their cars up my tailpipe are my speed these days. +4

Chip Chandler Been social distancing myself! Wiring and firing soon! Congrats Rob! [photo hot rod truck] +1

Steve Woodard Love those Smiths instruments. I was imprinted with my first car, a ‹65 MGB, which I ended up driving through 44 states and six provinces over a 5-year period. +1

Richard Belas So when are you going to sell it? Not to me, though.

David Weisman That's so much fun to watch! How would you compare to a Porsche 914 or Toyota MR2? +1

> **Rob Siegel** I've only driven one 914 and that was 30 years ago. I've never driven an MR2. So I can't offer any direct comparative observations. As I said in an earlier post, the Europa is a tiny, fragile, incredibly low, primitive, quirky, buzzy little car. There are so many reasons not to like it, and yet I just can't get enough of the little thing. I think a big part of the appeal is that it's the antithesis of a market-designed car. It's a car with a very consistent design philosophy: Lightness equals performance. +1

David Weisman Rob Siegel that's what makes it fun!

> **Brian Ach** Take a 914, then subtract 600lbs and all sense of German-ness, add some baked-in handling jazz and a ton of oh-shit-I'm-doing-90mph-in-a-car-made-of-tin-foil-and-spit and there you have it. +2

Rob Siegel that's probably not too far off +1

Ed MacVaugh In 1972, I had come back from SE Asia with a fistful (for the time) of money, and I had a choice of many cars, that in retrospect I should have bought, but didn't Astons, and Ferrari, and Jaguar, but the choice narrowed down to Europa versus the 914. The 914 was almost $2000 more money when $2000 was real money, but I still bought the car. It was substantially heavier, felt planted, was air cooled (which coming from Montana meant wouldn't freeze in the winter) and was unibody. It had some heat, but a removable roof. The Lotus was incredibly light, tossable, noisy in the right way, and you could hear the intake! Tires were half as expensive as the Porsche, but the brakes felt funny, the nose was too high up in the air, like it was going to launch, and women couldn't get in the car gracefully. +2

Vince Strazzabosco I keep getting flashbacks from Rendezvous when I watch this…

Elizabeth Marion Are you um, exceeding the speed limit? What? You didn't know because your speed-o-meter isn't working? Off to jail you go! +1

> **Rob Siegel** Mmmmmmaybe

Sam Schultz Elizabeth Marion that's just car control. Perfectly adjusting so the speed and rpm never change…

Dean Rozos Elizabeth Marion, he would never!! And I am sure that vehicle has a 5-star NHTSA rating!

Russell Baden Musta The top vid… every one of those cars coming the other way either didn't see you at all, or they all said, what the hell was that!? +1

Andrew McGowan Good stuff!! But in your rebuild, did it include a cylinder rehone/ rebore, and new rings? If so…this is where I disagree be on babying it. Your need to get on it to seal those rings! +1

Dean Rozos Nothing elitist about it! Pure reasonableness, as we expect from Siegel! I wish I knew what "completed winter projects" were.

David Kiehna The Europa is the best handling car you will ever drive with a roof. The Lotus Super 7 being the competition if you want a traditional front-engine rear-drive platform.

Lee Highsmith That thing is LOW. It looked like you were going to drive under the tires of that white SUV 😄

> **Rob Siegel** The spec is that the roof height is 42 inches, and that's about correct. It's off in the

photo because of the camera angle. The only car lower ever officially available in the US, I believe, was the Ford GT40, where the "40" is the 40-inch roof height. The nose, at the center of the front wheel, is about 27 1/2 inches. So, yeah, it's low :^)

Brian Ach I need to do this today +1

Rob Siegel We all do.

Scott Aaron I went yesterday. It's raining today and tomorrow here. I had my window and took advantage.

TLC Day 2182: Ice cream getter

MAR 27, 2020, 7:05 PM

The Lotus Chronicles, Day 2182:

Maire Anne baked brownies. I realized with horror that Ethan and I had already eaten all the ice cream. Drat! Fie! I can't order ice cream from Amazon, at least I don't think I can. Do I risk a trip to the grocery store, braving viral non-social-distance-obeying zombies, FOR ICE CREAM FOR BROWNIES?

Totally.

It's a strange new world, but it's easier when you're running errands in a freaking slot car.

116 LIKES, 37 COMMENTS

Mark Thompson Would it be a Zooropa if you owned this at UMASS? +1

Marc Touesnard You should wear the mask when driving the Lotus, even when there isn't a pandemic, you know, for attention. +1

  Layne Wylie And make a matching car bra for the Lotus. Which, now that I think about it, should really be called a "car mask." Since a car is a quadruped, the bra would go underneath somewhere. +1

Doug Jacobs DoorDash delivers from Ben & Jerry's stores. +4

Rob Koenig Wait, there are Ben & Jerry's STORES?!?!!!! +2

Tom Fota There used to be several Ben & Jerry's "Scoop Shops" in San Diego but now there's only one downtown in Seaport Village, a location only easily accessible to tourists. There's 10 in LA County.

Rob Koenig I love Seaport Village. We don't have any here in PA (that I'm aware of). 😔

Tom Fota My wife is the CFO of the company that owns Seaport Village! She was there weekly for a redevelopment meeting but that meeting is now Zoomed. +1

Rob Koenig I had family that lived in La Jolla. Visited many times over the years. I've always loved SD.

Doug Jacobs Rob, you're right, nothing in PA. Oh, wait, a couple in Philly.

Rob Koenig Hey, I had no idea! Thanks Doug. It's a hike, but if I'm already in Philly… +1

Janet Conway Did you pick up ice cream at the Star on Comm Ave? I was walking back from Tom's with our pizza for dinner when I heard/saw the brown Lotus turn onto Lexington St. Wasn't close enough wave but knew it was you. +3

Rob Siegel Yes!!

Tom Tate Janet Conway Right, how many Bozos can there be driving around in a brown bread van? 😂 +7

Janet Conway Tom Tate Just one special one. +1

Russell Baden Musta Also, you can pull the mask off and say, I'm busy block-sanding my fine British car, why are you all in wearing masks too? +3

Paul Wegweiser We made sure to stock up on ice cream yesterday…although we've already eaten a third of it. Wash all your groceries w diluted bleach and / or soapy water. Please. Stay healthy. +2

Neal Klinman Paul Wegweiser I just started doing this. [photo of fruits and vegetables in what looks like bleach]

Dohn Roush With the Lotus, are you stuck buying pints? +6

Michael Schwed Man, that headlight is the T to balance the heaven-sent A of the 3.0.

Harry Krix Are they brownies or are they BROWNIES? Real Brownies don't need no stinkin' ice cream and BROWNIES, well that's another story for a different channel when the kiddies are in bed… +2

Rob Siegel They're "brownies."

Eric King

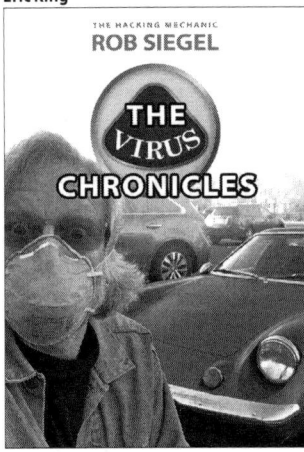

+13

Russell Baden Musta Eric King I'm saving this picture! 😂😂😂😂

Rob Siegel Russell, Eric King is the fellow who designed my last three books, and his parody covers like this are so good that they deserve their own series. +3

Russell Baden Musta Rob Siegel I love it!!!! This is just hilarious!! I'd frame this!

Stan Rose Amazon Fresh delivers Ice Cream, if Amazon Fresh is available where you are. I can't believe it wouldn't be available in Newton!

Neal Klinman I had fun shopping too. That's $250 worth in my old bike trailer. Old car would have been fun too. [photo of bike hauling small trailer] +1

Douglas Wittkowski A dangerous trip in the best of times, but in a Lotus? This guy right here, hero material! +2

Franklin Alfonso Well, I suppose ice cream is worth dying for...

Chris Mahalick I am laughing so hard at the thought of you ordering ice cream!! Unless it is packed in dry ice. Hmmmm.

Tim Salane LOL. Movie for the times, "The Omega Man," Charlton Heston in 1971. Also his early homage to the NRA.

Walter Jordan I like that just as we got speed cameras, masks are becoming socially acceptable. Although... I guess that doesn't apply to people in cars that only seem like they are speeding at just above walking pace. Good luck in your foraging endeavors!

Lid Elizabeth Adolph Reilly GoPuff

TLC Day 2183: 120 miles

MAR 28, 2020, 3:09 PM

The Lotus Chronicles (sort of), Day 2183:

Put 120 flawless miles on Lolita today running back and forth to Fitchburg. Other than my shit-eating grin, the only event was having one of the four pieces of windshield gasket-holding trim fly off. A damned shame, as new ones are expensive, the old ones all had matched patina, and at some point I will have to deal with replacing the cracked windshield and will need all the trim pieces.

But the purpose of the trip was really to get out to Fitchburg and put some miles on the four cars out there, get the fluids pumping and the seals exercised, level out the flat spots in the tires, frighten the rodents, etc.

--1979 Euro 635CSi: It's gorgeous, but it's at its best as a highway cruiser. These short hops in it don't really light my fire.

--1973 Bavaria: Even with the Bilstein HDs and the beefier sway bars, it's still a little floaty and boaty compared with the stiff 2002s, but it's just more fun than it has a right to be.

--Bertha: This is really the car that I should be driving around during the viral zombie apocalypse. It's so deliciously in your face and it doesn't give a fuck what you think about it. Nail the throttle, listen to the Webers, and forget all your worries.

--Z3 M Coupe: Whenever I toy with the idea of selling it, I drive it, and that thought immediately vanishes. By an order of magnitude, the best "sports car" of anything I own. And go-cart-wise, the only one that remotely compares with Lolita.

It certainly was a Saturday morning and afternoon that did not suck.

TLC Day 2630: New alternator

MAR 30, 2020, 8:22 PM

The Lotus Chronicles, Day 2630:

The original Lucas-Delco alternator in Lolita worked, sort of. It wouldn't output anything until the RPMs came up, then it would crank out over 15 volts, high enough for me to worry about boiling the battery. I'd need to always run with the blower fan and lights on to bring it down nearer to 14 volts. And the bearings were getting noisy. Finally replaced it with an inexpensive unit from DB Electrical that's listed as fitting Lotii, Triumph TR7, Ford Capri, and Massey Ferguson tractors. Go figure. Of course, the connectors are different. The original unit took two unique plugs from the car's wiring harness. I didn't want to cut them, so a few adapter cables later, and... voila! It now comes up reading about 14.2V and stays pretty close even with the accessories on. I won't need to continue to drive with one eye on the voltmeter. At least in theory.

I'm sure the car will continue to need constant attention—it IS, after all, a 46-year-old Lotus—but this is the last major item left on the punch list from when I began driving it last spring. Sure, there's still the cracked windshield, but that's mainly an inspection issue. So, oddly enough, after all the work this winter—the rebuilt front end, the adjustable shocks, the transaxle seal, the upgraded cooling system, the shimmed valve adjustment—THIS was the repair that, when completed and the car was driven, felt like a milestone.

Still, not exactly the go-to vehicle for escaping the viral zombie apocalypse.

92 LIKES, 77 COMMENTS

Rob Koenig A fine hack repair outcome! +1

Tom Melton How's the oil level staying?

    Rob Siegel So far so good. Thanks for asking. +4

Paul Wegweiser When did Lotus put the text "Abandon hope, all ye who enter here" on their volt meters? I'm guessing around 1970? +12

    Blair Meiser Paul, the day they signed the contract with Lucas! 😎 +4

    Blair Meiser I once had and sometimes enjoyed a 1967 Lotus Elan SE. I say sometimes because of Lucas! +1

    George Murnaghan I think it was when they gave up shillings. +2

    Paul Wegweiser George Murnaghan ...and yet... they still cling to Whitworth and WOODEN CARS! *shakes head* +1

    Blair Easthom Organic composites.

    George Murnaghan To be fair, they gave up on the English system of measurement, while we Muricans have clung to it +1

Josh Wyte How are you able to keep it plated/registered in VT?

Rob Siegel Gee, no one's ever asked me that :^) [link to article for Hagerty "Answering questions about The Vermont Loophole"] +2

> **Paul Wegweiser** Rob Siegel I thought it was because VT takes pity on you. I mean, why wouldn't they? +2

Tyrone Henriques Sweet Lotus! +1

Joshua Weinstein So now it's time to sell?

> **Rob Siegel** NYET!
>
> **Paul Wegweiser** It's WAYYYYYY past the ideal time to sell. That was... hrmmm... let me think... ah yes!... 2629 days ago. +3
>
> **Joshua Weinstein** If not earlier.

Chris Mahalick It is very cool to see you enjoying it so much. That is a really cool car. And totally well-sorted at this point. +2

> **Rob Siegel** It's not bad. +2
>
> **Dave Gerwig** Rob Siegel if the shoe fits, wear it. +1

Garrett Briggs Well *I* think it would be a good car for the zombie apocalypse. You could cut off the approaching zombies by hitting them right in the knees. Plus, it won't matter when they bounce off the windshield. It's already cracked. So you definitely shouldn't fix the windshield. A zombie will inevitably crack it again. +4

> **Rob Siegel** I like this approach.
>
> **Paul Wegweiser** Rob Siegel Don't worry. I won't say anything about the zombies eating brains and that driving an unreliable fiberglass car during the apocalypse probably takes you off the menu in the gray entree department. Why have I chosen this decision? Because that would be mean.
>
> **Rob Siegel** Your restraint is admirable. +1
>
> **Paul Wegweiser** Rob Siegel Admiral restraint reporting for duty, sir!
>
> **Dohn Roush** Gee, he's an admiral. Does the Lotus make you a midshipman?

Bronco Barry Norman Add a flame thrower or two and you will be set for the Z.A!

Russell Baden Musta Gonna start mine tomorrow to get out of the way to paint garage!!! +1

George Zinycz No Rob... I don't think this was a milestone—it was a portal into the next level of Lucas Hell. Now that the electrical system is getting a proper, steady voltage, the electrical components that got used to the sloppy voltage will now start spazzing out because of this change. I predict at least two electrical components will fry out during the next 5-6 drives. "Murphy was an optimist." +2

> **Rob Siegel** Bite my sparking fiberglass ass. +4
>
> **George Zinycz** Rob Siegel you'll curse me later, likely by the side of the road. And you know I'm right. Back me up Paul Wegweiser! +1
>
> **Paul Wegweiser** George Zinycz I'm still pondering the phrase "sparkling fiberglass ass." +4
>
> **Rob Siegel** Fine. "Well-patina'd fiberlass tushie." +2
>
> **Scott Linn** Paul Wegweiser "sparking" +3
>
> **Paul Wegweiser** Scott Linn Sparkling eh? I guess NOW we know who bought ALL the toilet paper!!!! +3
>
> **Mike Crane** Actually Rob, in case you don't know. The newer 3 pin connector on that ACR was a legacy issue re dynamos. They had a field coil wire and a power out, but when the alternator boosted the output, they fitted two larger spades to carry the extra current & two wires to take this, because of the wire capacity at the time & compatibility with old systems. 😊

Bill Mack Not that familiar with Europa electrics. But that looks not British at all.

> **Paul Wegweiser** Don't worry... we'll airbrush on some faux "scorch marks" and melted plastic bits to make it look more correct. +3
>
> **Bill Mack** Paul Wegweiser I was talking about the one he took off but that works for me!
>
> **Rob Siegel** Now, Paul, we TALKED about this... your assault license is only good for my GERMAN cars... +2

Luther K Brefo Can you have the original rebuilt?

> **Rob Siegel** I can, but one ear is cracked, and I'd sort of cajoled it into hanging there with some big-ass washers, so I'm not sure it's worth it. +2

Bruce Hoiseth Do you insure through Hagerty?

> **Rob Siegel** Bruce, yes. I also write for them. +2
>
> **Bruce Hoiseth** Rob Siegel I need to use them for my 1966 Mustang which is in Massachusetts and

has Massachusetts plates (low number plate my father got for me in 1977; want to keep that plate). +1

Patrick King Massey Ferguson. Hahah! 😂

    Rob Siegel I know, right?

    Rob Siegel [Amazon link to DB Electrical ALU0032 alternator]

    Rick Drost Maybe you wanna find yourself a tractor.

    Tim Warner Rob Siegel I have a Massey, a 1951, but it's only a 6 volt!

    Brian Aftanas And John Deere makes the perfect colors for a Lotus. +3

    Patrick King Brian Aftanas That's right! Same shades of green and yellow as Jimmy Clark's Lotus 49! Well no wonder there are interchangeable parts!

    George Murnaghan Beats the hell out of having a tractor engine in your British sports car.

Jack Fletcher You know by now than any old car that gets driven undergoes a "bug curve" of problems that are discovered/develop. And as driving hours accumulated it tapers off. Just like today's crisis, the curve flattens.

Rob Blake But when you want to escape the viral zombie invasion in STYLE it is. 😂 +2

Geoff Bartley OK so Rob… when can we go for a ride in it?

    Rob Siegel Geoff, when we can be within six feet of each other again, I'll be glad to take you for a ride :^)

Russell Baden Musta Here's mine from Summit Racing… one wire [photo of single-wire alternator] +1

Scott Linn So it bolted right in except for the wiring? Did you have to special source any connectors? Asking just in case for the future… Mine is still working after 9 years of ownership but you never know.

    Rob Siegel Yes, it bolted right in. I did have to buy some 3/8-in spade connectors. They were available on Amazon. +1

Walter Eschelbach Sure you have a voltage regulator or one built in that's working?

    Rob Siegel The voltage regulator in both the original alternator and in the new one is internal. I could've tried to open up the original alternator and replace the regulator, and the brushes, and since its bearings were noisy, I'd also need to do those too, but for $62, I preferred just to replace the whole thing.

David Kiehna There's a Bosch 55 amp model that's common for MG guys to swap into their cars. Just increase the size of the positive wire from the starter to the alternator.

TLC Day 2492: Cloverleafs

APRIL 5 AT 9:26 AM

The Lotus Chronicles, Day 2492:

Took Lolita out yesterday and rolled her over, in the one of the three connotations of that phrase that is neither sexual nor has an insurance claim and an ambulance ride associated with it. But with her now at a coming-out "You're not a kid anymore" 25,000 miles, I can't help hearing Lolita channeling Marcie Blane and saying "I want to be Robby's girl…"

Sorry. That was really bad. Even for me.

Oh, and yesterday we ran 12 cloverleafs. I wouldn't say I was cornering in anger, but I was definitely not grandmothering the ramps like I've been doing. I picked the speed up gradually, and intentionally lifted a little off the gas to feel if the rear end just twitched a bit or felt like was going to come around. Even with the not-yet-professionally-aligned front and rear suspensions—both of which have been re-bushed—and the Achilles tires that cost $241 installed for the set of 4, she feels (choose your metaphors carefully here, Siegel—you DID name the damned car "Lolita") like she knows what she's doing.

And about all these adolescent double entendres… look. This whole thing started because, as I've written multiple times, when I was 13, I worked at a stereo store for a guy who enjoyed cars, one day he came in driving a Europa, and a 40-something co-worker said the phrase that then lodged in my brain: "A car like that, you can get SEX out of." It took me decades to understand that he wasn't talking about the dynamic of men with cars attracting women—he was talking about THE SENSATION of owning and driving a wicked cool car. I can't say exactly why this particular batshit crazy fragile little 1600-pound car that took me six pain-filled years to get running so pushes these buttons for me, but she does. I joke that I'm in a destructive co-dependent wildly inappropriate relationship someone who screams and claws and threatens to tell my wife and smash all the windows in my house and I'M the one being abused and I can't get enough of it and keep coming back for more, but really, this is healthy. Especially in these Kafka-esque virustimes we find ourselves in, I'm grateful to have her. She's the best kind of distraction. And, as Townes Van Zandt sang about Loretta, "And I can have her any time."

She's just downstairs in the garage. She beckons. Damn it, girl, I'm WRITING!

Okay, maybe a quick 20 minutes before Maire Anne wakes up.

97 LIKES, 37 COMMENTS

    Brandon Fetch Aren't you too old for a midlife crisis? Or is this your second one?
        Robert Myles I think that in his case it's ongoing. +3
    John Morris Only the British would have a label in the tach reminding you how many cylinders you have. +8
        Bill Gau John, That's probably so they don't accidentally put the wrong tach in an Esprit. Who has time to read part numbers when you're slapping these things together ;^) +2
    Lee Highsmith When she finally does tell your wife, the reply will be "Oh I know. I've known all along. But it keeps him busy...". ;) +4
        Thomas Jones Took me a second to realize that's just a reflection and not the tach needle reading 6600rpm. +14
        Rob Siegel That is correct! +1
    Rob Koenig I thought the same thing! I thought Rob had suddenly decided to cast aside all inhibitions and wind it out whilst crawling down the road at 30 MPH. +3
    Rob Siegel Yes, 6600 rpm at 30 mph... hear'n it scream, livin' the dream... +5
    Jeffrey Miner Thomas Jones I'm surprised the redline isn't higher ... You'd think with a custom head, they would have beefed up the bottom end too, and 7k would be where it floats.
    Rob Siegel Jeffrey Miner This is actually a replacement tach out of an MG. The original tach needs to be repaired. I believe that the redline on the Europa TCS with the "Big Valve Twin Cam" engine is 6750 rpm.

> **Carl Witthoft** You mean Vida Loco
>
> **Franklin Alfonso** LOCA is correct. I am Cuban after all lol. (well, my parents were)
>
> **Robert Myles** He hasn't gone far, so it's living la vida LOCAL. I'll show myself to the door now... +3
>
> **George Murnaghan** La vida Lolita +2
>
> **Franklin Alfonso** Great sex with Lolita she bites you in the ass... again!

Eric King Tiny car bonus during this shutdown: less traffic!

Bob Sawtelle Anyone who quotes Townes Van Zandt is OK with me! Have a great day Pancho. +2

John McFadden I'm sure this was discussed here at the time of purchase, but "Achilles" tires? I guess the guys in Marketing weren't Greek mythology buffs :-) Funny thing is, they are probably light years better than anything available when the Lotus was new. +2

> **Bill Snider** I hear they're not really meant for Heel and Toe driving. Lol +6
>
> **Rob Siegel** BOOM! +1
>
> **John McFadden** My size 14s are my Achilles heel in that department!

Charles Morris Could have been worse... Icarus tires :D

Bill Gau Be careful not to blow that bastard up!

Brian Aftanas So tell us... when you were doing it with her on all those cloverleafs, did she twitch? +1

Brian Ach Geez I thought you literally "rolled her over" and wondered how fast you would have to be going in a Europa to roll one over

> **Rob Siegel** the future is rife with possibility +1

Scott Linn I only saw them in car magazines when I was maybe 15 or 16, thought they looked awesome. After getting mine mostly sorted, driving it is always an amazing experience. +1

Brian Aftanas My experience with rear mid-engine cars (er, my one car) is that the rear end does not "twitch" when you lift. They plow more than they drift. It's too easy to give too much grip to the rear tires. I once attended a BMW CCA autocross with the covert intention of spinning my 308. I couldn't do it: if I kept my foot in it, it would go increasingly sideways until the tires overwhelmed the engine and it stalled. +1

Ernie Peters She's not a kid anymore. You devil, you. +1

Mel Green I love to push the West Newton ramp coming in from Mass pike... my Yaris sedan corners like a bloody go-cart!! +1

> **Rob Siegel** Go Mel!

Lawrence P Beron What's the "getting in and out" process like? Do you use one of those foam rubber pads with the handles like gardeners use under their knees? +2

> **Rob Siegel** Well, it'd say a lot about which one of us in the relationship uses the knee pads... :^) +3

Brad Purvis You have to get sex out of it, because you certainly don't have room to have sex inside it. +3

> **Rob Siegel** Well, you SAY that...

Daren Stone It's pert near impossible to capture that Lotus feeling in words, but I'd say you nailed it in the post. Following along since Day 1, I would have bet you'd cut your losses somewhere along the way of putting it all back together again, but you didn't and so passed the painful part of owning a Lotus. And now you get to experience the happy part of owning a Lotus. Well done sir, well done. 😊 +2

TLC Day 2492: The wife and the mistress

APRIL 5 AT 2:50 PM

The Lotus Chronicles, Day 2492 (continued):

The wife and the mistress and I tangled it up together—the first time their paths had crossed since Maire Anne helped me roll the dead car into the garage after it was unloaded from the transport nearly seven years ago. This is as close to a threesome as I'm going to get in this lifetime.

And it was good.

(Actually it was very tame. I think Lolita was far more afraid of Maire Anne than vice versa. That's showing, I believe, the proper respect. And she is, after all, wearing a cap with a tarantula on it.)

168 LIKES, 27 COMMENTS

Matthew Zisk When does Maire Anne get to drive her?? +2
    Carl Witthoft Why would she want to? +3
    Rob Siegel That's actually a very accurate answer :^) +2
    Robert Myles Whenever she damn well pleases. Or whenever she loses her mind. Whichever comes first. +2
Steven Bauer We won't talk about devil's threesomes. +1
Christopher P. Koch Appropriate that she wore a spider in a spyder! +1
Doug Jacobs Man, this lockdown is not going well for you. I'm concerned. I think someone needs to step in here. I've never seen a weak metaphor beaten into the ground like that. Go for a walk my friend. +3
Paul Wegweiser When I fling Lily around with Wendy Burtner in the passenger seat, she mentions that the Recaros from the F Bomb DEFINITELY make the ride a little less tense. The factory seats realllly let you slide. It increases the feeling that the car is about to veer into a guardrail. +6
    Francois Doremieux Paul Wegweiser she does not take the driver seat? Your turn to slide around...
    Paul Wegweiser Francois Doremieux I keep trying to coax her into it... this summer I'll try harder... because I know she'd LOVE driving a 2002 as long as she was solo and could really wring it out without being self-conscious. She likes the fast. Went from a 330Xi to her current car: 2013 GTi. +3

Wendy Burtner Paul Wegweiser Yes, my only fear is scaring you. I'll drive Lily someday, For now... you are enjoying it way too much. 😊 +5

Brad Day Stock seats = Active Instability Control +1

Scott Aaron Paul Wegweiser my wife is Super-Meh on O2s. Always has been since the '80s. Only comments over the years were that the old-school alloys look better than the hubcaps, and the Recaros are an upgrade vs. the stock seats.

Rob Siegel If they like the car(s), that's great, if they don't, that's okay, but the fact that they tolerate and support us automatically elevates every one of them to goddess status. +5

Scott Aaron This is true. I'm lucky she likes cars. She'll drive the M3 occasionally, and I think her X5 is her "forever mobile." I've learned over the years to not bother her a whole lot with car talk. Least I can do.

Garrett Briggs Luckily my wife understands and also likes my old cars. (Ok at least the German One). She tolerates my old Toyota Pickup.

Joey Hertzberg Cute couple! +1

Frank Mummolo Watch out, Lolita! You've met your match (and more!) +3

Eric King

+8

Bill Snider Puts me very much in mind of a favorite French sports car The Matra Murena. You could tell it was French as it had 3 seats, 2 wide ones for driver and wife and a smaller one in middle for his mistress. ;-) [photo]

> **Rob Siegel** so... we are insinuating something about relative butt size of driver, wife, and mistress are we?

> **Bill Snider** Not me Rob, The French ;-) +4

Scott Chamberlain Just because you've survived a worldwide pandemic is not an excuse to take a road trip in a Lotus! +3

Dohn Roush Don't tell her you're trying to de-bug it... +3

Brian Ach awesomeness in all ways +1

Duncan Irving Never noticed before how sexy those blinker lights are. Practically indecent!

TLC Day 2498: A gallon of milk

APRIL 11 AT 8:54 AM ·

The Lotus Chronicles, Day 2498:

I hope everyone who runs out on a Saturday morning to buy a gallon of milk does so in a car that they love, and turns around and looks at it after they park it. Because this shit never gets old.

310 LIKES, 88 COMMENTS

Michael Otten When I close the garage door with my E24 inside, every time, I look at that beautiful 🍑 as the doors lowers. 😎 +1

    Maury Walsh You have to be careful with the tiny emojis. At first I thought it meant fish smell... 😄 +5

    Neal Klinman Michael Otten Yeah. And I still can't see what you're saying.

    Neal Klinman Michael Otten Oh. Now I get it.

    Rob Siegel "Fish dildo." what's unclear about THAT? +5

Mark Thompson It has elan...Oops, different car. +6

Doug Hitchcock When my 2002 is in the driveway, I always seem to step over to the window to check it out. +2

Buck Eddy How many cars do u have Rob? Favorite? +1

    Rob Siegel Counting the daily drivers and the Winnebago Rialta RV, 12. Although I'm very smitten with Lolita and will take any opportunity to experience the sensations she engenders, the queen of the roost is and will always be the 1973 3.0 CSi I've owned for 34 years. +9

    Russell Baden Musta Rob Siegel You are such a wordsmith! Beautiful!!! I believe you are correct... my "Lolita" needs fuel... I think maybe today it comes out for a trip to the gas station, and the long way home!! +2

    Rob Siegel Russell, it's funny. For years, I would look with some amount of interest at oddball VW-based kit cars on CL that were very low-slung and affordable, thinking "probably handles like a go-cart, probably fun." Now, when I stumble upon an ad for one, I think... "meh, I own the real thing." :^) +4

    Russell Baden Musta Rob Siegel that's absolutely correct! Back in 1970 there was a GT40-bodied VW in someone's yard on the way to the airport where I was learning to fly. I stopped one day and they let me drive it around their yard! Only thing I can remember was it was very very low.. Yes, we now own the real thing. I was probably in my Dad's Triumph Spitfire... +1

    Scott Linn I always thought my 68 Midget drove like a go cart. My friends too. Then I got the

Europa. Holy crap. +1

Larry Engel And hopefully we can all go to a convenience store that's 50+ miles away! +4

Gordon Parr on a twisty road! +1

Pete Sullivan beautiful

Bob Sawtelle If you don't look back at your car when you walk away are you even an "Enthusiast"? +4

> **Rob Siegel** Or you simply bought the wrong car 😊 +5
>
> **Patrick King** I was scanning to see if anyone said that before me. Saves me some typing. Except that I'm typing right now. 😊 +1

Dave Gerwig Great stance with the new front end! +2

> **Rob Siegel** Yes, I am very happy with the way that turned out. +1

Carl Witthoft I want a pink 57 sedan de ville +1

> **Rob Siegel** With Aretha in the back singing "Freeway of Love" +2

Don Stevenson Gotta love the patina on that copper paint!! +1

Doug Jacobs There sits poor Bluto [my X5], the ugly stepchild, never loved by his dad, the man who dotes on all the "pretty" kids. +1

Kathleen Wolf A thing of beauty. +1

Pete Lazier ...that car would look amazing in black and silver... 😊

Rob Siegel The sought-after John Player Special (JPS) version is in the John Player livery of black and gold. My car, a Twin Cam Special, is exactly the same car but in a different color.

> **Pete Lazier** Yes... I would paint the car to match in a second...
>
> **Wade Brickhouse** I always liked the JPS the MPs livery.
>
> **Scott Hopkinson** BMW Motorcycle made a JPS too: R100RS [photo]

Randy Yentsch As an added bonus, Lolita's color allows for infrequent washes. +1

> **Pete Lazier** I have never seen evidence that Rob washes his cars ever... might mess with the patina... +2
>
> **Rob Siegel** like my haircuts, once a year whether they need it or not :^) +5
>
> **Ian Sights** Back in the 60's before SUV was a term, my father's utility vehicles (International Harvester Travelall & Jeep Wagoneer) were tan to match the color of Missouri mud so they didn't require washing. The Lotus 7 and Morgan Super Sport got washed.

Alan Hunter Johansson And it has room to carry a gallon of milk! So utilitarian! +3

> **Scott Linn** Alan Hunter Johansson it also has a cold trunk and a hot trunk. +2

Marc Touesnard Yup [photo 2002] +3

Scott Poirier I took this to grab supplies the other day 😊 [photo E30 without hood] +2

Andre Brown Yes! Every time. +1

Dan McLaughlin My E39 M5 is blocked in my garage… sigh... by a M54B30 engine. At least we will have some great weather this week and I can finish the engine swap into my wife's E46. +1

Terry Shea Last night's supply run/ takeout pickup was in the '67 Barracuda. Who says lockdown has to be boring? +3

Bill Howard The Europa is two different cars—the second one starts aft of the front door.

> **Ben Greisler** I first saw one in the flesh sitting in the paddock at Lime Rock. This was my face: [gif child very hesitant] +2
>
> **Paul Wegweiser** Ben Greisler That gif never gets old. +1
>
> **Rob Siegel** Yeah, the worst view for that is dead side-on. But in the flesh (in the fiberglass?) it all is much more in context. I've never had someone come up to the car and say "DAMN that thing is even uglier in person than I THOUGHT it would be." Quite the opposite.
>
> **Rob Siegel** A few days ago at a traffic light, a guy on a motorcycle said "That's a Lotus Europa, isn't it?"
> "Yes, I said, "' '74 Twin Cam special."
> "You're a brave man," he said.
> I thought about it. "I suppose I am," I replied.
> Then I realized that this was a guy ON A MOTORCYCLE telling me that I'm a brave man. +8
>
> **Henry Noble** Rob Siegel and he was drunk and brave.

Alan Hunter Johansson On a serious note, I don't know how many times I've been saved from dead batteries because I've cast a fond look back and seen that I left the lights on. +4

> **Robert Boynton** Alan Hunter Johansson ...there's a Lucas joke lurking here. +1

Alan Hunter Johansson Robert Boynton I tried, but I Bosched it. +7

    Rob Siegel BOOM!

Sam Schultz Waiting to pick up curb side prescriptions. [photo E30 dashboard]

Paul Forbush It's amazing how the right light and camera angle can make a ratty car look so good. I know... I've owned a lot of ratty cars. I look at old pics sometimes and wonder why I got rid of them...

    Rob Siegel Are you insinuating something?

    Dohn Roush Yes, but Maire Ann will keep him anyway... +2

Francois Doremieux I do it in a car I can hope to bring more than a gallon of milk back in, and that's reasonably expected to make the trip back 😬 ... A 2002! +1

    Rob Siegel oh, snap! +1

Django Vaz I break my neck every time I park mine ♥ [photo Subaru WRX]

Eric King A gallon of milk and a... [insert pic: Europa bread van.jpg] +5

    Rob Siegel you suck

Jake Metz Eric King I need this livery to happen in real life +1

Coleman Maguire Yup do that all the time

Fred Bersot Yesterday, took the e46 for Home Depot run. Was first time driving it since October. +1

Johnny Schmitt ...have to admit I feel a little guilty driving my E9 or Targa around right now, as it makes me feel TOO good and we're supposed to be suffering in place, no? +4

    Dohn Roush Beautiful...

Jared Birk Yes! My son and I just got back from the store and hauled two gallons of milk home in the Corvette. As a bonus he got to drive it for the first time ever (he's 15) [photo] +3

Paul Forbush Not insinuating at all... being very direct. You have commented many times about the "patina" present on Lolita. My point is that this photo makes her look much more pristine than you have stated her to be. I was merely musing on how forgiving the camera can be. You and I have both travelled to look at cars on the strength of flattering photos, only to have them disappoint "in the flesh."

    Rob Siegel all good, just giving you a hard time. yeah, most of the patina is on the horizontal surfaces you can't see in these pics.

    Paul Forbush Lol. My autocorrect replaced "insinuating" with "insulting." Is there a message there? +2

    Paul Forbush Rob Siegel My e30s are lousy with "patina." I wish we worshipped that back in the day the way we do now...

    Paul Forbush Rob Siegel And I will be happy to return the favor.

    Paul Forbush Right now, I'm heading outside to hose off some of the spring pollen, the better to revel in the patina.

    Dave Gerwig Paul Forbush a perfect dust cover is easy to obtain during springtime. Took me all winter in the garage to get mine to lay down 😬

Chris Roberts Damn that looks good, Rob <3 +1

Richard Baptiste 💯 Rob else what's the point.

Roy Richard I hope you realize every vehicle in the world is sneezing down on you, in including the guy on the motorcycle. +2

George Whiteley On the way to the grocery store this morning for some necessary provisions. Just doing my part to keep the seals moist in these tough times. [photo E24 M6] +3

Jeff Dorgay So true!

David Weisman It must be a chick magnet +1

    Rob Siegel right... because that's what a happily-married 61-gear-old guy needs... +2

Duncan Irving Both [photo Porsche]

Mark Larson Purchased new in 2003. Have never tired of it. [photo E46]

Lee Perrault Lotus all admire Rob's vision + mechanical tenacity! Not so many can revel in self-sacrifice... 😎 +1

Scott Aaron Looks so much better w/the front lowered. That was so worth it. +1

Colin Tennessen I turn around to check the Subaru hasn't caught fire.

Chapter XIV: Adding Up The Costs

I'm neither a collector nor a dealer. I don't flip cars. I buy what lights my fire and hold onto some for longer than others. For cars that I keep over the long haul, I usually don't even try to keep track of the costs (or I'll accumulate them in a spreadsheet but never add them up), but for cars that I suspect are only sojourning for a bit with me as their automotive foster parent, I'll actively maintain a spreadsheet. However, it's less a profit-and-loss issue and more something I can use to try to calibrate my own behavior by seeing how badly I've misjudged things. I certainly didn't buy the Lotus to flip it; it was a straight-out passion purchase, a car I've been attracted to since I was in the 8th grade and worked for the guy who owned one. But it *is* instructive to know how much your passions are costing you. Plus, as I've long said, if I can't be a shining example, at least I can be a horrible warning.

Since I neither restore cars nor write big checks to someone else to do so, my goal with the Lotus was to leave the patina on the body intact and concentrate on mechanically reviving the fragile little fiberglass-bodied car as cost-effectively as possible. My mantra was that if a component *could* be reused, I would reuse it. I make no apologies for being relentlessly tight-fisted; I could never afford my seven vintage BMWs plus the Lotus plus the Winnebago Rialta RV without being so. Nonetheless, as this accounting exercise shows, once the meter starts running, it's impossible to shut it off and still walk through the stages from running to legal to well-sorted and enjoyable.

It's a given that almost every automotive project is more expensive than you'd planned. When I tally up expenses, I find it helpful to divide them into purchase and registration costs, costs due to a single large issue such as an engine rebuild, general sort-out expenses, and future projected costs. If there are non-essential expenses, I divide those out as well (here, the only real one was powercoating the valve cover). By dividing things up this way, I find that I at least stand a chance of learning something from it instead of simply recoiling in horror from the final number.

Initial costs: $8656

These are the front-end costs to purchase the car, ship it to my house, and square things with the state of Massachusetts (well, Vermont). So, in theory, if I bought a car that needed nothing, these would be the only costs. On paper, it sounds pretty good. The car cost me six grand. Enclosed point-to-point air-ride shipping from Chicago to Boston seemed appropriate for the long-dead fragile fiberglass-bodied car, and the $1220 cost from with Intercity Lines was pretty reasonable. As I wrote about in the Facebook posts, Vermont assesses sales tax like Massachusetts, which is not on what you actually paid, but on the basis of the low NADA value, which for a '74 Europa is $8650, so I didn't get soaked too badly.

Initially, I didn't include insurance costs with this tally, since, with Hagerty, the annual addition of a low- to mid-value car to my multi-car policy isn't much. However, I did insure the car when I purchased it in case something happened to it during shipping, and that was six years ago, so it added up. Even though it wasn't running, I kept it insured so I'd be covered if a tree or a meteor hit the garage. I never imagined that the car would sit in my garage, insured but immobile and taking up precious space, *for six years*. I'll admit that, during my darker moments, I thought "maybe I'll get lucky and a meteor will come through the window, take out the Lotus, and spare the garage." For grins, I called Hagerty. My annual cost of the policy on the Europa is $132. Like I said, not much. But over six years, it adds up, and if I'm going to look the beast in the face, I might as well see it snarling and with drool coming out of its mouth.

Purchase	$6,000
Shipping	$1,220
Tax	$539
Title and registration	$75
State inspection	$30
Insurance	$792
Subtotal	$8,656

Engine rebuild costs: $5910

As I've detailed, when I bought the Europa, I didn't have a clue that such a small number of Lotus-Ford Twin-Cam engines were built (about 34,000, only 4,700 of which have the Europa-specific front timing cover necessary to clear the right-hand leg of the car's Y-shaped frame), that used running twink engines are rare as hen's teeth, and that you certainly don't find them for BMW 2002 engine money. I removed and tore down the engine, unseized the stuck piston, and took the disassembled engine to The Lotus Engine God. The fact that the block was sleeved, and that the seized piston damaged its sleeve, was problematic, as the sleeves needed to be bored to remove the damage, and standard-sized pistons would no longer fit. I then met a fellow who'd had TLEG rebuild his motor, and learned that it cost him twelve grand. My cost-conscious self thought "I don't want to even pay a third of that." (I knew that I would, but that doesn't mean I wanted to.) I had TLEG clean and crack-check the block, head, crank, and rods, then pulled the engine out of his shop and took it to the more conventional machine shop who does my BMW engine work. The four-year timeframe notwithstanding, the shop reassembled the head, bored the block to accept the custom pistons, and re-bushed the rods. I reassembled the block, mated the troublesome aftermarket cartridge water pump and its timing covers, and attached the head. Considering that the machine shop bill included $600 for the custom pistons, and that I spent over a thousand bucks on the cartridge water pump kit and related parts and machining work, the $5910 total—about the cost of the purchase of the car itself—is the best I could've hoped for. Not surprisingly, there were times during whole water pump episode that I rued the day I'd committed to that path, but I've had zero issues with it, and I am happy knowing that I shouldn't have to pull the engine out and half apart if it ever needs replacing. Here's the breakdown:

Main machine shop bill	$3,165
Dave Bean water pump	$907
TLEG cleaning checking	$800
Main order engine parts	$552
Chain and tensioners	$158
TLEG cutting cover	$100
Adhesives	$50
Bean seal	$44
Twin-Cam rebuild book	$34
Threaded plugs	$64
Gaskets and bolts	$22
Oil tube	$10
Oil pump o-rings	$4
Subtotal	$5,910

Seven years prior, when I was trying to determine the best path for rebuilding the engine, I briefly considered an unseize-it-and-just-slap-it-back-together approach. The idea was that I'd never even driven a Europa, and this way I'd find out what the car was like and decide if I liked it and whether it was worth committing more money to. It sounds reasonable on paper, but it's really not practical. First, as my friend Lindsey Brown says, never start pulling apart an engine unless you're prepared to deal with whatever it is you find. Second, can you imagine having done the engine assembly work I did, only to have to do it again for the "real" rebuild? Third, as I said in the chapter on "The stages of sorting out a car," we usually want a performance car to *actually perform*. So, yes, the engine had to start and run, but also not sound like a tractor, have power, and not leave an oil fog behind it, and then other systems in the car—suspension, steering, brakes—needed to be brought up to similar snuff in order to get the car not only running and driving, but to the level where it could actually satisfy passion. Which brings us to the next two items:

First wave sort-out costs: $2982

The next set of expenses are comprised of nearly everything else that wasn't directly related to the engine rebuild I needed in order to get the car running and capable of driving to the Lotus Owner's Gathering event. It's too long to itemize, but it came to $2982—about half the cost of the purchase of the car—which isn't a shock when you consider that you expect a car that hasn't run for 40 years to need absolutely everything. The biggest individual items were the clutch, the components for the braking system, the exhaust, the tires, and road-force balancing.

Second wave sort-out costs: $1457

After the LOG event, I did a second round of sorting, much of which occurred over the winter. $508 for the front-end rebuild (done *very* cost-consciously using a used 35-year-old set of Spax adjustable shocks), $508 for the cooling system upgrades, $221 for the transaxle seal repair, and $223 for parts and tools related to the valve adjustment add up to $1457.

Projected costs: $1445

The cracked windshield will need to be replaced in order for the car to pass Massachusetts inspection (it's still wearing VT plates). Even if I can find a good used one, the cost of the sealing and trim pieces is even higher than the cost of the glass. I've installed traditional rubber-gasketed windshields, but this one is bonded in place. It needs a pro. The car also will also need a four-wheel alignment which involves inserting washers between the frame and the rear trailing arms, so there's shop labor over and above the standard alignment charge. Adding all this up, the estimated projected costs are $1445.

And the total is… oh crap

If you tally it all up—and lord knows I'd rather not, at least not without a Valium, some vodka, and a bucket within ready reach—this is what you get.

Initial costs	$8,656
Engine rebuild	$5,910
First sort-out	$2,982
Second sort-out	$1,457
Anticipated costs	$1,445
Total	$20,450

This isn't sticker shock; this is trauma. The only other car I've ever had twenty grand in is my 1973 BMW 3.0CSi, and that's having owned it since 1986 and given it an outer body restoration, a new drivetrain, interior, fuel injection, and air conditioning.

In a world where running project Europa Twin Cam Specials sell for eight to twelve grand, and sorted drivers go for fifteen, it's difficult for me to look at these numbers and claim with a straight face that it was worth it, even considering that I love cars with stories and patina and that this one has a particular, ahem, erotic component. Twenty grand is a shit ton of money to me. There are any number of cars I'd spend twenty grand on before electing to drop it on a Europa TCS. And with mine's distressed appearance, I could get it flawlessly sorted and its value probably wouldn't top twelve grand. Plus, well, let's just say it: It's brown. I *like* brown cars, but I'm not blind with regard to the effect on value.

So why do we *do* these things? And I say "we" because I am hardly alone. It's a combination of factors. Sometimes we simply don't know what we're getting into. Sometimes we know but we do it anyway, either for the challenge, or because the car speaks to us in some way. Sometimes it's the frog-in-the-pot analogy, not realizing the water is getting hotter and hotter until it's too late and you're cooked. All of those dynamics were definitely in play with the Lotus. I advise other people that they need to unflinchingly honest with regard to what a car needs and what it's worth once it gets it. Only then can you judge whether you should buy a car, and for what price. I completely ignored my own advice with the Lotus, and it shows.

The traditional wisdom is that there's nothing cheaper than someone else's money, that resurrection and restoration are for suckers, that it's always best to buy a car someone else has restored, resurrected, revived, call it whatever you will. For a number of reasons, I routinely reject that wisdom. I'm a bottom feeder who generally enjoys buying highly-challenged cars with good stories, paying little for them, and nursing them back to health. The fact that I also write about this process makes it almost seem like a rational endeavor. But even if I didn't regard it as grist for the journalistic mill, my middle-class roots make me sympathetic to the approach that paying little, adding your own sweat equity, and spreading the amount over a period of years is a de-facto way of financing a project that you'd never be able to afford otherwise.

But even still… twenty grand? For *this particular* Europa? Like, ouch. While I don't walk around moping that I'm "underwater" in the car, there are limits, and Lolita is clearly exceeding them.

In my first book, I talk quite a bit about how cars intersect with family dynamics. Some of those are financial. Passions are beautiful things, but I've never understood why some people feel that because a passion is automotive in nature, or because someone subscribes to the philosophy of "do it once do it right," that gives the person a right to spend limitless amounts of money and make unilateral decisions on things that should be family financial matters. While I appreciate that, for example, twenty grand spread over ten years is only about five bucks per day, and that's a reasonable cost for satisfying one's passion, this is a mechanism for denial, and part of the point of adding it up is to not be in denial.

I intentionally showed the spreadsheet to Maire Anne and told her the final number. She was as surprised and as concerned as I was, and she should be. As I've mentioned, one of my wife's passions is that she is an avid quilter, owns multiple sewing machines, and goes on quilting-related trips. I don't ask her what she spends any more than she asks me, but I think that, if that activity was draining twenty grand out of the family budget, I'd want to know. That's what a good relationship is all about.

To me, this is about more than just passion. It's about balance. Yes, we need passions as outlets to salve the pain of life, whether that's going into the garage for an hour at the end of a bad day, wanting to lay hands on tools, fix a problem, and *not* talk with anyone, or whether it's going to a Cars and Coffee on a Sunday to rub up against some other like-minded weirdos and get the scent on you. But we also need balance. We have to make choices. I don't have unlimited funds. That much money going into one car has to come from somewhere else. If that means I need to sell one of the other cars, I should know that, and that's a choice I should be prepared to make. I'm being intentionally brutal, but I think that's useful. I'd love to own an Elan 2+2 coupe. But now, if I see a long-dead project car with a seized engine for six grand, I know that there's little chance of getting it running and well-sorted for under twenty large, even without addressing interior, body, and paint issues.

One of the best pieces of advice my endlessly-wise mother ever gave me on parenting was this: "If your children show interest or passion in something, treat that like a flower, because if you don't, you'll kill it with neglect, or worse." If I've been guilty of anything, it's over-applying that advice to myself and using it to justify some of my own excesses. Maybe I was doing some of this because my mother was dying right in the middle of Lolita's rebirth. I couldn't save her, but I could resurrect the car. Well, no. Looking at the dates, I was already in the process of trying to start the engine when she first went into the hospital. The timing was coincidental. I don't know. I guess I'm reaching for something that isn't really there. Scratch that.

Look. We all stumble forward and figure these things out as we're going along. Hopefully we neither drive our families into bankruptcy nor crater our relationships with our family members as we're learning. But the "learning" part of it is important. There is a lesson to be gained from the Lotus' cost in terms of underestimating the time and the money involved, and I'd be a fool if I didn't learn it. It doesn't mean I won't ever invest time and money in a challenging project again, or that the flames of my automotive passion will be doused in any way. But a little (or a big) reality check every once in a while is a good thing.

I love the way the batshit crazy little thing screams and claws when I nail it. But actual illicit sex would've cost me a lot less than twenty grand.

On the other hand (as Tevye would say), one of my secrets to happiness is that I've gotten very good at not craving things I can never have. A vintage Italian mid-engine exotic will never grace my driveway. I'll never be able to hop in a Lamborghini Miura whenever I want, twist the key, baby it 'till it warms up, then wind it out on an open stretch of road. I'm not saying that the Europa is even remotely in that league, performance-wise, value-wise, anything-wise, but I am saying that, when I back it out of the garage and take it for a drive, that's the closest I'm ever going to get, and I'm completely centered in the experience rather than wishing it was something else. Maybe that *is* worth twenty grand.

As I think about it, it's oddly appropriate that the Europa, designed to be the first affordable mid-engine sports car that could give a non-trust-fund Joe or Jane a little taste of driving a Gran Prix racer, is doing *exactly* that for me, half a century later. God bless its quirky "breadvan" silhouette. If it looked like a Miura and had a V12 in it, twenty grand would be the cost of a valve cover, my garage wouldn't have one in it, and to get "sex out of" it, I'd be jerking off to a poster of the car and Tara King instead.

Epilogue

Well. Not a bad ride, huh? Let's wrap this up. I'll let my left brain go first. I've described:

- The excitement of stepping outside your comfort zone with an unfamiliar car and how you can almost feel new synapses forming as you learn new things and encounter and solve new problems (and want to kill the designers responsible for them).

- My generally good track record of understanding my limitations and not biting off more than I can chew.

- How not having a clue what I was getting into with this Lotus Twin-Cam engine up-ended my track record.

- How my relentless desire for cost-containment forced a change in the machine shop, and how that, coupled with my changing financial fortunes, ground the project to a nearly four-year halt.

- The forward motion that can be gained by doing one thing a night—time is going to pass anyway; you might as well get some work done while it's passing (he says with a straight face after doing fucking nothing for four years).

- How even though I'd shut down all spending on the car, there was no pretending that it wasn't costing me money since it was sitting in my garage and I pay money to rent other off-site garage space.

- How there came a point where I looked the project in the eye and faced the choice of selling it at a big loss as a bailed-out-of-it car with a disassembled engine that couldn't even be rolled out of the garage, or buckling down, assembling the engine, getting it into the car, and getting it running, knowing that that would be even *more* expensive.

- How, having done that, I had no idea at all whether I'd like, love, hate, or meh the car.

- How finding out that I loved this weird buzzy little thing was the surprise of my automotive life.
- How, once you begin pursuing a goal of making a car fully functional, it's really difficult to contain spending.
- But how glorious it is making that leap from functionality to performance so the Lotus could run and drive like it's supposed to.

I hoped that Lolita and I would have a more epic road trip, something where we'd truly pit ourselves against the elements and triumph over adversity. 50 miles there and back to LOG in Sturbridge with a 20-mile side jaunt to Thompson Motor Speedway in Connecticut ain't exactly *we were somewhere around Barstow on the edge of the desert when the drugs began to take hold*. But I'm not even doing those gonzo trips to my regular vintage BMW events; as of this writing, all are cancelled due to Covid-19.

So it's short social-distancing pleasure drives for me and Lolita—not far enough to make Maire Anne nervous that the car will die and I'll have to share a lengthy ride in the cab of a tow truck with a hygiene and PPE-impaired stranger.

Okay. Left brain, take a seat. Right brain, your turn. How does that make you *feel*, Rob?

You know what? It's not bad. In fact, it's great. Really. How lucky I am to have this impossibly cool adolescent lust car, to have bought it as a long-dead survivor, resurrected it *my* way making *my* choices, reached the point where it's fairly well-sorted, and be able to just hop in it, twist the key, and pleasure-drive it on low-speed twisties through leafy rural communities west of I95, then mid-speed two-lanes, then back home on the interstate. Especially now when we all need to find our happy places.

For many years I had a BMW Z3 roadster. Anyone who's owned a convertible knows about the whole body relaxation response they engender when you drop the top and drive. It's like rolling Xanax.

Lolita is *not* Xanax. She's Viagra. With a vibrator. And the Buzzcocks blasting "Orgasm Addict."

Back in the introduction to this book, I launched us all onto the plot arc where I described the event in my life where, when I was a 13-year-old kid, a 40-something year old man and I were ogling our boss' Lotus Europa, and he said "a car like that, you can get *sex* out of." It's not hyperbole to say that my thoughts about men, cars, passion, and physical sensation trace back to that seminal (heh) moment. Seven years ago I not only bought the Europa, I named the goddamned thing "Lolita," something so politically incorrect that I've never even explained the reference because it's absolutely indefensible. I started with simple dick jokes ("stressed member" nudge nudge), then went all-in (heh) and subjected you to 500 pages of salacious comments about jackshaft lubrication, the six-year-long tantric orgasm of getting Lolita running—I'm close, stick your finger in my ear and all that—and—finally—"consummating" the relationship with the car. I even named chapters where that happened *Sploosh* and *Barely Legal*. Subtle as a goddamn heart attack, this guy. And at the end of last chapter, I wrapped it up in pink paper and a little bow with "I love the way the batshit crazy little thing screams and claws when I nail it. But actual illicit sex would've cost me a lot less than twenty grand." But that misses the entire point.

Actual illicit sex probably *would've* cost me a lot less than twenty grand... *if I did such things*. But I don't. And that's the whole ball of wax right there.

You know what? I don't really want to go on a long road trip with Lolita. It was more something I felt that I *should* do. Sorted or not, she's loud, erratic, and demanding. Too many quirks. Too many rough edges. I think she'd drive me crazy. We're really best together on short stress-busting bursts. I have other cars that are much more reasonable comfortable appropriate traveling partners for long trips.

Jesus, Rob, just say the words: Fuck buddy.

Come on, Lolita. I've got 20 minutes. Let's go do something... sensational.

TLC Day 2629: The Last Fucking Post

AUG 20, 2020, 8:06 PM

The Lotus Chronicles, Day 2629:

In 2008, comedian Lewis Black was negotiating to do a televised special at the Kennedy Center when someone from the Center went to one of his live shows and counted the number of F-bombs he dropped. It was reportedly 42. The Kennedy Center declined to have him. When he talks about this in his routines, Black says "Who knew that The Kennedy Center has a five-fuck limit?"

When my first book was picked up by Bentley Publishers and I worked with them on the manuscript, I was in no way pressured to clean up the language, but both they and I wanted to know the F-number. It turned out to be six. So, one over The Kennedy Center's limit.

As *The Lotus Chronicles* nears publication, I became interested in what its F-number is and how that compares with my other books. I knew it was high, as it contains not only my colorful Facebook language, but other people's comments. Here's the F-body count of the books:

Memoirs of a Hack Mechanic: 6

Electrical book: 0

Vintage ignition book: 0

Ran When Parked: 5

Just Needs a Recharge: 1 ("Propane? In an air conditioning system? Are you fucking CRAZY? Do you WANT to die?")

Resurrecting Bertha: 13

The Lotus Chronicles:

Wait for it...

83.

Yep. It's not only a weighty tome, it's a salty one.

Okay, well, I thought, most of those can't be me, right? There must be some serious fucking load sharing with the Hack Mechanic faithful, right?

I rolled up my sleeves and did the, ahem, fucking statistics.

-- Paul Forbush: 1

--Russell Baden Musta: 1

--Collin Blackburn: 1

--Garrett Briggs: 1

--Craig Fitzgerald: 1 (shocking, I know)

--Mike Miller: 1

--Paul Wegweiser: 2

--Brian Ach: 2

--Brad Purvis: 2

--Doug Jacobs: 10

Leaving yours truly with 61.

Of course, given the pain the fucking car caused me, it makes perfect sense.

Extrapolating this pattern, my next book will consist entirely of F-bombs expressed as nouns, verbs, and gerunds.

I do know one thing: The Kennedy Center is very unlikely to have me read the audiobook live on stage, as I am 77 fucks over their limit.

But Lewis Black might let me open for him.

[Note: No, I have not included this post in the F-count. This being 2020, I'm concerned that doing so could cause the universe to end.]

117 LIKES, 82 COMMENTS

Doug Jacobs I'M THE WINNER!!!!!! +4
Craig Fitzgerald I regret that I had but just one fuck to give for my Robby. +4
Ken Doolittle How many "shits"?
    Rob Siegel I don't give a shit.
    (okay. 114. seriously. that's a lot of shit.) +9
    George Zinycz Rob Siegel well we come to you because you know your shit +2
Sean T. Going I can't fucking believe I didn't make this fucking list. Fuck. +3
    Rob Siegel you are literally fucking lame. +1
Ken Batts "77 fucks over the line, sweet Jesus..." +5
Craig Fitzgerald Jesus Christ, if there was any tome that should be festooned with a fanfare of fucks, it's the electrical book. +6
David Kemether It's a wonder how the British are so polite. +2
Steve Rapson In a similar vein, Mark Twain was known as serial profaner, often embarrassing his wife, Olivia, in polite company. As they prepared for a social event, and in an effort to shame her husband into realizing how crude he sounded, Olivia stood before him and rattled off every cuss word she'd ever heard.
Twain pulled on his cigar and said, "My dear, you know the words, but not the music."
So thanks for the music, Rob. +5
Brian Ach Where are my fucking royalties? +1

End Notes

Rob Siegel has been writing the monthly column *The Hack Mechanic*™ for BMW CCA *Roundel* magazine for over 30 years. He is the author of the books *Memoirs of a Hack Mechanic, The Hack Mechanic*™ *Guide to European Automotive Electrical Systems*, and *Mechanical Ignition Handbook: The Hack Mechanic*™ *Guide to Vintage Ignition Systems*, all from Bentley Publishers, and *Ran When Parked: How I Resurrected a Decade-Dead 1972 BMW 2002tii a Thousand Miles Back Home, and How You Can, Too, Just Needs a Recharge: The Hack Mechanic*™ *Guide to Vintage Air Conditioning,* and *Resurrecting Bertha: Buying back our wedding car after 26 years in storage*, all from Hack Mechanic Press. Rob lives in West Newton Massachusetts with his saint-like wife Maire Anne Diamond, two black cats, one black dog, a roomful of his wife's insects (don't ask), and as many cars and guitars as he can get away with. Given his druthers, he'd rather be a full-time singer-songwriter, but it's easier and more fun to continue to play the part of The Hack Mechanic. Rob says "Most days, I actually am that guy, but really, it's just fucking exhausting being me."

Other books by Rob Siegel

Memoirs of a Hack Mechanic: How Fixing Broken BMWs Helped Make Me Whole (a memoir with actual useful stuff)
Bentley Publishers, 2013
ISBN 978-0837617206

The Hack Mechanic™ Guide to European Electrical Systems
Bentley Publishers, 2013
ISBN 978-0837617510

Mechanical Ignition Handbook: The Hack Mechanic™ Guide to Vintage Ignition Systems
Bentley Publishers, 2017
ISBN 978-0837617206

Ran When Parked: How I Resurrected a Decade-Dead 1972 BMW 2002tii and Road-Tripped it a Thousand Miles Back Home, and How You Can, Too
Hack Mechanic Press, 2017
ISBN 978-0837617671

Just Needs a Recharge: The Hack Mechanic™ Guide to Vintage Air Conditioning
Hack Mechanic Press, 2018
ISBN 978-0998950716

Resurrecting Bertha: Buying back our wedding car after 26 years in storage
Hack Mechanic Press, 2018
ISBN 978-0998950723

Printed in Poland
by Amazon Fulfillment
Poland Sp. z o.o., Wrocław